International Conference on Smart Transportation, Energy and Power (STEP 2021)

Journal of Physics: Conference Series Volume 2195

Online
3 - 5 December 2021

ISBN: 978-1-7138-5162-2
ISSN: 1742-6588

Printed from e-media with permission by:

Curran Associates, Inc.
57 Morehouse Lane
Red Hook, NY 12571

CURRAN ASSOCIATES INC. proceedings.com

Some format issues inherent in the e-media version may also appear in this print version.

Institute of Physics
Dirac House, Temple Back
Bristol BS1 6BE UK

Phone: 44 1 17 929 7481
Fax: 44 1 17 920 0979

techtracking@iop.org

Additional copies of this publication are available from:

Curran Associates, Inc.
57 Morehouse Lane
Red Hook, NY 12571 USA
Phone: 845-758-0400
Fax: 845-758-2633
Email: curran@proceedings.com
Web: www.proceedings.com

TABLE OF CONTENTS

Journal of Physics: Conference Series **2195** (2022) 011001 doi:10.1088/1742-6596/2195/1/011001

Preface

Due to recent pandemic, the 2021 International Conference on Smart Transportation, Energy and Power (STEP 2021) which was planned to be held in Sanya, China, was held virtually online during December 03-05, 2021. The decision to hold the virtual conference was made in compliance with many restrictions and regulations that were imposed by countries around the globe. Such restrictions were made to minimize the risk of people contracting or spreading the COVID-19 through physical contact. There were 110 individuals who attended this online conference, represented many countries including Singapore, Australia, United States, Germany, and China.

The conference aimed to provide a platform for experts and scholars, engineers and technicians, R&D personnel engaged in intelligent transportation, environmental engineering, mechanical engineering, measurement and control technology and instrumentation science to share scientific research results and cutting-edge technologies, to understand academic development trends, to broaden research ideas, to strengthen academic research and discussion, and to promote cooperation in industrialization of academic achievements. We have invited experts, scholars, business people and other related personnel from domestic and foreign universities and research institutions to participate in the conference.

During the conference, the conference model was divided into three sessions, including oral presentations, keynote speeches, and online Q&A discussion. In the first part, some scholars, whose submissions were selected as the excellent papers, were given about 5-10 minutes to perform their oral presentations one by one. Then in the second part, keynote speakers were each allocated 30-45 minutes to hold their speeches. In the second part, we invited nine professors as our keynote speakers. Prof. Minghui Hong, National University of Singapore, Fellow of Singapore Institute of Engineering, performed a speech: *Laser Applications on Railways.* And then we had Prof. Qian Xiao, East China Jiaotong University. He presented a wonderful speech: *Research on vibration reduction of metro vehicle on-board equipment based on elastic body.* Prof. Minhao Zhu, Southwest Jiaotong University was invited to present his talk *Intelligent urban rail transit construction based on quality safety infrastructure.* Prof. Shanyong Zhang, Southwest University. He delivered a wonderful speech: *Advances in CrCoNi-based medium-entropy alloy coatings.* The others keynote speakers were as follow: Prof. Goh Hui Hwang, Guangxi University, China; Prof. Wenping, Anhui University; Prof. Xinlong Tian, Hainan University; Prof. Hongzhong Ma, Hohai University; Assoc. Researcher Juanxiu Xiao, Hainan University. Their insightful speeches had triggered heated discussion in the third session of the conference. Every participant praised this conference for disseminating useful and insightful knowledge.

The proceedings are a compilation of the accepted papers and represent an interesting outcome of the conference. Topics include but are not limited to the following areas: Electric vehicle technologies, Intelligent control systems, Signal and Communication Processing, Optimal Control, Nonlinear Systems and Control, Nuclear Energy and more related topics. All the papers have been through rigorous review and process to meet the requirements of International publication standard.

We would like to acknowledge all of those who supported STEP 2021. The help and contribution of each individual and institution was instrumental in the success of the conference. In particular, we would like to thank the organizing committee for its valuable inputs in shaping the conference program and reviewing the submitted papers.

We sincerely hope that the STEP 2021 turned out to be a forum for excellent discussions that enabled new ideas to come about, promoting collaborative research. We are sure that the proceedings will serve as an important research source of references and knowledge, which will lead to not only scientific and engineering findings but also new products and technologies.

The Committee of STEP 2021

Committee

Conference Chairman
Porf. Changming Li, Suzhou University, China
Prof. Hesheng Liu, East China Jiaotong University, China
Prof. YinQuan Yu, East China Jiaotong University, China

Academic Committee Chairman
Prof. Yongdong Li, Tsinghua University, China
Prof. Yijun Shen, Hainan University, China

Academic Committee Member
Prof. Shoudao Haung, Hunan University, China
Prof. Xiaofu Xiong, Chongqing University, China
Prof. Zhengyou He, Southwest Jiaotong University, China
Prof. JinSong Kang, Tongji University, China
Prof. Hongzhong Ma, Hohai University, China
Prof. Ronghai Qu, Huazhong University of Science and Technology, China
Prof. Tianhao Tang, Shanghai Maritime University, China
Prof. Fushuan Wen, Zhejiang University, China
Prof. XinZhang Wu, Guangxi University, China
Prof. Dehong Xu, Zhejiang University, China
Prof. Guoqing Xu, Shanghai University, China
Prof. Xing Zhang, Hefei University of Technology, China

Organizing Committee Chairman
Prof. Qian Xiao, East China Jiaotong University, China
Prof. Shengkui Zhong, Hainan Tropical Ocean University, China

Organizing Committee Member
Prof. Hua Geng, Tsinghua University, China
Prof. GohHui Hwang, Guangxi University, China
Prof. Ke Jin, Nanjing University of Aeronautics and Astronautics, China
Assoc. Prof. Min Chen, Zhejiang University, China
Assoc. Prof. Qiping Chen, East China Jiaotong University, China
Assoc. Researcher Juanxiu Xiao, Hainan University, China

International Technical Program Committees
Assoc. Prof. Abdul-Sattar Nizami, Pakistan
Assoc. Prof. Kishore Kumar Pedapenki, India
Assoc. Prof. Vineet Dahiya, India
Assoc. Prof. Muhamad Bin Mansor, Malaysia
Assistant Prof. Subhojit Dawn, India

Peer Review Statement

All papers published in this volume have been reviewed through processes administered by the Editors. Reviews were conducted by expert referees to the professional and scientific standards expected of a proceedings journal published by IOP Publishing.

- **Type of peer review:** Single Anonymous
- **Conference submission management system:** Morressier
- **Number of submissions received:** 102
- **Number of submissions sent for review:** 60
- **Number of submissions accepted:** 60
- **Acceptance Rate (Submissions Accepted / Submissions Received × 100):** 58.8
- **Average number of reviews per paper:** 2
- **Total number of reviewers involved:** 8
- **Contact person for queries:**
  Name: Xuexia Ye
  Email: xx.ye@keoaeic.org
  Affiliation: AEIC Academic Exchange Information Centre

Numerical Study on the Influence of Angles of Repose on the Stability of Cargo in a Cargo Hold

Zhanwei Wang[1a], Jianwei Zhang[1b*], Hang Zhou[1c]

[1]School of Naval Architecture and Maritime, Zhejiang Ocean University, Zhoushan 316022, China

[a]wzw0724@zjou.edu.cn , [b*]zhangjianwei@zjou.edu.cn , [c]zhouhang@zjou.edu.cn

Abstract—In recent years, cargo instability caused by fluidization is one of the main reasons for bulk carrier capsizing. Considering the complexity and diversity of the actual cargo, the mechanism of fluidization has not been fully revealed. Taking a single cargo hold of a 57,000 DWT bulk carrier as an example, a 3D cargo hold scaled model was established, and the movement state of the cargo in the cargo hold under different angles of repose is numerically simulated. Results show that under the precondition of model parameters, angle of repose of the cargo can be further modified by the friction coefficient between particles. The smaller of the angle of repose, the stronger of the fluidity of the cargo becomes. There exists a critical angle of repose, below which the center of gravity offset of the cargo becomes more obvious and the stability of the cargo drops sharply. Simulation model and numerical method involved in this research can provide some references for predicting the stability of the marine transportation of solid bulk cargoes.

1. Introduction

Fluidization, mainly refers to the special state that a type of bulk solid cargo is prone to occur during ship navigation. This type of cargo usually consists of a certain proportion of fine particles and moisture, such as laterite nickel ore and iron concentrate. Due to the continuous dynamic load, including the ship's own swaying motion and the strong wind and waves under severe sea conditions, the cargo gradually presents a fluid state, completely losing its strength and stiffness [1]. In severe cases, the ship may even capsize due to the instability of cargo caused by fluidization [2]. For this kind of problem, Guan et al. [3] analyzed the evolution process of nickel ore fluidization in detail through a six-degree-of-freedom motion platform test loaded with a nickel ore cargo hold model. It is concluded that roll motion plays a major role in cargo fluidization. Based on the indoor small-scale shaking table model test, Zhou et al. [4,5] studied the fluidization characteristics of iron ore concentrate and laterite nickel ore. Results all show that water migration is a key factor leading to cargo fluidization. Due to the poor repeatability of model test and many external interference factors, however, numerical simulation methods such as finite element method (FEM) and discrete element method (DEM) have been adopted by scholars to further study the movement law of easily fluidized cargoes in the cabin. Based on the finite element software FLUENT, Zhang et al. [6] established the numerical model of cargo sloshing in cabin and rolling coupling motion between cabin segments by using simplified surface nonlinear method and volume of fluid (VOF) method. It is further verified that higher wave amplitude and resonance frequency are the main factors leading to ship capsizing. Ju et al. [7] established a DEM numerical model to compare with the static tilt test of Japan Classification Society, and further verified that the friction coefficient can be equivalent to the effect of moisture between cargo particles.

Previous studies have shown that DEM is more suitable than FEM in studying the dynamic change process of multiphase, discrete and physical property dispersion of particle motion and particle flow characteristics [8]. Therefore, in view of the uniform loading of a single cargo hold of 57,000 DWT grade bulk carrier, a 3D cargo hold scale model is established based on the discrete element software PFC3D to simulate the movement of cargo particles in the cargo hold. Considering the complexity of the actual cargo properties, this study focuses on the influence of different repose angles on the movement of cargo particles in the cargo hold, in order to analyze and discuss the stability of the cargo in detail from the perspective of initial cargo properties.

2. Theoretical Background and Equations

The contact behavior of multiple particles, as the key issue of DEM, is determined by the selected contact model. In the contact model, particles are regarded as rigid bodies, and the contact between particles is a soft contact, that is, the particles are allowed to deform slightly at the contact point. When the contact behavior occurs between particles, the selected contact model would be activated. The force-displacement law provided by the contact model acts on the particles in contact to update the contact force and moment between the particles. The updated force and moment respectively act on the particles in contact, thereby updating the force of each particle and the current position change of the particle. Considering that the rolling resistance linear model has certain advantages in simulating granular materials, especially in the case of angle of repose formed by them. Therefore, the model is mainly used to complete the DEM modeling. Law of force-displacement in PFC3D are further described as follows.

In a timestep, the force-displacement law updates the relative displacement increment of the particles according to the change in the contact gap between the particles, that is, the following equations is satisfied:

$$\Delta\delta_n := \alpha\Delta\delta_n \tag{1}$$

$$\Delta\delta_s := \alpha\Delta\delta_s \tag{2}$$

$$\alpha = \begin{cases} \frac{g_s}{g_s-(g_s)_o}, & (g_s)_o > 0 \; and \; g_s < 0 \\ 1 & \end{cases} \tag{3}$$

where, $\Delta\delta_n$ is the normal relative displacement increment, $\Delta\delta_s$ is the shear relative displacement increment; g_s is the surface gap between particles; $(g_s)_o$ is the surface gap at the beginning of the time step. In addition, the force-displacement law updates the contact force and moment between particles as follows:

$$F_c = F^l + F^d \tag{4}$$

$$M_c = M^r \tag{5}$$

where, F^l is the linear force, F^d is the dashpot force, M^r is the rolling resistance moment. Generally, the linear force F^l can be decomposed into linear normal force F_n^l and linear shear force F_s^l, which satisfies the following equations:

$$F_n^l = \begin{cases} k_n g_s, & g_s < 0 \\ 0, & g_s \geq 0 \end{cases} \tag{6}$$

$$F_s^l = \begin{cases} (F_s^l)_o - k_s\Delta\delta_s, & ||(F_s^l)_o - k_s\Delta\delta_s|| \leq -\mu F_n^l \\ -\mu F_n^l[(F_s^l)_o - k_s\Delta\delta_s)/||(F_s^l)_o - k_s\Delta\delta_s||], & \text{otherwise} \end{cases} \tag{7}$$

where, k_n is the normal stiffness, $(F_s^l)_o$ is the linear shear force at the beginning of the timestep, k_s is the shear stiffness, $\Delta\delta_s$ is the adjusted relative shear-displacement increment of Eq. (2), and μ is the particle friction coefficient. The rolling resistance moment M^r is updated through the following steps.

First, the rolling resistance moment M^r is incremented as:

$$M^r := M^r - k_s\bar{R}^2\Delta\theta_b \tag{8}$$

where, $\Delta\theta_b$ is the relative bend-rotation increment, $\bar{R}$ is the contact effective radius, defined as: $1/\bar{R} = 1/R^{(1)} + 1/R^{(2)}$, $R^{(1)}$ and $R^{(2)}$ respectively correspond to the radii of the two particles in contact.

The magnitude of the updated rolling resistance moment M^r is then checked against a threshold limit:

$$M^r = \begin{cases} M^r, & ||M^r|| \leq \mu_r \bar{R} F_n^l \\ M^*(M^r/||M^r||), & \text{otherwise} \end{cases} \quad (9)$$

where, μ_r is the rolling friction coefficient.

3. Simulation of angle of repose

3.1. Basic model parameters

Angle of repose, as an important physical parameter to characterize the bulk cargo's looseness and fluidity, its study is of great significance to the safety of bulk cargo transportation. Before using contact model to simulate the angle of repose, it is necessary to calibrate the particle parameters [9]. Previous studies have shown that the main factors affecting the angle of repose are the friction coefficient between particles, friction coefficient in contact with the ground and the types of cargoes (especially the stiffness and modulus) [10-12]. Considering the diversity of influencing factors, this numerical simulation study takes nickel ore particles as the research object, unifies the relevant variable parameters, and only changes the friction coefficient between the particles to adjust the value of the angle of repose.

In addition, reference [13] points out that when the ratio of the average particle size of the cargo to the boundary size of the model is greater than or equal to 25, the size effect of the particle size in the model can be ignored. In this numerical study, the selected particle size is 0.01 m, and the ratio of particle size to model boundary size is greater than 25. Basic parameters of the model are shown in Table 1.

Table 1. Specific setting of rolling resistance linear model parameters.

Model parameter	Value	Model parameter	Value
Normal stiffness k_n (Pa)	1e5	Particle friction coefficient μ	Calibrated by simulation of different angles of repose
Shear stiffness k_s (Pa)	5e4	Rolling friction coefficient μ_r	
Normal-to-shear stiffness ratio $k^* = k_n/k_s$	2	Wall friction coefficient	0.3*
Effective modulus E^* (Pa)	1e6	Gravity field g (m/s^2)	9.81
Particle density D (kg/m^3)	2500	Time step (s)	1e-4

*Note: Bottom friction coefficient

3.2. Simulation model

Funnel method is usually used to measure the angle of repose of soil particles. However, once the friction coefficient or the viscous force between cargo particles is large, it is easy to block the discharge port. Therefore, considering the complex characteristics of the actual easily fluidized cargoes, the inclined plane method is mainly used for numerical simulation of angle of repose.

Comparing reference [8] and reference [10], a numerical simulation model for measuring the angle of repose of the cargo is designed. Numerical simulation of angle of repose is shown in Fig. 1, and the specific steps of using the model to test the angle of repose are as follows:

- Establishment of the cabin (see Fig. 1a). The baffle is used to divide the cabin, and the left cabin is used to load cargo.
- Cargo loading (see Fig. 1b). The loading depth is set to 0.37 m, and the total mass of the cargo is about 77.40 kg.

- Remove the middle baffle and wait for the cargo to reach a state of equilibrium (see Fig. 1c).
- After the cargo reaches equilibrium (see Fig. 1d), save the current view and data for further processing.
- Modify the friction coefficient between particles and repeat the above steps to obtain different angles of repose.

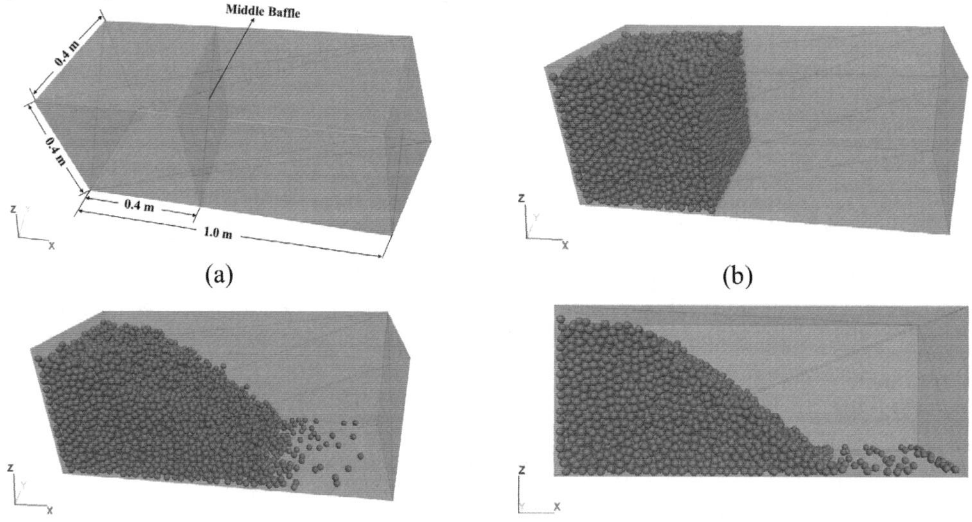

Fig. 1. Numerical simulation of angle of repose.

3.3. Simulation results processing

From the angle of repose simulation results, it can be seen that the cargo slides freely under the condition of their own gravity, and often present an uneven surface. To further solve the effective angle of repose, the boundary drawing method and Origin software are used for further processing.

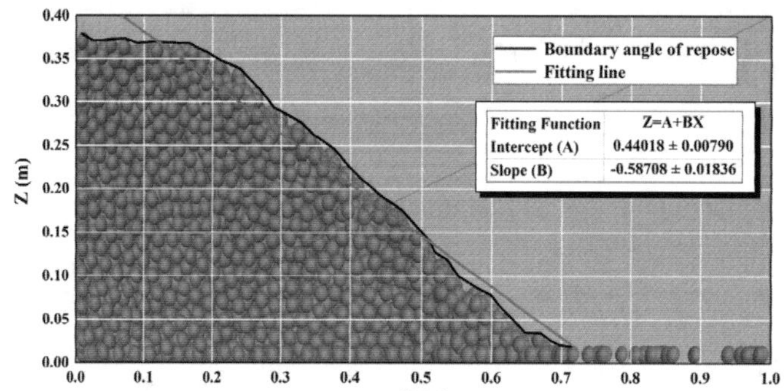

Fig. 2. Boundary fitting of angle of repose.

The specific steps are as follows:

- When the cargo reaches equilibrium, the cross section in x and z direction is taken, and the x and z coordinate points of boundary particles are extracted. Then, the plane rectangular coordinate system is reestablished to complete the drawing of the cargo repose angle boundary.
- Linear regression equation is used to fit the boundary line, and the corresponding curve fitting equation is obtained. Then, the effective angle of repose $\bar{\theta}$ is calculated by weighted processing according to the change interval of curve slope. Boundary fitting of angle of repose is shown in Fig. 2, and the data processing and analysis results are shown in Table 2.

From Table 2, it can be known that under the condition of calibrated model parameters, the angle of repose of the cargo can be corrected by the change of the friction coefficient between particles. At the same time, it is further verified that the friction coefficient between particles has a significant effect on the angle of repose.

Table 2. Data processing analysis results.

Serial number	Particle friction coefficient μ	Rolling friction coefficient μ_r	Maximum angle of repose θ_{max} (°)	Minimum angle of repose θ_{min} (°)	Effective angle of repose $\bar{\theta}$ (°)
1	0.90	0.80	31.19	29.63	30.41
2	0.60	0.45	29.16	27.68	28.42
3	0.40	0.20	25.53	24.90	25.22
4	0.35	0.18	23.57	23.06	23.32
5	0.24	0.17	21.28	20.96	21.12
6	0.22	0.16	20.91	20.49	20.70
7	0.20	0.15	19.34	19.11	19.23
8	0.18	0.10	17.63	17.37	17.50
9	0.12	0.08	15.04	14.76	14.90
10	0.10	0.05	13.85	13.60	13.73
11	0.05	0.02	11.46	11.11	11.29
12	0.01	0.01	9.69	9.16	9.43

4. Swing Table Test Simulation

4.1. Cargo hold model

Refer to reference [5], a 3D rectangular cargo hold is established in this simulation based on DEM. The similarity ratio between the size of the model and the actual cargo hold is 1:25, which can approximately simulate the uniform loading of a single cargo hold of a 57,000 DWT grade bulk carrier. Set the origin O (0,0,0) as the swing center, and the motion form of the cargo hold corresponds to the roll motion of the actual ship. Considering the complexity of actual sea conditions, the roll motion amplitude is set to 20° and the frequency is 1 Hz. In addition, similar to the angle of repose simulation test, the total mass of the cargo is about 77.40 kg, and the actual loading height is 0.22 m. The model frame and calculation model are shown in Fig. 3.

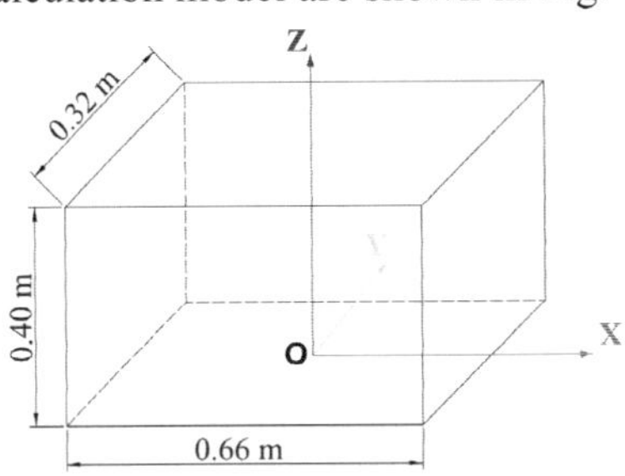

(a)

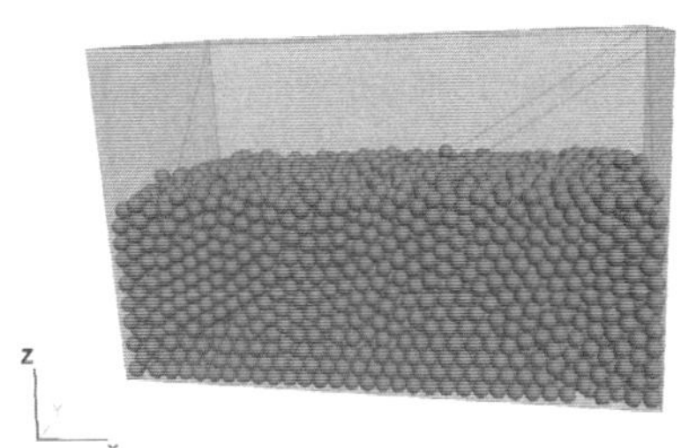

(b)

Fig. 3. Cargo hold model: (a) Model frame; (b) DEM calculation model.

4.2. Simulation steps

Based on the previous angle of repose model, six groups of angles of repose parameters are selected for simulation and analysis. The simulation mainly follows the following steps:

- Establish a cargo hold, complete cargo loading, and wait for the cargo to reach a balanced state.
- Roll motion simulation. The loading of roll motion is realized by applying sinusoidal cyclic buffer velocity to the cargo bulkhead.
- Data record. The traversal function and the command of monitoring are called to record the coordinate change of real-time movement of cargo particles and the change of contact force on cargo hold.

Journal of Physics: Conference Series **2195** (2022) 012001 doi:10.1088/1742-6596/2195/1/012001

- Data export and graphic post-processing. Change the corresponding parameters and continue to simulate the next angle of repose condition.

4.3. Result analysis

The movement of the cargo with angle of repose of 14.90° in one period is shown in Fig. 4. Whenever the cargo hold returns to the equilibrium state, the overall cargo is more likely to be tilted to one side, and the center of gravity alters obviously at this time.

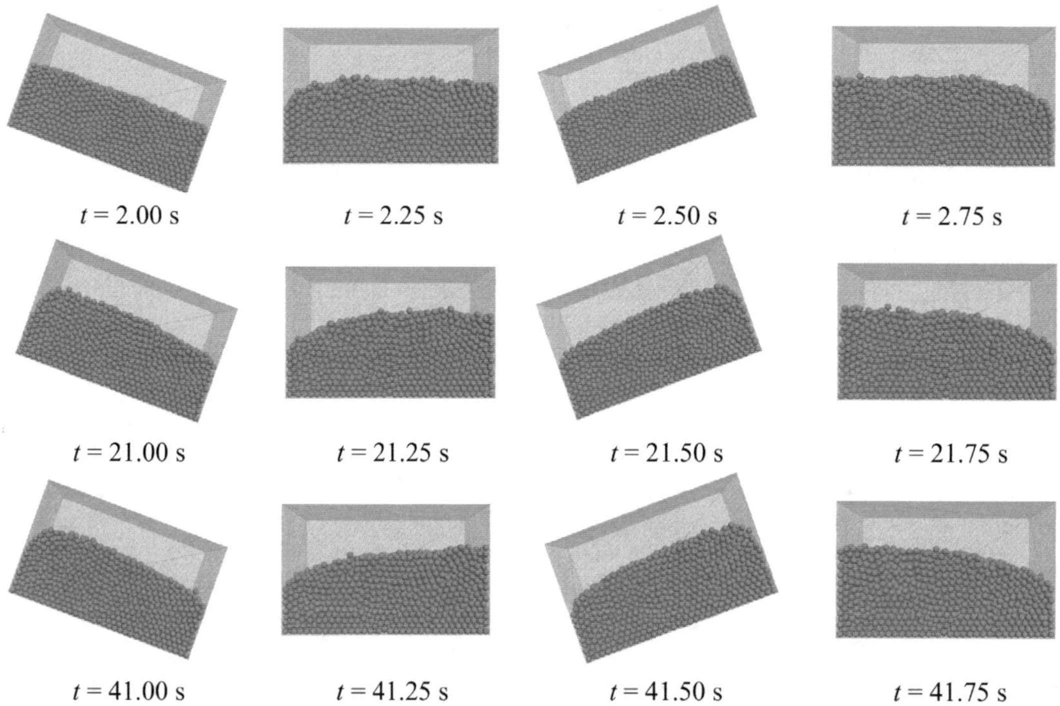

Fig. 4. Movement of the cargo with angle of repose of 14.90° in one period.

The center of gravity offset of the cargo under different angles of repose is shown in Fig. 5, in which the real-time loading curve of cargo hold roll motion is shown in Fig. 5a. It can be seen that with the decrease of the angle of repose of the cargo, the variation of the center of gravity offset of the cargo becomes more obvious. The variation of dynamic contact force on the cargo hold under different angles of repose is shown in Fig. 6. Among them, Fig. 6a shows the variation of the dynamic contact force on the cargo hold when the cargo angle of repose is 30.41°, while Fig. 6b shows the variation of the dynamic contact force difference of the other groups of angles of repose relative to the angle of repose of 30.41°. It can be seen that with the decrease of the angle of repose, the dynamic contact force on the cargo hold changes more and more large. The main reason is that the decrease of the angle of repose causes the shift of center of gravity of the cargo. The variation of dynamic contact force exactly corresponds to the variation of gravity center offset of the cargo. In general, the relationship between the center of gravity offset of the cargo and the angle of repose is mainly divided into the following three stages:

- When the angle of repose is 30.41°, the cargo has almost no center of gravity deviation.
- When the angle of repose is in the interval 23.32°~25.22°, the center of gravity of the cargo has a small deviation, which is specifically expressed as: the maximum center of gravity deviation in the x-direction is 0.00121 m, and the maximum center of gravity deviation in the z-direction is 0.00055 m.
- When the angle of repose is in the interval 14.90°~21.12°, the center of gravity offset of the cargo changes significantly.

Based on the above analysis, there should be a critical angle of repose in the interval of 21.12°~23.32°, which makes the cargo appear a large center of gravity deviation. Therefore, in order to ensure the stability of the cargo in the cabin, the angle of repose of the cargo should be greater than this critical value.

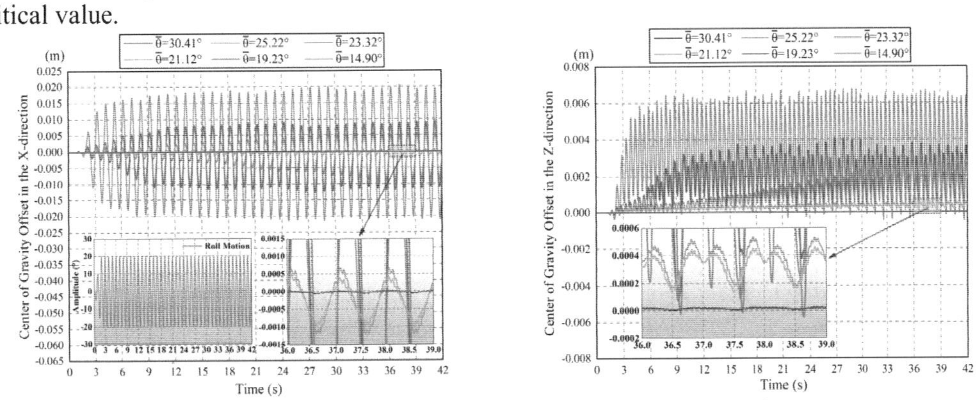

(a) (b)

Fig. 5. Center of gravity offset of the cargo: (a) X-direction; (b) Z-direction.

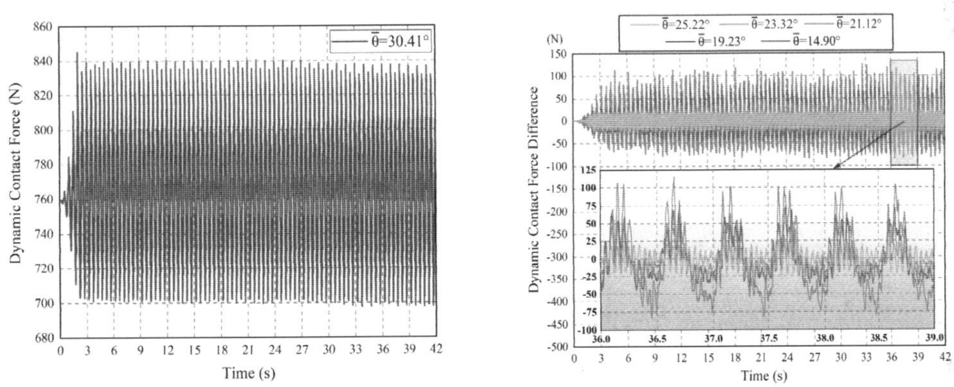

(a) (b)

Fig. 6. Variation of dynamic contact force on the cargo hold under different angles of repose: (a) $\bar{\theta}$=30.41°; (b) $\bar{\theta}$=25.22°, 23.32°, 21.12°, 19.23° and 14.90°.

5. Conclusion

Based on the rolling resistance linear model, firstly, a numerical simulation model for measuring the angle of repose of cargo is established on the basis of previous studies. Then, the movement of the cargo in the cargo hold under different angles of repose is simulated. The main conclusions are as follows:

- In the case of partial parameter calibration, the change of friction coefficient between cargo particles has a significant impact on the angle of repose. Different angle of repose can be obtained by changing the particle friction coefficient and rolling friction coefficient between particles.
- The angle of repose is proportional to the fluidity. The smaller the angle of repose is, the stronger the fluidity of cargo is, leading to the greater deviation of the center of gravity and the gradual decline of the stability of the cargo.
- Under the condition of this study, there exists a critical angle of repose: above this value, the center of gravity of the cargo is less affected and the cargo are basically stable; below this value, the center of gravity of the cargo is greatly affected, and the stability drops sharply.

Acknowledgments

This research was financially supported by the National Natural Science Foundation of China (Grant No. 51809237) and the Science and Technology Project of Zhoushan (Grant No. 2020C21022).

References

[1] ZHAO Yue-lin, MENG Shu-xian. Safety transportation of easily fluidized cargoes[J]. Journal of Dalian Maritime University, 2012,38(04):15-18.

[2] Zhang Jianwei. Numerical Study on the Effect of Liquefiable Cargo Sloshing on Motion and Stability of a Cargo Hold Section[D]. Dalian Maritime University,2016.

[3] GUAN Chen, DONG Guo-xiang, GAO Jia-yong, JIN Yun-long. Plateform experiment and research of nickel ore liquefaction process[J]. CHINESE JOURNAL OF HYDRODYNAMICS,2014,29(06):700-705.

[4] ZHOU Jian, JIAN Qiwei, WU Xiaohong, LI Ning, ZHU Yaomin. MODEL EXPERIMENTAL STUDY OF FLUIDIZATION OF IRON CONCENTRATE ORE IN BULK[J]. Chinese Journal of Rock Mechanics and Engineering,2013,32(12):2536-2543.

[5] ZHOU Jian, ZHU Yao-min, JIAN Qi-wei, JIA Min-cai, JIN Yun-long. Modeling Experiments on the Fluidization of Laterite Nickel Ore in Bulk[J]. Chinese Journal of Geotechnical Engineering,2014,36(08):1515-1520.

[6] ZHANG Jian-wei, WU Wen-feng, ZHU Fa-xin, LU Jin-shu. Numerical study on capsizing characteristics of a laterite nickel ore carrier, CHINESE JOURNAL OF HYDRODYNAMICS, 2019,34(01):77-84.

[7] Ju,Lei, Yanzhuo Xue, Dracos Vassalos, Yang Liu, and Baoyu Ni. Numerical Investigation of Rolling Response of a 2d Rectangular Hold, Partially Filled with Moist Bulk Cargo. Ocean Engineering 142 (2017): 348-62.

[8] Liu Hongjun, Zhang Wenyi, Ji Yao, Qi Bing, Li Kun. Parameter calibration of soil particles in annual rice-wheat region based on discrete element method[J]. Journal of Chinese Agricultural Mechanization,2020,41(12):153-159.

[9] Zheng Kan, He Jin, Li Hongwen, Diao Peisong, Wang Qingjie, Zhao Hongbo. Research on Polyline Soil-breaking Blade Subsoiler Based on Subsoiling Soil Model Using Discrete Element Method[J]. Transactions of the Chinese Society for Agricultural Machinery, 2016, 47(09):62-72.

[10] Wang Yunxia, Liang Zhijie, Zhang Dongxing, Cui Tao, Shi Song, Li Kehong, Yang Li. Calibration method of contact characteristic parameters for corn seeds based on EDEM[J]. Transactions of the Chinese Society of Agricultural Engineering,2016,32(22):36-42.

[11] Tian Xiaohong, Tan Bin, Liu Ming, Tan Hongzhuo, Liu Yanxiang. Study on Influence Factors of Grain Angle of Repose by Funnel Method[J]. Journal of the Chinese Cereals and Oils Association,2011,26(10):108-113.

[12] Hou Zhanfeng, Dai Nianzu, Chen Zhi, Qiu Yi, Zhang Xiwen. Measurement and calibration of physical property parameters for Agropyron seeds in a discrete element simulation[J]. Transactions of the Chinese Society of Agricultural Engineering, 2020, 36(24): 46-54.

[13] LI Wenjie, JU Lei. Numerical Simulation of Fluidization of Shipborne Solid Bulk Cargo[J]. SHIP ENGINEERING, 2019,41(12):64-70.

A novel solution for smart unmanned minibus based on intelligent fusion sensing technology

Wei Song[1] *

[1] Product Engineering Department of Nanjing Iveco Automobile Co., Ltd., Nanjing 210028, China

*Corresponding author's e-mail: song.wei@naveco.com.cn

Abstract. Thanks to recent developments in artificial intelligence and smart control technologies, unmanned vehicles are becoming pervasively adopted in a variety of scenarios. As a representative example, unmanned minibuses are integrations of vehicle level intelligence and internet of vehicles, where intelligent link mainly depends on the recognition of environment and precise control of drive-by-wire chastises. On top of these, decision-making systems for driving control can be fulfilled to achieve continuous improvement in safety and reliability. The most important parts of the telematics system include high precision map and positioning, which are the foundation for cooperation among vehicles, road-side control units, and cloud-side centers. In this work, we present a solution for unmanned minibuses that integrated smart features for unmanned driving scenarios. Both hardware design and software capabilities are presented. Prototypes have been manufactured and gone through multiple tests, which indicates feasibility of the proposed solution.

1. Introduction

In recent years, several countries have put forward timetables to stop further development of traditional fuel-powered vehicles. For example, the Netherlands will ban the sale of fuel vehicles in 2025, followed by several states in U.S. and German in 2030 [1]. Consequently, new energy vehicles (NEV) have become the priorities for most vehicle makers propelled by national strategies such as China's "Energy Conservation and New Energy", "Electric Vehicle Popularization Challenge Blueprint" from U.S, and "EU 2020 Strategic Innovation Plan". Along with development of NEVs, intelligent and internet-based functions are getting more and more pervasively adopted, leading to a global industrial change towards electrification, intelligence and network-based vehicles [2].

Buses are one of the most important vehicle types and are the major objective for traffic electrification. In most cases, buses operate on fixed routes at regulated intervals without adjustment of capacity to fit actual demand. Consequently, buses could be empty or overloaded with passengers, resulting in a waste of public resources and energy [3]. The situation tends to be especially severe for minibuses to transport the last three kilometers to final destinations. The main application scenarios of minibuses are enterprises, tourist attractions, university campuses, airports, and so forth. Characteristics of a typical minibus include simplicity in usage scene, relatively lower speed, and so forth.

Thanks to developments in artificial intelligence and electronic technologies, minibuses are getting more intelligent and upgraded with unmanned features. The core technology of an electric unmanned minibus is the integration of vehicle level intelligence and internet of Vehicles. Vehicle level intelligence mainly relies on two aspects, namely environmental perception system and onboard controllability. On the other hand, internet of vehicles system mainly covers high-precision positioning as well as efficient collaboration between vehicles and cloud-side centers. Integration between vehicle level intelligence and Internet of Vehicles technologies enables route planning, self-driving, remote control, as well as a variety of other cutting-edge technologies.

In this work, we propose a novel solution for unmanned minibuses, which covers both hardware and software parts. Prototypes have been manufactured and tested to verify performance of the proposed design. The rest of the paper is organized as follows. Section II presents the proposed solution from software and hardware perspectives. Prototypes are manufactured and tested in Section III. Section IV concludes the paper.

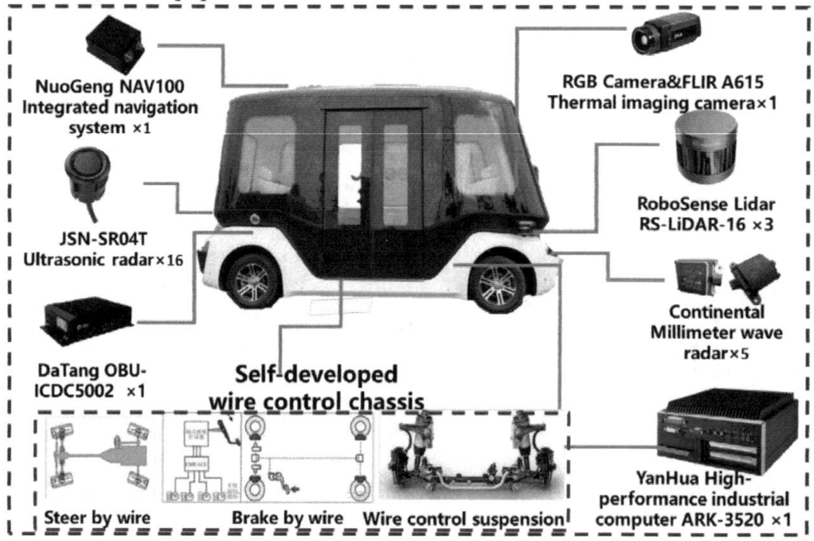

Figure1 core technologies of an electric unmanned minibus

2. Proposed unmanned minibus solution

2.1. A. Intelligent perception and control

Sensor fusion is a feasible solution to compensate for the shortcomings of relaying on a single sensor and is adopted to realize environmental perception on minibuses in complex traffic scenarios. On top of this, active obstacle avoidance, which is the prerequisite and foundation for ensuring the safe and reliable driving can be realized. When applied on minibuses, three major functions are meant to be fulfilled, namely joint calibration and communication agreement among heterogeneous sensors, target recognition, and state estimation that covers multi-target tracking and environmental map construction.

Accurate lane detection is the first step to realize these functions, which could be challenging in foggy or relatively dark areas and when multiple traffic lines coexist. In this work, an extraction-correction strategy is used, where lanes are first binarized into digital format and corrected to handle distortions. In this way, lanes can be detected through colour space and gradient threshold. Initial reference of a lane is critical for lane detection, autopilot, as well as other vehicle-level intelligent functions. In this work, initial reference of a lane is detected through weighted pixel statistics located close to the bottom of each frame that are assigned greater weight to reduce influences from noise. Results from this step help determine position of a rectangle that covers pixels on the lanes so that lane detection can be performed in a recursive manner. Coordinate points of the lanes can be calculated accordingly to facilities lane trajectory fitting, which formulates mathematical expression and helps calculate specific distances between vehicle and lanes or objectives, as shown in Figure 2.

Meanwhile, ultra-high-definition cameras with infrared systems are deployed to capture visual information of traffic signs and traffic lights, so that image reconstruction, positioning segmentation, object detection and recognition can be performed. In this work, various traffic signs are meant to be detected and sent through CAN buses to call corresponding algorithms to perform object recognition that facilities driving control.

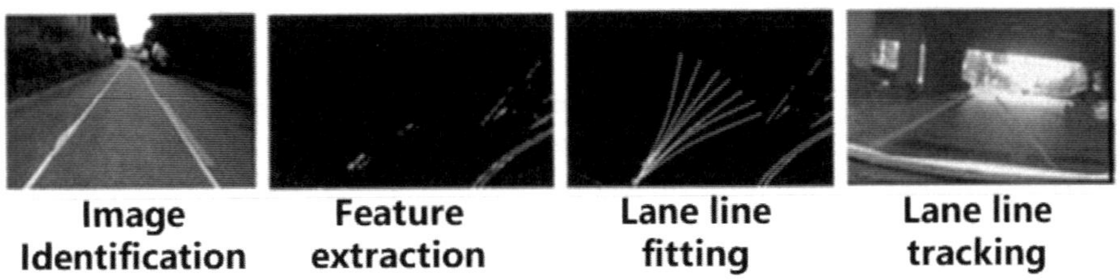

Figure 2 Coordinate point calculation for lane line tracking

As for object recognition, three different algorithms are established, namely dual-modal target sensing, dual-mode three-dimensional target sensing, and high-precision positioning. The three algorithms work in a collaborative way to achieve higher accuracy and to ensure better reliability. All the algorithms are embedded on board and are supported by cloud-side servers. Specifically, recorded pictures and videos are used for offline training through comparison and identification with help of a dual-mode data acquisition system, where over 60000 pairs of dual-mode images have been processed to accuracy of object detection especially under low illumination and low visibility conditions. Another part of image processing is conducted on cloud side using industrial computers and results transferred through 5G network. In this way, a dual-modal fusion can be performed to detect, segment and identify objectives.

Object detection is further enhanced through three-dimensional target formulation to ensure robustness against harsh conditions such as rainy, foggy, dark, and strong background noise. In this process, a three-dimensional grid is created on the sensing object. On top of this spatial grid, feature extraction is performed and transferred into a mathematical expression. The entire process, including classification, standardization and identification, is shown in Figure 3.

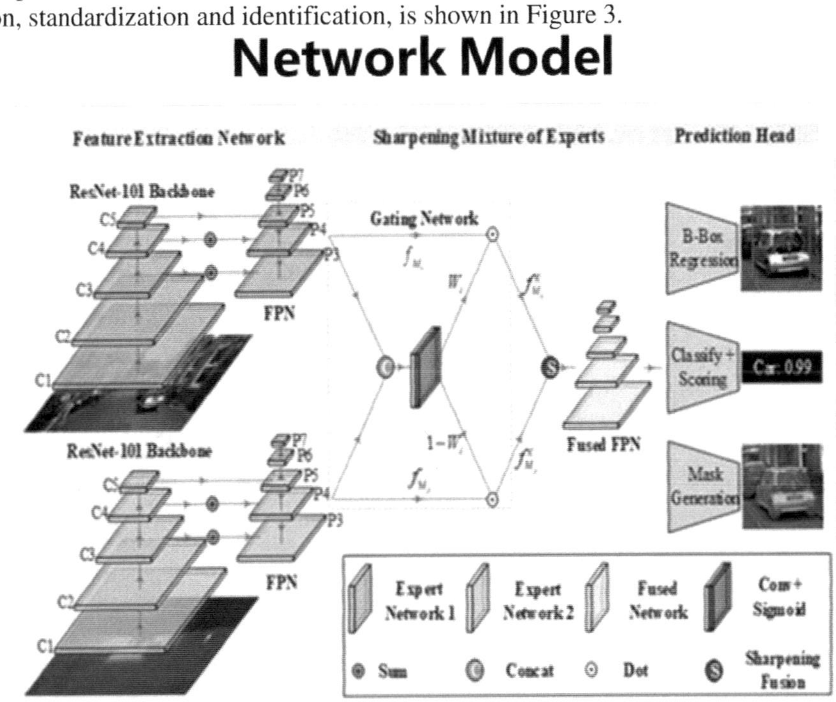

Figure 3 Entire processes of three-dimensional grid for enhanced object detection

Another critical function is high-precision positioning. In this work, positioning is achieved through integration of vision, lidar odometer, vehicle speed, as well as information from cloud side so that an optimal estimation of position can be obtained. As the first step, point distribution of voxel grid is calculated as

$$x \sim N(q, \Sigma) \tag{1}$$

$$q = \frac{1}{n}\sum_i x_i \qquad \Sigma = \frac{1}{n}\sum_i (x_i - q)(x_i - q)^T \tag{2}$$

where N is the collection of data points in a voxel grid, q stands for the center of each voxel grid, Σ denotes mean square deviation of the distance between each voxel grid point and center, x_i is the *i*th data point, and in total there are *n* points. In this way, the probability density for a single point locate in a given voxel grid can be expressed as

$$p(x) \sim \exp\left(-\frac{(x-q)^T \Sigma^{-1} (x-q)}{2}\right) \tag{3}$$

Similarly, the case for all data points can be given as

$$\text{score}(p) = \sum_i \exp\left(-\frac{(x_i' - q_i)^T \Sigma_i^{-1} (x_i' - q_i)}{2}\right) \tag{4}$$

Outputs from this function will be used as the evaluation standard for object detection.

2.2. Intelligent wire-controlled chassis

Intelligent chassis system is essential to work together with smart software, which is the foundation to incorporate intelligent network-based technologies. A typical wire-controlled chassis is an assembly of vehicle controller, braking system, electronic stability programmer (ESP), as well as a few other key components. In this work, a novel chassis design is proposed to realize functions of minibuses with unmanned features.

Unlike traditional hydraulic brake systems that use vacuum boosters for hydraulic lines to work, brake-by-wire systems convert hardware signals from brake pedals into digital format from the brake pedal in manual mode, or from embedded controller in unmanned mode. Consequently, a redundant braking system needs to be deployed for both safety and functionality. In this work, signals are sent to redundant brake controllers synchronously to control servo motors on wheels. A four-wheel hydraulic disc brake and an electric booster drum brake system are adopted, which also integrates electronic parking function, stopping vehicle from slipping or braking intervention. Smoothness is also guaranteed thanks to the two-way braking features.

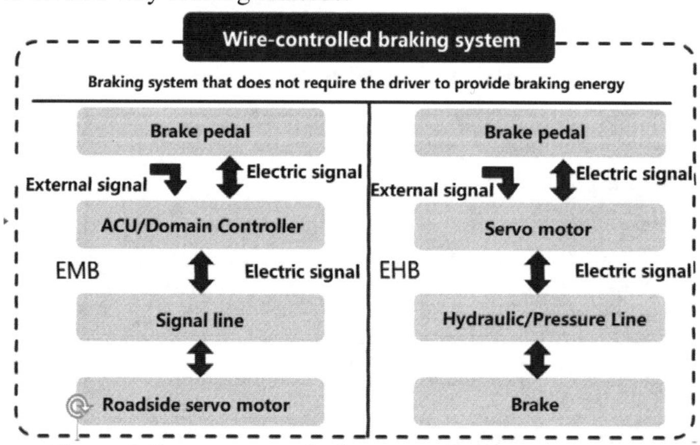

Figure 4 Components of wire-controlled braking system

The intelligent unmanned minibus adopts a wire-controlled hydraulic power steering control system, which retains the hydraulic steering system design of the traditional commercial vehicles and is controlled by a servo motor. One extra motor is added to the steering column to drive steering gear through a deceleration mechanism. Sensors are deployed under steering wheels to read rotation angle and torque synchronously. Electronic controller units will be activited accordingly, which helps determine the angle to be steered based on steering commands sent by the intelligent driving system. Meanwhile, speed is sent back as a feedback signal for realtime control. In this way, a closed controling loop is formed. Furthermore, a customized suspension system is designed to ensure a global optimization of stiffness, frequency deviation, weight, as well as other factors.

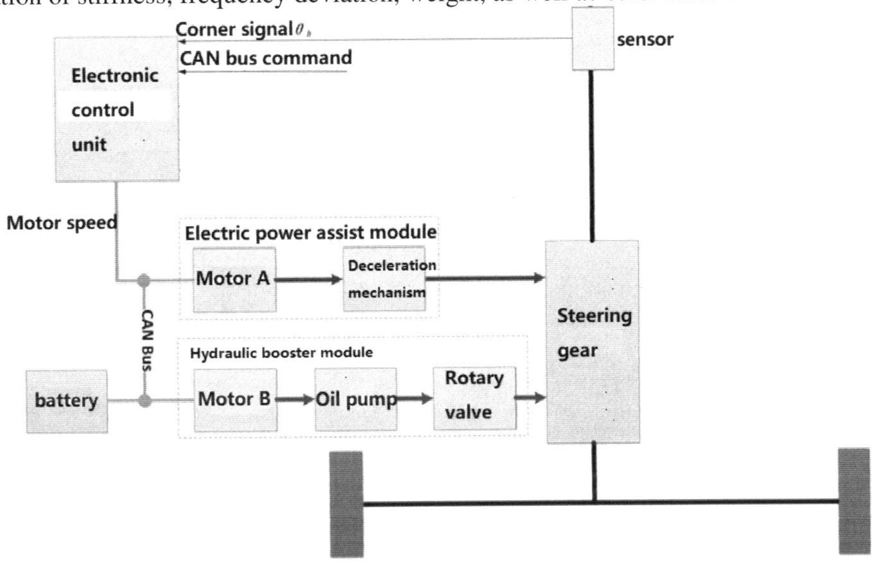

Figure 5 Wire-controlled hydraulic power steering control system for the intelligent unmanned minibus

3. Test Results

To verify the proposed minibus design, prototypes have been built and deployed in an industrial campus in Nanjing. Both vehicle hardware design and intelligent features are tested. Detailed results are given as follows.

3.1. Hardware capability

For unmanned minibuses, hardware design is the foundation of flexibility, smoothness and smart control. Meanwhile, the minibus is designed to meet requirements of comfort and manuverbility in both manned and unmannd modes [6]. Be more specifically, the minimum turning radius under low velocity scenario is around 4.6m when steering wheel is at 500 degree angle. With help of smart controlling software, steering error can be controlled within 1 degree. Mechanical delay for breaking is within 0.2 seconds. Ramp tests are carried out on an impending bench, where both step and ramp instructions with up to 10 Nm are executed. Test results show that ramp response delay is less than 2 seconds with an overshoot in torque response at 0.2 Nm. Steady state error is 0.1Nm and the maximum gradient of torque response is 7.6 Nm/s. As a brief conclusion, dynamic responses of the vehicle are stable and accurate.

3.2. Software and intelligent control

Software is meant to fully tap hardware capabilities. Besides aforementioned functionalities such as object detection and autopilot, the unmanned minibus is also designed to incorporate V2X Internet of Vehicles features, which perceives traffic signals. Communications among vehicles and with Road Side Units (RSU) are also enabled to realize vehicle identification, electronic deduction, billing, and so forth. Onboard sensors provide information such as the current position of the vehicle, vehicle speed, etc. On top of this, routing and charging guidance can be performed in unmanned mode. In manual mode, recommended route and speed will be displayed on user interface.

4. Conclusion

This paper presents an unmanned minibus solution in a comprehensive way to cover multiple key functions on both hardware and software design. Intelligent features of the vehicle, including object detection, identification, and autopilot, are fulfilled through software trained offline, whereas hardware design guarantees manoeuvrability, comfort, and robustness. Furthermore, the proposed design also incorporates internet-based features that allows for remote software upgrade. In the end, prototype vehicles have been manufactured and verified through operations in real driving conditions.

References

[1] Yu X.Y., Lan Y., Huang Y.Y., et al. (2019) Learning from South Korea's New Policy for Harnessing Haze to Strengthen my country's Management of Diesel Vehicles. Environment and Sustainable Development, 4.
[2] Bian M.Y., Li K.Q. (2018) The top-level design of the strategy of automobile power based on intelligent networked vehicles. China Engineering Science, 1.
[3] Perumal S., Dollevoet T., Huisman D., et al. (2021) Solution Approaches for Integrated Vehicle and Crew Scheduling with Electric Buses. Computers & Operations Research, 132(2):105268.
[4] Chen Q., Xie Y., Guo S., et al. (2021) Sensing System of Environmental Perception Technologies for Driverless Vehicle: A Review of State of the Art and Challenges. Sensors and Actuators A Physical, 319:112566.
[5] Wan S., Li X., Xue Y., et al. (2019) Efficient computation offloading for Internet of Vehicles in edge computing-assisted 5G networks. The Journal of Supercomputing, 4.
[6] Chen Z.P., Yang J.W., Qi K.G. (2020) Hardware requirements analysis of thermal management controller for hybrid electric vehicles[J]. Automotive Electrics, 5.

A ship target dynamic recognition method based on image recognition and VR technology

Peng Fei[1,a], Xiumin Chu[1,b], Xiongfei Geng[2,c], Jianan Luo[2]*

[1]School of Transport and Logistics Engineering, Wuhan University of Technology, Wuhan, China

[2]China Waterborne Transport Research Institute, Beijing, China

[a]abelfei@hotmail.com

[b]chuxm@whut.edu.cn

[c]gengxf@wti.ac.cn

* Corresponding author: luojianan@wti.ac.cn

Abstract—Effective identification of ship identity is an important means of water traffic management. Aiming at the technical limitations of AIS, radar and other means, a ship automatic recognition method based on machine learning is proposed, and a water traffic visualization system integrating multi-source recognition technology is realized based on VR technology. The test shows that the method and system can solve the technical problems of ship identification caused by data easy to be tampered with, difficult to distinguish massive targets and large obstacles, and provide an effective method for navigation support in key waters.

1. Introduction

With the rapid development of social economy and the deepening of water transportation, the number of ships is increasing, and the types are diversified. There are ship types such as oil tanker, bulk carrier, container ship and fishing boat, which makes the action coordination of ships difficult. At the same time, the ship is developing towards large-scale and high-speed, which not only increases the operation difficulty of the ship itself, but also puts forward higher requirements for the channel. It is necessary to provide a good traffic environment, special measures and services for its safe navigation. In addition, the increasing number of dangerous goods on board and the restriction of the controlled river section have led to the increase of navigation pressure of inland waterway, traffic congestion and hidden dangers of water traffic accidents, threatening shipping safety and river ecological environment, and restricting the normal operation of social economy [1]. Therefore, the intelligent tracking and monitoring of inland waterway, especially the navigation ships in the control river section, will help to improve the navigation command decision-making and improve the navigation status of the control river section.

Water traffic dynamic monitoring can find and coordinate maritime traffic targets in time. It is an important technical means to improve the efficiency and safety of ships entering and leaving the port, berthing and berthing, and navigation in the port channel. At present, VTS, AIS, radar, real-time monitoring video, GNSS positioning terminal, binding of mobile app and ship location, ship report, remote identification (RFID) are the main technical means of dynamic supervision of water traffic [2-3].

However, inland waterways face more technical obstacles, such as narrow waterways, many intersections, high ship density, many water navigation obstructions, large changes in channel water depth, complex navigation environment, long and wide navigation area, etc. Therefore, it is difficult to apply image recognition technology to inland ship tracking and monitoring. To sum up, the existing data fusion methods can not fuse the image with ship positioning, resulting in the decline of ship tracking accuracy at the port and poor navigation efficiency in the port channel.

This study uses modern information collection and AR monitoring technology to monitor ship traffic conditions, rectify ship traffic order and assist ship navigation. Through these measures, the establishment of good traffic order and the reduction of ship collision, grounding, reef and other maritime accidents are of great significance to improve the navigation capacity of inland waterways, reduce ship traffic accidents, reduce pollution and improve the information level. [4] At the same time, it helps to strengthen the law enforcement of waterways and improve the efficiency of waterway supervision. The research on port digital service technology can promote the application of land traffic tracking and monitoring and other video tracking and monitoring fields.

2. SHIP TARGET LOCATION AND RECOGNITION

2.1. Ship target detection in HSV color space

The detection of ship targets through HSV color space is different from frame difference detection method. Frame difference method belongs to continuous multi frame dynamic target detection, while HSV color space detection belongs to single frame static target detection [5]. Because the color texture of water is obviously different from the ship target, the hue, saturation and lightness of water in HSV color space image are different from the ship target.

HSV space ship target detection principle: through research and analysis, most of the upper part of the hull of many ship targets are brightly colored, and the s component is obviously different from that of water, so the s component is an important means to detect the hull; However, the lower part of the hull is in contact with the water surface for a long time. Generally, the color is gray and the lightness value is relatively low, especially for small cargo ships. Therefore, there is a certain difference between the V component of the lower part of the hull and the water, so it is more effective to detect the lower part of the hull; Then combine s and V component detection to get a more complete target. In addition, due to the shadow of shore objects or ship hull affected by light, the effect of target detection will also be affected. Through the monotonous hue H component of shadow and fixed within a certain range, the influence of shadow can be removed by finding out the relevant range value through experiments.

HSV space ship target detection process: since the measurement scene is in the water area, the ship targets on the water shall be separated. In order to eliminate the influence of the shore, a detection area shall be set in the water area. Since the water area in the detection area is generally larger than the ship target area, the maximum value of the statistical histogram obtained for the s and V components in the detection area is the water background reference value t of the component T(s) And T(v), and then detect the s and V components at the virtual coil position of the current frame detection area, which is consistent with the reference values T(s) and T(v) If the absolute value of the difference exceeds a certain threshold, the pixel is judged as 1, otherwise it is 0; then the absolute value diagrams of the two components are calculated or fused together to judge that pixel 1 is the target and pixel 0 is the water background; finally, the shadow is removed through the h-component value of the virtual coil position. When the h-component value of the pixel with target 1 is within the shadow fixed value range, the pixel is reset to 0.

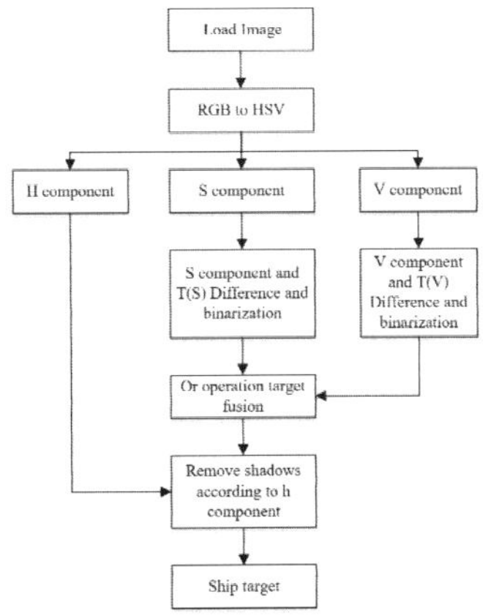

Figure 1. Basic technical process of ship target recognition method.

2.2. Virtual coil target capture

Several problems to be solved in ship target data acquisition and capture: First, capture when the ship target appears in the best position in the camera. Second, ensure that the capture of ship targets in the image is complete and the screen ratio meets the standard requirements. Third, the captured ship target shall be marked without repeated capture.

2.2.1.Setting of virtual detection coil and detection area: on the video image of the fixed gun, the virtual coil is set in the center of the water area, perpendicular to the navigation direction of the ship, and is composed of three rows of coil blocks side by side. Each coil block is adjacent to 5 to 8 coil blocks. The detection area is set to facilitate the use of HSV space ship target detection method. The detection area is set in the water surface area where ships other than the shore often appear.

2.2.2.Virtual coil block detection: each coil block, as an independent calculation unit, calculates whether the change of each pixel of the coil block exceeds the set threshold through the frame difference method or HSV color space detection method, and then counts the ratio of the number of pixels exceeding the threshold of the whole coil block to the total pixels of the coil block, If it is greater than 50%, it is judged that the target is detected by the virtual coil block and marked as 1. If it is less than or equal to 50%, it is marked as 0.

2.2.3.Target detection judgment: the rule for judging whether the virtual coil detects the target, and neighborhood retrieval is carried out for each coil block marked 1. If the number of neighborhood coil blocks marked 1 of a coil block marked 1 is greater than or equal to 3, it is judged that the virtual coil detects the ship target, If the number of adjacent coil blocks marked 1 of all coil blocks marked 1 is less than 3, it is judged that no ship target is detected.

2.2.4.Start capture: when the ship target triggers the virtual coil, the capture machine will capture the complete target through focusing, moving and other operations at the virtual coil position.

2.2.5.Waiting for the capture to be completed: waiting for the capture to be completed is a process of constantly reading video images. When the ship target is found to start the capture, the ship target will continue to enter the virtual coil until the stern leaves the virtual coil. This process is a process of waiting for the capture to be completed. When the ship continues to enter the virtual coil until it leaves, the virtual coil block will continue to detect the ship target until there is no ship target, indicating that it is waiting for the capture to be completed.

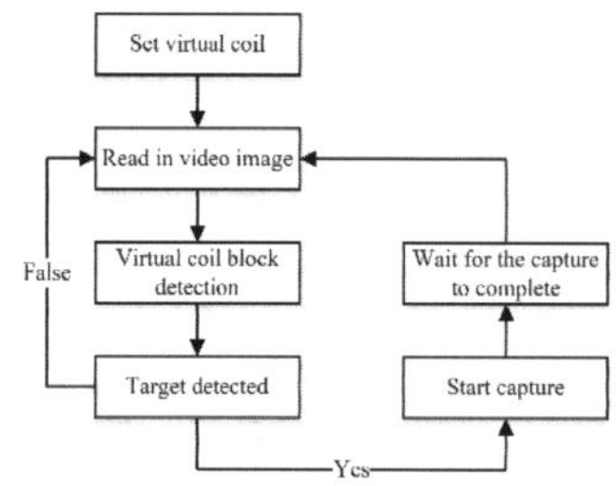

Figure 2. Ship target capture by CCTV.

2.3. Ship location and segmentation

The core task of ship target location and segmentation is to obtain the position information and contour information of all ship targets in a given image, so as to accurately separate the image of each ship target from the original image. The practical approach is to detect all ship targets in the image and frame the object with a rectangular frame. This is also a basic task in computer vision - target detection. This is a basic core module for us to build the whole system. Only by accurately framing all ship targets can we build an efficient ship identification and monitoring system. Target detection algorithms are roughly divided into traditional algorithms and algorithms based on deep learning.

Another category of target detection algorithms based on deep learning is non-region-based recommendation. The representative of this kind of algorithm is SSD algorithm [6]. The design idea of this kind of algorithm is end-to-end. The generation of candidate regions, feature extraction, location offset prediction and category prediction are encapsulated in the same network, which greatly improves the overall performance of the algorithm. SSD algorithm is one of the detection networks with the best real-time performance and the highest detection accuracy. Therefore, the construction of this system will be based on SSD algorithm.

In the training phase, the SSD algorithm framework is composed of input, feature extraction, default box generation, matching, prediction network and loss function, as shown in Figure 3 below. It is introduced as follows:

2.3.1.Input: the input information required by the SSD algorithm framework includes the training picture and the position information of all targets in the picture (i.e. the coordinates of the real frame) and category information (i.e. category number);

2.3.2.Feature extraction: it is composed of vgg-net (visual geometry group) and other neural network architectures with excellent performance to extract picture features and output feature maps extracted from different convolution layers. The selected convolution layer also becomes the feature extraction layer;

2.3.3.Default box generation and matching: according to the preset parameters, a certain number of default boxes with different aspect ratios are generated on different feature extraction layers for target box selection. After the default box is generated, it will be matched with the real box in the input information according to a certain matching strategy, and the matching result information will be sent to the prediction network;

2.3.4.The prediction network is composed of a group of convolution layers. After synthesizing the characteristic map and matching information, the target category prediction and position offset prediction are given;

2.3.5.The loss function layer will evaluate the gap between the predicted value and the real value, and the optimization of the loss function will adjust the weight of the network in back propagation, so that the network can better detect the target and fit the location information and category information of the target.

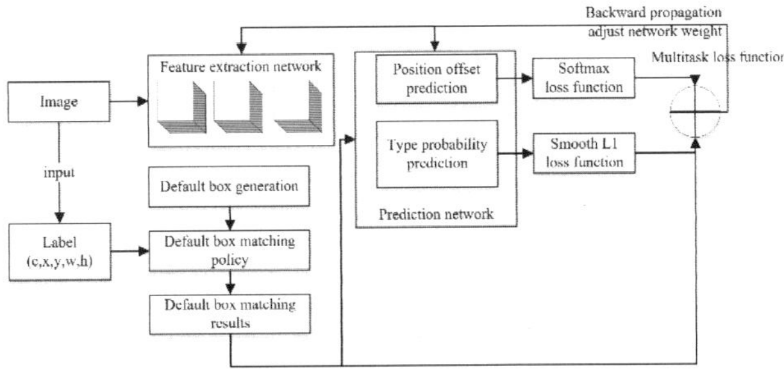

Figure 3. Framework and flow diagram of SSD algorithm training stage

SSD algorithm regards location prediction and category prediction as regression problems at the same time, so the loss function can combine location loss and category loss. Therefore, the form of loss function is multitasking. The loss function is an important driving force of neural network, because its optimization determines how the network learns the content. Therefore, SSD network adopts "smooth"_ L1 loss function is used as position loss function, and softmax loss function is used as confidence loss function, and then through a coefficient factor α Combine the two. The mathematical expression of the loss function in SSD algorithm is shown in formula (1).

$$L(x,c,l,g) = \frac{1}{N}(L_{conf}(x,c) + \alpha L_{loc}(x,l,g)) \quad (1)$$

Where n represents the number of matched candidate boxes. If n is equal to 0, l is considered_ loc=0, α It is a weight coefficient used to balance the losses of the two parts and avoid excessive proportion of the losses of one part. The expression of position loss function is shown in formulas (2) to (6).

$$L_{loc}(x,l,g) = \sum_{i\in Pos}^{N} \sum_{m\in\{cx,cy,w,h\}} x_{ij}^{k}\, smooth_{L1}(l_i^m - \hat{g}_j^m) \quad (2)$$

$$\hat{g}_j^{cx} = \frac{(g_j^{cx} - d_i^{cx})}{d_i^w} \quad (3)$$

$$\hat{g}_j^{cy} = \frac{(g_j^{cy} - d_i^{cy})}{d_i^w} \quad (4)$$

$$\hat{g}_j^{w} = \log(\frac{g_j^w}{d_i^w}) \quad (5)$$

$$\hat{g}_j^{h} = \log(\frac{g_j^h}{d_i^h}) \quad (6)$$

Where, x_{ij}^k indicates the matching condition. When it is 1, it means that the i_{th} default box matches the j_{th} real box, and the real category is k, otherwise it is 0. The position loss function adopts the position regression function smooth L1 loss, where l represents the predicted coordinate offset, $\hat{g}_j$ represents the coordinate offset corresponding to the real box on the matching of the current default box. d_i stands for the default box. The mathematical expression of the confidence loss function is as follows:

$$L_{conf}(x,c) = -\sum_{i\in Pos}^{N} x_{ij}^p \log(\hat{c}_i^p) - \sum_{i\in Neg} \log(\hat{c}_i^0) \quad (7)$$

$$whrere\ \hat{c}_i^p = \frac{\exp(c_i^p)}{\sum_p \exp(c_i^p)} \quad (8)$$

Where, C represents confidence, and the meaning of other variables is the same as above.

3. Ship Feature Extraction and Matching Method

3.1.Ship feature extraction

In this study, ORC (Oriented FAST and Rotated BRIEF) method is used to extract ship propagation features. ORB is a fast feature point extraction and description algorithm. The following are the steps of ORB feature extraction and corresponding feature matching [7].

ORB uses fast (features from accelerated segment test) algorithm to detect feature points. Taking the point to be detected as the center, the gray value is compared with the gray value of 16 adjacent pixel points. When compared with other pixels, if the gray difference is greater than a set threshold, it is considered that the point to be detected is different from the adjacent point. If 12 consecutive points are different from the point to be detected, the point to be detected is considered as a corner point, that is, a feature point.

In order to avoid taking too much time to calculate most non feature points, this method first calculates whether the gray values of 1, 5, 9 and 13 points are the same. Because if the point to be detected is a feature point, at least three of the four points are the same. Therefore, by comparing the above four points, the detection speed of feature points can be accelerated.

ORB uses BREF algorithm to calculate the descriptor of feature points. The core idea of BREF algorithm is to select n point pairs around feature point P with specific rules and combine the comparison results of these n point pairs as descriptors. Find the point pair around the feature points and obtain the 0 or 1 description of the point pair by comparing the gray value of the point pair. The same method is used for other point pairs to finally generate orb descriptors.

Specifically, the feature description area is framed with the feature point as the center and a certain length as the radius. Select 512 pairs of points and number them, P1 (a, b)...PN (a, b)... P512 (a, b). The descriptor of feature points is obtained by combining the calculated attribute results of multiple points. The characteristic descriptor of orb can be considered as a string of 0 and 1.

Since the descriptor of ORB feature points is a string of 0 and 1 binary strings, the matching of feature points can be transformed into string matching, which greatly saves storage and matching time. The matching of orb features is actually to calculate the Hamming distance between two feature vectors, that is, the number of different characters in the same position. It is defined as follows:

$$d(f^1, f^2) = \sum_{i=1}^{d} XOR(f_i^1, f_i^2) \qquad (9)$$

3.2.Ship feature matching

Performing the above processing on each ship target feature point to obtain a feature point set of an effective ship target image. Using FLANN matching algorithm, the first feature point set kp1 and the second feature point set kp2 are extracted from any two effective ship target images respectively [8]; According to the first feature point set kp1 and the second feature point set kp2, the corresponding first feature description set des1 and the second feature description set des2 are obtained by using the SIFT feature extraction method or surf feature extraction method. Then using the feature matcher FLANN, the first feature description set des1 and the second feature set des2 are matched to obtain multiple sets of feature matching point pairs, and the Euclidean distance between each set of feature matching point pairs is calculated.

$$\begin{gathered} A = (x_1, x_2, \dots x_n) \\ B = (y_1, y_2, \dots y_n) \\ AB = \sqrt{(x_1 - y_1)^2 + (x_2 - y_2)^2 + \cdots + (x_n - y_n)^2} \end{gathered} \qquad (10)$$

Where, A and B are two different feature matching points in a set of feature matching point pairs, x_n and y_n is the coordinate of the feature point corresponding to a feature matching point and the coordinate of the feature point corresponding to the B feature matching point in the nth group of feature matching point pairs, and ab is the Euclidean distance.

Select a set of feature matching point pairs with the smallest distance as the minimum distance and record it as the standard value. A set of feature matching point pairs with a minimum distance of less than 3 times are used as the matching degree. If the matching degree is less than the preset matching degree threshold, the matching feature corresponding to the matching degree is stored in the ship database as the nearest matching feature for subsequent search. If the matching degree is greater than or equal to the preset matching degree threshold, the matching degree is used for feature matching in the ship database to obtain the nearest neighbor matching feature. According to the nearest matching feature, obtain the corresponding electronic chart in the ship database; Using the ship automatic identification system, retrieve the identity of the target ship, determine the identity information of the target ship and calibrate it, compare the location information of the ship automatic identification system, obtain the AIS location information of the target ship, and then obtain the ship data;

Among them, the identity information of the target ship includes ship call sign, mobile identification code, course, speed, Captain, ship width, draft, port of destination, cargo type and sailing time. The ship data includes AIS data of each target ship and a set of feature matching point pairs corresponding to the nearest matching feature. Further, we coded the above method and applied it to the Lancang River AR [9-10] ship monitoring system in Yunnan Province, China, as shown in Figure 4.

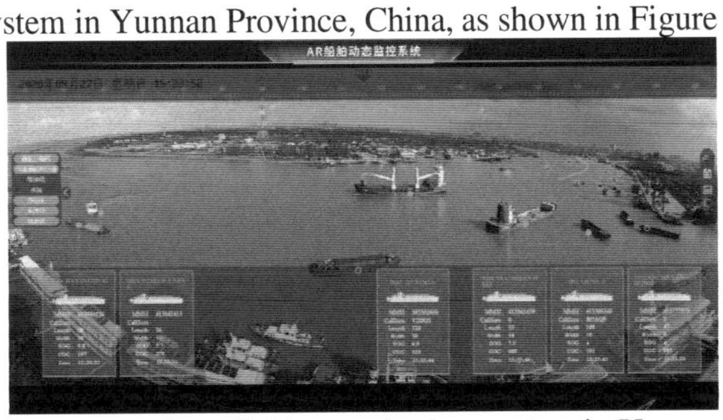

Figure 4. Demonstration of ship monitoring system in Yunnan Province

4. CONCLUSION

This study proposes an automatic ship identification method based on machine learning, which can automatically capture and identify ship objects through video system, and quickly identify ship identity and location by integrating ship data such as AIS and VTS. By further integrating AR application scenarios, the corresponding ship digital information can be superimposed on the real visual scene, and the information beyond vision can be superimposed to enrich the information perception of the channel. In the application scenario of ship dynamic monitoring in key waters, the use of AR technology can make managers more simply and intuitively observe the navigation environment information of the ship and the ship navigation in the port.

ACKNOWLEDGMENT

This work was supported by Green Intelligent Inland Ship Innovation Programme.

REFERENCES

[1] XU Wuxiong, CHU Xiumin, LEI Jinyu, Parallel-Control-Based Navigation Risk Management System for Busy Inland Waters, 3rd ed., vol. 2. Oxford: Clarendon, 2018, 41(01), pp. 64-69.
[2] Gangyan Xu,Chun Hsien Chen,Fan Li,Xuan Qiu. AIS data analytics for adaptive rotating shift in vessel traffic service[J]. Industrial Management & Data Systems,2020,120(4).
[3] Sun Yang, Peng Xinya, Ding Zexuan, Zhao Jiansen. An Approach to Ship Behavior Prediction Based on AIS and RNN Optimization Model[J]. International Journal of Transportation Engineering and Technology,2020,6(1).
[4] Bai X, Chen Z, Zhang Y, et al. Infrared ship target segmentation based on spatial information improved FCM[J]. IEEE transactions on cybernetics, 2015, 46(12): 3259-3271.

[5] Huang Z K, Liu D H. Segmentation of color image using EM algorithm in HSV color space[C]//2007 International Conference on Information Acquisition. IEEE, 2007: 316-319.
[6] Chen Q, Huang N, Zhou J, et al. An SSD Algorithm Based on Vehicle Counting Method[C]//2018 37th Chinese Control Conference (CCC). IEEE, 2018: 7673-7677.
[7] Shu C W, Xiao X Z. ORB-oriented mismatching feature points elimination[C]//2018 IEEE International Conference on Progress in Informatics and Computing (PIC). IEEE, 2018: 246-249.
[8] Gupta A, Chaudhary D K, Choudhury T. Stock prediction using functional link artificial neural network (FLANN)[C]//2017 3rd international conference on Computational Intelligence and Networks (CINE). IEEE, 2017: 10-16.
[9] Lvov M S, Popova H V. Simulation technologies of virtual reality usage in the training of future ship navigators[J]. 2019.
[10] Wu H, Yang J, Chen C, et al. Research of Virtual Ship Fire-fighting Training System Based on Virtual Reality Technique[C]//IOP Conference Series: Materials Science and Engineering. IOP Publishing, 2019, 677(4): 042100.

Research on Abnormal Event Classification of Expressway Tunnel based on Security Control

Jiangbi Hu[1], Zechao Zhang[1], Qingyun Cao[1], Shangwen Qu[1] and Ronghua Wang[1]*

[1]Faculty of architecture, civil and transportation engineering, Beijing University of Technology, Beijing, 100124, China

*Corresponding author's e-mail: wangrh@bjut.edu.cn

Abstract: Based on the perspective of tunnel traffic safety operation and control and the human, vehicle, road, and environment system in tunnel operation process, adopting event tree analysis method according to the different influencing factors, generation mechanism and performance characteristics of tunnel abnormal events, this research propose abnormal event classification tree diagram. According to the current abnormal event management and control mechanism in China, the administration departments and disposal facilities associated with different types of abnormal events in expressway tunnel operation are sorted out, which can provide decision-making support for the rapid treatment of abnormal events and realize intelligent control of tunnel.

1. Introduction

Accurate judgment of tunnel abnormal event type is the core technology and basic work of tunnel abnormal event control and emergency response capacity construction [1-4]. The establishment and improvement of a systematic and comprehensive tunnel abnormal event classification system plays a crucial role in the selection of abnormal event treatment scheme, allocation of resources, determination of rescue opportunity and other aspects, and is also an important step towards intelligent "one-click response" of tunnel.

In terms of classification types, Wang S.F. et al. classified traffic events affecting the normal operation of expressway tunnels into seven categories, including severe weather, traffic accidents, fire accidents, dangerous accidents, planned events, emergency events and other events [5]. Shen Y.J. et al. classified 121 road tunnel traffic accidents in China into rear-end accidents, collision accidents, vehicle ramming into tunnel wall accidents, fire accidents, vehicle rollover accidents and other accidents [6]. As to classification methods, based on the definition of the event, Wang Y. et al. extracted the characteristic attributes of the event from the definition of the event and classified the event using the characteristic attributes of the event [7]. Based on the evolution of emergencies, Zhang Y.Q. constructed an emergency case database from two perspectives: event classification system structure and event evolution law [8].

At present, there are some problems in the classification of abnormal events in expressway tunnel, such as unclear event classification, fuzzy boundary, or cross classification. There is still a large space for optimization and research necessity in the systematization and comprehensiveness of the classification system of abnormal events in expressway tunnel. In this paper, the mechanism of abnormal events generated by tunnel operation is analyzed based on the level of traffic system, and the event tree analysis method commonly used in risk analysis is adopted to analyze the event factors, event types, event treatment departments and facilities involved in the identification, judgment, and treatment of abnormal events in expressway tunnel. It is necessary to establish a systematic, clear, and practical classification system of abnormal events in tunnels, which is of great significance to the safety, efficient operation, and intelligent control of expressway tunnels.

2. Classification characteristics of tunnel abnormal event

2.1 Influencing factors and mechanism of abnormal tunnel events

The traffic operation system of expressway tunnel is composed of multiple factors such as human, vehicle, road, and environment. The incoordination and non-integration of any factor and link in the operation system will form potential safety hazards and even cause serious accidents. The correlation of each influencing factor is shown in Figure1. When different factors or different individuals of the same factor do not coordinate or even conflict with each other in the operation process, abnormal events will occur in the expressway tunnel.

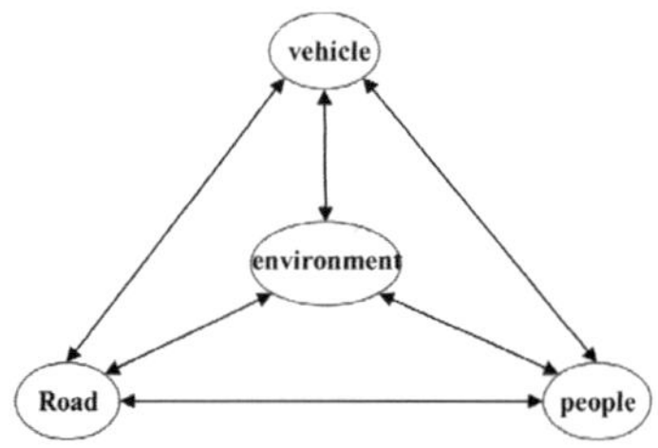

Figure 1. Influencing factors of abnormal tunnel events

The human factors in the tunnel can lead to casualties and traffic accidents if not managed well. Since the clearance in the tunnel is limited, once there are people in the tunnel, they will be very close to the driveway, which will easily lead to conflicts between people and passing vehicles, resulting in traffic accidents.

In abnormal events caused by vehicle factors, vehicle is an important component of tunnel traffic operation. Conflicts between single vehicles, vehicles and tunnels will be caused by uncoordinated operation between vehicles, large speed difference between vehicles, abnormal behavior of vehicles: changing the driving state of other vehicles, increased driving workload of drivers, vehicle faults and so on.

The abnormal events caused by tunnel factors are mainly structural damage, loss of function and obstacles of tunnel pavement, which directly lead to the vehicle running abnormally in tunnel.

Tunnel environmental factors have a certain influence on the occurrence of abnormal events, but environmental factors, personnel, vehicles, and tunnel factors will not produce abnormal events without incompatible operation or conflict. Therefore, environmental factors will not be considered separately in the classification process of tunnel abnormal events.

2.2 Analysis of tunnel abnormal event characteristics

In terms of the conflict between people and vehicles, it has been stipulated in The Road Traffic Safety Law of the People's Republic of China that the people who violate the tunnel are classified as abnormal personnel events [9], and the objects of control are the people who enter the expressway illegally. Abnormal event features are identified and judged by the characteristics of personnel.

In the conflict between vehicles and vehicles, abnormal event features are represented by vehicle features, and the object of abnormal events in operation control is the vehicle driven by the driver. During the process of driving in the tunnel, vehicles violate relevant regulations and increase the safety risk of other vehicles in the tunnel, including over speed driving, low speed driving, parking in lane, reversing driving, astern running, frequent lane changes, crossing the line driving abnormal traffic behavior events; If a vehicle has an accident in a tunnel, it may cause harm to people and property, etc. Once the event occurs, it is very easy to cause more serious secondary accidents if not timely treated, including multiple vehicles rear-end collision, scraping collision and single vehicle rollover, cartwheel, fire and other single-vehicle accidents, and events that affect the safe operation of other vehicles in the tunnel.

There are two aspects of conflict between vehicles and tunnels. The objects of abnormal events are vehicles, including collisions between vehicles and tunnels when vehicles hit the tunnel wall or the access road. The object of abnormal events control is the tunnel pavement, including the events that affect the normal running of vehicles in the tunnel, such as the discharge of tunnel pavement.

Through the analysis of the characteristics of abnormal events in expressway tunnels, 16 types of abnormal events in tunnels are obtained, as shown in Table 1.

Table 1. Abnormal events in expressway tunnel

Serial number	**Event**	**Serial number**	**Event**
1	Speeding	9	Roll
2	Low speed driving	10	Fire
3	Parking in lane	11	Rear-end collision
4	Reverse driving	12	Scraped
5	Astern running	13	Bump wall
6	Frequent lane changes	14	Collision repair way
7	Crossing the line	15	Abandoned object
8	Cartwheel	16	Pedestrian

3. Establishment of tunnel abnormal event classification system

3.1 Event tree analysis

Event Tree Analysis (ETA) starts from the initial state of the event and analyzes the results of various sequences that may be caused by the initial event in a certain order, thus qualitatively or quantitatively evaluating the characteristics of the system [10]. When abnormal events occur in expressway tunnels, abnormal events are treated according to the order of abnormal event characteristics and event handling mode. The evolution law of occurrence and development follows the basic principle of event tree analysis.

3.2 Tunnel abnormal event classification system

Tunnel abnormal events are judged by influencing factors and event characteristics. Therefore, according to the basic logic of event tree analysis, tunnel abnormal event tree is studied and compiled from two perspectives of influencing factors and event characteristics, and then tunnel abnormal events are classified. The influencing factors of the event include human factors, tunnel factors and vehicle factors. The characteristics of the event include abnormal traffic behavior, single vehicle accident, collision accident, tunnel road surface and personnel. Combined with the 16 types of abnormal events of expressway tunnel sorted out above, the classification tree of abnormal events of expressway tunnel is obtained as shown in Figure 2.

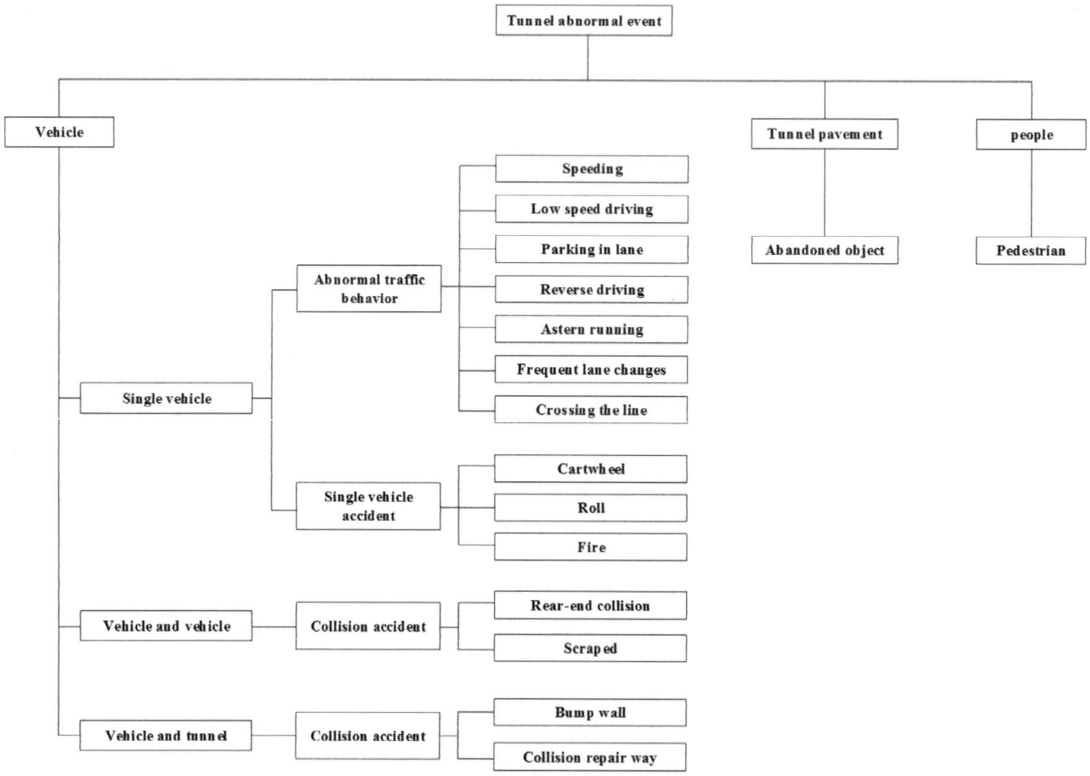

Figure 2. Classification tree of abnormal events in expressway tunnel

By using event tree analysis method, 16 types of abnormal tunnel events are divided into six categories from the influencing factors and event characteristics of abnormal tunnel events: abnormal vehicle traffic behavior events, single-vehicle accidents, vehicle-vehicle collision accidents, vehicle-tunnel collision accidents, tunnel pavement events and human events.

4. Management and control of tunnel abnormal events by category

Different types of abnormal events in expressway tunnels require different personnel, facilities and management units for event control and treatment. Abnormal events occurring in expressway tunnels are generally handled in coordination by tunnel management units and relevant departments according to the event type, severity, and department functions [11]. In terms of control and treatment of abnormal events in expressway tunnel and related facilities, the Specifications for Design of Highway Tunnels, Section 2, Traffic Engineering and Affiliated Facilities, stipulates the setting requirement [12] and functions of various facilities [13][14]. Combined with the influencing factors and event characteristics of abnormal events in the expressway tunnel, the classified management and control of abnormal events in the tunnel can be obtained as shown in Table 2.

Table 2. Classification control of abnormal events in expressway tunnel

Event	Type	Feature	Unit	Facility
Speeding	Single vehicle	Abnormal traffic behavior	Traffic police, road administrator	Traffic monitoring facilities, emergency calling facilities
Low speed driving				
Parking in lane				
Reverse driving				
Astern running				
Frequent lane changes				
Crossing the line				
Cartwheel		Single-vehicle accident	Traffic police, fire, road administration, medical treatment	Traffic monitoring facilities, emergency calling facilities
Roll				
Fire				Traffic monitoring facilities, ventilation facilities, lighting facilities, fire protection facilities, emergency calling facilities
Rear-end collision	Vehicles and Vehicles	Collision accident	Traffic police, fire, road administration, medical treatment	Traffic monitoring facilities, emergency calling facilities
Scraped				
Bump wall	Vehicles and tunnels	Collision accident	Traffic police, fire protection, road administration, medical treatment	Traffic monitoring facilities, emergency calling facilities
Collision repair way				
Abandoned object	Tunnel pavement	Tunnel pavement	Traffic police, road administrator	Traffic monitoring facilities, emergency calling facilities
Pedestrian	Personnel	Personnel		

5. Conclusion

Based on the tunnel operation system, the influencing factors and event characteristics of abnormal events in the expressway tunnel are analyzed. By using the event tree analysis method, the abnormal events in the tunnel are divided into six categories: vehicle traffic abnormal behavior event, single-vehicle accident, vehicle-vehicle collision accident, vehicle-tunnel collision accident, tunnel pavement event and personnel event. According to the logic of occurrence, characteristic judgment, and accurate identification of abnormal events in tunnel, the tree chart of abnormal events classification in expressway tunnel is drawn, and the classification of abnormal events in tunnel is realized systematically. Finally, according to the different types of abnormal tunnel events sorted out, the units and facilities involved in the treatment of abnormal events of different types of expressway tunnel are given, which provides theoretical and methodological basis for the realization of "one-click response" security control technology of abnormal tunnel events.

References

[1] Bai, Y., Han, Z., Li, N. (2008) Highway tunnel emergency capability evaluation. J. Highway traffic technology, S2:127-130.
[2] Chi, H. Ji, L. Chen, A., et al. (2006) Emergency management. Higher Education Press, Beijing.
[3] Yang, J., Chen, J.M., Zhao, H. (2005) Study on classification and classification of emergencies in emergency management. J. Management review, 017(004):37-41+8-64.
[4] Yu, L. J., Chen, S. N. (2001) Literature taxonomy. Wuhan University Press, Wuhan.
[5] Wang, S. F., Wang, H., Tu, Y. et al. (2013) Highway tunnel traffic incident classification and emergency management. J. Highway, 07:190-193.
[6] Shen, Y.J., Yang, Y., Zou, X. L., et al. (2018) Traffic accident statistics and casualty evaluation in

domestic highway tunnel operation period. J. Tunnel construction, 038(004):564-574.
[7] Wang, Y., Cao, C. G. (2020) Research on event classification based on event attribute. J. Journal of Chinese Information Science, 34(10):12.
[8] Zhang, Y. Q., Su, X. N., Tang, M. W. (2017) Construction of emergency case base based on emergency evolution model. J. Library and information work, 61(18):9.
[9] Central People's Government of the People's Republic of China. (2005) Road Traffic Safety Law of the People's Republic of China. http://www.gov.cn/banshi/2005-08/23/content_25575_3.htm
[10] Cui, Y.W., Gao, X.H., et al. (2011) Risk management Risk assessment techniques GBT27921-2011. Standards Press of China, Beijing.
[11] Zhang, W. (2018) Analysis of the emergency command system of Long Tunnel. J. Traffic informatization in China, 09:122-126.
[12] Ji, W.Y., Zou, X.C., Tu, Y., et al. (2014) Code for Design of highway tunnels Section II Traffic Engineering and ancillary facilities. People's Communications Publishing House Co., LTD, Beijing.
[13] Hu, J.C. (2021) Research on maintenance and Detection technology of mechanical and electrical facilities in Expressway Tunnel. J. Gansu science and technology, 37(04):61-62+26.
[14] Zhu, X.Q. (2021) Key points of detection technology for mechanical and electrical facilities of expressway tunnel. J. Traffic is the world, 07:139-140.

Study on the handheld monitoring equipment and the supervisory method for transportation loss of oil tank trucks

He Zhizhou[1] ,Wang Zuobi[2] ,Ma Hua[3] ,Shideyi[4] ,Hejingbin[5] ,Mu Zhaoxia[6] and Bai Yuanbo[7]

[1] Department of Computer Science, Xi'an Petroleum University, Xi'an City, Shaanxi Province, 710000, China.

[2] Changqing Oilfield No.2 Oil Production Plant, Qing'yang City, Gansu Province, 745000, China.

[3] Changqing Oilfield No.2 Oil Production Plant, Qing'yang City, Gansu Province, 745000, China.

[4]CNPC Chuanqing Drilling Engineering Company Limited, Xi'an City, Shaanxi Province, 710000, China.

[5]CNPC Chuanqing Drilling Engineering Company Limited, Xi'an City, Shaanxi Province, 710000, China.

[6] Changqing Oilfield No.2 Oil Production Plant, Qing'yang City, Gansu Province, 745000, China.

[7] Department of Digital Media, Xi'an Petroleum University, Xi'an City, Shaanxi Province, 710000, China.

[1]Corresponding author's e-mail: 478739064@qq.com

Abstract: At present, the oil of oil fields in remote areas and well sites without petroleum pipeline are mainly transported by oil tankers. Oil will spill and be stolen during the process of transportation, resulting in a large difference between the amount of oil loaded at first and the amount of oil unloaded in the end. The traditional method is to measure and print the bill when loading oil, not to re-measure when unloading oil, and to pay according to the bill, which causes great economic loss. Aiming at tackle this problem, this paper designs and realizes a handheld monitoring device and supervision method for the transportation loss of oil tanker. It integrates EEPROM, RTC real-time clock, PCF8563T memory, LED display lamp, LCD1602 display screen and speakers, and communicates and transmits oil quantity information with primary instruments of oil pumping equipment and oil unloading equipment, the information stored by the device can be read by the communication software of the host computer, and the information can be analyzed, calculated and settled accurately by the management software of the host computer. It can effectively prevent oil leakage and oil stealing during transportation.

1. Research on hand-held monitoring equipment

The objective of this study is to design and implement a hand-held monitoring device and supervisor y method for transportation loss of oil tank trucks, it integrates EEPROM(Electrically Erasable Progr ammable Read-Only Memory) , RTC(radiotelephone communication) real-time clock, PCF8563T memory, LED (light-emitting diode) display lamp, LCD(liquid crystal display)1602 display screen a nd speakers, and communicates and transmits oil quantity information with primary instruments of oi l loading and oil unloading equipment, the information stored by the device can be read by the commu nication software of the host computer;The information can be analyzed, calculated and settled accurat ely by the management software of the host computer. It can effectively prevent oil leakage and oil ste aling during transportation.

1.1 The basic structure and work flow of hand-held monitoring equipment
The hand-held monitoring device comprises a shell, wherein a processor which is installed inside the shell and electrically connected with a clock circuit, a reset circuit, a clock chip RTC and an EEPROM for accessing to time information , data storage real-time; The output terminal of the processor is electrically connected with an LED display tube, an LCD and a speaker for outputting the information of oil loading or unloading;

1.2 The basic workflow of hand-held monitoring equipment
The input end of the processor is electrically connected with a keyboard for setting device parameters and querying data records. What's more, the processor is electrically connected with a debugging and loading circuit for carrying out program downloading, debugging and switching; The shell surface is provided with a RS232 communication interface electrically connected with the processor, through which the RS232 communication interface is electrically connected with a PC, and a hand-held monitoring device which sends device recording data to the management software of the PC. The management software decodes the data frames and stores them in the database, and makes related inquiry, analysis and settlement of information involving personnel, meter and oil loading or unloading. The processor communicates in serial with a primary instrument of an oil loading or unloading device, which is used for measuring oil quantity. The hand-held monitoring equipment can automatically record work number, time and oil quantity when communicating with loading or unloading primary instruments to prevent human tampering with data.

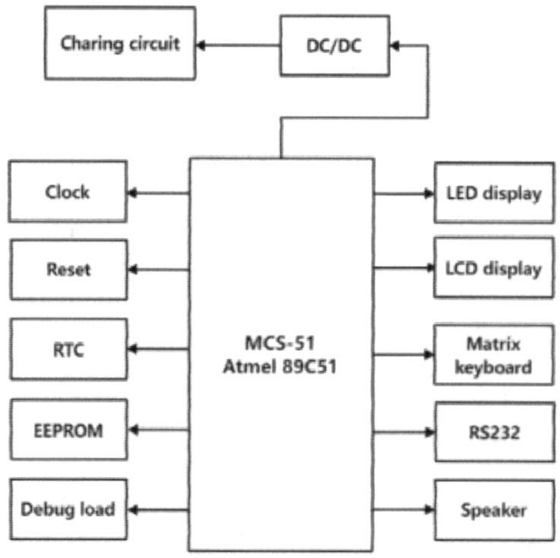

Figure 1. Schematic diagram of a hand-held monitoring device

Among them, the processor adopts MCS-51 series single-chip microcomputer, with cheap price and low power consumption. The EEPROM adopts the chip of AT24C64, and the date and time in the chip are modified when the system is used for the first time, powered by batteries of the device to ensure the accuracy of the time. The clock chip RTC adopts a PCF8563T chip with a storage capacity of 64K. The user defines the stored message format, which belongs to a private protocol to avoid information interference. Clock chip and memory are both 2-wire I^2C devices, with fewer wires between the device and the microcomputer. The LCD is LCD1602. The role of the MCS-51 processor is to control, store and transfer data of each functional module; The clock chip RTC provides the time for recording oil loading and unloading in the tank, which is compared with the time of the upper computer to prevent record cheating; The role of EEPROM is to store related information of oil quantity in real time; The configuration keyboard is used for setting initial use parameters of the device, downloading data, switching debug , and selecting display contents; The LED is used for displaying the working state of the device, and the speaker can emit a prompt sound during each operation. The LCD is used for displaying the relevant information recorded by the device, the operator information and oil quantity information. Besides, its backlight can be turned on or not to adapt to various circumstances and periods; The function of RS232 communication interface is to communicate with primary instrument or PC during the process of refueling or unloading oil. The equipment can accurately record oil quantity of tanker trucks, accurately carry out freight settlement, effectively eliminate oil leakage and oil theft.

The keyboard is a 2 × 4 matrix, where four keys are respectively connected to two external interrupts and two timing clocks of the processor, which works in an interrupt manner, while the remaining four keys work in query mode. The keys have an anti-jitter circuit for stable key reception. The eight keys can be combined to record the number of oil tanks, oil volume, query transport information, etc. Keyboard array makes full use of the external interrupt signal and timing interrupt signal of MCU. The whole system belongs to the minimum system of MCU. After the system is powered up and initialized normally, the display light is green; Then the speaker makes a prompt sound during each steps to ensure correct operation. The system is programmed with C51 language, which boasts flexible algorithm, tiny compiling code, and the program is stored directly in the internal program memory of MCS-51 thereby it is easier to upgrade. The driver can carry the device for its small size, accurately measuring ability. The internal 16K electric erasable memory storage program is chosen to store in the device, aiming to store information including oil volume and transport personnel, as well as data transmission procedures. The procedures is convenient for upgrading, too.

2.The research process for the handheld monitoring device is based on the following steps.

2.1. The supervision method of the hand-held monitoring equipment for the transportation loss of the oil tank truck

The regulatory method of the hand-held regulatory equipment for the transportation loss of oil tank trucks includes the following steps:

Step 1: Initiate the operation of the hand-held monitoring device, calibrate the current time, and input the oil tanker driver's job number;

Step 2: The hand-held monitoring device is connected with the primary instrument communication interface of the oil loading or unloading device. Activating the oil loading or unloading device and sending 2-byte 0xAA to the hand-held monitoring device through the serial port. Then the primary instrument transmits the oil loading quantity and time, or oil unloading quantity and time to the hand-held monitoring device for storage;

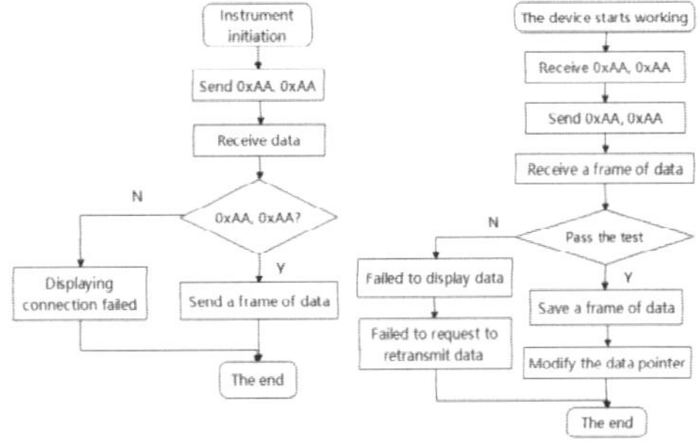

Figure 2. Workflow diagram of hand-held supervisory equipment

Journal of Physics: Conference Series **2195** (2022) 012005 doi:10.1088/1742-6596/2195/1/012005

Step 3: Connecting the communication interface of a PC with the hand-held supervisory device, transmitting the data in the memory of the hand-held supervisory device to the upper management software in the PC, and the upper management software automatically decodes the device data for data storage and automatic settlement;

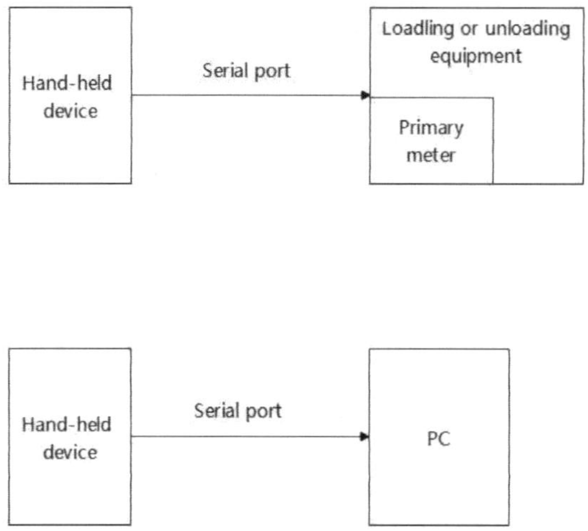

Figure 3. Hardware Connection diagram of hand-held monitoring equipment

Step 4: Check the loading and unloading data to get the information of oil fuel consumption in transportation.

2.2. Communication between the PC and the hand-held monitoring device

Moreover, when the PC reads the information of the hand-held monitoring device, it sends a 2-byte 0x55 to the hand-held monitoring device. After the hand-held monitoring device receives the marked data, the information stored in the device is sent to the management software in the PC, which analyzes the data and stores it in the database for data analysis, query and settlement.

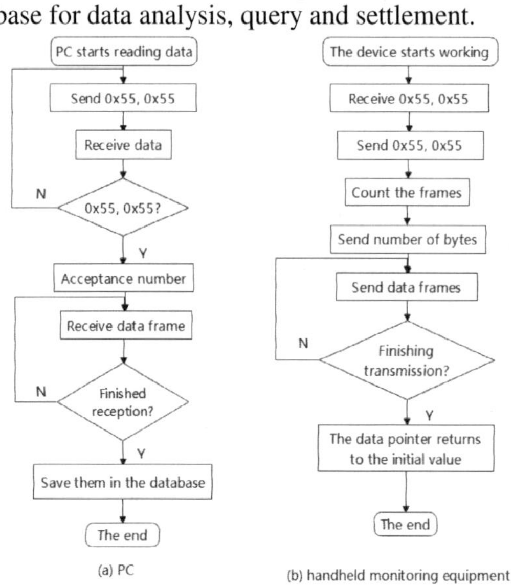

Figure 4. Flow chart of communication between hand-held supervisory device and PC

Furthermore, the PC stores information involving drivers, primary instrument and tankers; when the PC imports data of the hand-held monitoring device, the data is decoded and then stored transportation fuel consumption information. The PC performs conditional cascade information inquiry, calculation and settlement.

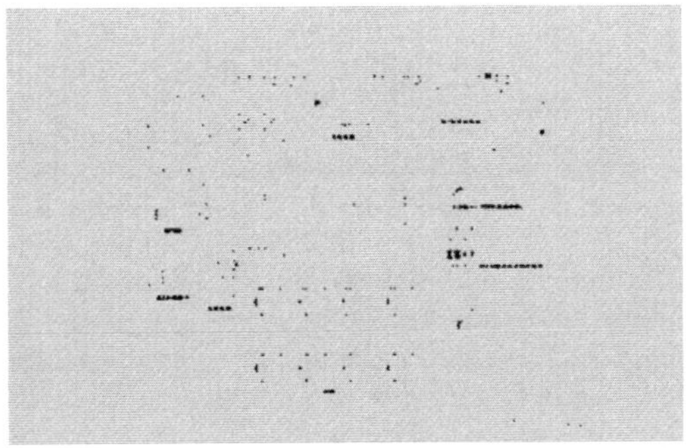

Figure 5. Circuit connection diagram for hand-held supervisory equipment

Display/modify	Job number/time	Start of oil extraction	End of oil extraction
Data increase	Cursor movement	Forward query	Backward Query

Figure 6. Key definition diagram for hand-held regulatory devices

3.Conclusion:

Compared with the existing technologies, the handheld monitoring device and supervisory method have the following significant advantages:

3.1. The first conclusion.

This study provides a hand-held monitoring device and supervisory method for the transportation loss of oil tanker. The interface between the hand-held monitoring device and the primary instrument of the oil loading or unloading device is the same, adopting the RS232 interface. The data format at both ends of the device is the same, avoiding data conversion. The communication protocol is completely identical and its data analysis is pretty simple.

3.2. The second conclusion.

This study provides a hand-held monitoring device and supervisory method for the transportation loss of oil tanker. The connection between the hand-held monitoring device's real-time clock, data memory, LED display interface and single chip microcomputer all uses I^2C bus. The keyboard is connected with the timer clock input through the external interrupt request of the single-chip microcomputer. The single-chip microcomputer works in a minimal way without bus expansion. Therefore, its size is small and power consumption is low.

3.3. The third conclusion.

This study provides a hand-held monitoring device and supervisory method for the transportation loss of oil tanker. When the device is activated or turned off, it can automatically read the current time, the oil quantity and saves. The management software can read the device data, analysis of the settlement while the driver can only view the oil information of the equipment and have no access to modify the information. All these measures and techniques can prevent artificial forgery efficiently.

Reference

[1] Lang Ye , Wang Wei , Wang Peng . Research on the application of computer simulation technology in extrusion forming process. Mechanical Engineering and Automation, 2017 (5): 123-124.

[2] Ding Li . Practical Application Analysis of Computer Simulation Technology in the Teaching of "E-Commerce Logistics" . Technology and Innovation, 2017 (3): 136-137.
[3] Shang Liang . A preliminary exploration of the teaching reform of optoelectronic information course based on computer simulation technology. China Modern Education Equipment , 2018,No.291 (11): 43-45.
[4] Zhang Peiqiang . The overall design of the automated logistics system based on computer simulation technology is discussed. Automation and Instrumentation , 2016 (8):202.

Optimal control of high speed unmanned vehicle path tracking

Shengwen Fan[1a*], Tao Xie [1b*], Zhigang Chen[1c], Xuan Hu[1d], Lei Gao[1e]

[1] Department of Electrical Engineering, North China University of Technology, Beijing 100144, China

[a]*fsw@ncut.edu.cn, [b]*137138634@qq.com, [c]czg0915@126.com,

[d]18811752696@163.com , [e]tyu1961@126.com

Abstract—In order to ensure the accuracy and stability of tracking control of driverless vehicle at high speed, an improved model predictive control method was designed. First of all, for the current high-speed unmanned vehicle tracking control field, there is little research and the tracking accuracy is poor. Starting from the three-degree-of-freedom vehicle dynamics model, by analyzing the vehicle yaw stability, the model is added with envelope constraints, and the linear time is derived. The variable path tracking controller model is designed with or without additional constraints.Through Matlab and CarSim co-simulation, it is shown that the MPC controller with envelope constraint has higher tracking accuracy than the ordinary MPC controller, and the vehicle has better robustness to the road adhesion coefficient, and the vehicle driving stability is significantly improved.

1. Introduction

Autonomous driving is one of the most promising development directions of the future automobile industry. Path tracking control is a key technology for unmanned driving research. It mainly controls the turning angle of the steering wheel of the vehicle to make the vehicle drive according to the desired trajectory and finally reach the target position.

The research on path tracking began in the 1950s. With the continuous development of control technology, there are more and more control algorithms for path tracking. The mainstream ones are: PID control, pure tracking control, optimal control, and model prediction Control, artificial neural network control, etc.

Literature [1] designed a vehicle lateral control strategy based on a model predictive control algorithm, and achieved a better tracking effect by controlling the steering wheel angle of the vehicle. Rafaila et al. studied the motion control strategy under the complex model. By establishing the nonlinear model of the vehicle, the design method of the nonlinear model predictive controller was given, and the tracking error was small in the low and medium speed conditions [2]. Literature [3] established a vehicle kinematics model based on a linear time-varying model predictive control method, and proposed a control strategy that only considers the front wheel steering. The lateral tracking effect is good under the double-line shifting condition. Liu Yang et al. used nonlinear model prediction to design the wheeled robot tracking controller, and considered terminal constraints [4]. Zhang Qian, etc., through the establishment of a vehicle dynamics model, the vehicle horizontal and vertical control issues are jointly considered, so that the controller has strong robustness [5]. Starting from the nonlinear vehicle model, Pan Shiju and others designed the linearization method and discretization method of the model, and then designed the linear model predictive controller and conducted experiments [6].

Many scholars have done in-depth research on path tracking control based on model prediction, and achieved good tracking accuracy. However, the tracking conditions are mostly concentrated in low-to-medium speed driving conditions, and there is little research on the tracking control of high-speed unmanned vehicles. In this regard, this paper designs a path tracking strategy based on model predictive control theory.

By analyzing the boundary of high-speed vehicle yaw stability, it adds envelope constraints to the controller to improve vehicle tracking accuracy and driving stability under high-speed conditions. The simulation results show that the maximum tracking error of this control method is only 0.6m when the vehicle speed is 80km/h, and the vehicle tracking accuracy and driving stability are better.

2. Vehicle dynamics modeling

To realize the motion control of unmanned vehicles, a suitable unmanned vehicle model must be established first. Considering that the vehicle is a complex nonlinear system with variable parameters, in order to reduce the computational complexity of the controller and ensure the real-time and reliability of vehicle tracking, the vehicle model should be simplified as much as possible. Therefore, assumptions are made: (1) Assume that the left and right wheels of the steering wheel have the same force, and the rotation angle and load are the same; (2) The vehicle has no load transfer; (3) The impact of vehicle aerodynamics is negligible; (4) Ignore the impact of tire return torque on the vehicle; (5) believe that the suspension has no impact on the vehicle [7].

Under the above idealized assumptions, the establishment of a three-degree-of-freedom vehicle dynamics model is shown in Figure 1.

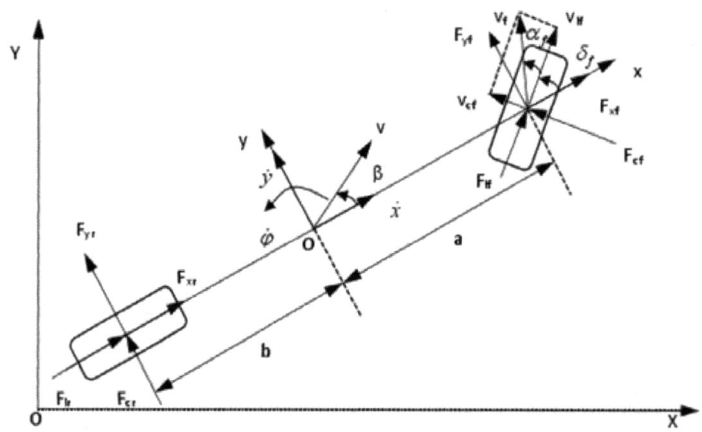

Fig.1 Three-degree-of-freedom vehicle dynamics model

Among them: a is the distance from the front wheel axis to the center of mass; b is the distance from the rear wheel axis to the center of mass; φ is the center of mass slip angle; δ_r is the front wheel angle; δ_r is the rear wheel deflection angle; F_{lf}, F_{lr} is the longitudinal force received by the front and rear tires; F_{ef}, F_{er} is the lateral force received by the front and rear tires; F_{xf}, F_{xr} is the x-axis force received by the front and rear tires; F_{yf}, F_{yr} is the y-axis force received by the front and rear tires.

Analyzing the force of the tire in the x-axis and y-axis directions, the relationship can be obtained:

$$F_{xf} = F_{lf} \cos\delta_f - F_{cf} \sin\delta_f \tag{1}$$

$$F_{xr} = F_{lr} \cos\delta_f - F_{cr} \sin\delta_f \tag{2}$$

$$F_{yf} = F_{lf}\sin\delta_f - F_{cf}\cos\delta_f \tag{3}$$

$$F_{yr} = F_{lr}\sin\delta_f - F_{cr}\cos\delta_f \tag{4}$$

Assuming that the front wheel deflection angle is small and the tire force is in the linear range, the nonlinear dynamics model of the vehicle can be obtained:

$$\begin{cases} m\ddot{y} = -m\dot{x}\dot{\varphi} + 2\left[C_{cf}\left(\delta_f - \dfrac{\dot{y}+a\dot{\varphi}}{\dot{x}}\right) + C_{cr}\dfrac{b\dot{\varphi}-\dot{y}}{\dot{x}}\right] \\ m\ddot{x} = m\dot{y}\dot{\varphi} + 2\left[C_{lf}S_f - C_{cf}\left(\delta_f - \dfrac{\dot{y}+a\dot{\varphi}}{\dot{x}}\right)\delta_f + C_{lr}S_r\right] \\ I_z\ddot{\varphi} = 2\left[aC_{cf}\left(\delta_f - \dfrac{\dot{y}+a\dot{\varphi}}{\dot{x}}\right) - bC_{cr}\dfrac{b\dot{\varphi}-\dot{y}}{\dot{x}}\right] \\ \dot{Y} = \dot{x}\sin\varphi + \dot{y}\cos\varphi \\ \dot{X} = \dot{x}\cos\varphi - \dot{y}\sin\varphi \end{cases} \tag{5}$$

Expressed by the state space equation:

$$\begin{cases} \dot{\xi} = f_{dyn}(\xi_{dyn}, \mu_{dyn}) \\ \eta = h_{dyn}(\xi_{dyn}) \end{cases} \tag{6}$$

Among them, the state quantity is $[\dot{y},\ \dot{x},\ \varphi,\ \dot{\varphi},\ Y,\ X]$, the control amount is δ_f.

3. Design of MPC

3.1. Linear error equation

Unmanned vehicles have strict requirements on the real-time performance of the controller at high speeds, and the nonlinear dynamic model is difficult to meet. The linearization method of the first-order Taylor expansion is used for the above formula to obtain the linear time-varying equation:

$$\dot{\tilde{\xi}} = A(t)\tilde{\xi} + B(t)\tilde{\mu} \tag{7}$$

At this point, the approximate linearization of the nonlinear system is completed, but since the system represented by the state equation is a continuous system, and the model predictive controller is based on a discrete system design, the system needs to be discretized, using the first-order difference quotient approach to approximate the discrete to get the equation of state:

$$\tilde{\xi}(k+1) = A_{k,t}\tilde{\xi}(k) + B_{k,t}\tilde{\mu}(k) \tag{8}$$

In: $A_{k,t} = I + TA(t)$, $B_{k,t} = TB(t)$, The same can be obtained:

$$\begin{cases} \xi(k+1) = A_{k,t}\xi(k) + B_{k,t}\mu(t) + d_{k,t} \\ \eta(t) = C_{k,t}\xi(k) + D_{k,t}\mu(t) + e_{k,t} \end{cases} \tag{9}$$

In: $\tilde{A}_{k,t} = \begin{bmatrix} A_{k,t} & B_{k,t} \\ 0_{m\times n} & I_m \end{bmatrix}$, $\tilde{B}_{k,t} = \begin{bmatrix} B_{k,t} \\ I_m \end{bmatrix}$ $\tilde{C}_{k,t} = \begin{bmatrix} C_{k,t} 0 \end{bmatrix}$.

Considering that the calculation at this time is still relatively complicated, in order to simplify the calculation, the following assumptions are made:

$$\tilde{A}_{k,t} = \tilde{A}_t, k = 1,\ldots,t+N-1$$
$$\tilde{A}_{k,t} = \tilde{A}_t, k = 1,\ldots,t+N-1 \tag{10}$$

The matrix form of the output of the available system:

$$Y(\mathrm{t}) = \Psi_t \xi(\mathrm{t}|\ t) + \Theta_t \Delta U(\mathrm{t}) \tag{11}$$

In:

$$Y(\mathrm{t}) = \begin{bmatrix} \eta(\mathrm{t}+1|\ t) \\ \eta(\mathrm{t}+2|\ t) \\ \cdots \\ \eta(\mathrm{t}+N_c|\ t) \\ \cdots \\ \eta(\mathrm{t}+N_p|\ t) \end{bmatrix} \quad \Delta U(\mathrm{t}) = \begin{bmatrix} \tilde{C}_t \mathrm{A}_t \\ \tilde{C}_t \mathrm{A}_t^2 \\ \cdots \\ \tilde{C}_t \mathrm{A}_t^{N_c} \\ \cdots \\ \tilde{C}_t \mathrm{A}_t^{N_p} \end{bmatrix} \begin{bmatrix} \Delta u(\mathrm{t}|\ t) \\ \Delta u(\mathrm{t}+2|\ t) \\ \Delta u(\mathrm{t}+3|\ t) \\ \cdots \\ \Delta u(\mathrm{t}+N_c-1|\ t) \\ \Delta u(\mathrm{t}+N_c|\ t) \end{bmatrix}, \quad \Theta_t = \begin{bmatrix} \tilde{C}_t\tilde{B}_t & 0 & 0 & 0 \\ \tilde{C}_t \mathrm{A}_t \tilde{B}_t & \tilde{C}_t\tilde{B}_t & 0 & 0 \\ \cdots & \cdots & \ddots & \cdots \\ \tilde{C}_t \mathrm{A}_t^{N_c-1}\tilde{B}_t & \tilde{C}_t \mathrm{A}_t^{N_c-2}\tilde{B}_t & \cdots & \tilde{C}_t\tilde{B}_t \\ \tilde{C}_t \mathrm{A}_t^{N_c}\tilde{B}_t & \tilde{C}_t \mathrm{A}_t^{N_c-1}\tilde{B}_t & \cdots & \tilde{C}_t \mathrm{A}_t\tilde{B}_t \\ \vdots & \vdots & \ddots & \vdots \\ \tilde{C}_t \mathrm{A}_t^{N_p-1}\tilde{B}_t & \tilde{C}_t \mathrm{A}_t^{N_p-2}\tilde{B}_t & \cdots & \tilde{C}_t \mathrm{A}_t^{N_p-N_c-1}\tilde{B}_t \end{bmatrix}$$

3.2. Objective function design

In order to reflect the accuracy of vehicle tracking, the variables of tracking anti-interference ability should be reflected in the design of the objective function. Reference [8] uses an objective function of the following form:

$$J(\xi(t),u(t-1),\Delta U(t)) = \sum_{i=1}^{N_p} \left\| \eta(t+i,t) - \eta_{ref}(t+i,t) \right\|_Q^2 + \sum_{i=1}^{N_e-1} \| \Delta u(t+i,t) \|_R^2 \tag{12}$$

In the above objective function, the first term is a reflection of the system tracking accuracy, and the second term is a reflection of the system tracking stability. $\eta_{ref}(t+i\,|\,t)$ is the reference output of the system and $\eta(t+i\,|\,t)$ is the actual output of the system. Np is the system prediction time domain, and Ne is the system control time domain. Q is the tracking accuracy weight matrix, and R is the tracking stability weight matrix.

Since there are many constraint conditions in the actual system, including constraints on the control quantity, control increment, and output quantity, it is necessary to consider the constraint conditions in the process of solving the objective function. The following formulae are the constraints on the system control quantity, control increment, and output quantity:

$$\begin{cases} u_{\min}(t+k) \le u(t+k) \le u_{\max}(t+k), k = 0,1,\cdots N_e - 1 \\ \Delta u_{\min} u_{\min}(t+k) \le u(t+k) \le u_{\max}(t+k), k = 0,1,\cdots N_e - 1 \\ y_{\min}(t+k) \le y(t+k) \le y_{\max}(t+k), k = 0,1,\cdots N_e - 1 \end{cases} \tag{13}$$

3.3. Yaw stability boundary

The analysis of vehicle handling and stability is an important means to ensure the safety of high-speed driving. The phase plane method is the most commonly used method in vehicle stability analysis. In order to reflect the vehicle's yaw characteristics and limit the tire's side slip angle, this paper analyzes the rear wheel sideslip and selects the yaw rate and lateral velocity as the phase plane parameters to constrain the vehicle to the maximum yaw rate and the tire's maximum side. Within the bias force, the driving stability of the vehicle is ensured.

When the vehicle yaw instability occurs, the rear axle will cause greater damage. According to the maximum cornering force constraint of the rear wheel, the control threshold of the yaw angle change rate can be obtained:

$$\dot{\psi}_{ss,\max}(k) = \frac{\bar{C}_{\alpha r}\alpha_{r,\,\lim}\left(1+l_r/l_f\right)}{mv_x(k)} \tag{14}$$

Constraining the slip angle of the rear wheels can be transformed into a constraint on the lateral speed v_y of the vehicle:

$$-\alpha_{r,lim} \le \frac{v_y - l_r\dot{\psi}}{v_x(k)} \le \alpha_{r,lim} \tag{15}$$

Combining equations (14) and (15), the vehicle stability constraint can be expressed as:

$$-G_{sh}(k) - S_{sh}(k) \le H_{sh}(k)\xi(k) \le G_{sh}(k) + S_{sh}(k) \tag{16}$$

In:

$$H_{sh}(k) = \begin{bmatrix} \frac{1}{v_x(k)} & -\frac{l_r}{v_x(k)} & 0 & 0 & 0 & 0 \\ 0 & 1 & \frac{g}{v_x(k)} & 0 & 0 & 0 \end{bmatrix}, G_{sh}(k) = \begin{bmatrix} \alpha_{r,\lim} \\ \frac{\bar{C}_r\alpha_{r,\lim}(1+l_r/l_f)}{mv_x(k)} \end{bmatrix},$$ $S_{sh}(k)$ is the relaxation factor.

Equation (16) forms a closed envelope, as shown in Figure 2:

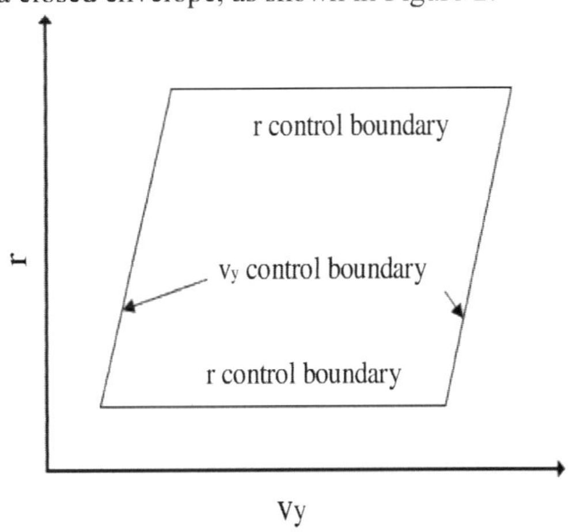

Fig.2 Yaw envelope

In:r is the rate of change of heading angle; v_y is the lateral velocity.

When the yaw rate and lateral speed of the vehicle are within the envelope, the stable driving of the vehicle can be ensured.

3.4. Controller solution

The objective function solution of the above formula generally adopts the method of converting the objective function into a quadratic programming problem. So the above system solution can be transformed into the following constraint problem solution:

$$\min_{\Delta U_t,\varepsilon}\left[\Delta U(t)^T,\varepsilon\right]^T H_t\left[\Delta U(t)^T,\varepsilon\right] + G_t\left[\Delta U(t)^T,\varepsilon\right] + P_t$$

$$s.t.\begin{cases} \Delta U_{\min} \le \Delta U(k) \le \Delta U_{\max} \\ U_{\min} \le u(t-1) + \sum_{i=1}^{k}\Delta U(i) \le U_{\max} \\ Y_{\min} - \varepsilon \le \Psi_t\xi(t|\,t) + \Theta_t\Delta U(t) \le Y_{\max} + \varepsilon \\ -G_{sh}(k) - S_{sh}(k) \le H_{sh}(k)\xi(k) \le G_{sh}(k) + S_{sh}(k) \end{cases} \tag{17}$$

By solving the above formula, the input of the system control increment can be obtained, and the control increment of the system at the next time can be obtained by the state quantity at the current moment. Repeating the solution process in each control cycle, the control increment sequence in the control time domain can be obtained as shown in the following equation:

$$\Delta U_t = [\Delta u_t, \Delta u_t + 1, ..., \Delta u_{t+N_e-1}]^T \tag{18}$$

The input increment sequence does not need to be all applied to the system. The system uses the first quantity in the sequence as the input increment of the system at the current moment, and can predict the output of the system at the next moment. Repeat the above process at the next moment to get the output of the system at each moment in the future, that is:

$$u(t) = u(t-1) + \Delta u_t^* \tag{19}$$

The system loops iteratively according to the above process, and finally realizes the control of the entire system.

4. Simulation analysis

In order to verify the performance of the controller, the double-line-shifting trajectory was selected as the reference trajectory, and the simulation platform was built based on MATLAB/CARSIM.

The main parameters of the vehicle model in the simulation are shown in Table 1:

Table 1 The main parameters of the vehicle

Parameter meaning	Parameter value
Wheel rolling effective radius（R）	0.22m
Vehicle quality（m）	350kg
Moment of inertia（Iz）	700kg/m2
Steering stiffness of front and rear wheels（Cf、Cr）	10000、16000N/rad
Distance from center of mass to rear axle（Lr）	1.335m
Distance from center of mass to front axle（Lf）	1.335m
Wheelbase of front and rear axles（w）	1.65m

The main parameters of the controller are shown in Table 2:

Table 2 The main parameters of the controller

Parameter meaning	Parameter value
Forecast time domain /Np	20
Control time domain /Ne	5
Extreme lateral sideways（ey,max）	10000、16000N/rad
Weight matrix（Q）	Diag（1000,600）
Weight matrix（R）	10
Relaxation factor ψ	1000

4.1. Different controller tracking performance

The adhesion coefficient is set to 0.8. At different vehicle speeds, the tracking effect of the controller designed in this paper and the traditional model predictive controller without constraints added is compared.

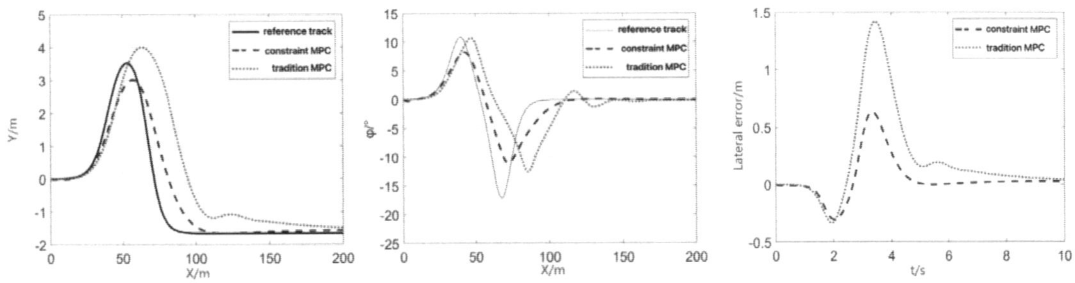

Fig.3 80km/h Simulation results

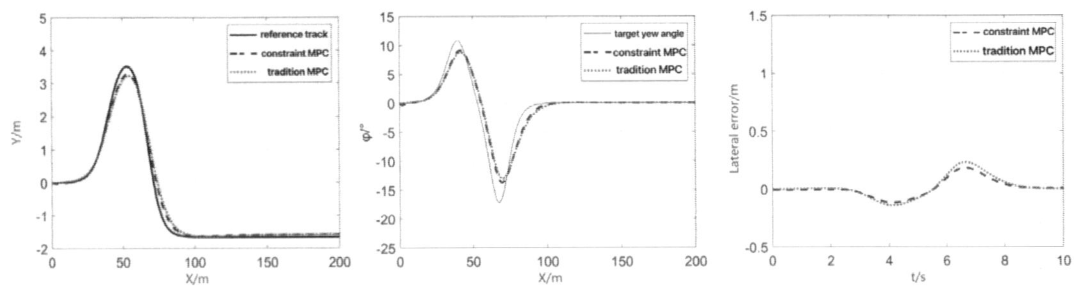

Fig.4 40km/h Simulation results

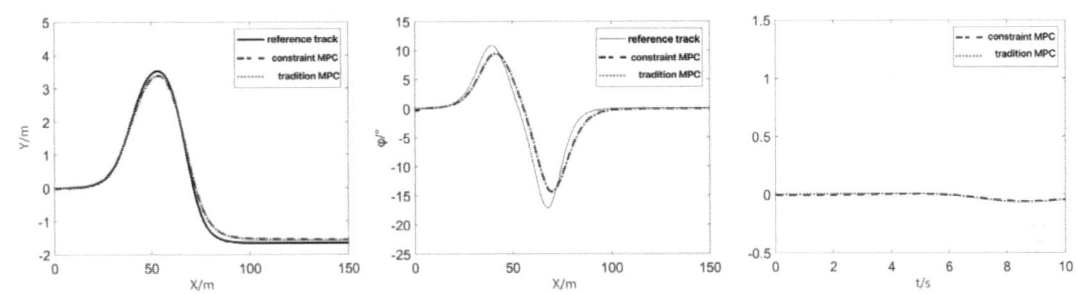

Fig.5 20km/h Simulation results

It can be seen from Fig. 3, Fig. 4, and Fig. 5 that under the conditions of low and medium speed, the tracking effect of the traditional model predictive controller and the optimized model predictive controller with constraints is not much different, and the tracking error is close. However, when the vehicle speed reaches 80km/h, the traditional model predicts that the controller has a large error. The maximum lateral error reaches 1.5m, and at 120km, the controller is slightly unstable. After optimization, the constrained model predictive controller can track the reference trajectory steadily throughout the entire process, and the maximum error is only 0.6 meters, and the optimization effect is obvious.

4.2. Controller tracking effect under different adhesion coefficients

The adhesion coefficients are respectively taken as: u=0.8 under good conditions and u=0.4 under low adhesion conditions. Other controller parameters are set as: prediction time domain Np=20, control time domain Ne=5, and vehicle speed is set to 40km/h. The simulation result is shown in Figure 6.

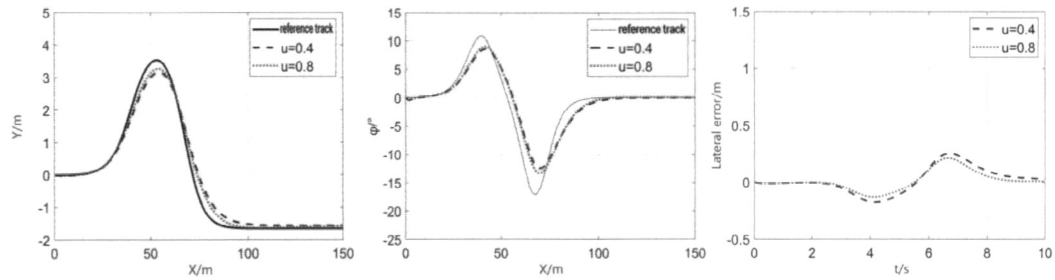

Fig.6 Tracking effect of different adhesion coefficient controllers

It can be seen that under the two adhesion coefficients, the vehicle trajectory tracking accuracy is higher, and the error is always within 0.3m, indicating that the optimized controller can better adapt to different road conditions and is more robust to changes in the road adhesion coefficient good.

5. Conclusion

Based on the results and discussions presented above, the conclusions are obtained as below:

(1) This control strategy can effectively improve vehicle tracking accuracy and tracking stability under high-speed operating conditions.

(2) The controller has strong robustness to the change of road adhesion coefficient, and has certain practicability.

Acknowledgments

Thanks to the platform support given by the NRT fleet of North China University of Technology.

References

[1] Falcone P,Borrelli F,Asgari J.Predictive active steering control for autonomous vehicle systems[J].IEEE Trans Contr Syst Tech,2007,15(3):566-580.

[2] Rafaila C , Livint G. Nonlinear model predictive control of autonomous vehicle steering[C] // 19th International Conference on System Theory, Control and Computing. Cheile Gradistei, Romania: IEEE, 2015: 466—471.

[3] Jian-guo Guo and Jun Zhou "Altitude Control System of Autonomous Airship Based on Fuzzy Logic" The 2nd International Symposium on Systems and ontrol in Aeronautics and Astronatics 10–12 December 2008.

[4] Yang Liu. Path tracking control of mobile robot based on model predictive control [D]. Jilin University, 2016.

[5] Qian Zhang. Research on trajectory tracking control method of intelligent vehicles [D]. Harbin Institute of Technology, 2015.

[6] Shiju Pan, Hua Li, Zhiyuan Su, Youchun Xu. Intelligent vehicle trajectory tracking method based on tracking error model [J]. Automotive Engineering, 2019, 41(09): 1021-1027.

[7] Liangxiu Zhang, Guangqiang Wu, Xiaoxiao Guo. Autonomous vehicle linear time-varying model predictive path tracking control [J]. Journal of Tongji University, 2016, 44(10): 1595-1603.

[8] Jianwei Gong, Yan Jiang, Wei Xu. Model predictive control of unmanned vehicles [M]. Beijing: Beijing Institute of Technology Press, 2014.

Numerical Study on the Effect of Transverse Baffles on the Stability of Iron Concentrate in a Cargo Hold

Anqi Wang[1a], Jianwei Zhang[1b*], Zhanwei Wang[1c]

[1]School of Naval Architecture and Maritime, Zhejiang Ocean University, Zhoushan 316022, China

[a]Z20086100020@zjou.edu.cn , [b*]zhangjianwei@zjou.edu.cn , [c]wzw0724@zjou.edu.cn

Abstract—Cargo fluidization is one of the main causes of ship capsizing accidents, which has been concerned by International Association of Classification Societies. Iron concentrate is regarded as the research object, its movement in the cargo hold is simulated using discrete element method. To effectively reduce the sloshing amplitude of cargo in the cargo hold, four conditions that considering the transverse baffles are designed. The stability of cargo is analyzed in detail from the offset of cargo gravity center and the contact force on the bulkhead induced by the cargo movement. Numerical simulation results show that the installation of transverse baffles can effectively reduce the sloshing amplitude of cargo in the cargo hold. Especially, it can better restrain the variation of center of gravity of cargo in the y-direction. The study focuses on the influence of baffles with different lengths on the stability of cargo in the cargo hold, which can provide a reference for the prevention of bulk carrier accidents.

1. Introduction

Iron concentrate, a kind of fluidizable cargo, is composed of a certain proportion of fine particles and moisture content. This kind of cargo is prone to fluidization once it is continuously affected by wind and wave load and ship's own motion during maritime transportation. The formation of fluidization makes the cargo easy to tilt to one side of the cabin, increases the inclination angle of the ship, and even the capsizing of the ship under severe sea conditions [1].

Recently, lots of studies on the transportation safety of easily fulidization cargo have been made. Based on the model test of indoor shaking table, Zhou et al. [2,3] analyzed the influence of different factors on the fluidizing characteristics of iron concentrate. Results show that moisture content is the key factor affecting the fluidization of iron concentrate. Ding et al. [4] established a three-dimensional numerical model of fluidized nickel ore, and analyzed the stability of the nickel ore in a cargo hold. Results show that there is a certain phase difference between the motion of liquefied nickel ore and ship motion, but the former is relatively regular. Through the finite element software Fluent, Zhao [5] conducted numerical simulation of roll motion in the process of transportation of easily fluidization cargo. Results show that the installation of the longitudinal baffles can greatly improve the safety of cargo transportation.

So far, researchers have done a lot of study on the fluidization mechanism of easily fulidization cargo, and some effective methods to ensure the stability of cargo have been put forward, for example, adding longitudinal baffles. Considering that the installation of longitudinal baffles would affect cargo handling to a certain extent, a new two-dimensional numerical model of cargo hold is established using discrete element software PFC2D. The model focuses on the influence of different lengths of transverse baffles on the cargo stability under the condition of sway motion, which can provide some references for improving the safe maritime transportation of easily fluidization cargo.

2. Linear contact model

Contact behaviour, the key problem of discrete element method, is mainly determined by the selected contact model. The force-displacement law provided by the contact model is used to transfer and update the forces between particles, then the position information of particles could be updated. In PFC2D, the linear contact model is applicable to study the macroscopic characteristics of soil particles [6]. It could well transfer the force between particles. Therefore, the linear contact model is used to analyze the movement of iron concentrate in the cargo hold.

The contact model is activated immediately when there is contact between particles, i.e. the gap between particles is no more than zero. The contact forces between particles mainly include two kinds, linear force and dashpot force. The former provides linear force with linear elastic (no-tension) characteristics, while the latter is used to provide viscous characteristics, as follows.

$$F_c = F^l + F^d \tag{1}$$

Where F^l is the linear force, F^d is the dashpot force. Linear forces and dashpot forces are decomposed in normal and shear directions:

$$F^l = -F_n^l + F_s^l \tag{2}$$

$$F^d = -F_n^d + F_s^d \tag{3}$$

where F_n^l is the linear normal force, F_s^l is the linear shear force, F_n^d is the dashpot normal force, F_s^d is dashpot shear force.

In this study, the viscous behavior between particles is not considered, so the linear contact model mainly updates the linear force. With the interaction between particles, the linear normal force and the linear shear force will be updated iteratively according to the variation of the clearance between particles, as shown in Eqs. (4) and (5) below.

$$F_n^l = \begin{cases} k_n g_s, g_s < 0 \\ 0, otherwise \end{cases} \tag{4}$$

$$F_s^l = \begin{cases} F_s^*, \|F_s^*\| < F_s^\mu \\ F_s^\mu (F_s^*/F_s^*), otherwise \end{cases} \tag{5}$$

Where k_n is the normal stiffness of particles, F_s^* is the trial shear force, F_s^μ is the shear strength, which are updated according to the following equations:

$$F_s^* = (F_s^l)_o - k_s \Delta\delta_s \tag{6}$$

$$F_s^\mu = -\mu F_n^l \tag{7}$$

where $(F_s^l)_o$ is the linear shear force at the beginning of the time step, k_s is the shear stiffness of particles, $\Delta\delta_s$ is the relative shear displacement increment of particles, and μ is the friction coefficient between particles.

3. Material property verification

Internal friction angle, an important property of soil, is mainly calculated and often obtained through direct shear test [7]. Reference [2] points out that when the particle size of iron concentrate is selected as: 7.2% with particle diameter greater than 0.5 mm, 32.6% with particle diameter greater than 0.25-0.50 mm, and 48.3% with particle diameter greater than 0.10-0.25 mm, the internal friction angle is generally about 34.5°.

3.1. Model parameter calibration

During the indoor direct shear model test, the size of the shear box is generally set as 6.18 cm * 4.00 cm [7]. Therefore, a shear box numerical model with the same size is established in this study. Reference [8] points out that when the ratio of model boundary size to particle radius is not less than 25, the influence of size effect can be ignored. To avoid the size effect, the ratio of boundary size to particle size is greater than 25. Based on the iron concentrate particle gradation given in reference [2], parameters of the linear contact model selected are mainly set as shown in Table 1 below.

Table 1. Linear contact model parameters.

Parameter	Value	Parameter	Value
Particle normal stiffness k_n (Pa)	1e8	Particle density $\rho\ (g/cm^3)$	2700
Particle shear stiffness k_s (Pa)	1e8	Particle friction μ	0.5
Stiffness ratio $k = k_n/k_s$	1	Acceleration of gravity g (m/s^2)	9.81

3.2. Direct shear test simulation

To effectively calibrate the internal friction angle, direct shear tests under confining pressures of 100 kPa, 200 kPa, 300 kPa and 400 kPa are completed. When the shear displacement reaches 4 mm or the shear stress reaches the peak, the shear strength of the sample can be obtained. The simulation of direct shear test is shown in Fig. 1.

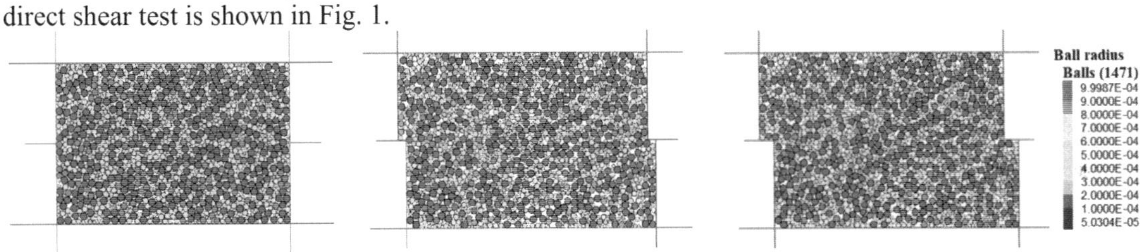

(a) Initial state (b) Shearing process (c) Shear completed

Fig. 1. Simulation of direct shear test.

Table 2. Shear strength under different confining pressures.

Normal stress (kPa)	Shear strength (kPa)	Normal stress (kPa)	Shear strength (kPa)
100	86	300	237
200	159	400	291

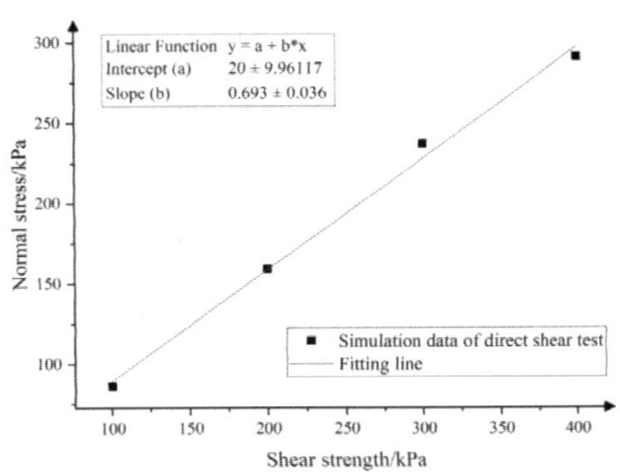

Fig. 2. Relationship between shear strength and normal stress.

According to the shear strength theory of soil, the shear strength of soil is expressed as a function of the normal stress on the sliding surface:

$$\tau_f = \sigma \tan \varphi \tag{8}$$

where τ_f is the shear strength of soil, σ is the normal stress, φ is the internal friction angle.

From Eq. (8), there exists a linear relationship between shear strength and normal stress. Therefore, the inclination of the straight line based on the linear fitting equation is the internal friction angle of the soil. The variation of shear strength of samples under different confining pressures are shown in

Fig. 2 and Table 2. According to the numerical simulation results, the internal friction angle between particles is 36°, which is about 4.17% error compared with the iron concentrate in the reference [2]. The main reason for this error is that the particle number generated in modelling are different from the actual iron concentrate particles.

4. Cargo hold sloshing test simulation

4.1. Model building

The size of the numerical model of the rectangular cargo hold established in this study is 0.06 m * 0.04 m. This is a 1:250 scale cargo hold model according to the actual cargo hold size of 57,000 DWT bulk carrier [3]. The loading height of cargo is set as 0.0195 m. To further study the movement of cargo in the cargo hold, the sway motion of the cargo hold is mainly discussed. The amplitude of sway motion is 0.003 m and the frequency is 1 Hz. Based on the particle size of cargo in the direct shear test above, the initial state of the numerical model is shown in Fig. 3, and the real-time loading curve of sway motion is shown in Fig. 4.

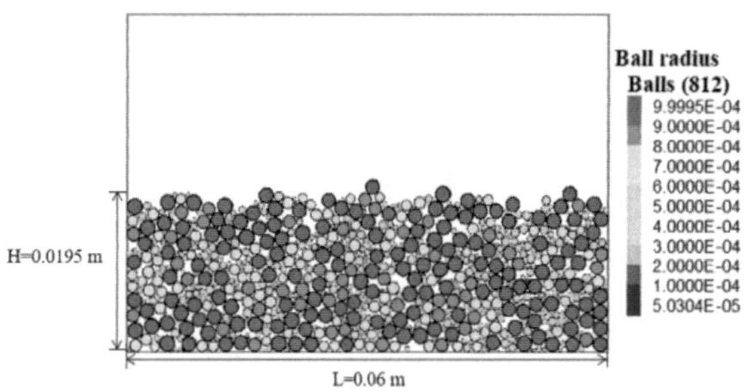

Fig. 3. Initial state of model.

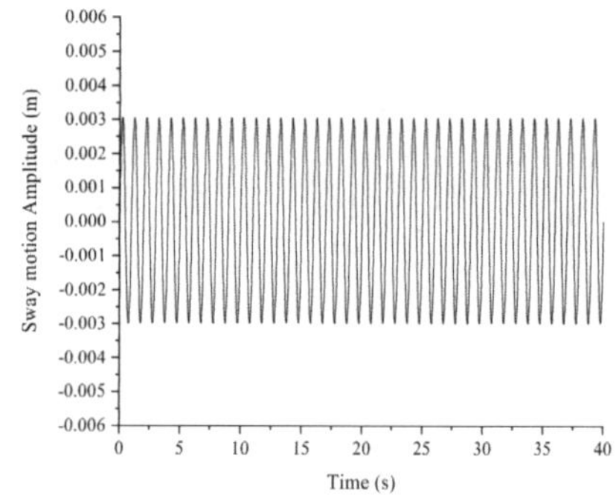

Fig. 4. Sway motion.

4.2. Cargo sloshing test simulation

The closer the installation position of cargo hold transverse baffles is to the surface of cargo, the more stable the cargo movement is [7,9]. Combined with engineering practice, discontinuous transverse baffle is selected, and then three groups of baffles with different lengths are simulated. Three groups of baffles with different lengths are: case 1: L1=0.02 m, case 2: L2=0.015 m, and case 3: L3=0.01 m. Meanwhile, the numerical simulation without baffles is used as the control case 0.

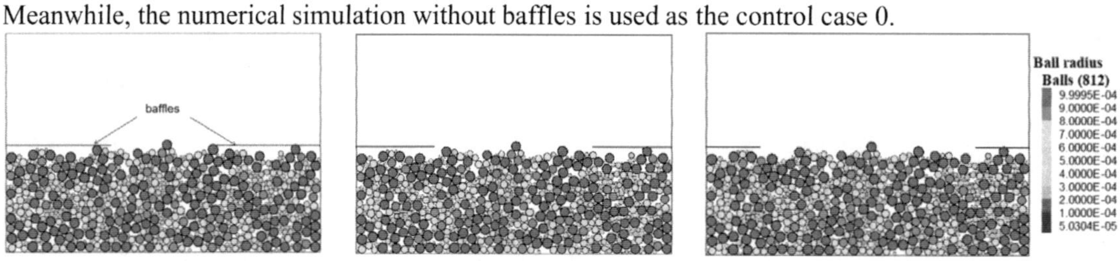

(a) case 1 (b) case 2 (c) case 3

Fig. 5. Initial state of cargo hold under different conditions

4.3. Result analysis

To better discuss the effect of transverse baffles of different lengths on the cargo stability, the variations of the center of gravity offset of cargo and the contact force are considered. Fig. 6 shows the variation of the center of gravity of cargo without baffle, while Fig. 7 shows the variation of the dynamic contact force on the bulkhead induced by the cargo movement.

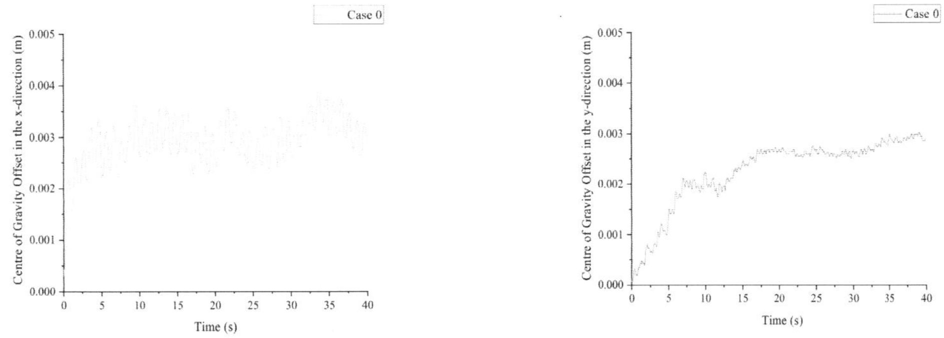

(a) (b)

Fig. 6. Center of gravity offset of cargo: (a) x-direction; (b) y-direction.

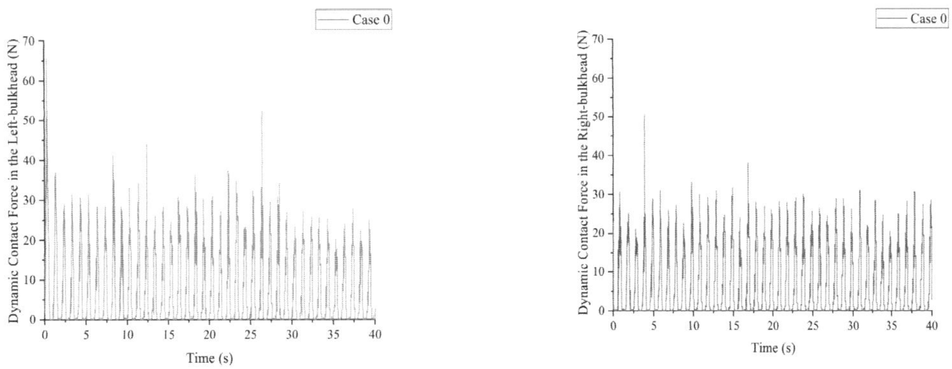

(a) (b)

Fig. 7. Dynamic contact force on the bulkhead induced by the cargo movement: (a) left side; (b) right side.

From Fig. 6, in the condition of no baffles, the offset of the center of gravity in the x-direction mainly fluctuates between 0.0022 m and 0.0038 m, while the offset in the y-direction increases with time. In this case, the stability of cargo in the cargo hold is very poor, and it is easy to slosh. From Fig. 7, there is a little difference in forces on the left and the right bulkhead. However, the force on the left bulkhead is slightly greater than that on the right. The reason for this phenomenon is that the initial movement direction of the cargo hold is to the left.

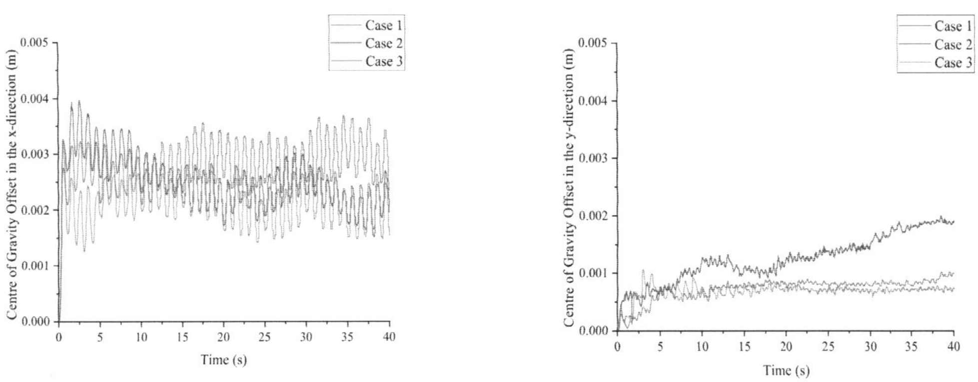

(a) (b)

Fig. 8. Center of gravity offset of cargo: (a) x-direction; (b) y-direction.

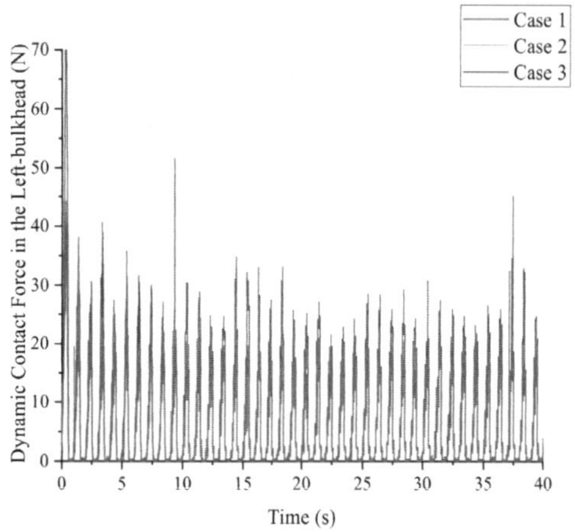

(a)

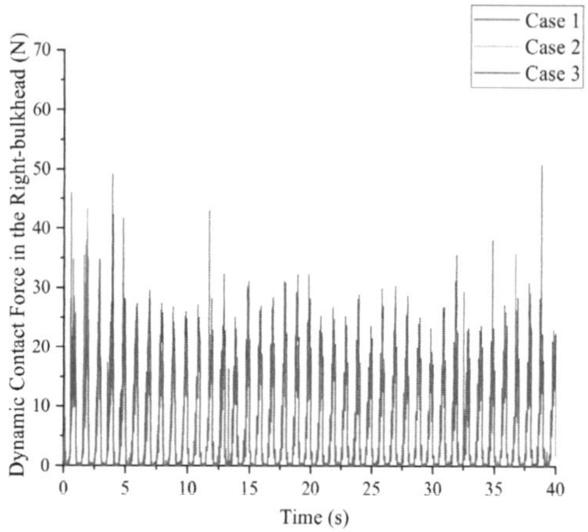

(b)

Fig. 9. Dynamic contact force on the bulkhead induced by the cargo movement: (a) left side; (b) the right side.

As can be seen from Fig. 8, when the length of baffle is 0.015 m, the center of gravity offset of cargo in the x-direction and y-direction is smaller than that in the cases of 0.02 m and 0.01 m. It can be seen from Fig. 9 that the forces on the left bulkhead and the right bulkhead are basically uniform in each case. However, when the length of baffle is 0.02 m, the bulkhead is subjected to greater force. Combined Figs. 6 and 7, the installation of transverse baffles in the cargo hold has a great influence on the center of gravity offset of cargo, but a small influence on the force on the bulkhead.

From the analysis and discussion above, it can be deduced that under the current working conditions, when the length of the transverse baffle is 0.015 m, the overall center of gravity offset of cargo is the smallest. This also shows that when the length of transverse baffle is about 1/4 of the length of the cargo hold, the good stability of cargo in the cargo hold could be ensured.

5. Conclusions

To effectively guarantee the stability of iron concentrate safe transportation, the movement characteristics of iron concentrate in the cargo hold is mainly studied in this study. And the influence of adding transverse baffles on the center of gravity offset of cargo and the dynamic contact force of bulkheads are considered. Then the influence of different lengths of transverse baffles on the stability of iron concentrate in the cargo hold is analysed in detail. The main conclusions are summarized as follows:

- The stability of iron concentrate can be effectively promoted by the installation of transverse baffles. The center of gravity offset of cargo in the y-direction is most affected by the arrangement of baffles.
- When the length of transverse baffle is 1/4 of the length of cargo hold, the stability of iron concentrate in the cargo hold is the best.

Acknowledgments

This research was financially supported by the National Natural Science Foundation of China (Grant No. 51809237) and the Science and Technology Project of Zhoushan (Grant No. 2020C21022).

References

[1] Jianwei Zhang. Numerical Study of the Effect of Liquefiable Cargo Sloshing on Motion and Stability of a Cargo Hold Section[D]. Dalian Maritime University,2016.

[2] Jian Zhou, Qiwei Jian, Xiaohong Wu, Ning Li, Yaoming Zhu. MODEL EXPERIMENTAL STUDY OF FLUIDIZATION OF IRON CONCENTRATE ORE IN BULK[J]. Chinese

Journal of Rock Mechanics and Engineering ,2013,32(12):2536-2543.

[3] Jian Zhou, Qiang Du, Qiwei Jian, Chen Li, Zhiqin Zhang. Discrete element numerical simulation of model experiments of iron concentrate ore fluidization during shipping[J]. Journal of Hohai University(Natural Sciences),2016,44(03):219-225.

[4] Junhong Ding, Yunlong Jin, Genguo Li, Hui Wang, Chen Guan. Numerical simulation and experiment study of sloshing characteristics of shipborne liquefied ore powder[J].SHIP SCIENCE AND TECHNOLOGY,2015, 37(05):36-41.

[5] Zihao Zhao. Numerical Study of the Effect of Longitudinal Bulkhead on the Liquefiable Cargo Hold[D]. Dalian Maritime University, 2018..Influence of the Micro Parameters of Discrete Element on the Macro Parameters of Sands [J]. NORTHWESTERN SEISMOLOGICAL JOURNAL,2011,33(S1):160-165.

[6] Jiangu Qian, Juyun Yuan, Chunfeng Zhao. Soil Properties and Soil Mechanics. China Communicatons Press Co.,Ltd,2015.

[7] Songqiang Gao. The Characteristics Analysis and Research of Sloshing Tank with Baffles[D]. Huazhong University of Science and Technology,2013.

[8] Wenjie Li, Lei Ju. Numerical Simulation of Fluidization of Shipborne Solid Bulk Cargo[J]. Ship Engineering,2019,41(12):64-70.

[9] Xin Luo, Dong Sheng. Numerical investigation on the mechanism of combined anti-slosh baffles[J]. Periodical of Ocean University of China,020,50(10):107-117.

A Feature Extraction Method for Non-intrusive Load Identification

Wei Wei[1,a], Tao Peng[2,b], Li Ye[3,c], Xianyu Feng[4,d]*, Zisong Jiang[5,e], Heyang Yu[6,f]

[1]State Grid Hubei Marketing Service Center Hubei Electric Power Co., Ltd Hubei, China

[2]State Grid Hubei Marketing Service Center Hubei Electric Power Co., Ltd Hubei, China

[3]State Grid Hubei Marketing Service Center Hubei Electric Power Co., Ltd Hubei, China

[4]Zhejiang Chint Instrument & Meter Co., Ltd Chint Group Hangzhou, China

[5]Zhejiang Chint Instrument & Meter Co., Ltd Chint Group Hangzhou, China

[6]College of Electrical Engineering Zhejiang University Hangzhou, China

[a]1036243440@qq.com, [b]2661900807@qq.com, [c]21515996@qq.com, [d]fxy@chint.com, [e]zisong.jiang@chint.com, [f]yuheyang@zju.edu.cn

Abstract—When some household appliances are running, the electrical quantities such as active power and reactive power will change slowly for a long time, which will affect the feature extraction of event-based non-intrusive load identification, resulting in low accuracy of load identification results. In view of this situation, a feature extraction method is proposed. Its core idea is that there is a certain law in the change of active power and reactive power in a very short time, and the curve can be fitted to speculate the values of these electrical quantities in the next time, so as to eliminate the influence. Based on this, a feature extraction method applied to non-intrusive load identification is realized. The test results show that this method can effectively improve the accuracy of non-intrusive load identification feature extraction data, and has good application value.

1. INTRODUCTION

With the development of the national economy, "power demand side management" has become increasingly important, which can give full play to the value of power companies and avoid unnecessary waste [1-2].

The use of load identification technology can finely analyze the characteristics of power consumption of power users, formulate energy-saving and supply-demand interaction strategies for the power grid, and provide a reference for the development of "power demand side management" [3-5].

The current load identification technology is divided into Intrusive Load Monitoring (NLM) technology and Non-Intrusive Load Monitoring (NILM) technology [6]. Compared with intrusive load identification technology, non-intrusive load identification technology has the characteristics of low cost and easy installation [7-10]. Only a device for collecting electricity consumption data is installed at the power entrance, and the electricity consumption data is analyzed through algorithms, thus acquiring the composition and operation of electrical equipment. In recent years, with the development of microelectronics technology, sampling sensing technology, and machine algorithms, the research on non-intrusive load identification technology has become increasingly widespread, mainly based on Hidden Markov Model (HMM) [11-13], deep learning neural network [14-16], event method [17-20], etc. Although there are many methods of NILM, event-based NILM, due to its strong operability, has been widely used in recent studies. The load decomposition of the event-based framework includes steps such as data measurement, data processing, event detection, feature extraction, and load identification. The effectiveness of the feature extraction step plays a key role in the accuracy of load identification. If the home appliances are running smoothly, when the NILM system detects the occurrence of an event (in this case, the start-up, state switching, and shutdown of an appliance are called events), the difference between the appliance feature quantity after the event ends and the appliance feature quantity before the event occurs can be extracted directly as the appliance feature extraction value. When household appliances are not all running smoothly, such as air conditioners, refrigerators, and other appliances, there will be a long-term slow increase or slow decrease in the amount of active power, reactive power, and other appliances (hereinafter collectively referred to as slow change). This will lead to deviations in the extracted values of the feature quantities, which will reduce the accuracy of subsequent steps such as load identification and matching.

Therefore, how to solve the problem of accurate feature extraction of non-intrusive load identification appliances with a slowly changing background has become an urgent problem. This paper introduces the compensation algorithm in the feature extraction step for the framework described above. The core idea is that there is a certain law of in the change of active and reactive power changes in a very short period of time when these slow transformers are operating, and a curve can be fitted to infer the values of these electrical quantities at the next time, so as to eliminate the effects.

2.Feature Extraction Principle of Slow Transformers

It is assumed that the feature extraction method of event-based non-intrusive load identification is to calculate the feature group ΔΩ1 of electrical appliance feature changes from the beginning to the end of the event, which mainly includes feature quantities such as active power, reactive power, and harmonic content.

Among them, the feature group ΔΩ1 is calculated by the following formula:

$$\Delta\Omega_1 = \Omega\left(t_{\text{event-off}}\right) - \Omega'\left(t_{\text{event-off}}\right) \tag{1}$$

$\Omega(t)$ refers to the feature group used for load identification that changes over time.

$\Omega(t_{\text{event-off}})$ is the actual value of the feature group at the end of the event $t_{\text{event-off}}$.

$\Omega'(t)$ represents the value of the feature group predicted by fitting the curve to the feature group data before the event that changes with time for load identification.

$\Omega'(t_{\text{event-off}})$ is the predicted value of the feature group at the end of the event $t_{\text{event-off}}$.

That is to say, in the context of a slowly-varying feature, when the event occurs, each appliance feature quantity is predicted by curve fitting of all appliance feature quantities to obtain the predicted value of each appliance feature quantity at the end of the event if the event does not occur. Finally, the actual operating feature quantity of the electrical appliance at the end of the event is compared with the predicted value of the appliance feature quantity, so as to compensate for the change of the electrical feature quantity during the event period under the background of the feature quantity when the slow transformer is operating, thereby improving the non-intrusive load identification of the electrical feature quantity and enhancing the identification accuracy of the overall algorithm.

3.FEATURE EXTRACTION ALGORITHM TO OVERCOME THE INFLUENCE OF SLOW TRANSFORMERS

Based on the above principles, this paper proposes the following algorithm:

Step (1): Sample the voltage and current at a rate of 256 points per cycle and calculate according to the sampling results, Thus, a feature sequence of active power, reactive power, and harmonic content is formed, which is combined into the above-mentioned $\Omega(t)$;

Step (2): Detect whether the feature sequence $R_1(t)$ is slowly rising or slowly falling (hereinafter collectively referred to as slow change). if it is determined that the feature sequence $R_1(t)$ is slowly changing, it is needed to record the starting time $t_{\text{slow-on}}$ of the feature sequence slowly changing, go to Step (3), otherwise go to Step (1).

Among them, the method of judging that the feature sequence $R_1(t)$ is slowly changing is as below: calculate the average value R_n of all the feature sequences of t_{increase} for a period of time, calculate the increment sequence ΔR_n of the average value, and when the average increment ΔR_n is greater than a certain threshold D_1, and If the number num is greater than a certain threshold $D_2(num>D_2)$, it is judged that the feature quantity is slowly increasing. The calculation method of the average value R_n of the feature quantity sequence and the increment sequence ΔR_n of the average value is as follows:

$$R_n = \frac{1}{t_{\text{increase}}} \sum_{t=n*t_{\text{increase}}+1}^{(n+1)*t_{\text{increase}}} R(t) \tag{2}$$

$$\Delta R_n = R_{n+1} - R_n \tag{3}$$

Step (3): Calculate the change $\Delta P_1(t)$ of active sequence P1(t) at time t, detect whether $\Delta P_1(t)$ is greater than the active event starting threshold P_{start}, and if it is determined that $\Delta P_1(t)> P_{\text{start}}$, it is needed to record the current moment as the event start moment $t_{\text{event-on}}$, and at the same time the moment is the end moment $t_{\text{slow-off}}$ of collecting the slowly-varying data interval of the active sequence. After the event starts, when detecting whether the active power sequence changes $\Delta P_1(t)$, $\Delta P_1(t\text{-}1)$, and $\Delta P_1(t\text{-}2)$ at time t, t-1, and t-2 are all less than the active event end threshold Pend, if yes, it is required to record the current time as the event end time $t_{\text{event-off}}$ and the actual value $\Omega(t_{\text{event-off}})$ of the feature group at this time, go to Step (4), otherwise continue to Step (3).

Step (4): Set the feature group of the slow change interval of the collected active power sequence to $\Omega_{\text{slow}}(t)$.

$$t \in \{t_{\text{slow-on}}, t_{\text{slow-on}}+1, \cdots, t_{\text{slow-off}}-1, t_{\text{slow-off}}\} \tag{4}$$

The variables of the feature are fitted with the following curve to obtain the curve group $\Omega_{\hat{y}}(t)$.

$$\hat{y} = \hat{a} + \hat{b}t \tag{5}$$

Where the total time length T_{slow} is calculated by the following formula:

$$T_{\text{slow}} = t_{\text{slow-off}} - t_{\text{slow-on}} + 1 \tag{6}$$

According to the least square method of curve fitting, it is needed to set a characteristic in the feature group $\Omega_{\text{slow}}(t)$ as $\omega(t)$, which is its corresponding fitting curve.

$$\hat{y} = p_{\omega}(t) = \hat{a}_{\omega} + \hat{b}_{\omega} t \tag{7}$$

$$\hat{b}_{\omega} = \frac{\sum_{i=t_{\text{slow-on}}}^{T_{\text{slow}}} t_i y_i - \sum_{i=t_{\text{slow-on}}}^{T_{\text{slow}}} t_i \cdot \sum_{i=t_{\text{slow-on}}}^{T_{\text{slow}}} y_i}{\sum_{i=t_{\text{slow-on}}}^{T_{\text{slow}}} t_i^2 - \left(\sum_{i=t_{\text{slow-on}}}^{T_{\text{slow}}} t_i \right)^2} \tag{8}$$

$$\hat{a}_{\omega} = \sum_{i=t_{\text{slow-on}}}^{T_{\text{slow-on}}} y_i - \hat{b}_{\omega} \cdot \sum_{i=t_{\text{slow-on}}}^{T_{\text{slow-on}}} t_i \tag{9}$$

Step (5): in the curve group $\Omega_{\hat{y}}(t)$, let $t = t_{\text{event-off}}$, then the predicted value $\Omega'(t_{\text{event-off}})$ of the feature group is obtained if the event does not occur at the end time of the event; the feature group $\Delta\Omega_1$ is obtained after the electric appliance characteristic extraction and compensation, which is the result.

The changes in active power, reactive power, and current harmonic amplitude changes in the feature group are calculated by the following formulas:

$$\Delta P_{\Omega-\Omega'} = P(t_{\text{event-off}}) - P'(t_{\text{event-off}}) \tag{10}$$

$$\Delta Q_{\Omega-\Omega'} = Q(t_{\text{event-off}}) - Q'(t_{\text{event-off}}) \tag{11}$$

$$\Delta i_{\Omega-\Omega'} = \sqrt{\begin{array}{l} I_{m,t_{\text{event-off}}}^2 + I'^2_{m,t_{\text{event-off}}} \\ -2 * I_{m,t_{\text{event-off}}} I'_{m,t_{\text{event-off}}} \\ \cos\left(\varphi_{k,t_{\text{event-off}}} - \varphi'_{k,t_{\text{event-off}}}\right) \end{array}} \tag{12}$$

The process steps are as follows:

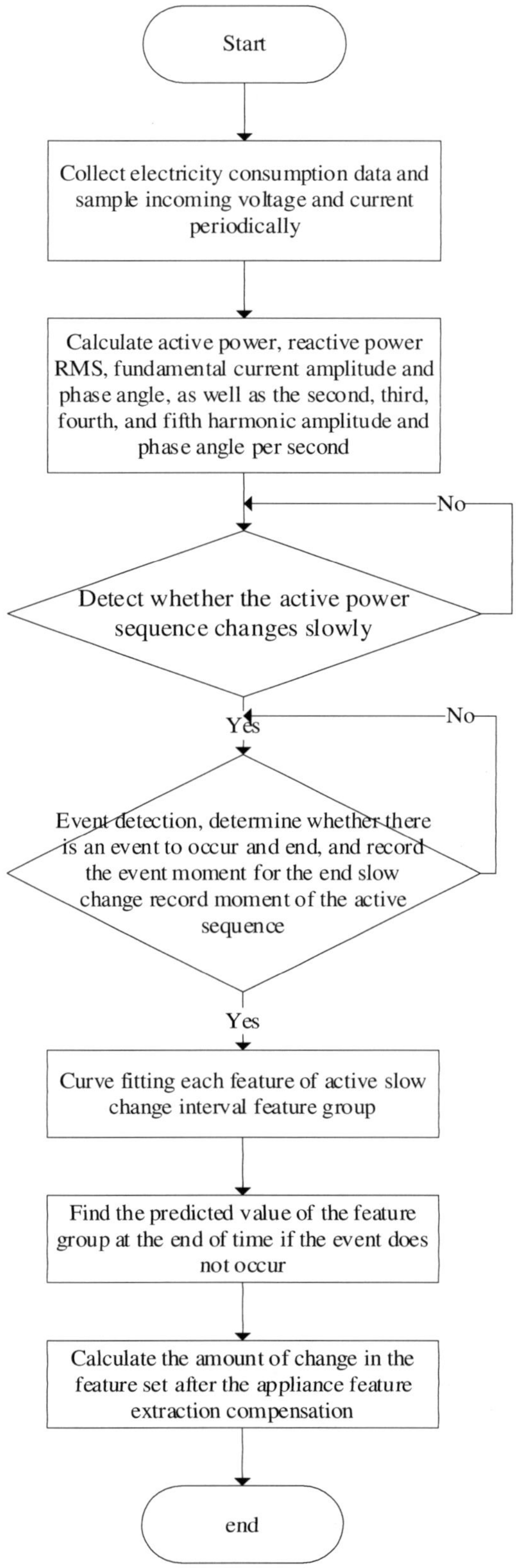

Figure 1. Flow chart of feature extraction of slow transformer

Journal of Physics: Conference Series **2195** (2022) 012008 doi:10.1088/1742-6596/2195/1/012008

4.Test and Verification

This paper uses Matlab 2019a development platform and m programming language to develop and implement the method and uses a PC with the main frequency of 2.6GHz CPU and memory of 4G to complete the test and verification of this method. The following will carry out technical verification around a household load operation scenario.

The verification background is that a fixed-frequency air conditioner in the home is turned on and the active power is slowly increasing. At the 10th second, the occurrence of the "rice cooker on" event is detected, and this time is recorded. At the 20th second, the "rice cooker on" event ends. The total active power of the household increases rapidly. At this time, it is the algebraic sum of the active power of the fixed-frequency air conditioner and the rice cooker. In the same way, the total reactive power also changes to the algebraic sum of the operating power of the fixed-frequency air conditioner and the rice cooker. Further, the change in the amplitude of the fifth harmonic is considered as the vector sum of the operating power of the fixed-frequency air conditioner and the rice cooker. The curve is used to fit each feature of the household load before the event, and predict the value of each feature at the end of the event, as shown by the thick solid line at the 20th second in Figure 2, Figure 3, and Figure 4 below.

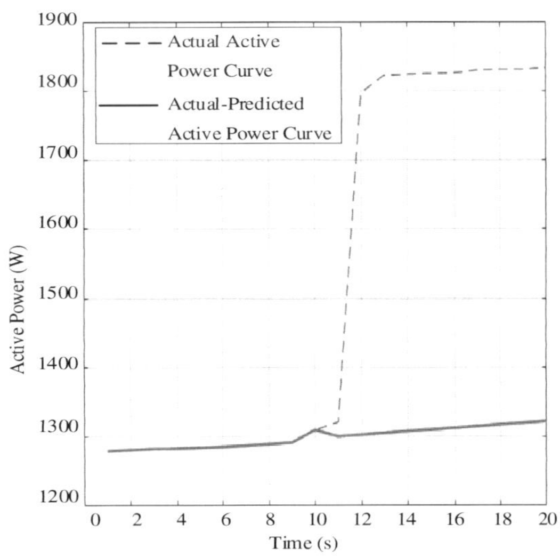

Figure 2. Active power characteristic fitting and actual comparison diagram

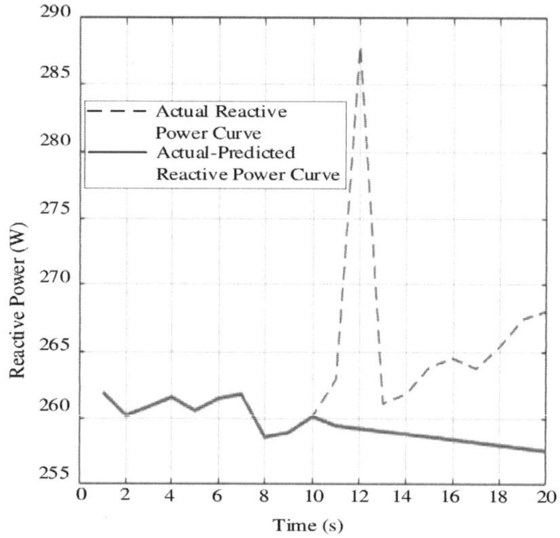

Figure 3. Reactive power characteristic fitting and actual comparison diagram

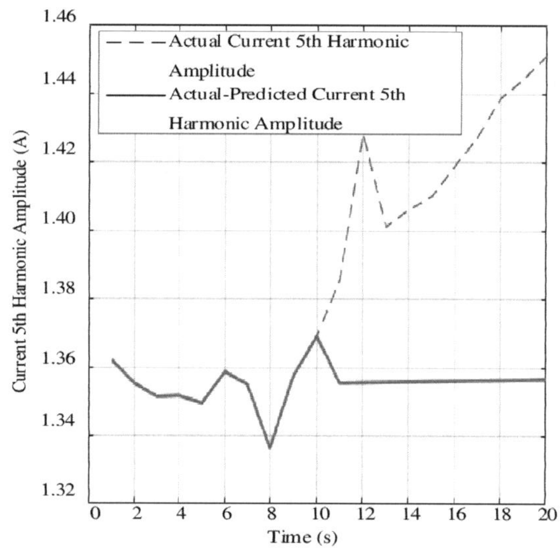

Figure 4. Current 5th harmonic amplitude fitting and actual comparison diagram

The change amount of the feature group after the electrical feature extraction and compensation is calculated, and then it is compared with the data of the rice cooker under stable conditions. The results are shown in Table 1.

TABLE 1. THE VALUE OF THE RELATED VARIABLES OF THE FEATURE GROUP BEFORE AND AFTER THE ELECTRICAL FEATURE EXTRACTION AND COMPENSATION AND THE AMOUNT OF CHANGE IN THE STEADY-STATE

Feature quantity	The actual value when the event starts	The predicted value when the event ends	The actual value when the event ends	Original change	Compensation change	The amount of change in steady state
a	1310.0	1322.4	1832.9	522.9	510.5	508.23
b	260.12	257.53	268.02	7.9	10.49	13.72
c	1.3690	1.3566	1.4511	0.102673	0.115535	0.112467
d	147.8109	147.6028	150.3175	———	———	———

Note: In the above table, a represents active power; b represents reactive power; c represents the amplitude of 5th harmonic of the current; d represents the phase angle of 5th harmonic of the current.

It can be seen from the above table that after compensation, the compensated eigenvalue is close to the eigenvalue in the steady state, which will improve the accuracy of non-intrusive load identification.

5.CONCLUSION

With respect to the load identification method based on events and probability, if the cut-off probability is increased, it is prone to misjudgment, and if the cut-off probability is reduced, it is subject to the missed judgment. Thus, this paper provides a method for accurately extracting feature quantities when a slow transformer exists. This method has obvious significance for reducing missed judgments and improving identification accuracy.

ACKNOWLEDGMENT

The authors would like to thank the State Grid Corporation of China for its great support of this research. This paper is funded by the key technology research of smart IoT meter for energy service (52153220001Y) of State Grid Hubei Electric Power CO., LTD.

REFERENCES

[1] Zou H P. Application analysis of power demand side management technology in electrical economic management [J]. China management informatization, 2019, 22 (12): 98-99.
[2] Xia X Z, Chen H, Li W F, et al. Practice of power demand side management based on improving customer response [J]. Electrical technology and economy, 2018 (01): 62-64.
[3] Zhang Y J, Liu S L, Jiang J Q, et al. Research review on low-voltage intelligent distribution network technology[J]. Guangdong Electric power, 2019, 32 (1): 1-12
[4] Zhou M, Song X F, Tu J, et al. Residential electricity consumption behavior analysis based on non-intrusive load monitoring [J]. Power System Technology, 2018, 42(10): 3268-3274.
[5] Li Y, Wang B B, Li F X. Outlook and thinking of flexible and interactive utilization intelligent power[J]. Automation of Electric Power Systems, 2015, 39 (17): 2-9.
[6] Hart G W. Nonintrusive appliance load monitoring[J]. 1992, 80 6(12): 1870-1891.
[7] Shi S B, Zhang H. CUSUM transient event detection algorithm based on compound sliding window[J]. Electrical Measurement & Instrumentation, 2019, 56 (17): 13-18.
[8] Li R Y, Zhang P, Liu Y G, Zhang H, Zhou D G. Non -intrusive household load identification method based on random forest [J]. Electrical Measurement & Instrumentation, 2021, 58 (04): 9-16.
[9] Yu D W, Liu M. A non-invasive load decomposition method based on deep circular convolution model[J]. Electrical Measurement & Instrumentation, 2020, 57 (23): 47-53.
[10] Cheng X, Li L Z, Wu Hao, et al. A survey of the research on non-intrusive load monitoring and disaggregation[J]. Power System Technology, 2016 (10): 3117.
[11] Kolter J Z, Jaakkola T. Approximate inference in additive factorial HMMs with application to energy disaggregation[C], La Palma, Spain, 2012. Microtome Publishing, 2012.
[12] Kong W, Dong Z Y, Hill D J, et al. A Hierarchical Hidden Markov Model Framework for Home Appliance Modeling[J]. IEEE Transactions on Smart Grid, 2018,9; 9(4; 4):3079-3090.
[13] Yu C, Qin Z J, Yang Y D. Non-intrusive Load Disaggregation with Improved Factorial Hidden Markov Model Considering ON-OFF Status Recognition[J]. Power System Technology, 2021, 45(11): 4540-4550.
[14] Kelly J, Knottenbelt W. Neural NILM: Deep neural networks applied to energy disaggregation[C]//, Seoul, Korea, Republic of, 2015. Association for Computing Machinery, Inc, 2015.
[15] Yan X F, Zhai S P, Wang Z H, et al. Application of deep neural network in non-intrusive load disaggregation[J]. Automation of Electric Power Systems, 2019, 43 (1): 126-132.

[16] Meng L, Yu C, Zhang X X, et al. Non-intrusive load disaggregation based on 1D convolutional neural network and self-attention mechanism[J]. Power Systems and Big Data, 2020, 23(10): 1-8.
[17] Zhao B, He K, Stankovic L, et al. Improving Event-Based Non-Intrusive Load Monitoring Using Graph Signal Processing[J]. IEEE Access, 2018, 6; 6: 53944-53959.
[18] Wang J, Geng G, Chen K, et al. Event-Based Non-Intrusive Home Current Measurement Using Sensor Array[J]. IEEE Transactions on Smart Grid, 2018, 9(6): 5878-5886.
[19] Wu X, Han X, Liang K X. Event-based non-intrusive load identification algorithm for residential loads combined with underdetermined decomposition and characteristic filtering[J]. IET Generation Transmission & Distribution, 2019, 13(1): 99-107.
[20] Zhang X X, Bao Y Q, Shao Q, Qian Y J, Jiang D Q. Research on non-Intrusive detection of multi-Rang household appliances[J]. Electric Power Information and Communication Technology, 2021, 19 (05): 89-94.

Division of Risky Sections of Railway External Environment Based on TOPSIS

Zibiao Pan

School of Transportation, Beijing Jiaotong University, Beijing,100044, China

Email: 19120880@bjtu.edu.cn

Abstract. With the rapid development of economy and society, the external environmental problems of railways have become more and more prominent, and external environmental accidents have occurred from time to time. The external environmental problems along the railway have become more and more prominent, which has become one of the important risks that endanger railway safety and affect the order of trains. By analyzing the external environmental risk profile, quantifying the impact of different types of hidden danger points along the line, dividing the hidden danger section of the line, dividing the risk section, and carrying out special rectification of the sections with different risk levels, it will be conducive to efficient and centralized rectification of the external environment and improve environmental governance Efficiency and risk control level.

1. Introduction

At present, the identification and analysis of section risk is more common in the road field. Zhang Tiejun extends the accident-prone area to the accident points at both ends until the accident-prone end point and the extension point do not meet the accident density index [1]. Based on the fixed-step cumulative frequency curve method for identifying accident-prone sections, Xie Lian et al. used an improved density clustering analysis algorithm, combined with the cumulative frequency curve, to realize the adaptive extraction of risky sections and the identification of different distribution lengths of accident-prone sections. Improve the accuracy of identification of multiple sections [2]. In the railway field, the grid management technology is used to divide the sections, and the line sections are divided based on a fixed length. The state change law of management objects in different line sections is studied from a spatial perspective, and the state of the risk section based on the spatial location is realized supervision [3], but the risk identification and analysis of the external environmental section of the railway line is still in the exploratory stage. By analyzing the influence range of hidden danger points on the route and the relationship between adjacent hidden danger points, dividing the hidden danger section of the route and distinguishing the risk of the section will be of great significance to the centralized improvement of the external environment.

2. Description

The hidden dangers of the external environment of the railway refer to the unsafe conditions and behaviors that are carried out by the personnel of the railway transportation enterprise or the personnel and units outside the railway that directly or indirectly affect the line in the railway safety protection zone and other areas [4].

Hidden hazards exist objectively. Without considering other factors such as height, terrain, etc., the schematic diagram of hidden hazards and routes is shown in Figure 1, where L represents the vertical distance of the hidden hazard from the line, R represents the radius of the hidden hazard's influence distance, and S represents the hidden hazard in The radiating section of the line.

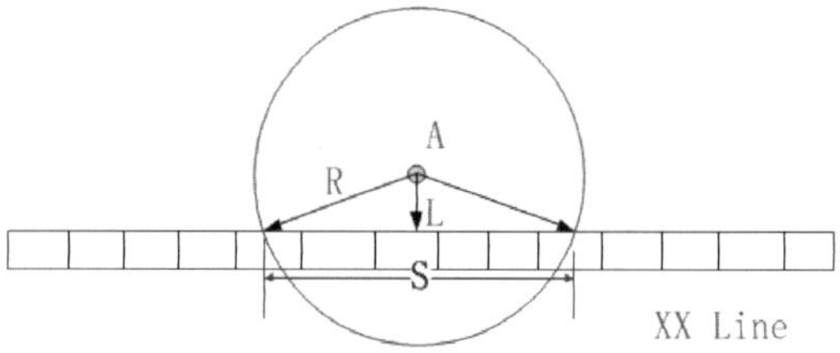

Figure 1 Schematic diagram of hidden danger line

In actual situations, different types of hidden danger points have different influence distances. There are four kinds of relationships between adjacent hidden dangers. As shown in Figure 2, the relationship set *T={tangency, intersection, right inclusion, left inclusion}*.

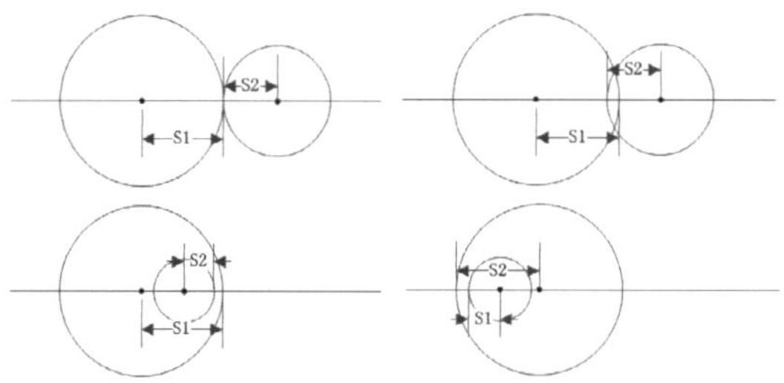

Figure 2 The relationship diagram of adjacent hidden danger points

3. Materials and Methods

Based on the relationship between adjacent hidden danger points, the relationship between the *i-th* hidden danger point and the *i+1-th* hidden danger point is represented by t_{ij} $(j=i+1)$, and the hidden danger points along the railway line are integrated, starting from the first Starting from the hidden danger point, divide the sections with $t_{ij}\notin T$ as the condition, where the hidden danger section H=$S_1 \cup S_2 \cup ... \cup S_i$, S_i is the radiation section of the *i-th* hidden danger point, when the number of hidden danger points in the hidden danger zone is low At 4 o'clock, the zone is a low-risk zone by default. Generally, the distance between the hidden danger and the line is much smaller than the hidden danger's influence distance radius, so when dividing the hidden danger zone, the influence radius *R* is used to represent the radiation zone *S*.

X1, X2, X3 indicate the number of hidden dangers in the evaluation index in the hidden danger zone, the number of hidden dangers per kilometer, and the length of the unit hidden danger from the line.

Standardization of negative indicators: $$\overline{X}_j = \frac{X_j^{\max} - X_j}{X_j^{\max} - X_j^{\min}}, j = 1,2 \quad (1)$$

Standardization of positive indicators: $$\overline{X}_j = \frac{X_j - X_j^{\min}}{X_j^{\max} - X_j^{\min}}, j = 3 \quad (2)$$

(1) and (2) where: X_j^{max} is the maximum value of the *j-th* index variable; X_j^{min} is the minimum value of the *j-th* index variable. After normalization, the data matrix is recorded as.

The positive ideal solution is denoted as C^+, the positive ideal solution of the *jth* index:

$$c_j^+ = \max_{1 \le i \le 23} b_{ij}, j = 1,2,3 \quad (3)$$

The negative ideal solution is denoted as C^-, then the negative ideal solution of the *jth* index：

$$c_j^- = \min_{1 \le i \le 23} b_{ij}, j = 1,2,3 \quad (4)$$

The distance from the *i-th* hidden danger zone to the positive ideal solution:

$$\lambda_i^+ = \sqrt{\sum_{j=1}^{3} (b_{ij} - c_j^+)^2}, i = 1,2,\cdots 23 \quad (5)$$

The distance from the *i-th* hidden danger zone to the negative ideal solution:

$$\lambda_i^- = \sqrt{\sum_{j=1}^{3} (b_{ij} - c_j^-)^2}, i = 1,2,\cdots 23 \quad (6)$$

The comprehensive evaluation value of the hidden danger zone is calculated as:

$$t_i = \frac{\lambda_i^-}{\lambda_i^+ + \lambda_i^-}, \quad i = 1,2,\cdots 23 \quad (7)$$

4. Results & Discussion

The total length of line A is 512.51km, with a total of 23 hidden danger sections, as shown in Table 1. Among them, H_S represents the start position of the hidden danger section, *H_E* represents the end position of the hidden danger section, *N* represents the number of hidden dangers in the hidden danger section, *P* represents the number of hidden dangers per kilometer, and *G_L* represents the length of the unit away from the route.

Table 1 A list of hidden danger sections of line A

Index	H_S	H_E	N	P	G_L	Index	H_S	H_E	N	P	G_L
1	762.08	765.749	5	1.36	16.20	13	1000.18	1003.48	4	1.21	5.00
2	764.8	777.505	48	3.78	16.94	14	1008.5	1012.65	13	3.13	18.46
3	777.55	780.65	9	2.90	10.11	15	1028.14	1032.05	4	1.02	25.25
4	779.38	802.75	197	8.43	24.34	16	1039.79	1046.87	12	1.69	19.08
5	802.826	857.948	264	4.79	42.64	17	1145.5	1153.2	22	2.86	13.41
6	869.738	878.8	13	1.43	26.15	18	1154.3	1172.51	31	1.70	20.16
7	878.865	889.56	8	0.75	26.88	19	1173.39	1182.33	6	0.67	20.00
8	907.25	911.4	9	2.17	32.78	20	1187.06	1192.98	8	1.35	15.25
9	926.55	933.3	19	2.81	50.47	21	1193.1	1203.61	16	1.52	59.06
10	937	970.742	173	5.13	81.99	22	1210.69	1228.21	19	1.08	27.95
11	974.55	980.14	10	1.79	39.40	23	1228.46	1241.38	12	0.93	46.83
12	985.392	991.653	12	1.92	57.33						

The calculated comprehensive evaluation value of each section is shown in Table 2.

Table 2 Comprehensive evaluation form of hidden danger zone

index	H_S	H_E	VALUE	index	H_S	H_E	VALUE
1	762.08	765.749	0.3876	13	1000.18	1003.48	0.4233
2	764.8	777.505	0.4784	14	1008.5	1012.65	0.4253
3	777.55	780.65	0.4460	15	1028.14	1032.05	0.3441
4	779.38	802.75	0.8015	16	1039.79	1046.87	0.3866

5	802.826	857.948	0.6470	17	1145.5	1153.2	0.4424
6	869.738	878.8	0.3519	18	1154.3	1172.51	0.3939
7	878.865	889.56	0.3345	19	1173.39	1182.33	0.3615
8	907.25	911.4	0.3360	20	1187.06	1192.98	0.3925
9	926.55	933.3	0.2726	21	1193.1	1203.61	0.1779
10	937	970.742	0.4317	22	1210.69	1228.21	0.3404
11	974.55	980.14	0.2941	23	1228.46	1241.38	0.2376
12	985.392	991.653	0.1986				

Use the natural breakpoint algorithm to classify the evaluation value, and divide the section into low-risk area, medium-risk area, and high-risk area. The evaluation value ranges are [0.1779,0.2941), [0.2941, 0.4784),[0.4784,0.8015]

5. Conclusions

In this paper, on the basis of the two-dimensional spatial distribution characteristics of hidden dangers in railway lines and their influence radius, the radiation section of hidden dangers on the line is obtained, and the hidden danger sections are divided by the distribution characteristics of adjacent hidden danger points. According to the characteristics of the hidden danger zone, the number of hidden dangers per kilometer, and the length of the unit hidden danger line as indicators, the risk classification of the divided hidden danger zones is carried out using TOPSIS and natural breakpoint methods, respectively. High risk section, medium risk section, low risk section.

References

[1] Zhang Tiejun, Tang Tengteng, Zhang Weihan. Bivariate interval filtering method for the identification of accident-prone sections[J]. Highway Transportation Science and Technology,2006, 23(3):139-142.

[2] Xie Lian, Wu Chaozhong, Lu Nengchao, et al. Research on the identification method of road traffic accidents based on improved clustering algorithm[J].Journal of Wuhan University of Technology (Transportation Science and Engineering Edition), 2014(4): 904-908.

[3] Wang Feng, Liu Yangkui. High-speed railway grid management theory and implementation technology [M]. Beijing: China Railway Press,2014: 67-70.

[4] Zhang Zhigang. Thoughts on the use of legal thinking and methods to rectify the hidden dangers of external risks in the railway safety environment [J]. Railway Economic Research,2015, 000(004):38-42.

Journal of Physics: Conference Series **2195** (2022) 012010 doi:10.1088/1742-6596/2195/1/012010

The Optimization Of The Distributed AI Embedded Cluster System

Zhiyuan Han[1,*], Jian Shen[1], Dandan Luo[1], Bo Gao[1], HuaQi Chen[1] and Jin Xie[1]

[1]Insigma Technology Co. Ltd., Hangzhou, China.

*Corresponding author email: hanzy@watone.com.cn

Abstract. This paper improves the core functions of the real-time video analysis system for edge computing which we proposed before. The improvement of this paper including distributed cluster status management and two-level horizontal expansion. The three-level task scheduling mechanism is realized, and the complete data link processing is realized. At the same time, RK3399pro is proposed to replace RK3399 as the hardware computing unit of the AI business. On the premise of satisfying real-time processing, the computing unit resources are saved. Compared with our previous system, the processing capacity of the new system is improved by 6-8 times.

1. Introduction

At present, in the analysis task of hundreds of video streams, the general processing method is generally to pull hundreds of video streams to the data center and use the GPU cluster for centralized analysis and processings[1]. However, the cost and power consumption of building a GPU cluster is relatively high.

In recent years, with the improvement of the processing capacity of Advanced RISC Machine (arm) chips, people have tried to use arm chips to build clusters[2]. The common way is to design multiple arm processor computing boards in a chassis, each computing board The card is a minimum system that can run independently, including arm processor, Memory, eMMC storage, network card, etc. Each computing board communicates with the outside world through Ethernet. In the same server chassis, all computing boards are interconnected through Ethernet and communicate with the outside world. All calculations are completed on the calculation board inside the chassis, and then the calculation results are transmitted through the network, to achieve the purpose of high performance and high concurrency.

In the previous various requirements based on the actual application scenarios of the smart transportation business scenario on the Hong Kong-Zhuhai-Macao Bridge, we found that a real-time video analysis arm cluster based on the edge of the distributed AI cluster was previously proposed[3]. In practice, a single insufficient processing capacity of the RK3399 computing unit leads to the need for more than ten computing nodes to work at the same time to meet the needs of real-time processing of a video stream. Therefore, in the next work, we used RK3399Pro instead of RK3399 to improve the processing capacity of a single node. At the same time, given the insufficient horizontal expansion capacity of the system, we further rationally planned the overall architecture to make it easier to meet the requirements of horizontal expansion and three-level task scheduling.

2. Core Hardware

In this paper, we chose Rockchip RK3399Pro as the core computing unit. Compared with RK3399, RK3399pro is additionally equipped with a Neural Processing Unit (NPU) module that supports 8bit/16bit operations and supports deep learning models under frameworks such as PyTorch and TensorFlow. deploy. In addition, after using the NPU for model loading and prediction, more abundant memory space can support the analysis and processing of more video streams.

As for the design of the AI cluster, we use 4 Gigabit switchboards, dual power supplies (one main, one standby), and the full load power consumption is 800W. Equipped with 8 hydraulic bearing cooling fans, the overall design of the chassis is 3u specifications.

3. System Optimization

3.1. Brief introduction of distributed system

Based on the previous ARM cluster state management, we added the output of AI data processing results and adds back-end management and application systems to complete the closed-loop of the overall landing scene. The state management part keeps the previous structure unchanged, and forms a cluster of nodes, reports the state to the PC server through TCP, and obtains the configuration information of each service from the server. Added a background management WEB system, you can view the status of devices and nodes, and manage and configure the services on the nodes. And the new MQTT message queue is added to receive the processing results of the AI service, the logic of the data processing layer is added to aggregate and logically process the AI processing results, and the in-depth data fitting and event alarms are carried out for the landing scene so that the AI processing results are original Data can truly serve the landing scene. Finally, the state data and AI processing output data will eventually be persisted in the database to complete the infrastructure construction of the entire platform. On this basis, the application layer design of the landing scene can be carried out(Fig. 1).

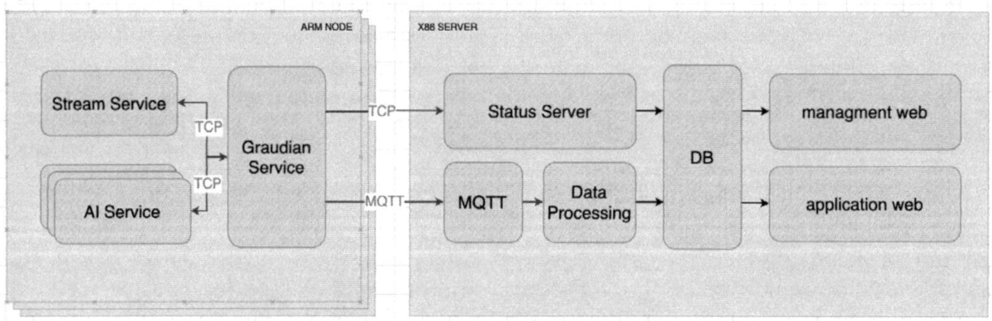

Figure 1. Structure Of AI Distributed System.

3.2. Three-Level Strategy for Distributed Computing

This distributed system uses 80 ARM plus 4 Gigabit switchboards as a basic cluster unit for task processing. Each ARM board has streaming and AI processing capabilities. According to the requirements of the computing scenario, the computing resources are deployed on a cluster basis and provide a three-tier progressive strategy of local priority, intra-cluster collaboration, and cross-cluster collaboration(Fig. 2).

Local first: ARM uses TCP protocol to connect the streaming service and AI service. When the processing capacity is sufficient, the image data obtained by the current ARM board will be processed first by the AI service on the same board, and the processed result will be reported through TCP.

Collaboration within the cluster: The cluster uses the GIGABIT protocol switchboard to form a mesh cluster to synchronize the task load and status using the GOSSIP protocol. After the local processing task exceeds the load, the image frame can be sent to other idle ARM boards in the cluster for AI processing and report the result data.

Cross-cluster collaboration: The overall strategy uses clusters as the smallest unit, and the load capacity of a single cluster will be fully considered during cluster deployment. When an emergency occurs and the cluster exceeds the processing load, the strategy will be activated for an upgrade, and the image frames will pass through the TCP of the x86 server. The server is sent to the cluster that handles idle.

And in practice, we can achieve a rapid improvement in the processing capacity and stability of the entire system by horizontally expanding the ARM cluster.

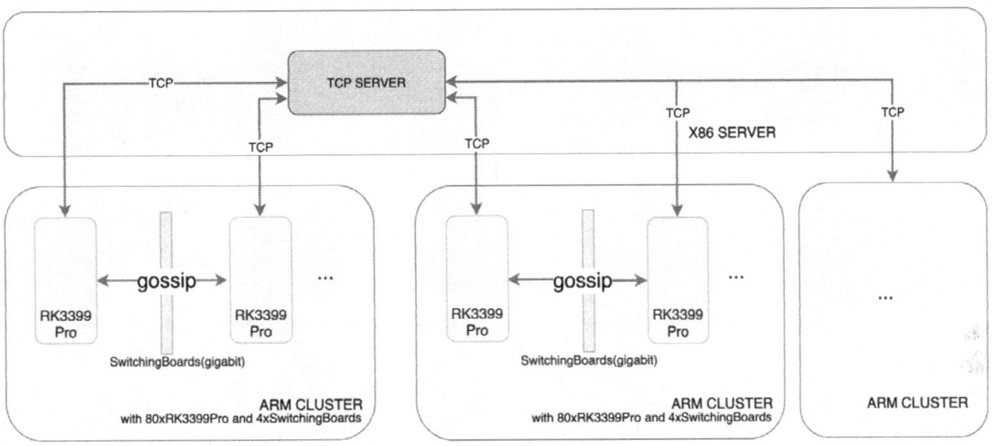

Figure 2. Three-Level Strategy For Task Processing.

3.3. Data-processing

The entire system uses a complete link for data processing, and the figure shows the complete data flow process within the entire system(Fig. 3). (1) RTSP stream analysis, after the image processing service outputs the AI service (AI Service B) designated by the processed picture and the image frame to be processed, the AI service is set by the server configuration. (2) The image frame to be processed is sent to the Guardian service on the ARM node by the TCP protocol. (3) After the Guardian service receives the image frame, it publishes the image frame to the topic of the service according to the specified AI service. (4) The corresponding AI service subscribes to its service topic and receives the image frames that it needs to process. (5) After the AI service receives the image frame and performs AI processing, it sends the processing result to the Guardian service. (6) After the Guardian receives the AI processing result data, it sends it to the MQTT Broker of the server. (7) The data processing service on the server receives the AI processing results for data integration analysis and produces event data.

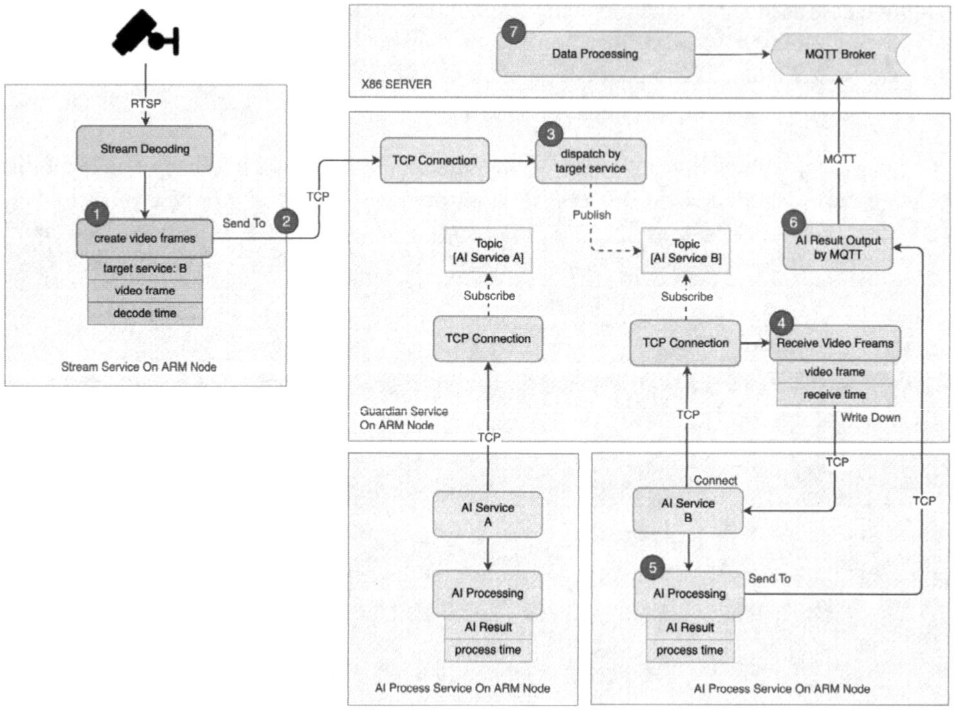

Figure 3. Data-Flow Introduction.

4. Experiments

First, we use mainstream frameworks and algorithms to verify the performance of RK3399 and RK3399Pro (NPU). There is some difference in model prediction capabilities between the two core computing unit. Secondly, we will compare the performance of the RK3399 cluster and RK3399Pro cluster in the 10/20/40/80 cluster mode.

4.1. Single board performance test

We use the official model[4] of YOLOv5 for performance testing. Experiments show that the prediction speed of RK3399Pro (NPU) under the small model (YOLOv5s) is about 20 times faster than that of RK3399. Under the larger model (YOLOv5x), the performance is also good. 15 times improvement (Table 1).

Table 1. The speed measures end-to-end time per 25 images(torch1.8.0 + torchvision0.9.0).

Detection Frameworks	**Model Size**	**Image resolution**	**RK3399 single board speed**	**RK3399Pro single board speed**
YOLOv5s	13.8M	640x480	47.3s	2.5s
YOLOv5m	41.1M	640x480	128s	7.6s
YOLOv5l	90.2M	640x480	215s	13.3s
YOLOv5x	168M	640x480	317s	20s

4.2. Cluster performance test

In the cluster performance test experiment, we use 10, 20, 40 and 80 clusters to conduct experiments to investigate the number of RTSP streams that can be processed in full frames at the same time in clusters of different sizes. The comparison experiment is divided into two groups, using the Yolov4-tiny model[5] and the Yolov5s model[4] for testing. According to the experimental data, the processing capacity of the RK3399Pro cluster is 6-8 times higher than that of the RK3399 cluster(Fig. 4), compared to Experiment 4.1 In the single-board test, we believe that the reason why the improvement is not as good as the single-board experiment may be that the task scheduling in the cluster caused part of the performance overhead.

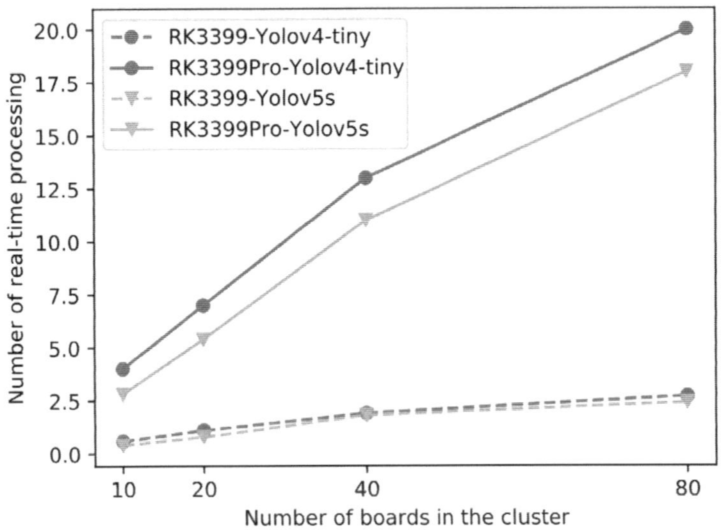

Figure 4. Number of rtsp streams processed by clusters of different scales.

5. Conclusion

A high-performance, practical system needs to be constantly updated and iterated for actual scenarios. The Distributed AI Embedded Cluster System we proposed based on RK3399Pro is optimized on the previous system. After a large number of experiments and project practices, this system has the following advantages over the previous system.

(1) This system can support real-time analysis and processing of 15-20 video streams at the same time.

(2) The system has a complete data analysis transmission link, from the original video stream to the output of the video detection results required by the application layer.

(3) The system supports flexible dynamic horizontal expansion, and under the premise of sufficient hardware, it can meet the real-time processing task of supporting thousands of video streams at the same time.

Acknowledgments

This work was supported by National Key R&D Project of China (No.2019YFB1600700).

References

[1] H. Shen et al., "Nexus: A GPU cluster engine for accelerating DNN-based video analysis", Proc. 27th ACM Symp. Oper. Syst. Princ., pp. 322-337, Oct. 2019.
[2] N. Ha, "Raspberry Pi Cluster for Parallel and Distributed Computing", MS Project Report, 2019.
[3] Wei Li et al., "Distributed AI Embedded Cluster for Real-time Video Analysis Systems with Edge Computing", 2021 International Conference on Mechanical, Automation and Computer Engineering., to be published.
[4] Ultralytics. [n.d.]. Yolov5. Retrieved Feb 20, 2021 from https://github.com/ultralytics/yolov5.
[5] J. Redmon. Darknet: Open source neural networks in c. http://pjreddie.com/darknet/, 2013–2016. 3.

Cooperative control mechanism of key objects and key nodes of dry bulk terminal

Hankun Shi[1], Xuelin Wang[2*], Zhaolun Zhang[3]

[1] China Waterborne Transport Research Institute, Beijing, China

[2] China Waterborne Transport Research Institute, Beijing, China

[3] China Waterborne Transport Research Institute, Beijing, China

[*]Corresponding author's e-mail: wangxuelin@wti.ac.cn

Abstract. Based on the operation process of dry bulk terminal and according to the unique process characteristics of dry bulk terminal, this study combs the specific distribution of key objects and key nodes of dry bulk terminal. Based on the analysis of the process mechanism and interactive information of dry bulk terminal, the collaborative control mechanism of key objects and key nodes is studied, which creates a certain theoretical framework for the automation and intelligent design of dry bulk terminal.

1. Introduction

Foreign ports have advanced research and application concepts in dry bulk intelligent technology, and introduce advanced computer and information technology to reduce logistics costs and improve operation efficiency. For example, Newcastle port in Australia is the world's largest coal export port[1]. Unmanned operation has been realized in the coal stacking link for a long time, so as to improve the coal stacking efficiency and reduce the operation cost. Rotterdam port in the Netherlands is committed to promoting the process of port digitization and improving port operation efficiency and service level through the application of modern information technology and artificial intelligence technology[2-3]. Rotterdam port uses Ethernet, cloud computing, mobile terminal equipment, Internet of things, GIS, video monitoring system, etc. to establish the port operation management system, realize the seamless connection between the information system instructions and the terminal machinery and equipment control functions, and make the efficient and reasonable allocation and scheduling of various port resources[4]. Shanghai Luojing phase II bulk cargo terminal has a set of full-automatic bridge grab ship unloader, full-automatic stacker reclaimer and full-automatic ship loader, all of which are remotely controlled through the central control room[5]. Huanghua port coal terminal has realized the operation of automatic coal terminal by integrating visualization system, monitoring and early warning system and emergency disposal system, and has become the first coal terminal in China to realize the whole process remote operation[6-7]. The remote operation system is introduced into the coal terminal of Huangye port to uniformly dispatch the dumper operation, stacker operation, reclaimer operation and loading operation of the coal terminal[8]. The Nanjiang bulk cargo terminal of Tianjin Port has completed the transformation of the automatic terminal, adopted a new automatic shipping system, scanned the hull through lidar, and then carried out the shipping work, so as to realize the automatic shipping operation under unmanned conditions[9-10]. It can be seen that in the future, the deep integration of process flow, equipment, system and new technology is the core path for the construction and development of automatic dry bulk terminal.

After comparing and analyzing the intelligent technology application of several dry bulk cargo terminals, it is analyzed that the problems existing in the current intelligent and automation direction of dry bulk cargo terminals mainly include the following two aspects: the first aspect is that bulk cargo terminals have the characteristics of wide variety of goods, huge quantity and complex operation process. Many links need to rely on manual experience, and the degree of intelligence is low, the upgrading of control technology can not catch up with the development of information technology. The second aspect is the lack of deep integration between new technologies and traditional systems. The integration of emerging technologies such as artificial intelligence, big data, blockchain and information physics system with port business scenarios and key equipment systems is far from enough to support the future development needs of container terminal intelligence and automation. Therefore, starting from the basic theory and overall architecture, it is very important to find the key objects and key nodes of dry bulk terminal, and deeply analyze the cooperative control mechanism between them.

2. Distribution of key objects and key nodes of dry bulk terminal

Dry bulk terminal has the characteristics of rich types of equipment and systems, non centralized goods, complex technology and so on. Therefore, before studying the cooperative control mechanism of dry bulk terminal, it is particularly important to sort out the key objects and key nodes in many equipment and systems, complex processes . The key objects of dry bulk cargo terminal are divided into five categories from the perspective of equipment types, namely terminal loading and unloading equipment, terminal loading and unloading equipment, terminal yard equipment, terminal horizontal transportation equipment and terminal operation auxiliary equipment. The equipment under these five types of equipment together constitute the key objects of dry bulk cargo terminal, as shown in Table 1. The organic cooperation between them determines the normal and orderly operation of dry bulk terminal production.

Table 1. Key objects of dry bulk terminal

Wharf type	Equipment type	Key object
Dry bulk terminal	Dock loading and unloading equipment	Mobile ship loader, swing ship loader, fixed ship loader, ship boom, ordinary gantry crane, bucket gantry crane, loading bridge Chain ship unloader, bucket wheel ship unloader, screw ship unloader, belt unloader, buried scraper ship unloader and pneumatic conveyor
	Dock loading and unloading equipment	Bridge crane grab, portal crane grab, loading and unloading bridge grab, single bucket loader, elevated bin hopper belt conveyor system, dumper, screw unloader, chain bucket unloader, bottom door dump truck, crane grab and simple push rod unloader
	Wharf yard equipment	Stacker, reclaimer, stacker reclaimer
	Wharf horizontal transportation equipment	Belt conveyor, clamp belt conveyor, chain conveyor, buried scraper conveyor, screw conveyor, bucket elevator, bracket elevator
	Auxiliary equipment for wharf operation	Electronic belt scale, funnel metering device, track scale, underground scale, water gauge measurement, bucket leveler, straight belt leveler, curved belt leveler, feeder, scraper, rake, shovel, bucket truck, wind and dust prevention net,

		sprinkler, hot air thawing device, steam heating pipe thawing device, gas or infrared thawing device, air blowing arch breaking device,arch breaking device of coal conveyor

The key nodes of dry bulk terminal are divided into four categories from the core system, namely automatic control system, management information system, remote control system and support system. The subsystems under these four types of systems together constitute the key nodes of dry bulk terminal, as shown in Table 2. The cooperation and interconnection between them assisted the normal operation of the key equipment of the dry bulk terminal, realized the business operation and the exchange and sharing of information and data, and improved the work efficiency of the dry bulk terminal.

Table 2. Key nodes of dry bulk terminal

Wharf type	**Core system**	**Key node**
Dry bulk terminal	Automatic control system	Automatic control system of bulk cargo handling machinery
		Automatic control system of bulk cargo conveying machinery
		Dust removal automatic control system
		Automatic lighting control system
		Other auxiliary systems automatic control systems
		Operation information monitoring system
		Inventory information monitoring system
		Equipment condition monitoring system
		Energy consumption monitoring system
	Management information system	Production operation management system
		Tally management system
		Stockpile management system
		Equipment management system
		Energy consumption management system
		Material management system
		Production and operation management system
		Gate control system
		Statistical analysis management system
		Basic information management system
	Remote control system	Remote control system of wharf ship handling equipment
		Remote control system of wharf yard equipment
	Support system	Wireless intercom system
		Port video surveillance system
		Ship dynamic monitoring system based on electronic chart
		Meteorological information management system
		Positioning and navigation system
		Online service system
		Large screen display system
		Data center system

3. Cooperative control mechanism of key objects and key nodes of dry bulk terminal

On the basis of focusing on the material movement process mechanism of dry bulk terminal and fully mastering the material movement interaction information of dry bulk terminal, this study combs out the control logic structure required to achieve the remote real-time online movement control function on the basis of ensuring the safe and stable operation of the terminal, starting from the level of key objects and key nodes, as shown in Figure 1.

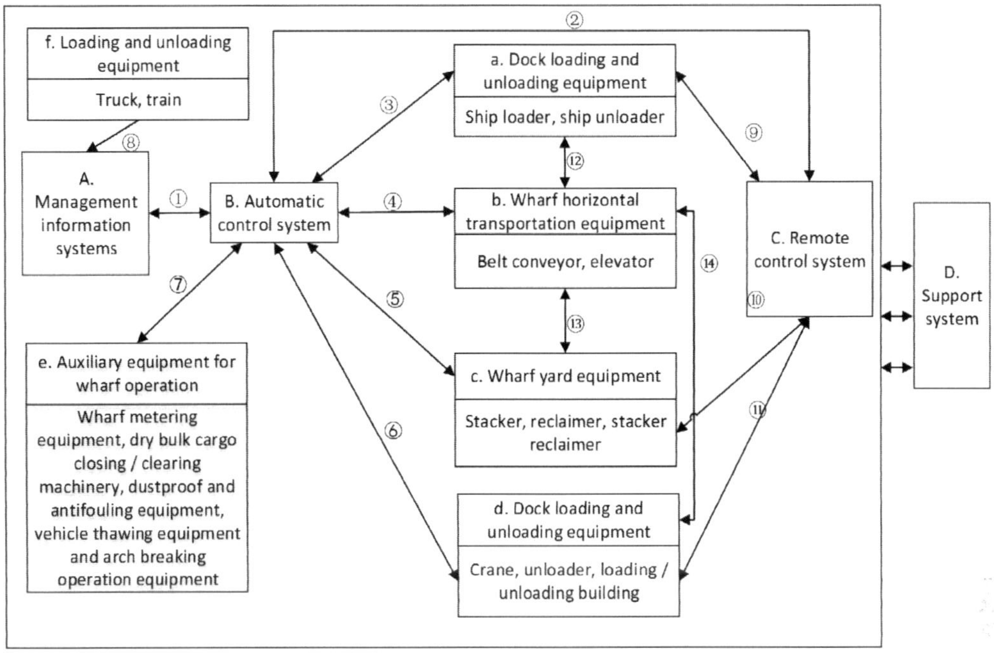

Figure 1. Remote real-time online material movement control logic of dry bulk terminal

Combined with the above control logic, the control mechanism of coordination and cooperation between key nodes and key objects of dry bulk terminal (between systems, between systems and equipment, between equipment and equipment) is shown in Table 3. The interaction mechanism between systems can open up the flow path of decision information between systems in the process of terminal operation, The interaction mechanism between system and equipment can integrate the internal relationship between information flow and physical flow, and the interaction mechanism between equipment can determine the optimal structure of equipment in geospatial layout.

Table 3. Cooperative control mechanism between key objects and key nodes of dry bulk terminal

Interaction category	**Information flow**		**Interactive information**
Interaction between systems	1	A-B	Work order
		B-A	Work order execution status
	2	B-C	Work order
			Work order execution status
		C-B	Work status
			Work intervention request
Interaction between system and equipment	3	B-a	Detailed work instructions
		a-B	Fault information
			Work status
			Work order execution status
			Abnormal operation instruction execution
	4	B-b	Detailed work instructions
		b-B	Fault information
			Work status
			Work order execution status
	5	B-c	Detailed work instructions
		c-B	Fault information
			Work order execution status

			Work status
			Abnormal operation instruction execution
	6	B-d	Detailed work instructions
		d-B	Fault information
			Work order execution status
			Work status
			Abnormal operation instruction execution
	7	B-e	Work order
		e-B	Fault information
			Work order execution status
			Work status
	8	f-A	Driver identification information, truck or train number
	9	C-a	Remote operation instruction
		a-C	Work status
	10	C-c	Remote operation instruction
		c-C	Work status
	11	C-d	Remote operation instruction
		d-C	Work status
Interaction between devices	12	a-b	Interlocking control signal
		b-a	Interlocking control signal
	13	c-b	Interlocking control signal
		b-c	Interlocking control signal

4. Conclusions

The analysis of key objects and key nodes of dry bulk terminal and the corresponding cooperative control mechanism can lay a theoretical foundation for the construction of dry bulk terminal control model and digital twin. On the basis of focusing on the material movement process mechanism and fully mastering the interactive information of material movement, combined with the actual needs and process characteristics of each dry bulk terminal, a system design scheme and engineering application guidance for automated dry bulk terminal can be created in the future.

This paper belongs to the research results of the national key research and development project "Theories and methods of super-large comprehensive port operation prediction based on cyber-physical integration" (2020YFB1710801)

References

[1] Ma Zuqi (translation). Australia: 1.2 billion national construction funds injected into railway construction [J]. World rail transit, 2009 (4): 2

[2] Zhang Zhiming. Key technology for design of 400000 t Ore Wharf in Dongjiakou port area of Qingdao port [J]. Water transportation engineering, 2013 (S1): 48-55

[3] Lei Wenwen. Design of intelligent control system for bulk cargo operation at dry bulk terminal [J]. Port loading and unloading, 2020 (4): 4

[4] Yuan Hang, Cao Yanjie, Yang Duobing. Research and development of intelligent reclaiming system for dry bulk terminal [C] / / Proceedings of application technology exchange meeting for automated container terminal. 2015

[5] Gao Yujun, Sun Bo, Chu Yangang, et al. Application of automatic operation technology of stacker reclaimer in Dongjiakou Ore Terminal [J]. Automation technology and application, 2019, 38 (11): 4

[6] Li Weibo. Research on the integration of management and control of coal terminal in Huanghua port [J]. Science and technology and innovation, 2016 (issue 16): 61-63
[7] Meng Wenjun. Development status and Prospect of dry bulk cargo terminals in Chinese ports [J]. Chinese ports, 2012 (10): 4
[8] Liu Lin. practice of intelligent transformation of coal port in Huanghua port [J]. Port science and technology, 2020 (12): 4
[9] Zhang Maolin, Qi Jian. Conception and practice of intelligent construction of bulk cargo terminal [J]. 2021 (2013-1): 31-36
[10] Shen Bo. Research and design of integrated production management system for bulk cargo terminal of Tianjin port [D]. Tianjin University, 2015

Journal of Physics: Conference Series **2195** (2022) 012012 doi:10.1088/1742-6596/2195/1/012012

Research on Multi-Objective Optimization Method of Edge Cloud Computing Virtual Machine Placement

Lingpeng Shi[1,a], Shida Lu[2,b], Tianbo Feng[3,c], Xiumin Zhao[4,d], Xiaolu Chen[5,e], Haoyang Cui[6,f*]

[1,2,3,4,5]State Grid Shanghai Electric Power Co. , Ltd. Information & Telecommunication Company, Shanghai, China

[6]Shanghai University of Electric Power, School of electronic and information engineering, Shanghai, China

[a]Lingpeng_Shi@163. Com, [b]Shida_Lu@126. Com, [c]Tianbo_Feng@163. Com

[d]Xiumin_Zhao@163. Com, [e]Xiaolu_Chen@163. Com, [f]cuihy@shiep. edu. cn

Abstract—With the emergence of demand for massive computing tasks in the edge cloud of the city, the disordered computing in the edge cloud leads to the high energy consumption of CPU computing and the problem of too long time delay caused by the blockage of computing tasks. This has become the first technical difficulty to solve in the construction of edge cloud. The virtual machine placement method is optimized by computing energy consumption and computing delay. First, the LRR physical host screening model is constructed to detect the status of the physical hosts and form a list of migrated physical hosts; secondly, the MMT time scale model is constructed to generate the list of migrated virtual machines; finally, the GA algorithm is used to place the virtual machines. The simulation results show that this algorithm can reduce the CPU energy consumption of the edge cloud center by 20. 21% and the time delay by 16. 11%. This optimization method has a good theoretical guidance effect on the construction of cloud computing.

1. INTRODUCTION

With the advent of the era of 5G, the Internet of Things, and the gradual increase in cloud computing applications, the pressure on data processing for cloud computing centers is becoming more and more intense [1]. If the resource allocation is unreasonable, in addition to the waste of resources, it will also bring greater computing pressure to the cloud computing center [2]. Therefore, it is of great significance to realize the multi-objective optimization of Virtual Machine Placement (VMP).

Virtual machine placement is an important step in virtual machine migration. Effective optimization of VMP can improve the utilization of physical resources, reduce energy consumption and achieve load balancing. From the perspective of the purpose of the virtual machine placement problem, there are currently three main types: energy consumption optimization technology for VMP [3]; VMP technology in the special architecture of the cloud data center; and the design of the placement method.

In recent years, the literature on the placement of virtual machines has focused on the following two aspects: changes in energy consumption models and improvements in intelligent algorithms. In terms of energy consumption model, it is divided into two types: multi-objective and single-objective [4]. In the early articles, when considering the energy consumption model of the data center, it was relatively single, and the typical goals were: energy consumption optimization, performance optimization, and resource allocation optimization [5]. In order to optimize the data center from the aspects of data center load balancing, reducing energy consumption, and improving resource utilization, literature proposed a series of intelligent algorithms based on the combination of multi-objective optimization problems and intelligent algorithms. Multi-objective solving algorithm. Literature [6] uses pheromone positive feedback mechanism and heuristic search to add ant colony algorithm to solve, realizes the integration strategy of virtual machine placement in cloud environment, and obtains results in cloud data center load balancing, reducing energy consumption, and improving resource utilization. Literature [7] first takes minimization of energy consumption and maximization of resource utilization as the optimization goals. By imitating the pheromone mechanism, the classical bat algorithm is discretized and improved, and the improved discrete bat algorithm is used to solve the model. Literature [8] sets the goal as the waste of physical resources in the data center and the time delay of the data center, and optimizes the two goals at the same time through an improved dual fitness genetic algorithm. Literature [9] established an optimization model to minimize energy consumption and resource waste, using Pareto dominance concept combined with ant colony algorithm to solve the problem. Literature [10] uses CPU, MEM and hard disk occupancy rates as evaluation indicators to describe the load of the system, and uses linearized evaluation to calculate the resource utilization of the host to maximize resource utilization. Literature [11] designed a multi-objective discrete differential evolution of network-aware virtual machine placement algorithm. Under the framework of differential evolution, based on the network-wide communication cost, maximum link utilization, hardware resource constraint violations, and link capacity constraints Violation degree comprehensively evaluates individuals and selects the best.

However, genetic algorithm has the inherent deficiency of premature convergence, and the current improvement schemes based on genetic algorithm pursue high intelligence, precision and reckless to avoid premature convergence, which is very easy to cause high operation energy consumption. In this paper, MMT and LRR algorithms are used to optimize the genetic strategy and improve the virtual machine placement process, which can not only avoid the problems of low efficiency and premature caused by the influence of decision coding, but also significantly reduce energy consumption and improve the sense of user experience.

2. MODEL BUILDING

2.1. The workflow of GA-VMP

The workflow specifically includes the following 4 steps:Step1 Complete physical host status detection based on the robust partial regression LRR method to form a list of candidate physical hosts for migration; Step2 Select the MMT algorithm based on the minimum migration time to complete the virtual machine selection and form a list of candidate virtual machines for migration; Step3 is based on the GA-VMP method to complete the virtual machine relocation; Step4 Repeat Step1 ~ Step3, set a cycle, and it will end when the time period is reached.

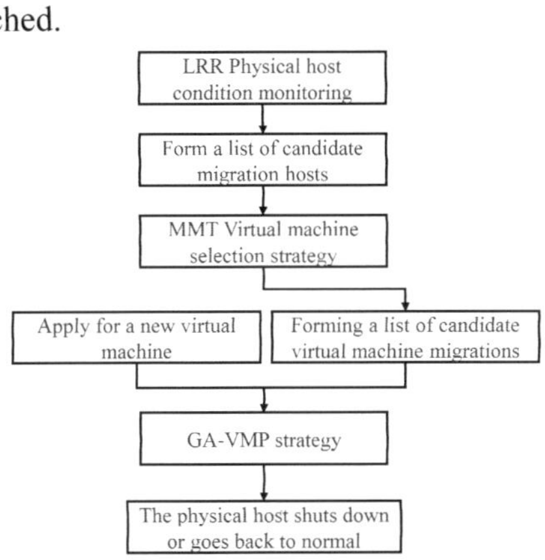

Figure 1. workflow of the GA-VMP

Figure 1 shows the workflow of the GA-VMP virtual machine placement strategy during the entire virtual machine migration process.

2.2.GA-VMP related terms

1) Cloud data center energy consumption model

The cloud data center is mainly composed of a large number of physical hosts, so its energy consumption is mainly composed of the energy consumption of all components of the physical host. Literature [12-15] believes that the energy consumed by a physical server (dual-core CPU, 4 memory, a disk, 2 PCI slots, a motherboard, etc.) is approximately: CPU accounts for 41%, memory accounts for 18%, and disk accounts for about 41%. 7%, PCI slots accounted for 22%, and motherboards accounted for 12%. According to this idea, this paper designs the energy consumption of a single physical server as:

$$E(U_{cpu})=E_{idle}+(E_{max}-E_{idle})U_{cpu} \tag{1}$$

$$U_{cpu}(t)=\sum_{c=1}^{PE_j}\sum_{i\in r_j(t)}\frac{mips_{i,c}}{MIPS_{j,c}} \tag{2}$$

$$E(U_{mem})=E_{idle}+(E_{max}-E_{idle})U_{mem} \tag{3}$$

$$E(U_{disk})=E_{idle}+(E_{max}-E_{idle})U_{disk} \tag{4}$$

$$E(U_{bw})=E_{idle}+(E_{max}-E_{idle})U_{bw} \tag{5}$$

$$E_{total}=E(U_{cpu})+E(U_{mem})+E(U_{disk})+E(U_{bw}) \tag{6}$$

Among them, $r_j(t)$ represents the index collection of virtual machines M_j allocated to the physical host. $U_{cpu}(t)$、$U_{mem}(t)$、$U_{disk}(t)$、$U_{bw}(t)$ indicates the CPU usage, memory usage, disk usage, and network bandwidth usage of the physical host at time. $0\%\leq U_{cpu}(t)$、$U_{mem}(t)$、$U_{disk}(t)$、$U_{bw}(t)\leq 100\%$. Indicates the energy consumption of the physical host CPU, memory, disk space, and network bandwidth when it is idle, that is, when $U_{cpu}(t)=0\%$、$U_{cpu}(t)=0\%$、$U_{mem}(t)=0\%$、$U_{disk}(t)=0\%$、$U_{bw}(t)=0\%$. Indicates the energy consumption of the physical host at full load, that is, when $U_{cpu}(t)=100\%$、$U_{mem}(t)=100\%$、$U_{disk}(t)=100\%$、$U_{bw}(t)=100\%$.

$Mips_{i,c}$ is the mips request situation of the c processing element of the *i* virtual machine. Is the overall MIPS computing power of the *c*-th processing element of the *j*-th physical host.

The number of MIPS requested by a virtual machine changes as the application changes, so the resource usage rate of the physical host should also change as the application changes. Therefore, the energy consumption of the physical server must be counted within a certain period of time. So according to formula (1) can be evolved into formula (7):

$$E_{total}(t)=E(U_{cpu}(t))+E(U_{mem}(t))+E(U_{disk}(t))+E(U_{bw}(t)) \tag{7}$$

In this way, the total energy consumption of the j-th physical host in the time period can be calculated according to formula (8):

$$E_j=\int_{t_0}^{t_1}E_{total}(t)\,dt \tag{8}$$

The energy consumption of the entire cloud data center is formula (9), where M is the number of physical hosts.

$$E=\sum_{j=1}^{M}E_j \tag{9}$$

2) SLA violation Time per Active Host (SLATAH).

When a cloud client submits a job to the cloud computing platform, the lack of resources will cause SLA violations. In the process of virtual machine allocation, an important performance indicator is the SLA time per host of each physical host. It reflects the online time of the physical host with high quality of service.

$$SLATHA=\frac{1}{M}\sum_{i=1}^{M}\frac{T_{s_i}}{T_{a_i}} \tag{10}$$

The number of hosts and virtual machines in the cloud data center are denoted by M and N respectively, where is the time when the CPU utilization of the physical host reaches 100%. T_{a_j} is the time that the physical host is online and active.

3) Performance Degradation due to Migrations (PDM) after virtual machine migration.

$$\text{PDM}=\frac{1}{N}\sum_{j=1}^{N}\frac{C_{d_i}}{C_{r_i}} \tag{11}$$

Where C_{d_i} is the estimated value of performance degradation caused by virtual machine migration, and C_{r_i} is the total CPU MIPS computing power during the entire time period during which the virtual machine is requested.

4) The combined index (ESV) of energy and SLA violations.

The SLA violation rate is calculated as formula (12):

$$\text{SLA violation}=\text{SLATAH}*\text{PDM} \tag{12}$$

The indicators of overall energy consumption balance are:

$$\text{ESV}=\text{E}*\text{SLA violation} \tag{13}$$

3. GA-VMP'S VIRTUAL MACHINE PLACEMENT ALGORITHM

3.1. The specific process of GA-VMP virtual machine placement

Suppose there is a collection of N virtual machines

$\{VM_i(pe_i, mips_i, ts_i, d_i)|i=1, \ldots, N\}$ VM physical hosts to be reassigned to the cloud data center during the virtual machine placement phase$\{M_i(PE_i, MIPS_j)|j=1, \ldots, M\}$ run on. Each virtual machine requests pe_i processing elements, the request includes $mips_i$ computing requirements, the start execution time of VM is ts_i, and the completion time is ts_i+d_i, The migration operation was not interrupted or performed within d_i a period of time. This article does not put the migration factor only on the single dimension of the CPU usage of the physical host, but expands the dimensions of physical resources, including memory size and disk space size, Network bandwidth, etc.

Assuming that each physical host can accommodate any type of virtual machine, and the energy consumption model of the physical host has a linear relationship with the utilization rate of various physical resources, then the goal of the virtual machine placement is to minimize the energy consumption of the cloud data center and make the most of it. It is possible to complete the execution of more virtual machines, so the objective function is:

$$G=\min\sum_{j=1}^{N} E \tag{14}$$

$$E_j=\int_{t_0}^{t_1} E_{\text{total}}(t)\,dt\ ,\ \ E=\sum_{j=1}^{M} E_j \tag{15}$$

As mentioned earlier, virtual machine placement is to complete the mapping of virtual machines to physical hosts, which is a multi-objective optimization problem. Therefore, GA-VMP introduces a multi-objective genetic algorithm to solve the virtual machine placement problem.

The specific meaning of the genetic algorithm will not be described in detail due to the space relationship. The specific steps of the GA-VMP solution are as follows:

The chromosome encoding creates an initial population for s chromosomes, and s is the size of the population, which realizes the mapping from the virtual machine number to the chromosome.

Calculation of fitness fitness function Fitness is converted from the objective function. Under the conditions of a given population, the evolution value of each chromosome is calculated.

The calculation of the new family population completes the creation of a new family population by performing the following steps and completes the update.

The selection stage is based on the evolution value of the chromosome, and an individual with two parents is selected from the current population.

The crossover stage uses the crossover probability to create a new offspring by modifying the chromosomes of both parents.

In the mutation stage, the mutation will be completed in a certain position of the chromosome through the mutation probability.

The acceptance phase in the current situation, a new child will be produced from the next generation.

The replacement phase completes the replacement operation by assigning the current generation of individuals to the next generation. If the termination condition is met, the algorithm will stop execution and return an individual with the highest evolution value. Otherwise, the algorithm returns to the fitness calculation phase to continue the cycle. implement. Figure 2 shows the flow of GA-VMP.

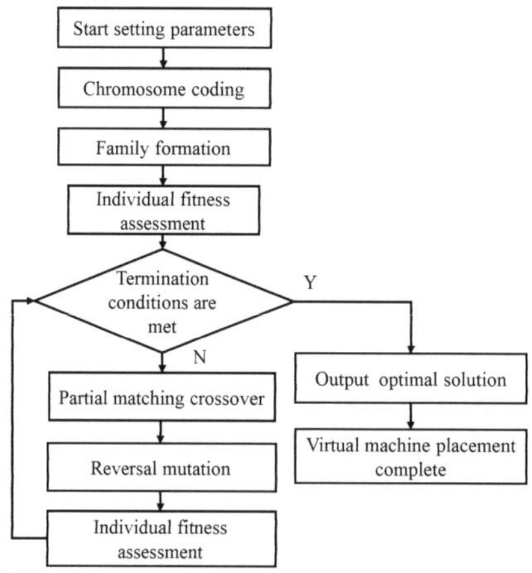

Figure 2. the flow of GA-VMP

The surface is the pseudo code of the virtual machine placement process of GA-VMP genetic algorithm. The code can be implemented in Java language with a small amount of modification.

3.2.Implementation of GA-VMP

GA-VMP is a genetic algorithm, which is mainly used to solve the multi-objective optimization problem of virtual machine placement. The objective function of GA-VMP is to minimize the energy consumption of the cloud data center. This article uses a hierarchical structure to reflect the chromosome relationship between individuals. This hierarchical structure has 3 levels:

Level 1 consists of initialized individuals.

Level 2 consists of a series of nodes, representing a group of virtual machines in the cloud data center.

Level 3 consists of a series of nodes, representing a group of physical hosts running on physical hosts.

Through this hierarchical structure, the placement of virtual machines to physical hosts is reflected, and the fitness function completes the calculation of the evolution value of each chromosome. This process is performed in parallel for virtual machine placement, which can reduce the calculation time for virtual machine migration. For the pseudo code of the fitness function, see Algorithm1.

Algorithm1 Calculation of the fitness value of each chromosome
Input: Chromosome
Output: Fitness of the chromosome
1 Power Of Datacenter = 0
2 For each host ∈ collection of hosts do
3 utilization Mips=host. get Utilization Of CPU
4 utilization Mem = host. get Utilization Of Mem
5 utilization Disk = host. Get Utilization Of Disk
6 Utilization B w = host. Get Utilization Of B w
7 Power Of Mips = get Power (host, utilization Mips
8 Power Of Mem = get Power (host, utilization Mem)
9 Power Of Disk = get Power (host, utilization Disk)

10 Power Of B w = get Power (host, utilization B w)
11 Power Of Host = power Of Mips + power Of Mem + power Of Disk + power Of B w
12 Power Of Datacenter = power Of Datacenter + power Of Host
13 End For
14 Evaluation value(chromosome) = 1. 0/power Of Datacenter

The GA-VMP virtual machine placement method in this article combines the existing robust local regression LRR physical host state detection method and the minimum migration time MMT virtual machine selection strategy in Cloudsim, forming a new type of virtual machine migration model called LRR -MMT-GA.

4. Experimental Results and Analysis

This article analyzes the performance of the GA-VMP algorithm proposed in this article from the aspects of virtual machine placement energy consumption and delay, and compares it with the following three algorithms:

1. Optimization Sequence Algorithm (Optimization Sequence Algorithm, OSA): A virtual machine sequence placement algorithm optimized by the idea of interval bubbling.
2. Simulated Annealing (Simulated Annealing, SA): A Monte-Carlo iterative stochastic optimization algorithm based on the similarity between the annealing process of solid substances in physics and general combinatorial optimization problems.
3. Particle Swarm Optimization (PSO): A self-adaptive multi-strategy heuristic algorithm that uses multiple particles to form a population for iterative calculation to find the optimal solution.

The simulation in this paper is implemented in Cloudsim and MATLAB R2019b simulation platform, and genetic algorithm is introduced as the benchmark algorithm. The initial size of the population is 20 and the number of iterations is 200. This algorithm is unique in terms of optimal configuration. The experiment used 100 physical nodes and 200 virtual machines. The CPU resources requested by the virtual machines were uniformly set to (0. 25, 0. 5, 1, 1. 5, 2, 2. 5, 3, 4) to simulate the different requests of the virtual machines. At the same time, the memory is also set to (0.25,0.5,1,1.5,2,2.5,3,4) the power consumed by the physical node when the CPU utilization is 0% and 100% is 175W and 250W respectively.

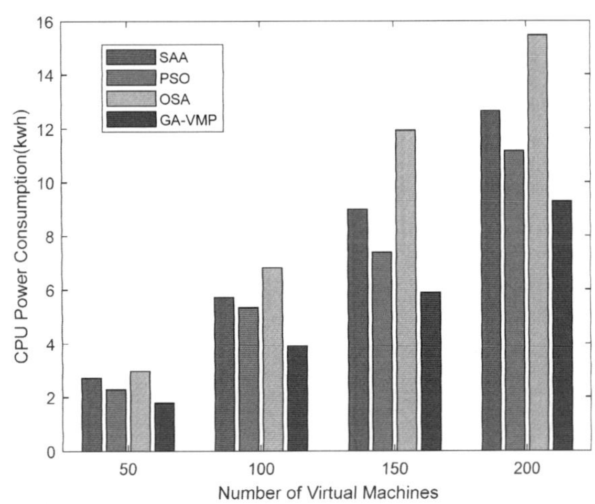

Figure 3. CPU energy consumption

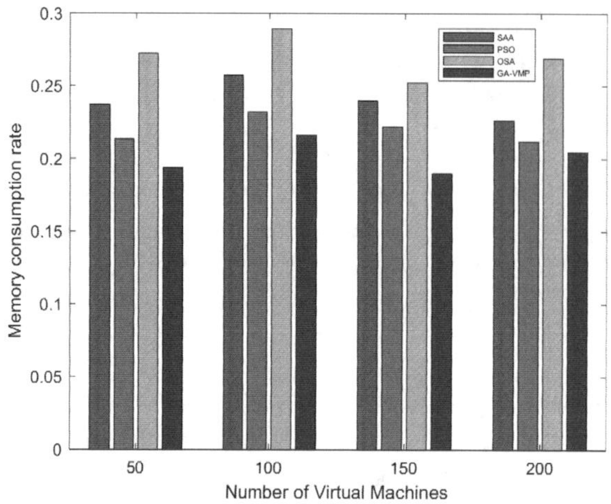

Figure 4. memory consumption

As shown in Figure 3, under the same virtual machine load, the comparison of CPU energy consumption occurs under four different algorithms. It can be seen that with the increase in the number of virtual machines placed, the CPU energy consumption of the four placement algorithms is on the rise. The OSA algorithm has the highest CPU energy consumption, SAA and PSO are relatively low, and GA-VMP has the lowest CPU energy consumption. Compared with the PSO algorithm, the CPU energy consumption of the algorithm in this paper is reduced by 20. 21% on average. The proportion of CPU memory consumption fluctuates with the increase in the number of virtual machines. As shown in Figure 4, there is no significant increase in memory consumption, indicating that the server has a strong virtual machine placement capacity. In addition, the memory consumption rate of the GA-VMP virtual machine placement algorithm is slightly lower than the other three algorithms, and it has stronger placement capabilities.

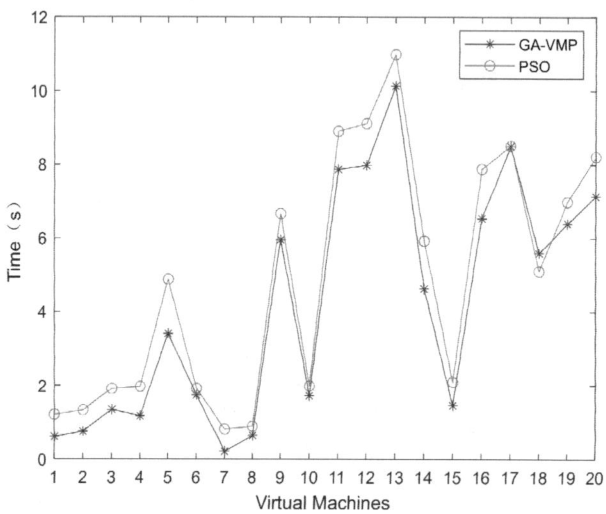

Figure 5. Comparison of the latency of virtual machine placement

Taking the placement of 20 virtual machines as an example, the particle swarm optimization algorithm with the best delay performance is selected and compared with the algorithm in this paper. The delay effect is shown in Figure 5, and the average delay is reduced by 16. 11%. Since the particle swarm optimization algorithm is prone to local optimal solutions, it needs to perform secondary placement and migration according to the allocation rate of virtual machines, which consumes a certain amount of time. The GA-VMP virtual machine placement algorithm mentioned in this article avoids localization to a certain extent through genetic iteration. The frequent appearance of the optimal solution can effectively improve the delay performance and enhance the user experience. In addition, the consumption rate of CPU and memory has been reduced to varying degrees.

5. Conclusion

This paper proposes a GA-VMP virtual machine placement method, which uses MMT and LRR algorithms to optimize genetic strategies. While improving the virtual machine placement process, it can reduce the energy consumption of virtual machine placement and enhance the user experience. Based on the future research of this article, the following two aspects can be considered: (1) Consider combining genetic algorithm with particle swarm algorithm, ant colony algorithm and other methods to design a method that can better jump out of the local optimal solution. (2) The granularity of the target virtual machine can be refined and analyzed to obtain the impact of different granularity on the placement performance of the virtual machine.

Acknowledgment

This study was supported by the Science and Technology Project of State Grid Shanghai Electric Power Co. , Ltd(52090F20007L). The authors gratefully acknowledge the anonymous reviewers for their valuable comments.

References

[1] SHANG P J, WANG R J, WANG Q D. A novel power management for CMP systems in data-intensive environ- ment ［J］. IEEE Transactions on Computers, 2016, 65 (6): 1816-1830.
[2] MAO Y Y, ZHANG J, LETAIEF K B. A Lyapunov Opti- mization Approach for Green Cellular Networks with Hy-brid Energy Supplies ［J］. IEEE Journal on Selected Are- as in Communications, 2015, 33(12): 2463-2477.
[3] BELOGLAZOV A, BUYYA R. Optimal online determin- istic algorithms and adaptive heuristics for energy and performance efficient dynamic consolidation of virtual ma- chines in cloud data centers ［J］. Concurrency and Com- putation: Practice and Experience, 2012, 24(13): 1397-1420.
[4] GU Y H, YI P. Method of Service Function Chain Ad- justment Based on Multi-VM Live Migration. Journal of Chinese Computer Systems, 2017, 38(5): 1022-1027.
[5] Wang JV, Cheng CT, Tse CK. A power and thermal-aware virtual machine allocation mechanism for cloud data centers [C]//IEEE International Conference on Communication Workshop. IEEE, 2015:2850-2855
[6] Zhang Congyue, Fu Xiong, Qiao Lei. Research on virtual machine placement based on multi-objective optimization in cloud computing environment [J]. Computer Applications and Software, 2021, 38(03): 32-38.
[7] Lin Kaiqing, Li Zhihua, Guo Shujie, Li Shuangli. Multi-objective optimization virtual machine placement algorithm in cloud environment[J]. Computer Applications, 2019, 39(12): 3597-3603.
[8] Huang Zhaonian, Li Haishan, Kui Qiang. Research on virtual machine placement based on improved algorithm in cloud computing [A]. Electronic Technology Academic Committee of China Shipbuilding Engineering Society. Paper of Equipment Technology Development Forum 2017 of Electronic Technology Academic Committee of China Shipbuilding Engineering Society[C]. Electronic Technology Academic Committee of China Shipbuilding Engineering Society: China Shipbuilding Engineering Society, 2017:5.
[9] F. Farahnakian et al. , "Using Ant Colony System to Consolidate VMs for Green Cloud Computing, " in IEEE Transactions on Services Computing, vol. 8, no. 2, pp. 187-198, 1 March-April 2015 , doi: 10. 1109/TSC. 2014. 2382555.

[10] Yan Jianen, Zhang Hongli, Xu Haiyan. Virtual machine placement method based on load optimization [J]. Intelligent Computers and Applications, 2019, 9(06): 178-180+183.
[11] Zang Weifei, Lan Julong, Hu Yuxiang, Chen Wentao. Network-aware virtual machine placement algorithm based on multi-objective discrete differential evolution[J]. Computer Engineering, 2019, 45(06): 96-102.
[12] Li Shuangli, Li Zhihua, Yu Xinrong. Virtual machine placement method based on multi-objective optimization [J]. Journal of Chongqing University of Posts and Telecommunications (Natural Science Edition), 2020, 32(03): 356-367.
[13] Zhou Zhen, Wang Hui, Li Junfeng. Virtual machine placement strategy based on family genetic algorithm[J]. Computer Engineering and Design, 2021, 42(02): 482-488.
[14] Xu Shengchao. Using genetic algorithm to complete the optimization of virtual machine placement strategy[J]. Computer and Modernization, 2020(12): 25-31+42.
[15] Liu Dan, Sui Xin, Li Li. Virtual machine placement energy-saving optimization algorithm based on cloud computing [J]. Journal of Changchun University of Science and Technology (Natural Science Edition), 2015, 38(06): 150-153.

Automatic partial discharge detection method of traction transformer based on wavelet transform

Jiyong Liu[1*], Xin Wang[1], Shengchun Yan [1], Yan Fan [1], Changjin Hao [2]

[1] Shuohuang Railway Development Co, Ltd, Shuozhou 036002, China

[2] Institute of Energy Sensing and Information, Sichuan Energy Internet Research Institute, Tsinghua University, Chengdu 610000, China

1981ljy@163.com

Abstract. Partial discharge parameter acquisition and storage of transformer lacks the setting of time threshold, which leads to the low reception rate of transformer discharge signal. An automatic detection method of partial discharge of transformer in substation based on wavelet transform is studied. By analyzing the ultrasonic waveform of discharge signal, the partial discharge parameters of transformer in high voltage substation are extracted. According to the principle of wavelet transform, the time threshold is set, the extracted parameters are collected and stored in sections, and the automatic detection algorithm of partial discharge parameters is established. By deriving the results, the automatic detection of partial discharge is completed. At this point, the design of automatic detection method for partial discharge of transformer in substation based on wavelet transform is completed. Analysis of experimental results: Through comparative experiments, it is found that the highest discharge signal receiving rate of this method is 87.5%, which is superior to other methods and has high application value.

1. Introduction

In the traction power supply system, the transformer plays an important role in the stable output of power, and its stable operation is of great significance to the safety of the power system. Statistics show that transformer faults are mostly caused by insulation problems[1]. Partial discharge of transformer is an important cause of insulation failure. Partial discharge will cause aging or breakdown of transformer insulation parts, which will greatly threaten the service life and operation reliability of transformer, and then affect the stability and reliability of power transmission in the power system. If the partial discharge of the transformer can be found in time, the cause and location of the fault can be determined, and the transformer can be repaired in time, the operation safety and reliability of the transformer can be greatly improved, and the transformer accidents caused by transformer insulation failure can be avoided[2].

At present, there are non-electric measurement and electric measurement methods to locate the discharge position. When partial discharge occurs in the high voltage transformer, it will be accompanied by abnormal phenomena of sound and heat. Parameters can be obtained by ultrasonic sensor, based on which partial discharge location can be carried out[3-4]. Electric measurement method mainly uses pulse current test to locate, although this method has high accuracy but is susceptible to environmental interference[5].As for the localization algorithm of partial discharge of transformer, literature [6] proposed that in the substation environment with more uncontrollable interference, the optimal identification parameters were obtained by using the node training of network layer to realize partial discharge identification of substation by utilizing the self-organizing competitive identification and strong anti-interference characteristics of Kohonen network. Literature [7] proposes to use generalized regression neural network and received signal amplitude intensity fingerprint for partial discharge location. Firstly, the fingerprint is established and solved by generalized regression neural network to complete detection and location. However, due to the lack of time threshold for the segmented acquisition of transformer partial discharge parameters, the reception rate of transformer discharge signal becomes low. This paper studies the automatic partial discharge detection method of high voltage transformer, and proposes an automatic partial discharge detection method of transformer using wavelet transform method.

2. Design of automatic partial discharge detection method for transformer

2.1. Partial discharge parameter extraction

There are many partial discharge signal parameters of high-voltage transformer. The amplitude, time interval, energy, frequency and other parameters of parameter waveform can be directly obtained from the ultrasonic waveform of discharge signal[8]. However, the availability of the parameters obtained by the waveform itself is poor. When the partial discharge intensity is large, the time interval of the waveform will be affected. Therefore, this type of parameter has certain limitations. This paper will not consider using this type of parameter for data analysis. Some potential parameters of ultrasonic can also be obtained by indirect means, and the waveform of Partial Discharge Ultrasonic of substation transformer can be converted and calculated to read[9]. For example, through spectrum calculation, parameters such as power spectrum value, frequency of power spectrum, average value of power spectrum, median value of power spectrum and so on of Partial Discharge Ultrasonic of substation transformer can be obtained, Parameter clusters such as square root, square difference and root mean square are obtained by statistical algorithm[10], which can avoid the impact of time interval and improve the availability of data.

By using the principle of wavelet transform, the local discharge times of transformer can be easily extracted. Every impact of ultrasonic wave is caused by discharge[11], the number of discharge times can be counted by counting the impact time of ultrasonic wave. Partial discharge diagram of transformer is shown in Figure 1.

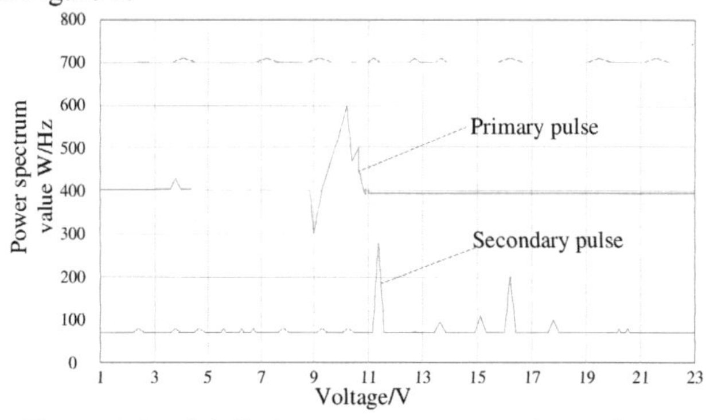

Figure 1. Partial discharge concept map of transformer

According to Figure 1, when the impact time of discharge ultrasonic exceeds the pulse cycle of partial discharge, two discharge reactions will occur, but only one set of parameters will be generated. When the impact time of discharge ultrasonic is less than the pulse period of partial discharge, it will also produce two discharge reactions, but two groups of parameters will be generated. Therefore, the same number of discharge signals may contain different groups of parameters, which means that the length of impact time determines the extraction accuracy of ultrasonic parameters, and the less discharge parameters will be obtained with the increase of impact time. Therefore, setting the exact impact time is of great significance to the extraction of discharge parameters. Table 1 shows the corresponding relationship between the modulation value of wavelet transform and the number of discharges.

Table 1 Partial discharge times

Modulation value	Effective discharge times	Invalid discharge times
1	324	271
2	312	254
3	276	76
4	259	57
5	237	21

According to the data in Table 1, the larger the set value of ringing, the greater the proportion of effective discharge. If conditions permit, continue to increase the set value of ringing count, the proportion of effective discharge will continue to increase. However, if the setting of ringing times is too large, it may lead to some unnecessary errors, so the ringing count is modulated to 5. Partial discharge parameters are a relatively stable signal in a small space, but when ultrasonic testing the location of partial discharge, due to the strong variability and randomness of partial discharge, in order to reduce the difficulty of extracting partial discharge parameters, control the spectral distribution of partial discharge and make it tend to a stable state. The partial discharge parameters of transformer in high voltage substation are extracted and processed based on wavelet transform, and then the discharge parameters are stored in sections..

2.2. Partial discharge parameter extraction

The types of signals that can be detected due to partial discharge parameters are called Duoduo, such as ultrasonic signals, high-frequency signals, UHF signals, etc. The above signals have different frequency bands and sampling frequencies. The discharge parameters of these signals are collected and stored. Some signals only need very low sampling frequency, and some signals need very high sampling frequency. Various types of discharge parameters can be processed based on wavelet transform. When collecting discharge parameters at low speed, there is little demand for data storage. At this time, the time threshold can be set. If the sampling time triggers the threshold, the collection will stop. Then it is necessary to process the above collected discharge parameters, continue to collect the discharge parameters, and continue the cycle. Finally, the collection of all partial discharge parameters will stop. Partial discharge will lead to high-frequency electromagnetic wave at the local position of the transformer, up to hundreds of MHz. When the instantaneous voltage to ground in the transformer is generated in the metal enclosure, the electromagnetic wave signal is difficult to capture and collect. Therefore, the signal shall be captured through a special capacitor and detector, and the detector shall be installed on the outer surface of the collector. According to the detector, the transformer discharge signal is captured and the detection medium in the transformer is analyzed.

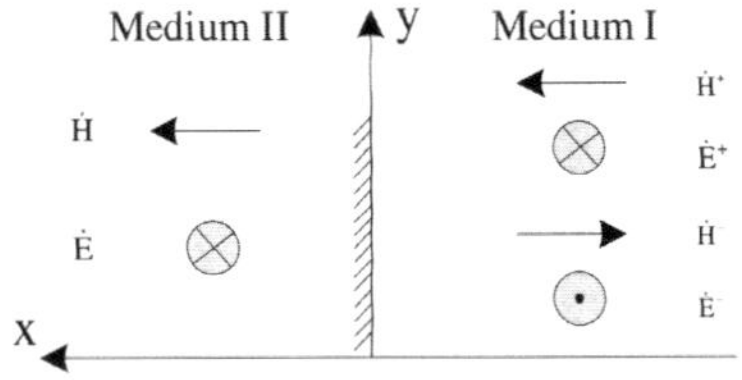

Figure 2. Effect of transient on partial discharge

There are two kinds of media in the transformer, I and II. Medium II represents the metal conductor material, and the interior of the transformer is divided into two parts through the metal inner wall. The process of forming transient effect on partial discharge is shown in Figure 2.

In Figure 2, H^{+} And H^{-} represents incident electromagnetic wave and reflected electromagnetic wave respectively. The electric field intensity of incident wave and reflected wave is E respectively $^{+}$ and E^{-}. The two values are opposite in phase and the absolute values are not equal. So we know the critical surface (X=0)E^{+}+E=0. Magnetic field H (x, t) at position x can be expressed as:

$$H(x,t) = H^{+}(x,t) + H^{-}(x,t) \tag{1}$$

Where, $H^{+}(x, t)$, $H^{-}(x, t)$ respectively represent incident and reflected magnetic fields, and the relationship between them and electric field intensity E is:

$$\begin{cases} H^{+}(x,t) = \dfrac{E^{+}(x,t)}{Z_{02}} \\ H^{-}(x,t) = -\dfrac{E^{-}(x,t)}{Z_{02}} \end{cases} \tag{2}$$

Where, Z_{02} is the wave impedance of the medium.

The discharge pulse current is:

$$I = I_0 \exp\left(-\frac{4\pi(t-t_0)^2}{t^2}\right)(H^{+}(x,t) - H^{-}(x,t)) \tag{3}$$

Where, φ is a constant; I_0 is the pulse peak value, which is the maximum value. At this time, $t = t_0$.

Partial discharge signal has periodic characteristics. When the signal period is less than the sampling time, the characteristics of the signal can be fully collected. When the period is greater than the sampling time, the integrity of the collected signal is insufficient. If the partial discharge signal meets the requirements of acquisition frequency, the continuous signal will have a large amount of acquisition data, and has high requirements for the capacity of storage equipment. Its cost will be very high, resulting in lack of economy.

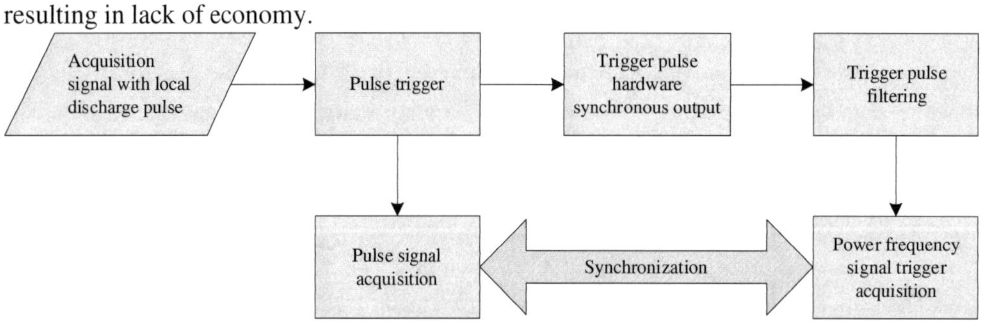

Figure 3. The process of subsection collection and storage of discharge parameters

There are mainly two methods to collect and process partial discharge signals: the first is to collect signals according to the form of signal envelope, so it is necessary to further compress the collected high-frequency signals, so as to obtain the sampled low-frequency signals. Although this method can solve the problem of too low storage capacity, the collected signal is missing some information, that is, after envelope acquisition, it will cause the loss of some important information. The second method is to use ultra-high speed subsection acquisition technology to collect UHF signals in partial discharge. The partial discharge signal obtained by this acquisition method can collect and process the collected matching signal, and then store the partial discharge parameters in this section, which can fully show the characteristics of the original signal. If it is necessary to collect and store segmented data, data can be obtained on a first in first out basis. The implementation process is shown in Figure 3.

2.3. Establish automatic detection algorithm for pd parameters

According to the collected and stored partial discharge parameters, based on wavelet transform and the theory of synchronous acquisition technology, the initial phase can be obtained by automatically detecting and calculating the synchronously collected partial discharge parameters, and the partial discharge parameters can be further obtained. After setting the number of partial discharge pulses of the system, start the synchronous power frequency sampling operation to locate the position of each power frequency sampled partial discharge pulse. Then, the position is used to divide the sampled partial discharge pulses, corresponding to different calculation parts. Calculate the local pulse group of each section according to the phase angle of the first pulse signal, check the time difference between the first pulse and a single pulse, and select the check standard as the power frequency of the measured pulse group. Calculate the phase value of a single pulse:

$$\varphi_i = 2\pi f(\Delta t) + \varphi_0 \tag{4}$$

Where, φ_i is the phase angle of the ith pulse signal, f is the actual power frequency voltage frequency of the site, t_i represents the initial moment of the ith pulse signal in the paragraph, t_0 represents the initial moment of the first pulse signal in the segment, φ_0 represents the phase angle of the first pulse.

In the low-frequency signal, the power frequency sampling signal is processed by filtering, and the zero crossing detection method is used to detect the zero crossing in ten cycles, and the corresponding zero crossing cycle is obtained. Therefore, the real-time frequency of power frequency signal can be obtained. Finally, according to the principle of wavelet transform, the initial phase value of power frequency signal is obtained. When sampling and processing the voltage value, the voltage function is assumed to be periodic, and only the sinusoidal odd harmonic part is considered. The voltage value is expressed by wavelet transform. Because the voltage value obtained by sampling processing is in a closed linear space, after using the partial discharge parameter self-test method, the above variables become a group of orthogonal bases and are in a closed linear space, that is, the algorithm of automatic partial discharge detection can be obtained. According to the self-test algorithm, the automatic partial discharge detection results are calculated, and the calculated self-test results are derived from the constructed path.

2.4. *Build the detection result export path*

After the calculation of the automatic calculation algorithm of partial discharge parameters, the calculated automatic partial discharge detection results are exported. Firstly, the result export path should be constructed. It is known that when partial discharge occurs in the transformer, the current is relatively large, resulting in electromagnetic waves and pulse signals generated by partial discharge at each port of the transformer. These signals will propagate in four directions in the complex structure inside the transformer. The electromagnetic waves and pulse signals in the transformer are induced by the propagation law of partial discharge signal, An ultrasonic sensor is installed in the external structure of the transformer to detect the discharge parameters. The discharge parameters, the propagation time of ultrasonic wave and the propagation speed in the transformer shell and medium are counted and calculated, so as to realize the positioning of partial discharge.

Therefore, the propagation characteristics of longitudinal wave and transverse wave of transformer ultrasonic wave are used for detection, combined with the physical properties of propagation medium, the propagation velocity of ultrasonic wave in the inner wall is measured, and the difference between the propagation velocity of longitudinal wave and transverse wave is calculated. According to the signal transmission interval of longitudinal wave and transverse wave, judge the sequence of signals received by the partial discharge sensor outside the transformer. Considering the attenuation effect of propagation in different media and rapid attenuation in the inner wall, combined with this property, the automatic detection of partial discharge can be realized. The export path of partial discharge self-test results is shown in Figure 4.

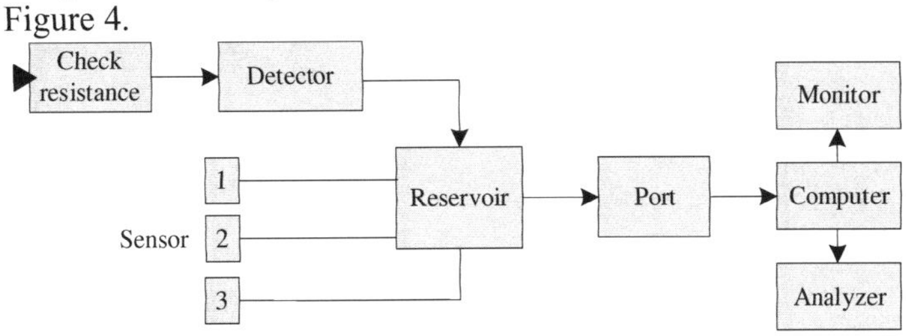

Figure 4. Derivation path of partial discharge self test results

When the ultrasonic sensor receives the partial discharge signal, the converter is used to convert the discharge signal into position signal, and the display is used to display the signal and display the partial discharge detection results, so as to realize the purpose of transformer partial discharge self-test

Journal of Physics: Conference Series 2195 (2022) 012013 doi:10.1088/1742-6596/2195/1/012013

by wavelet transform, and complete the method and function design of transformer partial discharge self-test based on wavelet transform.

3. Experimental Analysis

The wavelet transform based partial discharge automatic detection method and the traditional method 1 are used respectively[8]and traditional method 2[9]. The advantages of the automatic detection method are verified by comparing the receiving rate of transformer discharge signal between the three

3.1. Experimental preparation

Velocity of propagation of partial discharge measurement in the first place, and in the complex environment of transformer, the spread of the discharge signals will alternate in solid media and liquid media transmission, the isolation of two kinds of medium stage will cause the change of discharge signal propagation speed, the sound propagation speed is 1.42 km/s discharge signals receiving rate calculation. Since the conversion time of discharge signal between medium and medium can not be calculated accurately, the receiving rate error of discharge signal will be generated. The error range of the positioning method can be calculated by using the positioning equation.The three-dimensional coordinate was established. Assuming the location of partial discharge as the origin, a point in the coordinate was taken as the position of the ultrasonic sensor. The three-dimensional coordinate system was regarded as a cube, and then the cube was divided into 100 equal parts.Establish the function as follows:

$$V_s = \sum_{l=1}^{10} \left(\frac{\sqrt{S_1 - S}}{Vl}\right)^l \tag{5}$$

Where, l is the point variable, S_1 is the right x-axis coordinate value of the discharge coordinate point, S is the left x-axis coordinate value of the discharge left point, and V is the Y-axis coordinate.

The partial discharge phenomenon is detected according to the phase distribution and band characteristics of UHF. When the signal is detected, the interference noise signal needs to be separated. Then UHF and ultrasonic methods are used to accurately determine the receiving rate of discharge signal, or multiple groups of acoustic emission sensors are used to accurately locate the source. The defect type can be determined by using the phase characteristics, and the reception rate of discharge signal can be determined by combining the discharge type, discharge position, signal strength and severity. The automatic detection method proposed in this paper is used to receive and process the same discharge signal respectively as traditional method 1 and traditional method 2, and the reception rate of discharge signal is calculated.

3.2. Comparison of receiving rate of transformer discharge signal

The receiving rate of transformer discharge signal designed by traditional method 1, traditional method 2 and automatic partial discharge detection method of transformer based on wavelet transform is shown in Figure 5.

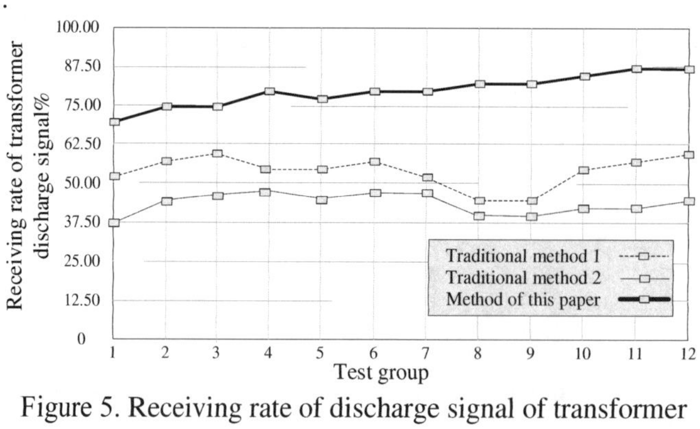

Figure 5. Receiving rate of discharge signal of transformer

It can be seen from the data in Figure 8 that during the self inspection of transformer discharge signal in traditional method 1, the reception rate of discharge signal is generally above 37.5%, the highest reception rate of discharge signal is 49%, and the lowest reception rate of transformer discharge signal is 37.5%. In traditional method 2, when performing self inspection of transformer discharge signal, the receiving rate of discharge signal is generally above 50%, the highest receiving rate of discharge signal is 62.4%, and the lowest receiving rate of transformer discharge signal is 48%. When the automatic partial discharge detection method of substation transformer based on wavelet transform is used for self inspection of transformer discharge signal, the receiving rate of discharge signal is generally above 75%, the highest receiving rate of discharge signal is 87.5%, and the lowest receiving rate of transformer discharge signal is 68.25%. Therefore, the method designed in this paper has superior performance.

4. Conclusion

The method presented in this paper can fully process the discharge signal data, effectively eliminate the relevant interference on site, and improve the quality and efficiency of partial discharge charged detection.But partial discharge detection is only a means of detection.Based on the theory and standard management method, the abnormal state of the equipment is systematically analyzed, and the important significance of partial discharge detection in the early warning and forecast of power system is realized from the serious influence caused by the abnormal equipment.

References

[1] Hou Rui, Yang Xuebao, Zhao Jianli. On line detection of partial discharge on insulated surface of cables in switchgear[J]. Jiangxi Electric Power, 2020, 44(12): 20-22.
[2] Yan Shuai, Li Pengyu, Wang Gaojie, et al. Discharge location method of open substation based on UHF wireless intelligent sensor array[J]. Electric Power, 2021, 54(02): 52-57.
[3] Zhou Bo, Gao Jun, Yang Hongquan, et al. Case Analysis on Suspension Discharge Defect of 220 kV GIS Disconnecting Switch[J]. High voltage apparatus, 2020, 56(12):313-317.
[4] Wei Wenting, Li Yue, Zhou Yu, Chen Yunfei, Wei Zhanpeng. The Partial Discharge Detection and Failure Analysis of 35 kV XLPE Cable[J]. Shandong Electric Power, 2020, 47(10): 39-44.
[5] Liu Hongliang, Ding Bin, Gao Shuguo, et al. Multi-terminal PD Detection Technology of Transformer Based on Wide Bandwidth Pulse Current Method[J]. High Voltage Apparatus, 2018, 54(11):241-246+252.
[6] Jiang Jiebo, Chen Ke, Shi Yonggui, et al. Partial discharge recognition of typical insulation defect based on Kohonen network[J]. Electric Power Engineering Technology, 2020, 39(05):43-48.
[7] Yu Qichen, Luo Lingen, Wu Fan, et al. UHF Partial Discharge Localization Methodology Based on Generalized Regression Neural Network[J]. Electric Power, 2021, 54(02): 11-17.
[8] Liu Qing, Chang Dingge, Deng Junbo. Optimal Selection on Spatial Spectrum Estimation Algorithms for UHF Direction Finding of Partial Discharge in Substation[J]. Transactions of China Electrotechnical Society, 2020, 35(16): 3551-3560.
[9] Li Ning, Hao Shuxing, Lei Yaoxu, et al. Research on partial discharge detecting repetitive-path problem in robot inspection[J]. Chinese High Technology Letters, 2020, 30(06): 615-625.
[10] Li Jie, Wang Lei, Wang Yingfang, et al. Research on partial discharge monitoring and localization method in an automatic directional way in substations[J]. Electrotechnical Application, 2020, 39(05): 23-28.
[11] Yu Qichen, Luo Lingen, Jia Yanbo, et al. Partial Discharge Positioning Technology in Substation Based on Received Signal Strength Indicator Statistical Analysis of UHF Signal[J]. High Voltage Engineering, 2020, 46(12): 4163-4171.

Optimal Dispatching of Integrated Energy System Based on Multi-Energy Complementation

Jiawei Xing*, Shumin Sun, Yan Cheng, Yiyuan Liu, Peng Yu, Yuejiao Wang, Shibai Wang, Xingyou Zhang, Guangqi Zhou, Nan Wang, Yifei Guan,

State Grid Shandong Electric Power Research Institute, Shandong Jinan 250002, China

*573602466@qq.com

Abstract—To raise the energy cascade use ratio and economic benefit, the multi-energy complementation-based integrated energy system (IES) has been considerably developed in recent years, so studying its internal optimal dispatching strategies will be of great importance. In this study, an optimal dispatching strategy applicable to the operation of typical IES was put forward. First, the classical IES architecture and the operating characteristics of each piece of equipment in it were explored. Second, a mixed-integer programming model with uncertainties was constructed for three energy systems—cooling, heating and power—and each equipment in these systems. In the end, the demonstration park of a major project in China was analyzed using the proposed strategy. The effectiveness of this strategy was verified by the simulation results.

1. Introduction

Given the limitation of all kinds of energy sources themselves and the diversity of human energy demands, it is hard for any energy source to undertake the heavy burden of energy transformation independently, so it is necessary to take full advantage of different technical, economic and environmental characteristics of various energy sources to realize their complementation within different temporal and spatial scales and continuously optimize the energy structure. Therefore, an integrated energy system (IES) came into being in response to the proper time and conditions [1]. The optimal dispatching of IES has been investigated a lot, mainly including specific optimal dispatching strategies and uncertain optimal dispatching strategies. In literature [2], an efficient-variable heat and power cogeneration model was proposed, which improved the operating performance of heat and power cogeneration and elevated the multi-energy coupling use ratio. In literature [3], the original algorithm was improved, and an integrated energy multi-agent interest balancing and optimal dispatching method based on the improved nondominated sorting genetic algorithm-II (NSGA-II) was proposed, which realized the multi-agent interest balance in the park. In literature [4], a distributed optimization strategy based on a master-slave game was put forward, and the uniqueness of Stackelberg's balance of this strategy was proved. This method effectively increased the revenues at the power supply side and power use side. In literature [5], a combined bidding and clearing strategy based on three agents—the superior power grid, operator and user group—in an IES was proposed, and the simulation results revealed that this strategy could effectively guide the three agents to keep well-functioning under competitive market environment. In literature [6], a multi-objective optimal dispatching strategy was proposed to minimize the economic and environmental dispatching goal of the system, thus coordinating and optimizing power-heating energy sources. However, uncertain factors were not introduced into the strategies of the abovementioned literature. As new energy sources continuously infiltrate systems, uncertain factors should be introduced into models to implement feasible and practical optimal dispatching strategies. In literature [7], a long-term programming method considering wind, power and load uncertainties was proposed to solve the capacity configuration problem in an IES. In literature [8], stochastic optimization models under the independent operation mode and the cooperative operation mode were constructed based on the uncertainties of clean energy considering three major agents: external backup system, user and park IES. The effectiveness of the models was verified by the simulation result. In literature [9], an optimal equipment configuration method for an IES considering the uncertainties of comprehensive demand response was put forward, and it was proven that this model showed stronger risk resistance. The literature [10] introduced the carbon trading mechanism, and a two-stage robustness optimization model was established based on the uncertainties of power load, heating load, and power-heating transferable load. The results revealed that this model could enhance the low-carbon quality and economic efficiency of the system. In literature [11], a mixed time-scale economic dispatch model considering characteristic energy difference was constructed, and the electricity and gas with a certain confidence level were set for standby application to cope with the uncertainties of new energy output. The results showed that this model was of great importance for dispatching IESs.

Based on the above background, certain and uncertain factors in the IES were simultaneously considered in this study to put forward a multi-energy complementation-based optimal dispatching strategy. First, the framework of typical IES was established, and each equipment therein and its characteristics were expounded. Second, the optimization model of IES was established, including objective functions and the constraint conditions for each equipment and the system. In the end, an integrated energy demonstration park in a major domestic project was analyzed, and the effectiveness of this strategy was verified.

2. Integrated Energy System (IES)

A typical IES consists of combined cooling, heating and power (CCHP) set, energy storage equipment, electrical refrigeration equipment and external power distribution network to which gas-fired boilers are connected. The concrete structure is as shown in Figure 1:

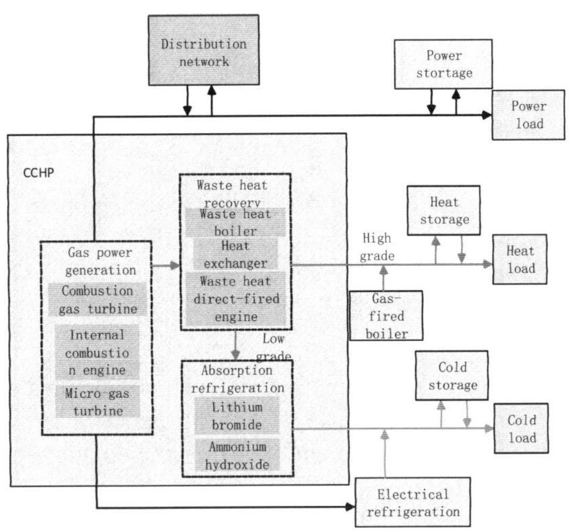

Figure 1: Typical IES structure

Where the black, red and blue solid lines with arrows represent the electric power transfer directions, the heat energy transfer directions, and the cold energy transfer directions, respectively. The yellow, green and purple boxes represent the equipment in the IES, load and external power distribution network, respectively.

2.1 CCHP set

Combined cooling, heating and power (CCHP) set, the power supply in an IES, is composed of gas power generating equipment, waste heat recovery plant and absorption refrigeration equipment. The gas power generating equipment (including the combustion gas turbine generator, the internal combustion turbine generator, the micro-gas turbine generator, etc.) can simultaneously generate electric energy and heat energy. Namely, it can be regarded as both a power source and a heat source. The electric energy emitted by the gas power plant can be directly supplied to electricity users, and the waste heat exhausted after the power generation can be treated by the waste heat recovery plants. The high-grade heat energy is directly supplied to heating loads after being treated by waste heat recovery plants (including the waste heat boiler, the heat exchanger and the waste heat direct-fired machine). In contrast, low-grade heat energy is supplied to cooling loads via absorption refrigeration equipment (e.g., the lithium bromide machine set) after being treated by the waste heat recovery plants.

2.2 Energy storage equipment

The energy storage equipment consists of electric energy storage, heat energy storage and cold storage plants. The electric energy storage plant, which is operated in the form of charging and discharging, can mitigate the uncertainties brought by the new energy output and enhance the stability of the power grid. Similarly, the heat storage and cold storage plants also operate in the form of storing and releasing cold and heat energy, i.e., storing energy in case of excess energy in the system and releasing energy in the event of insufficient energy. The energy storage equipment can be regarded as a load and power source in the system when storing and releasing energy, respectively.

2.3 Electrical refrigeration equipment

Just as absorption refrigeration equipment, electrical refrigeration equipment is a cold source in the IES. The difference lies in that the energy emitted by the electrical refrigeration equipment is converted from electric energy, while that of absorption refrigeration equipment is converted from the low-grade heat energy of the waste heat recovery plant.

2.4 Gas-fired boiler

The gas-fired boiler, the heat source in the IES, can directly supply heat energy to users. Compared with other boilers, it has better economic efficiency, so it has been widely applied to IESs.

3. Optimization Model

3.1 Objective function

To reduce the total cost of planning, investigation, design and construction in the electric power system and the expenses consumed in the service life cycle [12], and facilitate the dispatch department to coordinate the operation subjects in the IES, the minimum energy use cost of the whole system was taken as the objective function, as seen in Equation (1):

$$\min C = C_d + C_r + C_l \tag{1}$$

Where *Cd*, *Cr* and *Cl* denote the power use cost, heat use cost and cold use cost of the whole system, respectively, which are calculated through the following equations:

$$C_d = \sum_{t=1}^{24} [C_{CCHP,d}(t) \cdot P_{CCHP,d}(t) + C_p(t) \cdot P_p(t) + C_{cd}(t) \cdot P_{cd}(t)] \tag{2}$$

$$C_r = \sum_{t=1}^{24} [C_{CCHP,r}(t) \cdot P_{CCHP,r}(t) + C_{gl}(t) \cdot P_{gl}(t) + C_{cr}(t) \cdot P_{cr}(t)] \tag{3}$$

$$C_l = \sum_{t=1}^{24}[C_{CCHP,l}(t)\square P_{CCHP,l}(t) + C_{dzl}(t)\square P_{dzl}(t) + C_{xl}(t)\square P_{xl}(t)] \tag{4}$$

Where *CCCHP,d (t)*, *CCCHP,r (t)* and *CCCHP,l (t)* represent the cost spent by the CCHP to emit electric energy, heat energy and cold energy at time *t*, respectively. *PCCHP,d (t)*, *PCCHP,r (t)* and *PCCHP,l (t)* denote the electric power, heat power and cold power emitted by the CCHP at time *t*, respectively. *Cp (t)* stands for the electricity price of the power distribution network at time *t*. *Pp (t)* is the electric power exchanged by the power distribution network and IES at time *t*. When it is positive, the power distribution network supplies power to the system; when it is negative, the power distribution network purchases electricity from the system. *Ccd (t)* is the cost of electric energy storage at time *t*. *Pcd (t)* stands for the charging/discharging power of electric energy storage at time *t*; positive power denotes charging and negative power represents discharging. *Cgl (t)* and *Pgl (t)* represent the heat release cost and heat release rate of the gas-fired boiler at time *t*, respectively. *Ccr (t)* denotes the heat storage cost at time *t*. *Pcr (t)* is the heat stored and released by the heat storage device at time *t*; positive power represents heat energy storage, and negative power indicates heat energy release. *Cdzl (t)* is the operating cost of electrical refrigeration equipment at time *t*, and *Pdzl (t)* stands for the cold energy generated by the electrical refrigeration equipment at time *t*. C_{xl} *(t)* denotes the operating cost of the cold storage device at time *t*. P_{xl} *(t)* stands for the cold energy stored and released by the cold storage device; positive power indicates cold energy storage, and negative power represents cold energy release.

3.2 Constraint conditions

(1)CCHP operating constraints

The operating constraints of the CCHP set include power constraint, ramp constraint and thermoelectric coupling constraint, as seen in Equation (5):

$$\begin{cases} 0.4P^{N}_{CCHP}s_{CCHP}(t) \le P^{i}_{CCHP}(t) \le P^{N}_{CCHP}s_{CCHP}(t) \\ -\Delta P^{X}_{CCHP} \le P^{i}_{CCHP}(t+1) - P^{i}_{CCHP}(t) \le \Delta P^{S}_{CCHP} \\ P^{i}_{CCHP}(t) = G_{dr}(P^{i}_{CCHP,r}(t) + P^{i}_{CCHP,l}(t)) \end{cases} \tag{5}$$

Where P^{N}_{CCHP} denotes the rated operating power of the CCHP set, $S_{CCHP}(t)$ represents the operating state ($s_{CCHP}(t) \in \{0,1\}$) of the set. $P^{i}_{CCHP}(t)$ stands for the power of set *i*. ΔP^{X}_{CCHP} and ΔP^{S}_{CCHP} represent the lower limit and upper limit of ramp rate of the set, respectively. G_{dr} is the conversion coefficient of electrothermal power. $P^{i}_{CCHP,\ r}(t)$ stands for the heat release power of set *i*. $P^{i}_{CCHP,\ l}(t)$ is the refrigeration power of set *i* at time *t*.

(2)Energy storage constraint

The energy storage constraint includes energy storage and release constraint, ramp constraint, SOC constraint and energy balance constraint, as seen in Equation (6)-(8).

1) Energy storage

$$\begin{cases} P^{out}_{d} \le P_d(t) \le P^{in}_{d} \\ \Delta P^{out}_{d} \le P_d(t+1) - P_d(t) \le \Delta P^{in}_{d} \\ 0.05 \le S_{soc,d}(t) \le 0.95 \\ \sum_{t=1}^{24} P_d(t) = 0 \end{cases} \tag{6}$$

2) Heat storage

$$\begin{cases} P_r^{out} \le P_r(t) \le P_r^{in} \\ \Delta P_r^{out} \le P_r(t+1) - P_r(t) \le \Delta P_r^{in} \\ 0.05 \le S_{soc,r}(t) \le 0.95 \\ \sum_{t=1}^{24} P_r(t) = 0 \end{cases} \tag{7}$$

3) Cold storage

$$\begin{cases} P_l^{out} \le P_l(t) \le P_l^{in} \\ \Delta P_l^{out} \le P_l(t+1) - P_l(t) \le \Delta P_l^{in} \\ 0.05 \le S_{soc,l}(t) \le 0.95 \\ \sum_{t=1}^{24} P_l(t) = 0 \end{cases} \tag{8}$$

Taking electric energy storage as an example, P_d^{out}, P_d^{in}, ΔP_d^{out}, ΔP_d^{in}, and $S_{soc,d}(t)$ represent the maximum charging power, maximum charging power, maximum discharging ramp power, maximum charging ramp power and energy storage SOC at time t, respectively. The lower limit and upper limit of energy storage SOC are 5% and 95%, respectively. $P_d(t)$ is the energy storage power at time t. The sum is taken as 0 within 24 h to realize the charging/discharging balance.

(3) Electrical refrigeration equipment constraint

The constraint of electrical refrigeration equipment includes power constraint and energy balance constraint, as seen in Equation (9):

$$\begin{cases} 0 \le P_{dzl}(t) \le P_{dzl}^N \\ P_{dzl}(t) = K_{COP} P_{d1}(t) \end{cases} \tag{9}$$

Where P^N_{dzl} is the maximum power output of the electrical refrigerating machine set. K_{COP} is the electrical refrigeration performance coefficient, and a higher coefficient indicates better refrigeration performance. $P_{dl}(t)$ represents one part of electric energy emitted by the CCHP set at time t.

(4)Gas-fired boiler constraint

The gas-fired boiler constraint includes power constraint and ramp constraint, as seen in Equation (10):

$$\begin{cases} 0 \le P_{gl}^i(t) \le P_{gl}^N \\ \Delta P_{gl}^X \le P_{gl}^i(t+1) - P_{gl}^i(t) \le \Delta P_{gl}^S \end{cases} \tag{10}$$

Where P^N_{gl}, ΔP^X_{gl}, ΔP^S_{gl} represent the maximum power output of the gas-fired boiler, the lower limit and upper limit of the ramp power of CCHP set i, respectively.

(5)System constraint

The system constraint includes the equation of cold and heat power equilibrium expressed by Equation (11):

$$\begin{cases} P_{CCHP,d}(t) + P_p(t) + P_{cd}(t) = P_{s,d}(t) \\ P_{CCHP,r}(t) + P_{gl}(t) + P_{cr}(t) = P_{s,r}(t) \\ P_{CCHP,l}(t) + P_{dzl}(t) + P_{xl}(t) = P_{s,l}(t) \end{cases} \tag{11}$$

Where $P_{s,d}(t)$, $P_{s,r}(t)$ and $P_{s,l}(t)$ represent the power load, heating load and cooling load at time t, respectively.

4. Analysis of Calculated Example

4.1 Basic data

The cold and heating loads in the park of a major project were analyzed, which were the typical loads in this park in summer. The load curves, which referred to the literature [13], are as shown in Figure 2:

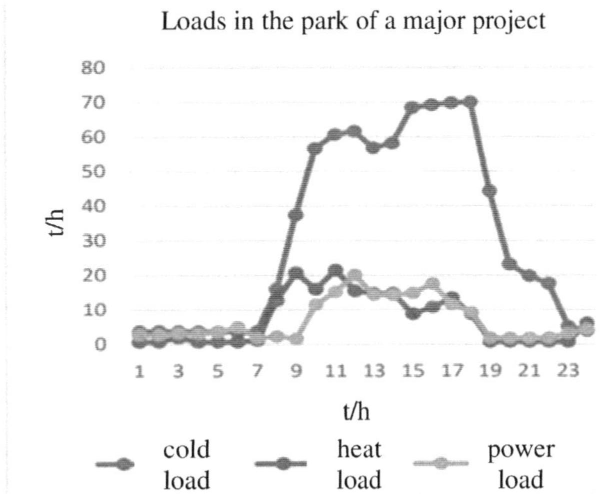

Fig.2 load curve of the park

In this park, the unit prices (time-of-use electricity price) of cold energy and heat energy were taken as RMB 0.587/kWh and RMB 0.563/kWh, respectively. The concrete time division and the corresponding prices are presented in Table 1:

Table 1: Time of Use Pricing

	1	2
Time frame	22:00-6:00	6:00-22:00
Price	0.517	1.071

There were two CCHP sets in this park. The gas-heat conversion coefficient of the waste heat boiler was taken as 4, the conversion efficiency of the lithium bromide machine set as 0.8, and the purchasing price of natural gas for CCHP sets as RMB 4.03/m^3. The operating characteristics of CCHP sets are shown in the following figure, and their energy use costs are displayed in Figure 3 and Figure 4, which directly came from the literature [14] and were acquired through the experimental data of CCHP units under standard operating conditions.

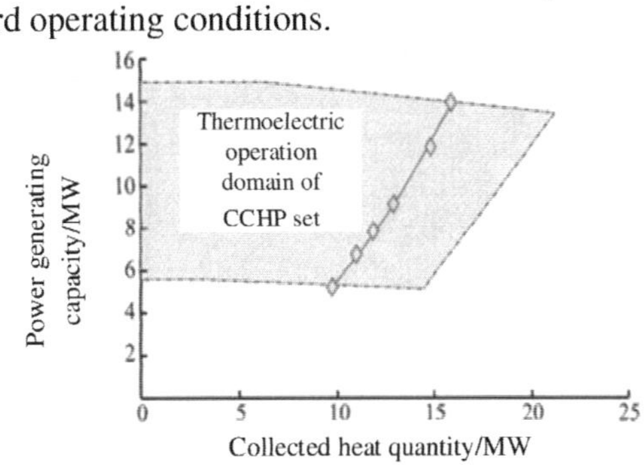

Figure 3: Thermoelectric Operating Curve of CCHP Set

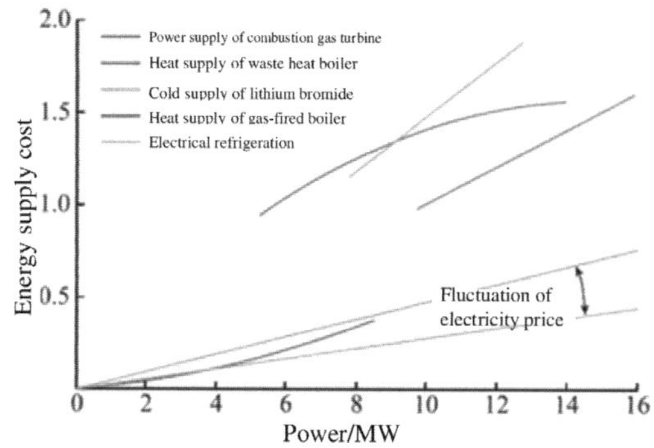

Figure 4: Energy Cost of CCHP

The parameters of the other machine sets in the park were described as follows: The coefficient of performance (COP) and the maximum power of electrical refrigeration equipment were set as 4.2 and 50 MW, respectively. The purchasing price of natural gas for a gas-fired boiler was RMB 3.03/m^3. The correlation coefficient between the cost and power of the gas-fired boiler was taken as 6.4. As for electric energy storage, the capacity, maximum charging power and maximum discharging power were set as 35 MWh, 5 MW and 5 MW, respectively. In terms of heat energy storage, the capacity, maximum heat storage power and maximum heat release power were taken as 20 MWh, 5 MW and 5 MW, respectively. In the cold storage part, the capacity, maximum cold storage power and maximum cold energy release power were selected as 40 MWh, 10 MW and 10 MW, respectively. In addition, there was also a photovoltaic power station in this park, with a maximum power of 8 MW. The typical power output curve of this photovoltaic power station is displayed in Figure 5:

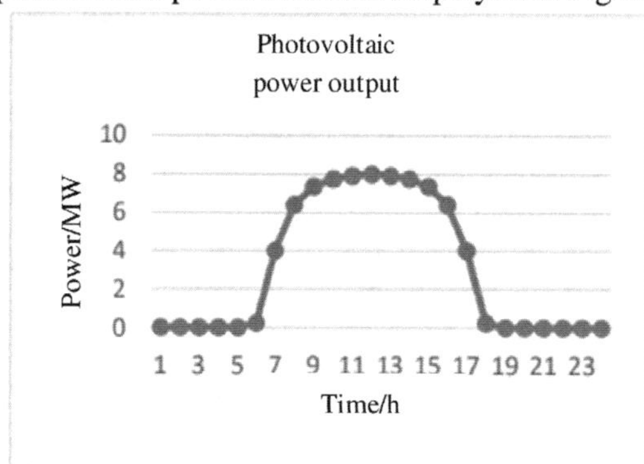

Figure 5: Power Output Curve of Photovoltaic Power Station

4.2 Simulation results

The simulation was performed by invoking yalmip optimization toolbox and using MATLABR2019a on a computer with the processor of intel (R) Core ™ i7-10170U CPU@ 1.10 GHz 1.61 GHz. The program was implemented through mixed-integer linear programming (MILP). The dispatch results are presented in Figure 6, Figure 7 and Figure 8.

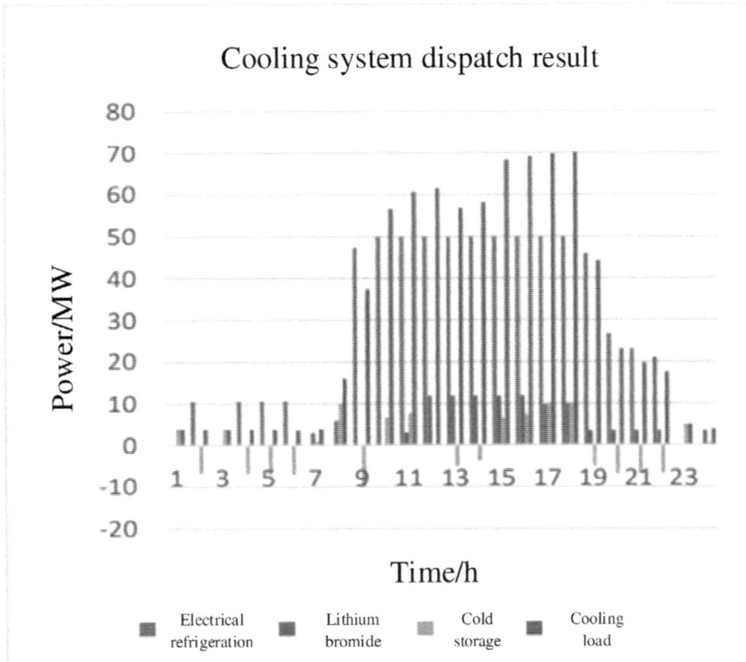

Figure 6: Cold Energy Dispatch Result

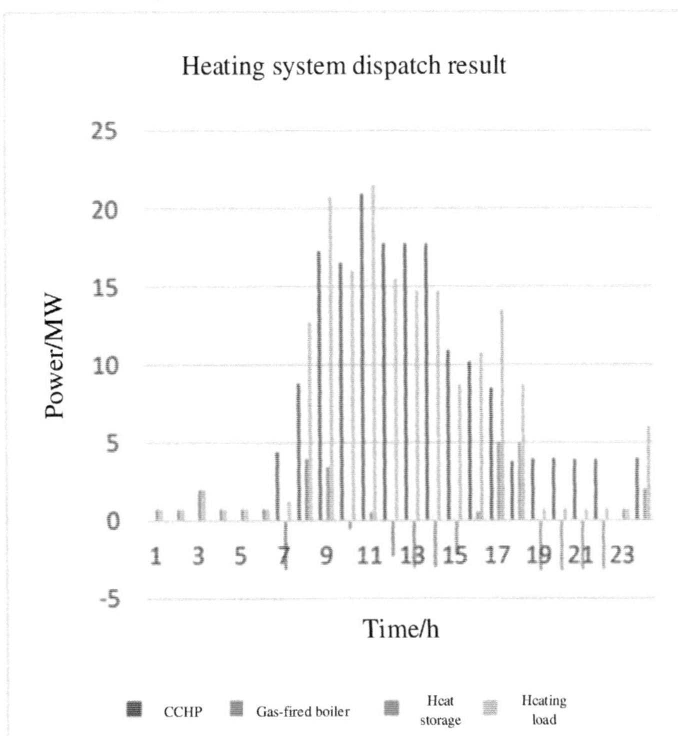

Figure 7: Heat Energy Dispatch Result

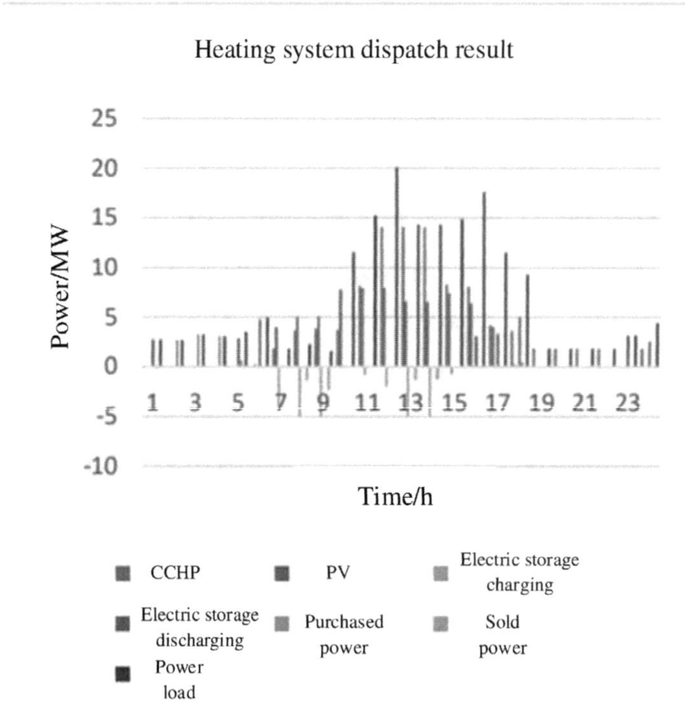

Figure 8: Electric Energy Dispatch Result

The above dispatch results could be analyzed from two angles: system and equipment.

4.3 System-side analysis

(1)Cooling system

The electrical refrigeration equipment, which was the main power source of the cooling system, balanced most of the cooling loads nearly in all time frames. The power it provided was 656.5549 MW, accounting for 78.48% of total power. The power provided by the lithium bromide machine set in the CCHP set was 107.8651 MW, accounting for 12.89% of total power. The power stored and released by the cold storage equipment was 72.1872 MW and 72.1871 MW, respectively, i.e., the power it absorbed as a load was 72.1872 MW, and the power it provided as a cold source was 72.1871 MW, accounting for 8.63% of the total load. The output power of electrical refrigeration equipment + output power of lithium bromide + output power of cold storage equipment = power absorbed by cold storage equipment + cooling load, which released the cooling power balance of the IES.

(2)Heating system

Unlike the function exerted in the cooling system, the CCHP set was the primary heat source of this system, and the power it provided was 174.4999 MW, accounting for 86.60% of total power. For the heat storage equipment, the power absorbed by it as a load was 27.0025 MW and the power released by it as a power source was 27.0026 MW, accounting for 13.40% of total power. The total power provided by the gas-fired boiler was 0 MW, and the heating load was 174.5 MW. The output power of CCHP + output power of heat storage equipment + output power of gas-fired boiler= power absorbed by heat storage equipment+ heating load, which achieved the heat power balance in the IES.

(3)Electric system

The electric power provided by the CCHP set was 96.7672 MW, accounting for 47.54% of total power, and that provided by the photovoltaic power station was 69.0988 MW (33.95%). Both discharging power and the charging power of electric energy storage were 27.6049 MW (13.56%). The power purchased by the IES from the power distribution network was 10.08 MW (4.95%), and that provided by it to the power distribution network was 6.2612 MW. The electric power output by

CCHP + electric power output by photovoltaic power station + output power of electric energy storage + power provided to power distribution network = power absorbed by electric energy storage+ power load + power provided by the power distribution network, which contributed to the electric power balance in the IES. It was noteworthy that the electric power provided by the power distribution network to this park accounted for a small proportion of the total load; namely, the power loads in the park were supplied with power mainly by the machine sets in the park. In case of a serious fault of the power distribution network, the park could be directly disconnected from the power distribution network to realize the island operation.

4.4 Equipment-side analysis

(1)CCHP set

The CCHP sets were the sole equipment that could simultaneously provide cold, heat and electric energy in the park. It could be known from the dispatch results that the two CCHP sets supported nearly all the heating loads and nearly half of the power loads in the park, as well as a small number of cooling loads. The reason was that the thermoelectric coupling operation mode of the CCHP units substantially improved the energy supply efficiency, being able to supply electric energy and heat energy simultaneously by consuming the same amount of energy, which reduced the energy cost. The operating cost of electrical refrigerating units (maximum unit price: RMB 1.071) was much lower than the refrigeration cost (unit price: RMB 4.03) of CCHP sets, and the cold energy provided by the CCHP sets was restricted by the heat energy. To maximize the revenue, the electrical refrigeration equipment was required to provide as much cold energy as possible. In addition, the CCHP sets operated mainly from 7:00 to 22:00, during which the electricity price was higher, i.e., the revenue of the CCHP sets was higher.

(2)Energy storage equipment

It could be discovered by comparing the energy storage curves in the above three figures that the charging power was small in all three forms of energy storage, with the proportion occupied in the total load not exceeding 14%. The operating cost of energy storage in the park was high, and the storage capacity was not large, so the energy stored in the park mainly served as standby energy, which was put into use when the other equipment failed normal operation or the capacity reached the limiting value.

(3)Electrical refrigeration equipment

This equipment was one of the two cold sources in the park. Since its operating cost was lower than the lithium bromide machine set in CCHP, it undertook the cooling loads in the park within most of the time frames. It could be seen from the dispatch results that from 10:00 to 18:00, the electrical refrigeration equipment realized full-power operation due to the large cooling loads.

(4)Gas-fired boiler

The equipment was one of the heat supply equipment in the park. Given that the gas purchasing price (RMB 3.03) of this equipment was lower than that (RMB 4.03) of CCHP sets, it should have served as the main heat source in the park. However, it could be seen from the operating results that the output power of the gas-fired boiler was always 0 MW; that is to say, it did not participate in the practical heating system dispatching. This can be explained by the following two reasons: first, the capacity of CCHP sets was enough to support the heating loads in the park; second, although the operating cost of CCHP sets was high, it could simultaneously provide electric energy and heat energy and generate revenues in two systems, so its net revenue was much greater than that of the gas-fired boiler. The gas-fired boiler was not needed to participate in the dispatching, given that the CCHP units were capable of sufficiently supporting the heating loads in the park.

5. Conclusions

A multi-energy complementation-based IES optimal dispatching strategy was proposed in this study to realize the cooling, heating and power balance. The following conclusions were drawn by combining the concrete simulation results:

1) CCHP sets are the primary power supply and heat supply equipment in the system, with their energy supply efficiency and economic efficiency higher than other equipment.

2) The electrical refrigeration equipment, as the leading cold supply equipment in the system, can balance the cooling loads in most time frames.

3) Gas-fired boiler and energy storage equipment service serve as standby equipment in consideration of their lower net profit than the other equipment, and they will be put into use in case of energy shortage.

4) The multi-energy complementation operation mode substantially improves the energy utilization ratio and enhances the economic benefits.

Acknowledgments
This work was financially supported by the State Grid Corporation Science and Technology Project (5206002000D9).

References

[1] Li P, Wu D F, Li Y W, Yin Y X, Fang Q, Chen B. Multi-objective combined optimal configuration of a multi-microgrid integrated energy system based on negotiation game [J/OL]. Power System Technology: 1-10 [2020-07-22].

[2] Cui Y, Yan S, Zhong W Z, Wang Z, Zhang P, Zhao Y T. Optimal thermoelectric dispatching of a regionally integrated energy system containing power-to-gas device [J/OL]. Power System Technology: 1-12[2020-07-23].

[3] Wang Y C, Zeng C Y, Xun J J, Tan X Y, Yu M Z. Multi-agent interest balancing and optimal dispatching of an integrated energy system based on improved NSGA-II [J]. Electric Power Automation Equipment, 2020, 40

[4] Wang H Y, Li K, Zheng C H, Ma X. Distributed collaborative optimal operating strategy of a community-integrated energy system based on master-slave game [J/OL]. Proceedings of the CSEE: 1-11 [2020-23].

[5] Cui Y, Yan S, Wang Z, Wang M C, Zhao Y T. Day-ahead iterative clearing strategy of integrated energy system considering three-agent bidding [J/OL]. Power System Technology: 1-11[2020-07-23].

[6] Zheng C M, Huang B N, Wang Z X et al. Multi-objective optimization dispatch for integrated electro-heating systems including network transmission losses [J]. Power System Technology, 2020, 44 (1).

[7] Zhao J, Yong J, Xun J J et al. Stochastic programming of park integrated energy system based on long time scale [J]. Electric Power Automation Equipment, 2020 (3): 10.

[8] Yang X B, Tan Z F, Lin H Y, Degeryzhv, Zhou F A.
Stochastic optimization of park integrated energy system and revenue distribution model considering cooperation alliance [J/OL]. Automation of Electric Power Systems: 1-18 [2020-07-23].

[9] Liu W X, Li Z Z, Yang Y et al. Collaborative optimal configuration for integrated energy system considering uncertainties of demand response [J]. Automation of Electric Power Systems, 2020, 44 (10): 41-49.

[10] Zhang X H, Liu X Y, Zhong J Q. Integrated energy system planning considering a reward and punishment ladder-type carbon trading and electric-thermal transfer load uncertainty [J/OL]. Proceedings of the CSEE: 1-13 [2020-07-23].

[11] Li R, Sun F, Liu H L, Ding X, Han Y, Yan J R. Mixed time-scale economic dispatch of user-level integrated energy system considering characteristic energy difference [J/OL]. Power System Technology: 1-10 [2020-07-23].

[12] Gu H F, Yu J, Li Y et al. Double-layer combined optimal economic dispatch of new towns containing multifunctional parks under environmental constraint [J]. Proceedings of the CSEE, 2020 (8): 2441-2452.

[13] Zhang J H, Huang W. Microgrid operational control and protection technology [M]. China Electric Power Press, 2010.
[14] Hao R, Ai Q, Jiang Z Q. Bi-level game strategy for multi-agent with incomplete information in regionally integrated energy system [J]. Automation of Electric Power Systems, 2018, 42 (004): 194-201.

Dynamic regulation method of distributed generation in distribution network under 5G network

Duanyun Chen[1], Yuanjiu Li[2], Jie Bai[3], Hongpo Zhang[2], Shimulin Xie[4*], Zeyi Tang[4] and Xujie Liu[4]

[1]State Grid FuJian Electric Power CO.LTD., Fuzhou 350001, China

[2]State Grid Fujian Quanzhou Electric Power Supply Company, Quanzhou 360000, China

[3]State Grid Information & Telecommunication Group Co., Ltd., Beijing 100052, China

[4]State Grid Info-telecom Great Power Science and Technology Co.Ltd., Fuzhou 350001, China

[*]Corresponding author's e-mail: baijie2@sgitg.sgcc.com.cn

Abstract. Aiming at the problem of uneven power flow distribution caused by distributed power supply access, which leads to high voltage deviation and active network loss, a dynamic control method of distributed power supply in 5G network is proposed. The distributed topology structure of the distribution network under 5G network is generated, and the static topology information is extracted to establish the power partition. The distribution network in each area is connected to its own bus through different feeders. The distribution network reconstruction period is divided in the partition, and the network reconstruction period and the corresponding switch action scheme are determined. The state of distributed power supply was estimated according to reconstruction period and active/reactive power balance index, and the stability of dynamic control was determined. Based on the stability control, the dynamic control model of distributed power supply is established, and the network voltage can reach the preset target by adjusting the power of power switch. The simulation results show that the average voltage deviation of the proposed control method is 1.976, which is 4.431 and 3.783 lower than those based on data drive and flexible load regulation, respectively. The active power network loss of the control method in this paper is 0.342 MW, which is 0.385 MW and 0.346 MW lower than that of the control method based on data drive and flexible load regulation, respectively. Therefore, this method can improve the absorption capacity of distributed power supply and promote the power flow reciprocity of distribution network.

1. Introduction

The development of energy technology drives the progress of human society, promotes the innovation of industrial technology, and is the cornerstone of global economic development. In the past few decades, the rapid progress of the energy industry has provided a strong guarantee for China's industrial upgrading and helped China's economy achieve leapfrog development. The traditional distribution network lags in construction, has limited control means and unreasonable structure, which restrict the flexibility of distribution network operation control [1]. With the massive access of distributed power supply, the urban power distribution network has been transformed from single-terminal power supply to multi-terminal power supply, making the urban power grid topology undergoing fundamental changes, further affecting the size and direction of fault current in the network, and bringing severe challenges to the protection and control methods of the existing urban power grid [2]. In addition, the increase of permeability of new energy and the increase of the proportion of nonlinear and impact loads put forward higher requirements for the related technologies to ensure the reliability of power supply and improve power quality [3-4]. Large-scale grid integration of new energy generation will profoundly change the dispatching control pattern of the main network[5]. Therefore, this paper proposes a dynamic control method of distributed power supply in the distribution network under 5G network to coordinate distribution network layout and capacity allocation with regional planning. Through coordination with active management and other means, the distributed power supply of the distribution network under 5G network is optimized and planned to improve users' economic benefits and social environmental benefits, so as to promote the construction of smart grid.

2. Dynamic regulation method of distributed generation in distribution network under 5G network

2.1. Generate the distributed topology of the distribution network under 5G network

Because there is no centralized master station to collect information, the network topology information in the distributed system is stored in the monitoring nodes of the urban distribution network. On the basis of static information, more abundant dynamic network topology information is identified through effective coordination among distributed generators in distribution network, and then these information are used to complete advanced automation functions such as fault recovery and island detection. The set composed of sections is called section set. In order to improve the feeder load balance, this paper solves the specific zoning scheme. The objective function of distribution network zoning method is:

$$\alpha_{\max} = 1 - \frac{\sum_{x=1}^{s_1} \sqrt{\sum_{y=1}^{s_2} \left(p_{xy} - \overline{p}\right)^2 / s_2}}{s_1} \quad (1)$$

In formula (1), $\alpha_{\max}$ represents the maximum feeder load balancing degree; s_1 and s_2 represent the number of feeders and the number of calculation points of feeder load balance respectively; x and y represent two different moments respectively; p_{xy} is the feeder load rate at a certain time; $\overline{p}$ is the average feeder load rate. When a distributed generation partition receives the switch number sent from two adjacent areas or the power side receives the switch number information sent from the downstream direction, and one of the switches controlled by itself is in the off state, it is determined to be the contact node position.

2.2. Division of distribution network reconfiguration period

After being put into operation for the first time, the distributed generation partition can independently obtain and update the dynamic topology information according to the changes of static topology information and system operation state. In order to complete the dynamic regulation function of power supply, the intelligent distribution terminals shall be able to complete information exchange through peer-to-peer communication network. Because the number of switch actions is related to the amount of load cut-off and the number of node cut-off, one of them can be selected as one of the power supply restoration objectives. When the reactive power adjustable capacity is insufficient, the active power cutting strategy is carried out. The update method of the set value can be expressed as:

$$\lambda(a+1) = \frac{\lambda(a) m_1 \delta}{m_2} \quad (2)$$

In formula (2), $\lambda(a)$ and $\lambda(a+1)$ correspond to the maximum values of range and standard deviation before and after updating respectively; m_1 and m_2 represent the number of iteration and expected reconstruction periods; δ is the random number conforming to the normal distribution with mean value of 1 and standard deviation of 0.1. Based on the division results of reconstruction period, network reconstruction can be divided into the following two steps.

2.3. Estimating distributed power status

The distribution network system is generally composed of distributed generation, load and energy storage devices. Each distributed generation transmits energy to the load through the power line. In order to adapt to the characteristics of decentralized structure and plug and play, a small signal model of distribution network is established based on the droop control principle. The expression of the active power injected into the microgrid AC bus by each distributed power converter in the distribution network is as follows:

$$W_1 = \frac{uv\sin\theta}{r} \quad (3)$$

In formula (3), W_1 represents active power; u represents the output voltage of distribution network converter; v represents common bus voltage; θ indicates phase difference; r is the equivalent output impedance. When the active power is constant, the unit amount of reactive power is injected, and the change of voltage amplitude is only related to the reactive power sensitivity matrix. Similarly, the reactive power injected into the microgrid AC bus by the distributed generation converter can be expressed as:

$$W_2 = \frac{uv\cos\theta - v^2}{r} \quad (4)$$

In formula (4), W_2 is reactive power. When the reactive power is constant, the unit amount of active power is injected, and the change of voltage amplitude is only related to the active power sensitivity matrix. Since the line impedance is usually much smaller than the load impedance, it is considered that the phase difference is very small. If the communication delay is less than the system sampling interval, it will not affect the data transmission; Otherwise, it will lead to packet loss. The data packet loss model of communication link is established based on Beili model, as follows:

$$\beta'(n) = \eta\beta(n) + (1-\eta)\beta(n-1) \quad (5)$$

In formula (5), β' represents the latest data value; β represents the initial data value; η is a random variable, taking values between 0 and 1; n indicates time. Bring the communication link data into the Kalman filter estimation equation to obtain the state estimation value at the sampling time, which can be expressed as:

$$\vartheta(n) = k\vartheta'(n-1) + h\varepsilon(n-1) \quad (6)$$

In formula (6), $\vartheta(n)$ represents the state estimation value at the current time; $\vartheta'(n-1)$ is the next forecast estimate; ε represents Kalman filter gain matrix; k and h are the estimation coefficients, respectively. If there is only one boundary section in the island area, it is impossible to directly connect with the pure load branch outside the island, and there are no other nodes downstream of the island.

2.4. Design the dynamic regulation model of distributed generation

The connection of distributed generation to distribution network will change the single source radial structure of traditional distribution network and change the voltage of each node in the distribution network. In order to solve the related power flow and voltage problems caused by distributed generation access, a dynamic regulation model for maximizing the consumption of distributed generation is proposed in this paper, so as to enhance the consumption ability of distribution network to distributed generation. The dynamic regulation model of distributed generation is shown in Figure 1.

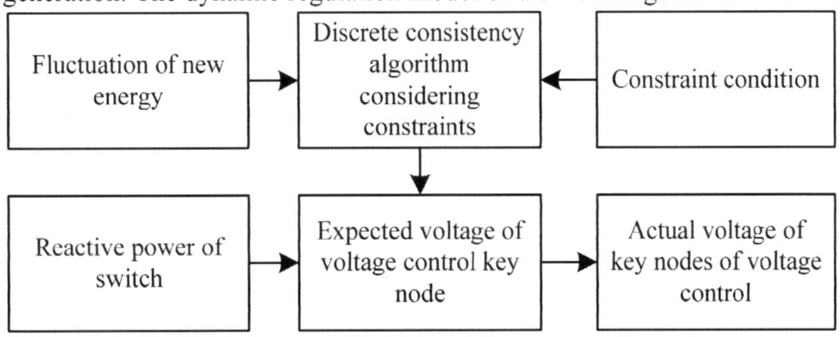

Figure 1 Dynamic regulation model

The inverter of distribution network works in constant power control, and the distributed generation is equivalent to a special load with negative power. The fluctuation of load rate on feeder is defined as:

$$\Delta\tau = \frac{q_1 - q_2}{c} \quad (7)$$

In formula (7), $\Delta\tau$ represents the load rate wave momentum; q_1 represents the power flowing through the feeder; q_2 represents the power flowing through the feeder when there is no load or new energy fluctuation; c is the feeder capacity. Generally, the new energy access node or feeder end node can be regarded as the key node of voltage control. The objective function of voltage adaptive control is:

$$y = \min \sum_{j=1}^{o} \left(F_j - E\right)^2 \quad (8)$$

In formula (8), y represents the objective function; o is the total number of voltage control key nodes; j is the key node; F is the actual voltage; E is the desired voltage. The expected voltage is calculated by the consistency algorithm to achieve the consistency of voltage changes of all key nodes. The input of dynamic regulation of distributed generation is the reactive power of flexible multi state switch, and the output is the actual voltage of key nodes of voltage control. Reactive power control can make the actual voltage of the key node of voltage control track the expected voltage well, so as to realize the distributed adaptive control of voltage. So far, the design of dynamic regulation method of distributed generation in distribution network under 5G network is completed.

3. Example analysis and verification

3.1. Simulation example

In order to verify the performance of the dynamic regulation method of distributed generation in distribution network under 5G network designed in this paper, simulation analysis is carried out based on IEEE 33 node distribution system to verify the consumption capacity of distributed generation. System parameters and power switch parameters are shown in Table 1.

Table 1 Example parameters

Parameter	Numerical value	Parameter	Numerical value
System rated voltage	15kV	Node load size	100 kVA
Line impedance of adjacent nodes	2Ω	Distributed power capacity	2000 kVA
Switching capacity	1500 kVA	Upper limit of charge and discharge power	1MW

In the IEEE 33 node distribution system, the distributed power supply connected to nodes 9 and 10 is wind power generation, and the maximum active power output is 0.8MVA. The distributed power supply connected to nodes 13, 14, 16, and 18 is photovoltaic, and the maximum active power output is 1.1MVA. It is assumed that except for the adjustable load, the load power of other nodes remains unchanged during the whole recovery process, and the constraint on the coverage of fluctuation range of each node is set to 100%. The power switch connects to the 10kV distribution network from the end of the feeder, and sets the power factor of the distributed power supply connecting to the power grid to 1. The feeder switch operates for a maximum of 4 times, and the minimum state transition interval is 2.

3.2. *Results and analysis*

Based on the above simulation examples, the consumption capacity of the distributed generation dynamic regulation method designed in this paper is tested. This method is taken as an experimental group to compare with the dynamic regulation methods of distributed generation in distribution network based on data-driven and flexible load regulation. In order to evaluate the maximum consumption capacity of distributed generation, voltage deviation and active network loss are selected as evaluation indexes for test in this example analysis. The voltage offset reflects the system voltage deviation caused by the access of distributed generation. Therefore, the smaller the voltage deviation index, the better. The minimum active network loss is usually the daily scale function of reactive power optimization power flow. Taking the minimum network loss as the objective function can reduce the system active power loss. Therefore, the smaller the active network loss index, the better. The comparison results of voltage deviation and active network loss of each distributed power generation dynamic regulation method are shown in Table 2 and table 3.

Table 2 Voltage deviation comparison

Number of tests	The dynamic regulation method of distributed generation in this paper	Dynamic regulation method of distributed generation based on data drive	Dynamic regulation method of distributed generation based on flexible load regulation
1	2.156	7.042	6.804
2	2.208	6.018	5.069
3	2.062	7.214	6.087
4	1.139	5.125	5.362
5	2.315	6.637	5.475

Table 3 Comparison of active power network loss (MW)

Number of tests	The dynamic regulation method of distributed generation in this paper	Dynamic regulation method of distributed generation based on data drive	Dynamic regulation method of distributed generation based on flexible load regulation
1	0.315	0.704	0.744
2	0.328	0.618	0.618
3	0.362	0.763	0.626
4	0.338	0.802	0.782
5	0.365	0.748	0.669

When the system voltage deviation is too large due to the excessive output of distributed power generation, the safe and stable operation margin of the system will be reduced. When the system is disturbed, it is easy to produce the problem of excessive local voltage deviation and cause damage to electrical equipment. According to the comparison results in Table 2, the average voltage deviation of the regulation method in this paper is 1.976, which is 4.431 and 3.783 lower than the regulation methods based on data drive and flexible load regulation respectively. Therefore, the regulation method in this paper has more stable operation ability. The distribution network has low active power network loss, which can not only increase the economy of system operation, but also improve the voltage quality of the system. According to the comparison results in Table 3, the active power network loss of the regulation method in this paper is 0.342 MW, which is 0.385 MW and 0.346 MW lower than the regulation methods based on data drive and flexible load regulation respectively. The regulation method in this paper has lower network loss. Based on the above results, the dynamic regulation method of distributed generation designed in this paper can suppress or reduce voltage deviation and active power network loss, improve the consumption capacity of distributed generation, and has strong engineering practicability.

4. Conclusion

The development trend of distribution network in the future is the modernization and automation of distribution system. It not only includes advanced power equipment, monitoring, regulation, communication technology and large-scale distributed energy access, but also represents the transformation of distribution mode and operation management mechanism. After the application of 5G network technology, a peer-to-peer communication network is established between intelligent distribution terminals to analyze and judge the fault situation and restore the power supply in non fault sections. Under the above background, this paper proposes a dynamic regulation method of distributed generation in distribution network under 5G network. Using the distributed control mode, the whole power supply process can be completed by the coordination between intelligent terminals. This method can suppress or reduce voltage deviation and active power network loss and improve the consumption capacity of distributed generation. In addition, the method of robust optimization can be considered, and the operation scheme obtained can be compared with the operation scheme obtained based on probability theory, so as to achieve a better balance in the applicability and economy of the operation scheme.

Acknowledgments

The study was supported by "Science and Technology Project of State Grid FuJian Electric Power Co.Ltd, China (Grant No. SGFJ0000DKJS2000260, Research on Key Technologies of 5G mobile edge computing platform for distribution network differential protection application scenario)".

References

[1] GAO Chong, HUANG Yuan-sheng, MA Hai-jie. (2020) Simulation of Spatial Load Density Distribution in Medium Voltage Distribution Network[J]. Computer Simulation, 37(3):56-60.
[2] Yan Xiangwu, Xu Yun, Li Ruojin, et al. (2019) Multi-Time Scale Reactive Power Optimization of Distribution Grid Based on Model Predictive Control and Including RDG Regulation[J]. Transactions of China Electrotechnical Society, 34(10):2022-2037.
[3] LI Siyu, ZHANG Bide, PENG Liwei, et al. (2019) Dynamic Island Partitioning Strategy for Active Distribution Network Considering Adjustment Ability of Flexible Load[J]. High Voltage Engineering, 45(6):1835-1842.
[4] Kouveliotis-Lysikatos, Iasonas, N, et al. (2019) A Double-Layered Fully Distributed Voltage Control Method for Active Distribution Networks[J]. IEEE Transactions on Smart Grid, 10(2):1465-1476.
[5] Askarpour M , Aghaei J , Khooban M H , et al. (2019) Voltage control of critical and non-critical loads in distribution networks with electric spring[J]. Electric power systems research, 177(Dec.):105988.1-105988.11.
[6] YANG Zhenquan, XIANG Ji, LI Yanjun. (2019) Active Control Strategy of Distributed Generations for Utility Grid Cooperation in Distribution Network[J]. Proceedings of the CSEE, (11):3176-3185.

Coordinated operation method of cascade hydropower stations considering runoff error

Miao You[1,2*], **Kongfei Hu**[1,2]

[1] Dadu River Hydropower Development Co., Ltd., Chengdu 610041, China

[2] Tellhow Software Co., Ltd., Nanchang, Jiangxi 330096, China

[*]Corresponding author's e-mail: ymvip163@163.com

Abstract. The traditional collaborative scheduling method of cascade hydropower stations group has the problem of imperfect regression model, which leads to excessive water consumption rate. A collaborative scheduling method of cascade hydropower stations group considering runoff error is designed. The influence factors of runoff forecast were obtained, the network parameters were called, and the regression model was built based on the runoff error according to the total power generation of cascade hydropower stations in the degree period. The objective function of hydropower system with hydropower stations as the center was established to optimize the cooperative dispatching mode. Experimental results: The average water consumption rates of the coordinated operation method of cascade hydropower stations group in this paper and the other two kinds of coordinated operation method of cascade hydropower stations group are 34.86%, 57.26% and 59.47% respectively, indicating that the coordinated operation method of cascade hydropower stations group with runoff error is more perfect.

1. Introduction

China has a large number of rivers, abundant runoff and huge drop, which is rich in hydropower resources, ranking first in the world. Since the construction of the first hydropower station, China's hydropower development has gone through a hundred years, and gradually formed the pattern of thirteen hydropower bases. Hydropower is a kind of clean renewable energy, which has the advantages of low power generation cost, strong peak regulation ability, stable operation of units, etc., and has become the first choice of energy development at home and abroad [1-2]. According to the arrangement of hydropower station group, the structure of hydropower station group can be divided into three types: cascade (series) hydropower station group, parallel hydropower station group and series-parallel hybrid hydropower station group. In the joint operation of cascade hydropower stations, the difference in the regulating performance of each reservoir can be used to make the control hydropower stations with better regulating performance change their operation mode to cooperate with the hydropower stations with poor regulating performance, so that the seasonal electric energy can be changed into long-term reliable electric energy, and the total guaranteed output of the system can be improved. Therefore, according to the characteristics of large-scale cascade hydropower stations, more effective and novel optimization theories and methods are needed to provide more scientific theoretical and methodological basis for the operation and management of cascade hydropower stations.

2. Coordinated operation method of cascade hydropower stations considering runoff error

2.1. Obtaining the influencing factors of runoff prediction

According to the layout of the hydropower station group, the structure of the hydropower station group can be divided into three forms: cascade (series) hydropower station group, parallel hydropower station group and series and parallel combination of hybrid hydropower station group [3]. Users can choose different models and set corresponding parameters to forecast runoff according to different basins and different stations. In the known state space, the element distribution mode of hydropower station group in this state space is obtained, and the specific expression formula is as follows:

$$P=\frac{1}{\{L(\delta)|\delta\in C\}} \quad (1)$$

In formula (1), L represents the state space, δ represents the element, and C represents the set of integers.Runoff connection is reflected in the downstream discharge of the upstream hydropower station as a component of the inflow flow of the next hydropower station, and the hydraulic connection is reflected in the utilization of cascade head and the influence of the connected water level of the next hydropower station on the working head of the upper hydropower station [4-5].The atmospheric circulation and solar activity directly affect the weather conditions of the basin, and play a decisive role in the precipitation and evaporation of the basin, and then affect the formation of runoff, while the underlying surface of the basin directly affects the process of runoff. If any variable of the element in the space satisfies formula (1), the sequence of random variables is said to have no aftereffect. Then the calculation formula for the transfer probability of discrete runoff series in space is as follows:

$$Q=\frac{G}{\varepsilon-1} \quad (2)$$

In formula (2), G represents part of the observed value, and ε represents the state frequency. Cascade hydropower station group compensation regulation is a branch of optimal operation development of hydropower station group, its core is to establish a mathematical model according to the demand, and to find a method to solve the model. In the subsequent runoff forecast of this power station, only each parameter of the network needs to be called and given input, a forecast value can be obtained by using the forecast model. When the runoff data is large enough, the statistical expression formula is as follows:

$$\eta^2=2d\frac{\gamma}{|\log_2 d|} \quad (3)$$

In formula (3), d represents the significance level and γ represents the marginal probability. The underlying surface conditions affect the process of runoff production and confluence in the basin, and the actual runoff and the average actual runoff in the early period can reflect the situation of water retreat in the basin comprehensively, and can show the influence of the underlying surface on runoff formation in the basin.

2.2. Runoff error regression model is constructed

The formation process of runoff is a complex nonlinear dynamic system, which is not only affected by natural factors such as rainfall, temperature and underlying surface, but also by human activities, even atmospheric circulation and astronomical factors [6]. The sources of runoff prediction errors mainly include model structure error, measurement data error and system initial state error, etc. The size of prediction error will directly affect the scheduling mode of hydropower stations [7-8]. By comparing and analyzing the output plan of each cascade power station and the actual output in the power system, it is not difficult to find the important position of peak load balancing. According to scheduling demand, the mean value of runoff data is:

$$Y = \frac{1}{H}\sum_{e=1}^{n} k_e - 1 \quad (4)$$

In formula (4), H represents the runoff coverage area, k represents the real interval, and e represents the variation range of runoff. In view of the operational flexibility of hydropower stations, the task of peak-shaving is generally given priority, secondary to system stability and flood control requirements. According to the cooperative dispatching object, the model objectives of hydropower station group dispatching can be divided into flood control, power generation, ecology, etc. On the basis of Formula (4), the expression formula of regression model is obtained, which is expressed as follows:

$$\varphi = \frac{\mu}{\left[l(\alpha+1)\sum_{j=1}^{i} \beta_j \right]^{\frac{1}{2}}} \quad (5)$$

In formula (5), α represents the arrangement of runoff data, β represents the real interval, μ represents the propulsion scale, and i, j represents the output coefficient respectively. Such as changes in river morphology and riverbed geomorphology, artificial natural hydrological cycle, biodiversity reduction, etc. In this mode, power generation plan is made based on the known total load task assigned by the power grid. Therefore, the selection of optimization criteria should not only ensure the benefit but also conform to the actual situation, but also reflect short-term benefits and ensure long-term benefits. Therefore, multi-objective optimal operation of cascade hydropower stations based on ecological dispatching has become an important research topic at present, especially for the construction of ecological civilization in China and ensuring the health of river ecosystem. Based on the above, the steps of constructing regression model are completed.

2.3. Optimizing the cooperative scheduling mode

The current scheduling concept of cascade hydropower stations mainly emphasizes the social and economic benefits brought by water resources utilization, but neglects the changes of watershed ecological environment caused by the changes of natural runoff caused by hydropower stations regulation. Power generation is one of the important tasks that large-scale hydropower stations need to undertake. Especially considering the characteristics of load curve, that is, the power station output process should be as consistent as possible with the load process change of system instructions. It can be seen from the above that the model is a multi-objective problem under various constraints. The formula for the minimum deviation sum of the output process of each hydropower station and the expected process of hydropower station group is as follows:

$$g = \min\left\{ \sum_{t=1}^{i} \left(S_t - \frac{1}{V} \right)^2 \right\} (6)$$

In formula (6), S represents the actual output in a fixed period, V represents the actual peak load output of each hydropower station, t represents the indicated output of the power load curve, and g represents the sum of residual errors between the actual output process of each cascade power station and the power load curve process. Due to its inherent limitations, the conventional operation of hydropower stations can no longer adapt to the requirements of economic and social development. When the abnormal extra abandoned water exceeds the standard threshold, Formula (6) becomes the following form:

$$\lambda = \max \sum_{t=1} \left(S \sum_{m \in Z} V - m \right)^2 \quad (7)$$

In formula (7), m represents the generation flow, Z represents the average generation head, and other variables have the same meanings as formula (6). Swarm optimization scheduling of hydropower stations is in conventional scheduling and optimization of systems engineering theory, developed on the basis of optimization scheduling is based on the theory of system engineering, establish a hydropower station group of target function of the hydropower system as the center, through modern computing technology and optimization method consists of the objective function and constraint condition of system of equations, The optimal scheduling method satisfying the scheduling principle is sought. This part of compensation benefit is the externality generated by the construction and operation of upstream hydropower stations to downstream hydropower stations. It is necessary to quantify the compensation benefits of construction and operation of controlled hydropower stations without considering joint operation:

$$h=\frac{1}{V}\left[m-n\left(1+\sigma^{2}\right)\right] \quad (8)$$

In formula (8), n represents outbound flow and σ represents inbound and outbound flow. In the model, the incoming water of hydropower stations is described by a deterministic process, which is usually predicted according to the historical runoff data and meteorological information. Because the deterministic model can make good use of all kinds of forecast information, and can integrate the experience of the scheduler, it is convenient and flexible, so the deterministic method is mostly used in the middle and long term optimization scheduling research at present. Based on the above description, the steps of optimizing the cooperative scheduling mode are completed.

3. Case analysis

3.1. Instance background and data

A hydropower station group located in X City was selected as the experimental object, and the average daily runoff of hydropower stations provided by the platform of hydropower station group could be collected. In order to predict runoff on a weekly scale, the average weekly runoff needs to be collected. According to the existing historical operation data of each cascade power station, the clustering indexes of daily load rate, daily peak-valley difference, number of peak value of sunrise force, time of peak appearance of sunrise force and number of sunrise force cycles were calculated, and the scheduling decision library of hydropower stations selected in the experiment was constructed to obtain the calculation results of typical output curves of each power station. Based on sampling survey and historical runoff data of hydropower stations, it is found that the difference between the daily average runoff in one week and the weekly average runoff is negligible. At the same time, the contour coefficient method is used to determine the optimal clustering number of sunrise force feature vector of each power station. The average runoff of each Sunday approximately represents the average runoff of this week. In addition, during data collection, m3 / s was taken as the unit, and it was assumed that the sediment concentration in the water at the time of measurement was not considered. Basic parameters of cascade hydropower stations are shown in Table 1:

Table 1 Basic parameters of cascade hydropower stations

name	A	B	C	D
Dead water level (m)	1206	1305	1148	982
Normal high water level (m)	1160	1223	976	814
The installed capacity （MW）	1025	960	878	658
Efficiency coefficient	9.0	8.5	9.0	8.5
Regulated storage capacity （m^3）	0.36	76.2	115.1	69.7

Journal of Physics: Conference Series **2195** (2022) 012016 doi:10.1088/1742-6596/2195/1/012016

According to Table 1, the basic parameters of the hydropower station group can be obtained, and the cooperative scheduling effect is tested by dynamic programming solution.

3.2. Effect analysis

The cooperative operation method of cascade hydropower stations group based on load distribution and cluster analysis are selected to make experimental comparison with the cooperative operation method of cascade hydropower stations group in this paper. The water consumption rates of three hydropower stations were compared under different runoff conditions. The smaller the value is, the better the performance of the hydropower station is. The experimental results are shown in Figure 1-2:

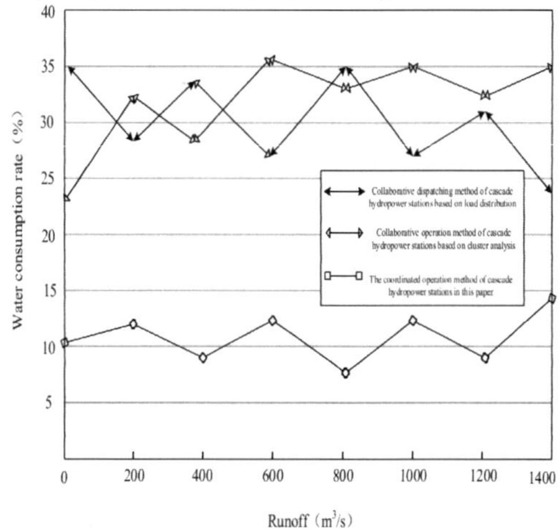

Fig. 1 Water consumption rate of 1200 (m^3/s) runoff (%)

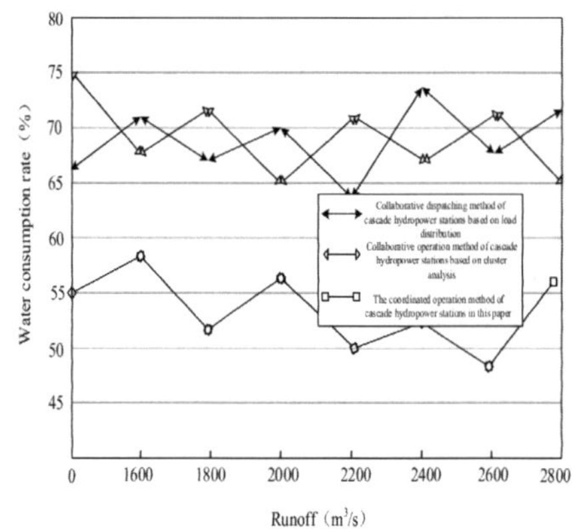

Fig. 2 Water consumption rate of 2600 (m3/s) runoff (%)

According to Fig. 1 and Fig.2, the average water consumption rates of the cooperative operation method of cascade hydropower stations group in this paper and the other two cooperative operation methods are 34.86%, 57.26% and 59.47% respectively. The cooperative operation method of cascade hydropower stations group in the illustration has a lower water consumption rate and is more suitable for practical operation tasks.

4. Conclusion

In view of the problem that the operation characteristics of hydropower stations are not considered in the process of traditional dynamic planning and dispatching of hydropower stations, based on the traditional model, the in plant economic operation of hydropower stations is introduced, the output of hydropower stations in the traditional dynamic planning is improved to the output of optimized operation in the plant, and the improved dynamic planning model of Hydropower Station Dispatching of hydropower stations is established. The model is applied to the optimal operation of the hydropower station group, and a good operation effect is obtained. In terms of scheduling mode, it adopts the mode of joint decision-making of library group scheduling diagram and load distribution algorithm. Due to the limited research conditions, the precision of collaborative dispatching method of cascade hydropower stations in this paper is not comprehensive enough, and will be strengthened in the future.

References

[1] Feng G L , Feng X T , Chen B R , et al. (2020) Characteristics of Microseismicity during Breakthrough in Deep Tunnels: Case Study of Jinping-Ⅱ Hydropower Station in China[J]. International journal of geomechanics, 20(2):04019163.1-04019163.11.
[2] Xu X , Hu W , Cao D , et al. (2020) Optimized sizing of a standalone PV-wind-hydropower station with pumped-storage installation hybrid energy system[J]. Renewable Energy, 147:1418-1431.
[3] Renardy S , Benitez J P , Tauzin A , et al. (2020) How and where to pass? Atlantic salmon smolt's behaviour at a hydropower station offering multiple migration routes[J]. Hydrobiologia, 847(2):469-485.
[4] Zare M , Pakparvar M , Jamshidi S , et al. (2021) Optimizing the Runoff Estimation with HEC-HMS Model Using Spatial Evapotranspiration by the SEBS Model[J]. Water Resources Management, (4).
[5] Abdulla F , Al-Shurafat A W . (2020) Rainfall-Runoff modeling for Semi-arid and trans-boundary Yarmouk River Basin[J]. Procedia Manufacturing, 44:180-188.
[6] Salim I , Sajjad R U , (2019) Paule-Mercado M C , et al. Comparison of two receptor models PCA-MLR and PMF for source identification and apportionment of pollution carried by runoff from catchment and sub-watershed areas with mixed land cover in South Korea[J]. The Science of the Total Environment, 663(MAY 1):764-775.
[7] Tiwari J , Yu B , Fentie B , et al. (2020) Probability distribution of groundcover for runoff prediction in rangeland in the Burnett - Mary Region, Queensland[J]. The Rangeland Journal, 42.
[8] Adib A , Lotfirad M , Haghighi A . (2019) Using uncertainty and sensitivity analysis for finding the best rainfall-runoff model in mountainous watersheds(Case study: the Navrood watershed in Iran)[J]. Journal of Mountain Science, 16(003):529-541.
[9] Ysa B , Yang X B , Ling S C , et al. (2019) A method of direct, real-time forecasting of downstream water levels via hydropower station reregulation: A case study from Gezhouba Hydropower Plant, China[J]. Journal of Hydrology, 573:895-907.
[10] Fu X , Wang H , Wang Z . (2020) Research on block-chain-based Intelligent Transaction and Collaborative Scheduling Strategies for Large Grid[J]. IEEE Access, PP(99):1-1.

Study on Thermal Runaway Behavior of Lithium Ion Battery under Overcharge Using Numerical Detecting Method

Ao GAO [1]*, Wenhui DONG [1], fang XU [1], Xiangyang XU [2], Lei FAN [2]

[1]National Engineering Laboratory for Fire and Emergency Rescue, Shenyang Fire Science and Technology Research Institute of MEM, Shenyang, Liaoning, 110034, China

[2]Nanjing NengQiNeng(NQN) Electronic Technology Co., Ltd. Nanjing, Jiangsu, 210032, China

* Corresponding author: gaoao@syfri.cn

Abstract—In order to provide support for the early warning model of thermal runaway of lithium-ion battery, the thermal runaway behavior of power lithium battery is studied by analyzing the charging current, voltage and other parameters of lithium-ion power battery under overcharge condition. In addition, the morphological changes of battery before and after thermal runaway are compared. It is found that under the conditions of this study, overcharging for 58 minutes will lead to thermal runaway of full load lithium-ion battery. In the early stage of thermal runaway, the leakage gas in the battery box will continue to increase, which is 1339 ppm in the First Level early warning, 3405 ppm in the Second Level early warning and 8292 ppm in the Third Level early warning.

1. INTRODUCTION

Ensuring the safe use of lithium battery pack is the basic requirement for battery pack of electric vehicle. Since the lithium battery pack of electric vehicle is composed of multiple single batteries in series, when the battery pack without battery management system is directly used, the difference between the single batteries in the battery pack will gradually increase during use and cannot be adjusted. That will be resulting in single overcharge, over discharge a series of faults, such as over-current and high ambient temperature of the battery pack, reduce the service life and performance of the whole battery pack, scrap and explosion in serious cases. In recent years, lithium battery energy storage industry has developed rapidly. Lithium battery energy storage is an energy storage technology that uses electrochemical reaction to realize the mutual conversion between electric energy and chemical energy [1-7]. It is the most rapidly developed energy storage technology at present. In particular, lithium-ion battery is most widely used in various battery energy storage technologies because of its high energy density and long cycle life [8-9]. Lithium iron phosphate has stable crystal structure and is not easy to decompose even at high temperature or overcharge. Therefore, it is commonly used as the cathode material of lithium ion battery and has a good development prospect [10-12].

Some scholars have studied the runaway behavior of lithium-ion battery in different external environments, simulated the thermal runaway behavior of battery under the conditions of extrusion, overcharge, over discharge and heating, and obtained a series of laws and conclusions [13-15]. Some scholars have established relevant thermal models of lithium-ion batteries by using simulation software, such as ANSYS, FLUENT or COMSOL, to simulate and analyze the thermal behavior of lithium-ion batteries under different abuse conditions [16-21]. The characteristic parameter alarm threshold of most models is lack of sufficient experimental data. Therefore, there are some problems, such as high false positive and false negative rate, lack of wide applicability and so on. In this paper, the behavior of single lithium-ion power battery before, during and after thermal runaway under overcharged condition is studied.

2. EXPERIMENT

Simulate the real situation, place a power battery pack inside the power battery box, overcharge one of the batteries, resulting in battery thermal runaway, simulate the scene of battery thermal runaway in the battery box in practical application, and actually test the early warning and fire control functions of the single power battery fire early warning and control device. The test battery box adopts the standard micro electric vehicle battery box. The dimensions of the battery box are 1100mm long, 800mm wide and 190mm high. 39 battery packs are arranged in the battery box according to the real working state, of which 9 are working battery packs, each consisting of 10 fully charged 12Ah ternary soft packed lithium batteries. One of them is overcharged, and the other 8 groups are arranged close to the periphery of the overcharged battery pack to test the thermal runaway of the overcharged battery pack, which leads to the thermal runaway of the adjacent battery packs and the diffusion of thermal runaway. The remaining 30 groups are battery pack models, which simulate the structure of the real battery box. The battery pack and battery pack models are arranged in 5 rows. The fire early warning control device is arranged in the gap 3 rows apart from the overcharged battery pack to test the early warning effect to the greatest extent. The fire extinguishing device is arranged at the far end from the overcharged battery pack to test the fire extinguishing effect to the greatest extent. The overcharge power supply adopts HT10020 equipment with constant wing energy as the charger, with constant current of 20A and constant voltage of 8V. The overcharge battery pack is charged directly through the charging cable. The monitoring equipment adopts notebook computer, which is connected to the fire early warning control device through the harness, and the communication mode adopts CAN bus to read the monitored status information in real time. A Nikon SLR camera is used for on-site shooting and recording at a fixed position, as shown in Figure 1. The specific physical diagram of the experiment is shown in Figure 2.

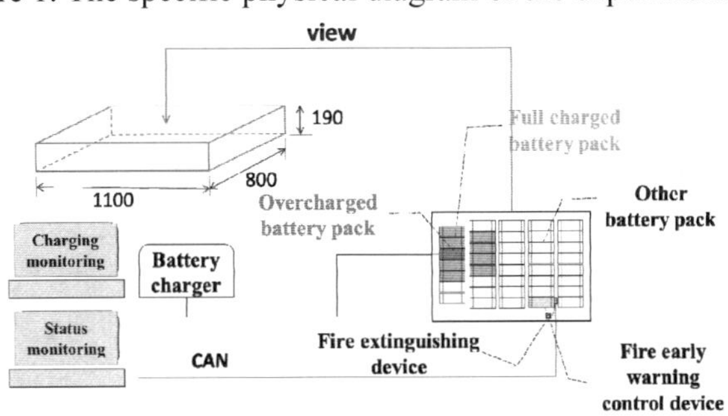

Figure 1. Schematic diagram of experimental system

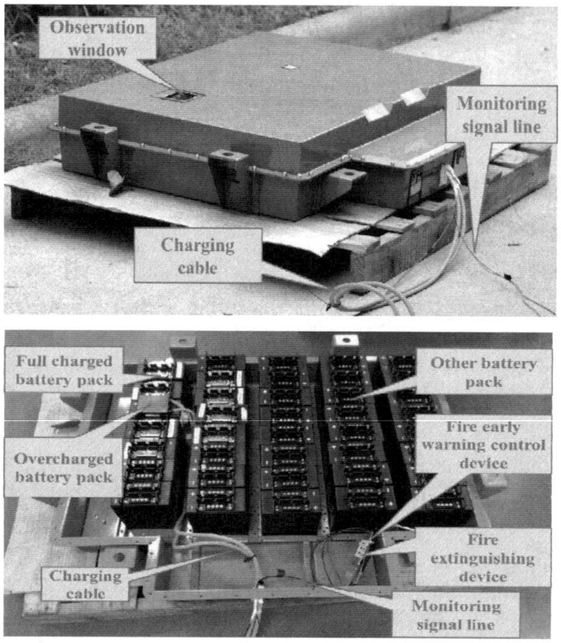

Figure 2. Physical diagram of experimental system

3. Results and Discussion

In the early stage of the charging process, the voltage rises slowly, as shown in Figure 3. When the overcharged battery pack is charged at constant current for about 58 minutes, the internal heat of individual battery cells is out of control, resulting in gas. The internal pressure increases, the package cracks, and the characteristic gas concentration in the box suddenly increase, triggering a First Level early warning. At this time, the gas smoke data rises rapidly to 1339 ppm, and the temperature does not change significantly, as shown in Figure 4.

After the First Level early warning, due to continuous overcharge, more battery cells were thermally out of control, the battery soft package further burst, the characteristic gas concentration further increased sharply, and the Second Level early warning was triggered after 2 minutes and 20 seconds. The gas smoke data rapidly increased to 3405 ppm, and the temperature did not change significantly, as shown in Figure 5. In this process, it can be seen that the charging voltage decreases, which is in line with the characteristics of thermal runaway of lithium-ion battery. As 10 overcharged batteries are packaged into a group, heat out of control and heat accumulation occurred successively. Even in case of power failure, it is inevitable that the temperature will rise further and burn. About 3 minutes and 30 seconds after the Second Level early warning, more batteries burst and combustion occurred. The Third Level early warning was triggered. The gas smoke data rapidly increased to 8292 ppm, and the temperature did not change significantly, as shown in Figure 6. The fire extinguishing device was automatically started to suppress the fire. The open fire was suppressed, but the heat runaway reaction continued and the heat was still gathering, After 8 minutes, the packaging of other battery cells in the overcharged battery pack burst, releasing a large amount of electrolyte and electrode reaction gas. However, due to the filling of aerosol and inert gas sprayed by the fire extinguishing device, there was no secondary combustion in the battery box.

As the thermal runaway occurs at one end of the battery box, the thermal runaway does not diffuse, and the battery box is flat, which is unfavorable to heat conduction, resulting in temperature imbalance inside the battery box. The ambient temperature on the test day is relatively high, and the temperature in the box before the test has reached 50 degrees. The fire early warning control device is placed in the gap 3 rows away from the overcharged battery pack, which is far away. The temperature change of thermal runaway fails to cause obvious monitoring temperature change, and the effect of temperature monitoring is not obvious under the specific conditions of the test.

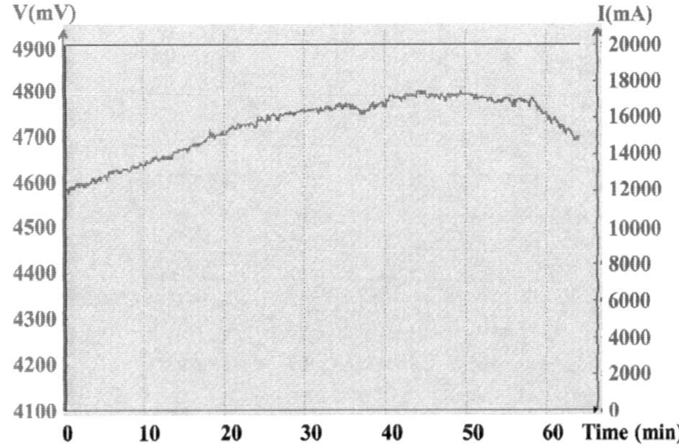

Figure 3. Change of current and voltage during charging

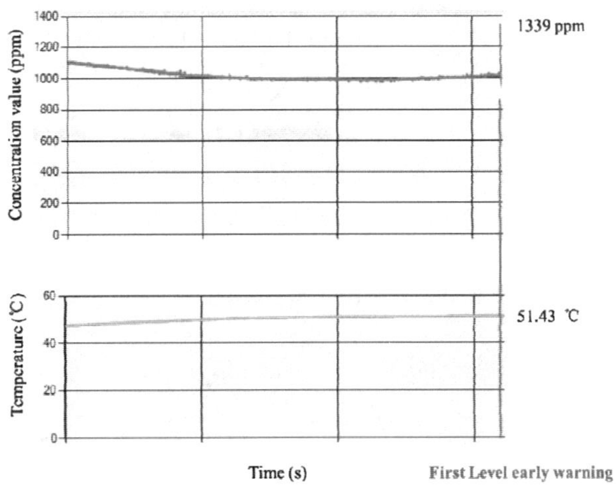

Figure 4. First Level early warning parameter change

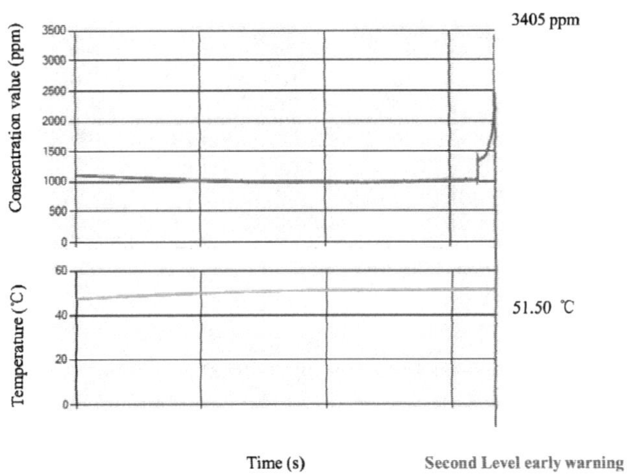

Figure 5. Second Level early warning parameter change

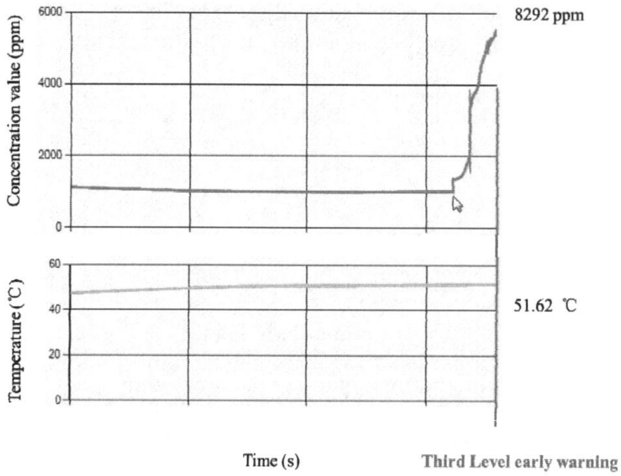

Figure 6. Third Level early warning parameter change

After 2 hours, the battery is unpacked and observed. After unpacking, no smoke is observed and there is no sign of re ignition. The overcharged battery pack has stopped the thermal runaway reaction, the overcharged battery pack has obvious signs of combustion, and the adjacent battery packs has no material damage and no uncontrolled thermal diffusion. The fire extinguishing device has been started, the fire extinguishing agent has been sprayed, and there is no obvious trace of fire extinguishing agent in the box.

It can be seen that the overcharged battery pack is completely burned, the combustion does not spread to the adjacent battery pack, there are burning marks on the appearance of the adjacent battery pack, and the structure is intact.

Figure 7. Battery pack status before and after the experiment

Overcharge causes thermal runaway of the soft pack battery in the battery pack. In the early stage of the thermal runaway, the fire early warning control device sends an early warning signal in time when the observable signs of thermal runaway are not obvious. The continuous overcharge caused the thermal runaway of 10 soft pack batteries in the battery pack and caused combustion. The fire extinguishing device was started in time, successfully suppressed the fire and prevented the thermal runaway from spreading to the adjacent battery pack.

4. CONCLUSION

By studying the thermal runaway behavior of lithium-ion battery under overcharge condition, this paper actually tested the fire early warning and control of the thermal runaway in the battery box with a length of 1100mm, a width of 800mm and a height of 190mm. When the thermal runaway of lithium-ion battery occurs after overcharging for 58 minutes, the leakage gas in the battery box will continue to increase, 1339 ppm in the First Level early warning, 3405 ppm in the Second Level early warning, 8292 ppm for the Third Level early warning. At this time, the temperature does not change rapidly.

Comparing the state of battery pack before and after the experiment, it is found that the state of thermal runaway battery is significantly different from other batteries. The quantitative determination of thermal runaway state is one of the key problems in battery safety research. This paper studies the thermal runaway behavior of lithium ion battery under overcharge quantitatively, which provides a technical basis for relevant research. In the future, more relevant research work should be carried out under other conditions.

ACKNOWLEDGMENT
This work is supported by National Key Research and Development Program (2019YFC0810703), Key Projects of Ministry of Public Security's Criminal Technology Double-Ten Program (2019YFC0810703), Liaoning Provincial Natural Fund Guidance Program Project (2019-ZD-0866) and Open Project of National Engineering Laboratory for Field Evidence Traceability Technology (2019NELKFKT03).

REFERENCES

[1] WU Shuang, JI Cong, SUN Guoqiang. Multiple objection operation strategy optimization research of distribution network including distributed energy storage[J]. Electric Power Engineering Technology, 2018, 37(2): 20-26.
[2] WU Ming, SUN Lijing, KOU Lingfeng, et al. Capacity configuration method for second-use electric vehicle batteries of active distribution network based on demand side response[J]. High Voltage Engineering,2020, 46(1): 71-79.
[3] YU Changhai, WU Jiping, YANG Haijing, et al. Research on control strategy of large-scale energy storage system participating in power grid frequency modulation[J]. Power Engineering Technology, 2019, 38(4): 68-73, 105.
[4] CAI Jilin, XU Qingshan, YUAN Xiaodong, et al. Configuration strategy of large-scale battery storage system orienting wind power consumption based on temporal scenarios[J]. High Voltage Engineering, 2019, 45(3): 993-1001.
[5] LIU Dahe, HAN Xiaojuan, LI Jianlin. Energy storage economy analysis of cascade cells based on photovoltaic power station scenario[J]. Electric Power Engineering Technology, 2017, 36(6): 27-31, 77.
[6] LEI Yong, LIN Xiaodong. Application of hybrid energy storage system based on SMES and BESS in smoothing the power fluctuations of wind farms[J]. High Voltage Engineering, 2019, 45(3): 983-992.
[7] SU Hao, ZHANG Jiancheng, WANG Ning, et al. Energy management strategy of large-scale hybrid energy storage system based on layered optimization[J]. High Voltage Engineering, 2018, 44(4): 1177-1186.
[8] WU X S, XIA S X, HUANG Y Q, et al. High-performance, low-cost and dense-structure electrodes with high mass loading for lithium-ion batteries[J]. Advanced Functional Materials, 2019, 29(34): 1903961.1-1903961.7.
[9] FOAD H, GANDOMAN, J J, SHOVON G, et al. Concept of reliability and safety assessment of lithium-ion batteries in electric vehicles: Basics, progress, and challenges[J]. Applied Energy, 2019, 25(1): 113343.
[10] ZHANG Yang. Preparation and modification study of LiFePO4 cathode material[D]. Xi'an, China: Xi'an University of Architecture and Technology, 2017.
[11] TAN Yiyong. Brief analysis of lithium iron phosphate battery[J]. Information Recording Materials, 2019, 20(6): 50-51.
[12] Sahraei E, Campbell J, Wierzbicki T. Modeling and Short Circuit Detection of 18650 Li-ion Cells Under Mechanical Abuse Conditions[J]. Journal of Power Sources, 2012, 220(4): 360-372
[13] Cai W, Wang H, Maleki H, et al. Experimental Simulation of Internal Short Circuit in Li-ion and Li-ion-polymer Cells[J]. Journal of Power Sources, 2011, 196(18): 7779-7783

[14] Wang H, Lara-Curzio E, Rule E T, et al. Mechanical Abuse Simulation and Thermal Runaway Risks of Large-format Li-ion Batteries[J]. Journal of Power Sources, 2017, 342: 913-920

[15] Spotnitz R, Franklin J. Abuse Behavior of High-power Lithium-ion Cells[J]. Journal of Power Sources, 2003, 113(1): 81-100

[16] Kim G H, Pesaran A, Spotnitz R. A Three-dimensional Thermal Abuse Model for Lithium-ion Cells[J]. Journal of Power Sources, 2012, 170(2): 476-489

[17] Baba N, Yoshida H, Nagaoka M, et al. Numerical Simulation of Thermal Behavior of Lithium-ion Secondary Batteries Using the Enhanced Single Particle Model[J]. Journal of Power Sources, 2014, 252: 214-228

[18] Melcher A, Ziebert C, Rohde M, et al. Modeling and Simulation the Thermal Runaway Behavior of Cylindrical Li-ion Cells—Computing of Critical Parameter[J]. Energies, 2016, 9(4): 292

[19] Coman P T, Veje C T, Darcy E C, et al. Modelling Li-Ion Cell Thermal Runaway Triggered by an Internal Short Circuit Device Using an Efficiency Factor and Arrhenius Formulations[J]. Journal of the Electrochemical Society, 2017, 164(4): A587-A593

[20] Zhao W, Luo G, Wang C Y. Modeling Nail Penetration Process in Large-Format Li-ion Cells[J]. Journal of The Electrochemical Society, 2014, 162(1): A207-A217

[21] Ohsaki T, Kishi T, Kuboki T, et al. Overcharge Reaction of Lithium-ion Batteries[J]. Journal of Power Sources, 2005, 146(1): 97-100

Optimal operation method of integrated energy system considering thermal inertia

Yongli Wang [1], Zhen Liu[1*], Chengcong Cai[1], Hekun Shen[1], Xin Chen [1] and Yang Yang[2]

[1] School of Economics and Management, North China Electric Power University, Changping District, Beijing 102206, China

[2] Economic and Technological Research Institute of State Grid Hebei Electric Power Co. Ltd, Hebei 050000, China

[*]Corresponding author's e-mail: lz_15737038162@163.com

Abstract. This paper proposes a comprehensive energy system operation optimization method considering the thermal inertia of the heating system. Firstly, based on the energy network of the integrated energy system, establish the thermal inertia and heat storage model of the heating pipeline; secondly, the goal is to minimize the total operating cost of the electricity-heat-gas integrated energy system. Considering constraints such as power balance and safe operation of equipment, an integrated energy system operation optimization model based on thermal inertia is established; the results show that the model can better balance scheduling cost and reliability.

1. Introduction

With the development of economy and society, the connection of multiple energy sources has become increasingly close. In this context, an integrated energy system came into being. Integrated energy system refers to a new type of energy system formed by coupling multiple energy systems such as electricity, heat, cooling, and gas in the links of energy production, transmission, and consumption, which can realize the scientific dispatch and energy cascade among multiple energy sources use. However, the transportation characteristics and time scales of different energy sources in the integrated energy system are quite different, which brings great challenges to the synchronization and coordination of source and load in the integrated energy system.

Scholars at home and abroad have conducted a lot of research on the thermal inertia and dynamic physical processes of thermal systems .The literature [1-2] studied the time lag of the heating network and the loss of the transportation process, and established the operation optimization model of the heat storage of the heating network and the thermal inertia of the building. Literature [3] considers the multiple thermal inertia of the thermal system, and proposes a dual-time-scale electric-heat coordination optimization method. Literature [4-5], in order to deal with the challenge of high proportion of renewable energy access, proposed an urban integrated energy system planning model that considers the reliability of energy supply and the uncertainty of wind and solar; Literature [6-8] considers the load side Regulate resources, propose a multi-energy system optimal dispatching plan that integrates demand response and game; Literature [8] introduces electric heat storage boiler device, establishes an energy flow model based on heat storage and heat transfer charge heat leakage dynamic process, and analyzes thermal characteristics Impact on the operation of the combined electric and heating system;

Based on the thermal system, this paper studies the operation optimization method of the integrated energy system considering thermal inertia. Establish a typical architecture based on IES simple topology to model the characteristics of the thermal system. Construct an optimal scheduling model for IES operation. Taking an IES in a certain place in Northeast China as a simulation example, the results verify the feasibility of the model.

2. Integrated energy system considering thermal inertia

2.1. Typical architecture of integrated energy system

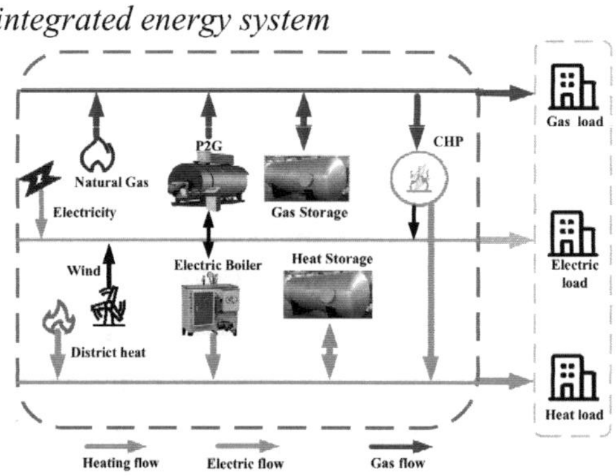

Fig.1 Schematic diagram of integrated energy system.

The conventional multi-energy system is shown in Figure 1. The energy matrix can describe the energy production, conversion, storage and consumption of the integrated energy system:

$$L = CP + S \tag{1}$$

In the formula, C is the energy conversion matrix of the integrated energy system; P 、 L 、 S are the input vector, load vector and energy storage vector of the integrated energy system;

$$\begin{bmatrix} L_G \\ L_E \\ L_H \end{bmatrix} = \begin{bmatrix} \alpha_1 & \beta_1\eta_{P2G} & 0 \\ \alpha_2\eta_{CHP,E} & 0 & 0 \\ \alpha_2\eta_{CHP,E} & \beta_2\eta_{EB} & 1 \end{bmatrix} \begin{bmatrix} P_{G,ex} \\ P_{E,ex} + P_{wind} \\ P_{H,ex} \end{bmatrix} + \begin{bmatrix} S_G \\ 0 \\ S_H \end{bmatrix} \tag{2}$$

In the formula, L_E 、 L_H and L_G are the electricity, heat, and gas loads in the integrated energy system, respectively; α_1 and α_2 are the distribution coefficients of natural gas to gas loads and gas turbines, respectively; β_1 and β_2 are electrical energy input to electricity to gas and electric boilers The distribution coefficient of; $P_{E,ex}$ 、 $P_{H,ex}$ and $P_{G,ex}$ are the energy interaction between the multi-energy system and the power grid, heating network, and gas grid respectively; P_{wind} is output for wind power; S_G and S_H are the output of gas storage and heat storage devices respectively; η_{P2G} and η_{EB} are power conversion The conversion efficiency of gas and electric boilers; $\eta_{CHP,E}$ and $\eta_{CHP,E}$ are the gas-to-electricity and gas-to-heat efficiencies of the gas turbine, respectively.

2.2. Thermal system inertia model and energy storage characteristics

The district heating network has large time lag and non-linear characteristics. Due to the influence of parameters such as the flow rate of the heat medium and the length of the heating pipe, the temperature change from the heat source to the user will have a certain time delay. This time lag is regarded as the thermal inertia of the heating network. At the same time, the heating network contains huge heat energy, which can be regarded as a kind of heat storage device. The two characteristics are analyzed below.

(1) Thermal inertia of heating pipes

Affected by the heating medium, the temperature change at the inlet of the heating pipe needs to be transmitted to the outlet of the pipe after a period of time. Its thermal inertia time can be expressed as:

$$\tau_{pipe} = \frac{\pi \rho_w l d_0^2}{4m} \tag{3}$$

In the formula, τ_{pipe} is the thermal inertia time of the pipe; l is the length of the pipe: ρ_w and m are the density and flow rate of the heating medium respectively; d_0 is the inner diameter of the pipe.

(2) Energy storage characteristics of heating network

Since the dynamic process of thermal energy transmission has a large time lag, and the hot water heat energy at the inlet and outlet of the pipeline at the same time is inconsistent, the heating network can be regarded as a special energy storage device that can provide energy buffering and delayed response. Therefore, the heat storage of DHN should be the difference between the heating output and the load of the system:

$$Q_{DHN}(t) = L_{(H)}t - (Q_{CHP}(t)) + Q_{EB} + Q_{HS}(t)) \tag{4}$$

In the formula, $L_{(H)}t$ is the heat load of t period; Q_{DHN} is the energy storage power of the heating network. Q_{CHP}, Q_{EB}, and Q_{HS} are the heat output power of gas turbine, electric boiler and heat storage in t period, respectively.

In this paper, the heating pipe network is regarded as a special kind of energy storage, so the temperature range of the supply and return pipes is the adjustable ability of the heating network and can be matched with the thermal inertia of the heating network.

$$\begin{cases} T_{pipe,\sup}(t) \le T_{pipe,\sup,\max} \\ T_{pipe,back}(t) \ge T_{pipe,back,\min} \end{cases} \tag{5}$$

In the formula, $T_{pipe,\sup}(t)$ and $T_{pipe,back}(t)$ are the pipe temperature of the water supply and return water respectively; $T_{pipe,\sup,\max}$ is the upper limit of the temperature of the water supply pipe; $T_{pipe,back,\min}$ is the lower limit of the temperature of the return water pipe.

3. Operational optimization model of integrated energy system considering thermal inertia

3.1. Objective function

This article aims to minimize the cost of integrated energy system scheduling, and establishes a two-stage optimal robust scheduling model of integrated energy system considering thermal inertia. The day-ahead scheduling stage optimizes the start and stop costs of all controllable equipment; the real-time scheduling stage adjusts the output of each equipment in the system based on the actual output of the wind and solar based on the known output of the day-a-day scheduling. The optimization goals are as follows:

$$C = \min_x \left\{ f_1 + \max_w \min_y f_2 \right\} \tag{6}$$

In the formula, C represents the total operation cost of the integrated energy system dispatch; f_1 is the total start-up and shutdown cost of each equipment, f_2 is the sum of the integrated energy system energy purchase cost and the wind curtailment cost, and x corresponds to the controllable equipment (CHP, P2G, electricity Boiler, heat storage and gas storage) start and stop status, 1 represents the 0-1 variable of the equipment operating status; y is the output corresponding to each equipment, represents the equipment (CHP, P2G, electric boiler, heat storage, gas storage, and Heating network (heat storage) is a continuous variable of output; w is the actual thermal inertia time of the heating pipeline in the project.

3.2. Restrictions

(1) Power supply balance constraint

$$L_E(t) = P_{CHP}(t) + P_{wind,s}(t) - P_{P2G}(t) - P_{EB}(t) + P_{ex,E}(t) \tag{7}$$

In the formula, $L_E(t)$ is the electric load of t period; $P_{CHP}(t)$ is the output electric power of the gas turbine in t period; $P_{P2G}(t)$ and $P_{EB}(t)$ are the electric power of the electric to gas and electric boiler in t period.

(2). Heat exchange constraints

$$Q_{CHP}(t) + Q_{EB}(t) = c_{water} m(t)[T_{pipe,\sup}(t) - T_{pipe,back}(t)] \tag{8}$$

In the formula, Q_{CHP} and Q_{EB} are the heat supply of the cogeneration unit and the electric boiler; $T_{pipe,\sup}(t)$ and $T_{pipe,back}(t)$ are the temperature of the supply and return water at the heat source, respectively.

(3) Equipment output and climbing constraints

$$\begin{cases} P_{i,\min} \le P_i(t) \le P_{i,\max} \\ -R_i^{down}\Delta t \le P_i(t) - P_i(t-1) \le R_i^{up}\Delta t \end{cases} \tag{9}$$

In the formula, $P_{i,\min}$ and $P_{i,\max}$ are the minimum and maximum output of the first device i, $P_i(t)$ is the output of the device i during t period, R_i^{down} and R_i^{up} are the climbing power and landslide power of the device i.

3.3. Model solving

Genetic algorithm is a computational model that simulates the biological evolution process of natural selection and genetic mechanism of Darwin's biological evolution theory. It is a method of searching for the optimal solution by simulating the process of natural law.

In this paper, a genetic algorithm is used to solve the model, and the process is shown in Figure 2.

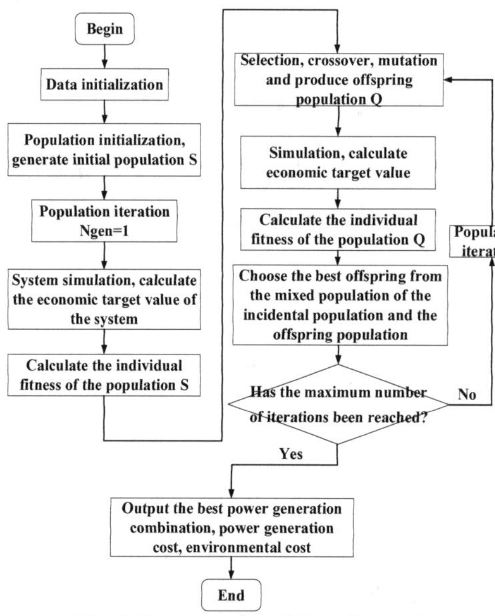

Fig.2 Genetic algorithm flow chart.

4. Numerical example simulation

4.1. Basic data

Taking an integrated energy system in a certain area of Northeast China as a simulation example, the specific parameters are shown in Table 1, Table 2, Figure 3 and Figure 4.

Table.1 Heating network pipeline parameters.

Pipeline	Head end note	End note	Length/km
1	N1	N2	2.2
2	N2	N3	3.6
3	N3	N4	2.5

Table.2 Equipment parameters.

Type	Power/MW		Adjustment rate/(MW/min)	
	Lower limit	Upper limit	Lower limit	Upper limit
CHP	3	10	0.2	0.5
EB	0	10	0.6	1
P2G	0	10	0.5	1
HS	0	5	0.4	1
GS	0	5	0.5	1

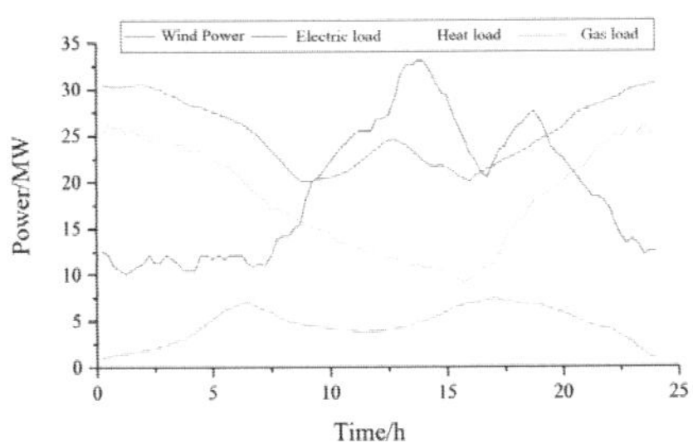

Fig.3 Wind power and multi-energy load forecast.

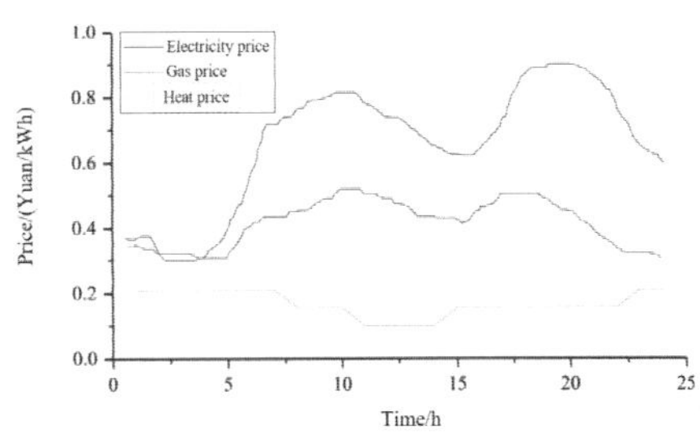

Fig.4 Price of electric, gas and heat.

The following two scenarios are used to analyze the dispatch operation of the integrated energy system.

Scenario 1 is the dispatch of the integrated energy system without considering the inertia of the thermal system.

Scenario 2 considers the thermal inertia and heat storage capacity of the heating network to test the operating economy at this time.

4.2. Simulation results

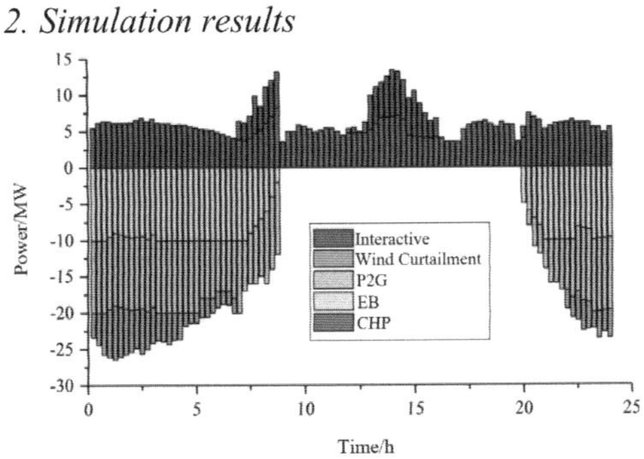

Fig.5 The output curve of each device under electrical balance (Scenario 1).

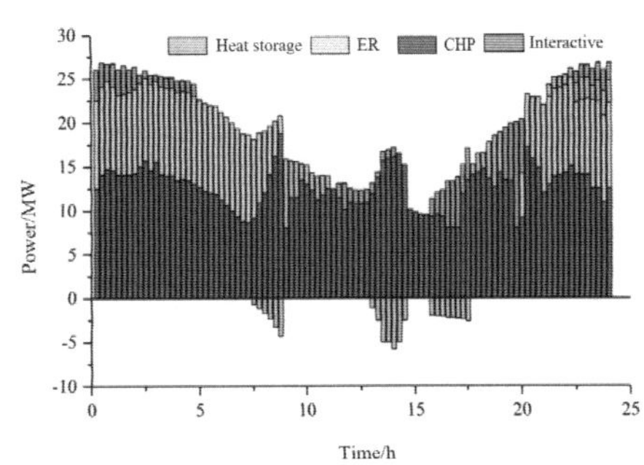

Fig.6 The output curve of each device under heat balance (Scenario 1).

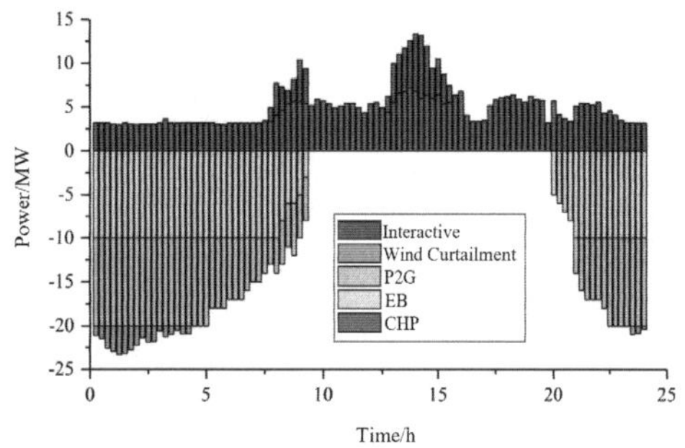

Fig.7 The output curve of each device under electrical balance (Scenario 2).

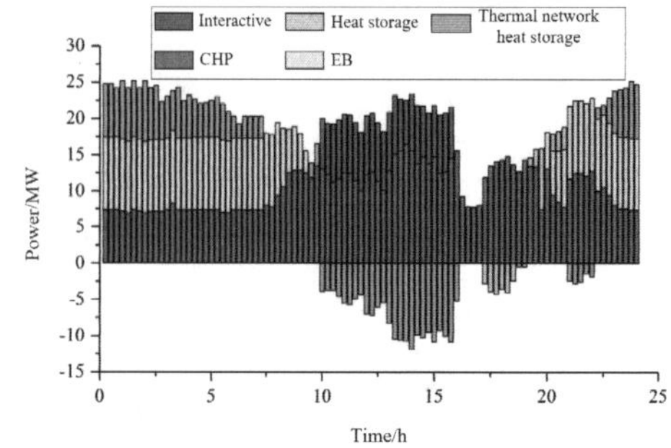

Fig.8 The output curve of each device under heat balance (Scenario 2).

In Scenario 1, the electric boiler and P2G consume this power. Due to the high wind power output and high thermal load, the wind abandonment is serious, as shown in Fig. 5 and Fig. 6. Scenario 2 after considering the thermal inertia, the output of the CHP unit is significantly reduced at night, reducing the wind abandonment phenomenon, as shown in Fig. 7 and Fig. 8.

5. Conclusion

The heat load demand of the heating network is mainly supplied by CHP and electric boilers. At night, the limited capacity of the heat storage device in scene one makes it difficult to reduce the output of the CHP unit. In Scenario 2, thermal inertia is considered. The waste heat generated by CHP is stored in the heating network first, and is released during peak heat load hours to increase the space for wind power grids. The simulation results show that: the coordination and optimization of the thermal inertia and energy storage characteristics of the thermal system is beneficial to reduce the CHP strong electrothermal coupling, and can effectively improve the wind power absorption capacity of the system and improve the flexibility of the integrated energy system.

References

[1] YI Zhongkai, LI Zhimin, Combined Heat and Power Dispatching Strategy Considering Heat Storage Characteristics of Heating Network and Thermal Inertia in Heating Area[J]. Power System Technology. 2018, 42(5):1378-1384.

[2] WANG Mingjun, MU Yunfei, MENG Xianjun, et al. Optimal Scheduling Method for Integrated Electro-thermal Energy System Considering Heat Transmission Dynamic Characteristics[J]. Power. System Technology. 2020, 44(1): 132-142.

[3] LIN Li, GU Jia, WANG Ling. Optimal Dispatching of Combined Heat-power System Considering Characteristics of Thermal Network and Thermal Comfort Elasticity for Wind Power Accommodation[J]. Power System Technology, 2019, 43(10): 3648-3661.

[4] ZHOU Xianzheng, CHEN Wei, GUO Chuangxin. An urban multi-energy system planning method incorporating energy supply reliability and wind-photovoltaic generators uncertainty[J]. Transactions of China Electrotechnical Society, 2019, 34(17): 3672-3686.

[5] JIN Hongyang, TENG Yun, LENG Ouyang, et al. Non-waste urban multi-energy coordinated energy storage model based on source-charge uncertainty state perception [J]. Journal of Electrical Engineering and Technology: 2020, 35(13): 2830-2842.

[6] XU Yeyan, LIAO Qingfen, LIU Dichen. Multi-player Intraday Optimal Dispatch of Integrated Energy System Based on Integrated Demand Response and Games[J]. Power System Technology, 2019, 43(7): 2506-2518.

[7] Nguyen D T, Le L B. Optimal bidding strategy for microgrids considering renewable energy and building thermal dynamics[J]. IEEE Transactions on Smart Grid, 2014, 5(4): 1608-1620.

[8] Liu X, Wu J, Jenkins N, et al. Combined analysis of electricity and heat networks[J]. Applied Energy, 2016, 162: 1238-1250.

Research on Optimization of Regional Integrated Energy System Based on Thermal Inertia of Building Clusters

Xiaodong Liu[1], Weiyuan Wang[2], Hengjun Zhou[1], Gang Wu[1], Yongjin Tao[3*]

[1] State Grid Nanjing Power Supply Company

[2] State Grid Jiangsu Electric Power Co., Ltd.

[3]Sichuan Energy Internet Research Institute, Tsinghua University

[*]Corresponding author's e-mail: taoyongjin@tsinghua-eiri.org

Abstract. With the increase in the share of energy from new energy sources such as photovoltaics and wind power, the development and application of integrated energy systems are becoming more and more rapid. However, if energy is not coordinated and optimized, wind and solar abandonment may occur. This is to achieve a variety of energy sources. The integration, complementarity and flexible regulation of the company have brought great challenges. The fifth-generation district heating and cooling system, as a new direction of comprehensive energy development, can greatly improve the energy efficiency of sources and networks and the convenience of user-side energy consumption, but its adjustment potential for the user-side is insufficient. This paper explores the thermal inertia of the existing infrastructure in the system, that is, the user-side building cluster in the regional thermal system, and uses the thermal inertia of the building cluster to improve the overall potential flexibility of the regional-level integrated energy system and realize the optimized operation of the regional integrated energy system, then improve the flexibility and economy of system operation.

1. Introduction

In order to cope with the challenges of limited fossil fuel resources and global environmental issues, renewable energy power generation technologies have developed rapidly on a global scale. As a new generation energy system that is conducive to promoting the consumption of new energy, the integrated energy system has received more and more attention and its related research has been developing rapidly[1]. The integrated energy system refers to linking these different energy entities more closely with each other through coupling technology, in which the interaction is mainly carried out through the energy conversion devices between the different energy entities to provide services and manage each energy entity in the best way[2]. With the increasingly in-depth research on the integrated energy system, the park-level integrated energy system is developing rapidly on a global scale[3]. The park has the characteristics of high energy density, high load utilization hours, and diversified forms of energy production and use. It is the best entry point to promote the development of an integrated energy system.

The user's demand for secondary energy is mainly cold, heat, and electricity. The common cold, heat and electricity integrated energy system mainly considers the complementary characteristics of the source side to provide users with energy supplies of different attributes. The fifth-generation district heating and cooling system is a new energy supply system concept, which can better integrate various energy forms at the hub, and provide the cold and heat of the energy hub to the users through the low-temperature two-way pipe network. Therefore, it meets the needs of different users, realizes the upgrade from the complementary utilization of energy to the integration of energy, and greatly reduces the network loss, and realizes the efficient and coordinated utilization of the source network[4].

Around 2017, Europe took the lead in proposing the concept of the fifth-generation district heating and cooling system(5GDHC)[5]. 5GDHC basic definition is: 5GDHC network is a brine or water as the carrier medium, the terminal sub-station network with energy supply source heat pump[6]. Its working temperature is very close to the ambient temperature, which is not suitable for direct heating and cooling. The low temperature heat/cold is supplied to the water source heat pump of the sub-station through the carrier medium to meet the cooling and heating needs of users. The low temperature of the carrier medium determines that it can directly use industrial and urban waste heat and renewable energy containing low-grade heat. The feature that the substations can supply energy in the reverse direction allows them to use the same pipeline to meet the heating and cooling needs of different buildings at the same time. Compared with the traditional heating and cooling network, 5GDHC technology has the advantages of efficient energy utilization, intensive space resources, user-friendly access, multi-energy interaction and coordination, and smart scheduling and operation. It can enhance the coupling of different networks such as heating and cooling in the integrated energy system, and greatly improve the efficiency of energy utilization. It is one of the hot research fields in recent years.

However, 5GDHC technology pays more attention to the integration of the source and the low temperature and high efficiency of the network, and there is not enough attention to how the user side participates in improving the overall energy efficiency. In practice, due to the time transmission delay of the heating pipe network, it is a high inertia system, and the user itself can be regarded as a heat storage equipment cluster. Literature[7] scheduling model to study the dynamic behavior cogeneration constraint model constraints and the heat storage device and heat pipe network meter. This model not only considers the economy of the system, but also improves the level of wind power consumption. However, the above research ignores the thermal inertia of the building clusters on the demand side, that is, due to the thermal insulation properties of the building clusters[8], the indoor temperature of the building clusters will not change drastically in a short time when the heat is absorbed, which allows the heating area located in the building cluster to act as a corresponding heat storage unit. In the literature[9], based on the heat storage capacity of the district heating system, the optimal dispatch model of the integrated energy system is studied. But how to take advantage of the thermal inertia of the district heating network and how to promote the integration of wind power in the electricity transmission network and not studied. In the literature [10-11], considering the thermal insulation characteristics of the building, according to the user comfort zone on the side of the cluster building, reduce the heat source output during the peak hours of wind power, and improve the peak shaving capacity of the cogeneration unit. In literature [12-13], In order to increase the flexibility of cogeneration units in promoting the integration of wind power, the heat storage characteristics of the central heating system are studied, and the output demand of cogeneration units is reduced when the heat load peaks. However, the detailed thermal model was not considered in the above research. Previous studies have carried out optimal scheduling based on the heat storage characteristics of the thermal system, but the building thermal inertia modeling is not detailed enough.

In this paper, a detailed model analysis based on the thermal inertia of the building cluster is carried out and applied to the 5GDHC system to establish a regional energy storage model based on the detailed building thermal inertia model, aiming at the optimal system operating cost, and comprehensively considering various equipment models in the 5GDHC and energy constraints, establish an optimal scheduling model, and realize regional-level energy coordination management. Taking a park as an example for simulation analysis, compared with the traditional optimization model, the user-side building response potential is further tapped, effectively reducing operating costs, verifying the effectiveness of the model, and improving the flexibility of the system.

2. Thermal inertia modelling of building clusters

The RC thermal network model of the building area is composed of thermal resistance with heat transfer capacity and heat capacity with heat storage capacity [28]. In the node method adopted, there are two types of nodes, namely, room nodes and wall nodes, and each node in the building area is grounded through a heat capacity node and connected to each other through a thermal resistance. The RC thermal network model specifically describes a heating/cooling area, while the building model is a combination of multiple similar structural areas. The heating/cooling areas in each building are roughly similar to each other.

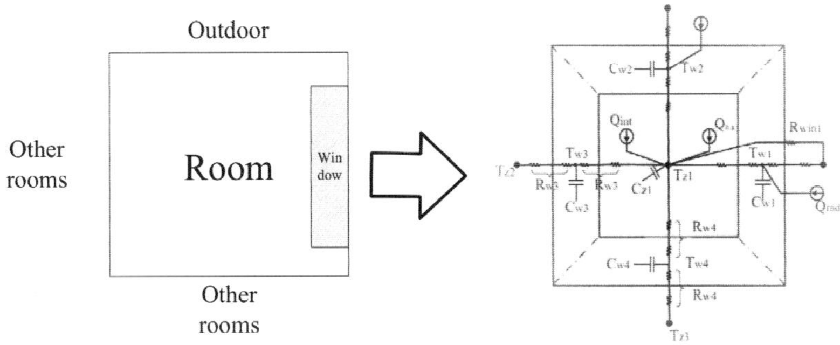

Figure 1. RC thermal network model

The specific mathematical model of the RC thermal network model in the heating/cold zone is described as follows:

$$C_{wj}\frac{dT_{wj}}{dt} = \sum_{s\in N_j^w} \frac{T_{sj} - T_{wj}}{R_{wj}} + r_j\alpha_j A_j^W Q_j^{rad} \tag{1}$$

$$C_{zi}\frac{dT_{zi}}{dt} = \sum_{j\in N_i^W} \frac{T_{wj} - T_{zi}}{R_{wj}} + \pi_k \sum_{c\in N_i^W} \frac{T_{cj} - T_{zi}}{R_{win}} + \dot{Q}^{int} + Q_{R,i} + \pi_k \tau_j^W A_j^{win} Q_j^{rad,win} \tag{2}$$

Where T_{zi} is the temperature of the i-th room, C_{wj} is the heat capacity of the j-th wall, N_j^w is the set of all adjacent points of the j-th wall, T_{sj} is the value of all the adjacent nodes of the j-th wall temperature, T_{wj} is the temperature of the j-th wall, R_{wj} is the thermal resistance between the center line and the j-th wall. If the j-th wall is not exposed to sunlight, $r_j = 0$, otherwise, $r_j = 1$. α_j is the absorption coefficient of the j-th wall, A_j^W is the area of the j-th wall, and Q_j^{rad} is the light intensity in the corresponding direction of the j-th wall.

Where C_{zi} is the heat capacity of the i-th room, T_{zi} is the temperature of the i-th wall, N_i^W is the set of all adjacent wall nodes in the i-th room, T_{cj} is the ambient temperature where the external node of the j-th wall with windows is located, $Q_{R,i}$ is the heat/cold power consumed by the building group provided by the energy supply device. If there is no window in this room, $\pi_k = 0$, or $\pi_k = 1$. τ_j^W is the transmittance of the window on the j-th wall, A_j^{win} is the total area of the window on the j-th wall, $Q_j^{rad,win}$ is the light intensity corresponding to the direction of the window, and $\dot{Q}^{int}$ is the i-th Heat source inside the room.

Taking into account the slow dynamic process of building heat dissipation/cooling and temperature changes, the building heat balance equation expressed by the differential equation is further differentiated, so as to achieve a simple and effective solution to the economic optimization problem of building clusters.

For the room in Fig.1, the model is established as follows:

$$\begin{aligned} C_{w1}R_{w1}T_{w1}(t+1) - (C_{w1}R_{w1} - 2) * T_{w1}(t+1) \\ = \Delta t * (T_{z1}(t) + T_{out}(t) + r_1\alpha_1 A_1^W Q_1^{rad} * R_{w1} \end{aligned} \tag{3}$$

$$\begin{aligned} C_{w2}R_{w2}T_{w2}(t+1) - (C_{w2}R_{w2} - 2) * T_{w2}(t+1) \\ = \Delta t * (T_{z1}(t) + T_{out}(t) + r_2\alpha_2 A_2^W Q_2^{rad} * R_{w2} \end{aligned} \tag{4}$$

$$\begin{aligned} C_{w3}R_{w3}T_{w3}(t+1) - (C_{w3}R_{w3} - 2) * T_{w3}(t+1) \\ = \Delta t * (T_{z1}(t) + T_{out}(t) + r_3\alpha_3 A_3^W Q_3^{rad} * R_{w3} \end{aligned} \tag{5}$$

$$C_{W4}R_{W4}T_{W4}(t+1)-(C_{W4}R_{W4}-2)*T_{W4}(t+1) \\ =\Delta t*(T_{z1}(t)+T_{out}(t)+r_4\alpha_4A_4^WQ_4^{rad}*R_{W4} \tag{6}$$

$$C_{z1}*\left(T_{z1}(t+1)-T_{z1}(t)\right) \\ =\Delta t*(\sum_{j=1}^{4}\frac{T_{wj}(t)-T_{z1}(t)}{R_{wj}}+\frac{T_{out}(t)-T_{z1}(t)}{R_{win1}}+Q_{R,i}+\dot{Q}_1^{int} \\ +\tau_1^WA_1^{win}Q_1^{rad,win}) \tag{7}$$

3. Modelling of the integrated energy system of the park

3.1. Objective function

The optimized operation of the integrated energy system in the park is aimed at economy, and the energy cost of the system is reduced by rationally arranging the unit's output plan and controllable load operation interval. Therefore, the objective function of the proposed optimal scheduling model is the lowest comprehensive operating cost of the park, which mainly includes three parts: power purchase cost, gas purchase cost, and operation and maintenance cost.

$$min\sum_{i=1}^{N_t}(C_{g,sum}+C_{gas,sum}+C_{e,sum}) \tag{8}$$

N_t is the number of days before the optimization period.

(1) Power purchase cost. The electricity purchase cost in the park is obtained by multiplying the electricity purchased by the grid $P_{grid,t}$ by the real-time electricity price λ_t^{TR}.

$$C_{g,sum}=\lambda_t^{TR}P_{grid,t} \tag{9}$$

(2) Gas purchase cost. The cost of gas purchase in the park is obtained by multiplying the power $P_{MT,t}$ of the combined heat and power by the coefficient η_{gas}^P of the combined heat and power, and then multiplying it by the converted natural gas price λ_{gas}^{MT}.

$$C_{gas,sum}=\lambda_{gas}^{MT}\eta_{gas}^{P}P_{MT,t} \tag{10}$$

(3) Operation and maintenance costs of cold, heat and gas supply.

$$C_{e,sun}=\sum_{m=1}^{N_{MT}}C_{MT,m}+\sum_{e=1}^{N_{ES}}C_{ES,e}+\sum_{b=1}^{N_{br}}C_{br,b}+\sum_{C=1}^{N_{AG}}C_{Ag,c}+\sum_{s=1}^{N_{is}}C_{is,s}+\sum_{a=1}^{N_{Ar}}C_{Ar,a}+\sum_{p=1}^{N_{pump}}C_{pump,p} \tag{11}$$

Where $C_{MT,m}$, $C_{ES,e}$, $C_{br,b}$, $C_{Ag,c}$, $C_{is,s}$, $C_{Ar,a}$, $C_{pump,p}$ is the equipment maintenance costs for micro gas turbine, accumulator, electric boiler, industrial air conditioning, ice storage device, absorption chiller and pump. The operation and maintenance cost of each equipment C_{gen} is the unit maintenance cost $C_{gen,unit}$ multiplied by the equipment's operating power $C_{gen,t}$.

$$C_{gen}=C_{gen,unit}C_{gen,t} \tag{12}$$

3.2. Energy supply equipment constraints

(1) Electric boiler

$$H_{br,t}=P_{br,t}\eta_{COP}^{br} \tag{13}$$

Where $H_{br,t}$ and $P_{br,t}$ are the heating power and power consumption of the electric boiler respectively, η_{COP}^{br} is heating efficiency of the electric boiler.

(2) Micro gas turbine

$$P_{MT,t}=H_{MT,t}R_{MT} \tag{14}$$

Where $P_{MT,t}$ and $H_{MT,t}$ are the electric power and heating power of the micro gas turbine, respectively, R_{MT} is the heat-to-electricity ratio of the micro gas turbine.

(3) Industrial air conditioners and absorption chillers

$$Q_{Ag,t} = P_{Ag,t}\eta_{COP}^{Ag} \tag{15}$$

$$Q_{Ar,t} = H_{Ar,t}I_{COP}^{br} \tag{16}$$

Where $Q_{Ag,t}$ and $P_{Ag,t}$ are the refrigeration power and power consumption of industrial air conditioners, respectively. η_{COP}^{Ag} is the refrigeration energy efficiency of industrial air conditioners. $Q_{Ar,t}$ and $H_{Ar,t}$ are the cooling power and heat consumption power of the absorption chiller, respectively. I_{COP}^{br} is the refrigeration energy efficiency ratio of the absorption chiller.

(4) Ice storage device

$$Q_{is,t+1} = (\eta_{is}^{s}P_{is,t} - Q_{is,t}/\eta_{is}^{r})\Delta t + (1 - \sigma_{is})M_{is,t} \tag{17}$$

$$M_{is}^{min} \leq M_{is,t} \leq M_{is,t}^{max} \tag{18}$$

$$P_{is}^{min} \leq P_{is,t} \leq P_{is,t}^{max} \tag{19}$$

$$Q_{is}^{min} \leq Q_{is,t} \leq Q_{is,t}^{max} \tag{20}$$

Where $M_{is,t}$ is the capacity of the ice storage device. $P_{is,t}$ and $Q_{is,t}$ is the ice making power and melting power of the ice storage device, respectively. η_{is}^{s} and η_{is}^{r} is respectively the ice making and melting efficiency of the device during cold storage, τ is the self-loss coefficient of the ice storage device. M_{is}^{min} and $M_{is,t}^{max}$ is the upper and lower limits of the capacity of the ice storage device, P_{is}^{min} and $P_{is,t}^{max}$ are respectively the upper and lower limits of ice making power, Q_{is}^{min} and $Q_{is,t}^{max}$ is respectively the upper and lower limits of melting power.

(5) Power storage device

$$S_{ES,t+1} = [\eta_{ES}^{s}P_{ES,t}^{s} - P_{ES.t}^{s}/\eta_{ES}^{r}]\Delta T + (1 - \sigma_{ES})S_{ES,t} \tag{21}$$

$$S_{ES}^{min} \leq S_{ES,t} \leq S_{ES}^{max} \tag{22}$$

$$P_{ES}^{min} \leq P_{ES,t}^{s} \leq P_{ES}^{max} \tag{23}$$

$$P_{ES,r}^{min} \leq P_{ES,t}^{r} \leq P_{ES,r}^{max} \tag{24}$$

Where $S_{ES,t}$ is the battery energy storage capacity, $P_{ES,t}^{s}$ and $P_{ES.t}^{s}$ is the charging and discharging power of the storage device, η_{ES}^{s} and η_{ES}^{r} is the charging efficiency and discharging efficiency of the power storage device, σ_{ES} is the self-loss coefficient of the power storage device, $S_{ES}^{min}$and S_{ES}^{max} is the upper and lower limits of the energy storage capacity of the power storage device, P_{ES}^{min} and P_{ES}^{max} is the upper and lower limit of the charging power of the storage device, $P_{ES,r}^{min}$ and $P_{ES,r}^{max}$ is the upper and lower limits of the discharge power of the power storage device.

3.3. 5GDHC constraints

(1) Heat pump unit, HP

$$P_{HP,i,t} \leq P_{HP,i}^{capacity} \tag{25}$$

$$COP_{HP,i} = \frac{Q_{R,i}}{P_{HP,i,t}} \tag{26}$$

$$P_{HP,i,t} + H_{HP,in,i,t} = Q_{R,i} \tag{27}$$

Where $P_{HP,i,t}$ is the heat pump power of the i-th room at time t, $P_{HP,i}^{capacity}$ is rated power of the heat pump in the i-th room, $H_{HP,in,i,t}$ is heat input from the system to the heat pump. $COP_{HP,i}$ is the energy efficiency ratio of the heat pump in the i-th room.

(2) Pipeline constraints

$$P_{pump,i,t}\eta_{pump} = H_{pump,in,i,t} * L * c_{pump} \tag{28}$$

$$\sum_{i}[P_{pump,i,t}] \leq P_{pump}^{cap} \tag{29}$$

Where $P_{pump,i,t}$ is pump power of the i-th room at time t, η_{pump}, $H_{pump,in,i,t}$, L, c_{pump}, P_{pump}^{cap} respectively represent the total efficiency of the pump, the heat flowing out of the energy supply pipe network, the distance from the pump to the room. According to Darcy's formula, the constant fitted according to the actual pipe diameter and temperature difference and the maximum installed pump power.

3.4. Multi-energy balance constraint

(1) Cooling balance

$$\frac{Q_{is,t}}{\eta_{is}^{r}} + Q_{Ar,t} + Q_{Ag,t} + \sum_{i}^{N_c} P_{HP,i,t} = \sum_{i}^{N_c} Q_{R,i} \tag{30}$$

Where N_c is the number of cooling rooms, η_{STD}^{r} is the heat exchange efficiency of storage device?

(2) Heating balance

$$H_{MT,t} + H_{br,t} + \sum_{i}^{N_c} P_{HP,i,t} = \sum_{i}^{N_c} H_{R,i} + H_{Ar,t} \tag{31}$$

Where $H_{R,i}$ is the heat load of the i-th room.

(3) Power supply balance

$$\begin{aligned} P_{MT,t}+P_{WT,t}+P_{PV,t} + P_{ES,t} + P_{ES,T}^{s} - P_{ES,T}^{r} + P_{grid,t} \\ = P_{HP,i,t} + P_{is,t} + P_{pump,i,t} + P_{Ag,t} + P_{br,t} + P_{load,t} \end{aligned} \tag{32}$$

Where $P_{load,t}$ is the total electric load value required by the park.

4. Example analysis

4.1. Basic data of the calculation example

This section takes an industrial park as an example to analyse the optimal scheduling model proposed in the previous section. In the industrial park, electric power is supplied by the grid company and the fans and photovoltaics in the park at the same time; and for heating, the park is equipped with a micro gas turbine with a rated power of 1200kW and an electric boiler with a rated heating power of 600kW. At the same time, the cooling system is equipped with an industrial air conditioner with a rated electrical power of 1,000 kW, an ice storage system with a rated electrical power of 500 kW, and an absorption chiller with a rated cooling power of 500 kW. On the user side, in addition to the energy demand of industrial users, 5GDHC also provides cold and heat supply for the surrounding 5 residential buildings. Each building has 5 floors, each floor is simplified into 4 heating and cooling areas, each area is 8m long, 8m wide, and 3m high.

In this case, a typical day in the cooling season is selected, and the operating parameters of each building and equipment are as follows：

Table 1. Industrial park building parameter table

Parameter	Numerical value
Window area A^{win}/m^2	4
Wall heat capacity $C_{wall}/(J\cdot K^{-1})$	$7.9*10^5$
Wall thermal resistance $R_{wall}/(K\cdot W^{-1})$	0.06
Window thermal resistance$R_{win}/(K\cdot W^{-1})$	0.02
Heat capacity of side wall with window $C_{wall(win)}/(J\cdot K^{-1})$	2.6*107
Room heat capacity $C_r/(J\cdot K^{-1})$	2.5*105

Table 2. Equipment economic parameters

Parameter	Numerical value
Micro gas turbine C_{MT} /[yuan·(kW·h)$^{-1}$]	0.059
Electric boiler C_{br} /[yuan·(kW·h)$^{-1}$]	0.026
Industrial air conditioning C_{Ag} /[yuan·(kW·h)$^{-1}$]	0.016
Ice storage device C_{is} /[yuan·(kW·h)$^{-1}$]	0.02
Absorption chiller C_{Ar} /[yuan·(kW·h)$^{-1}$]	0.018
Battery device C_{ES} /[yuan·(kW·h)$^{-1}$]	0.01

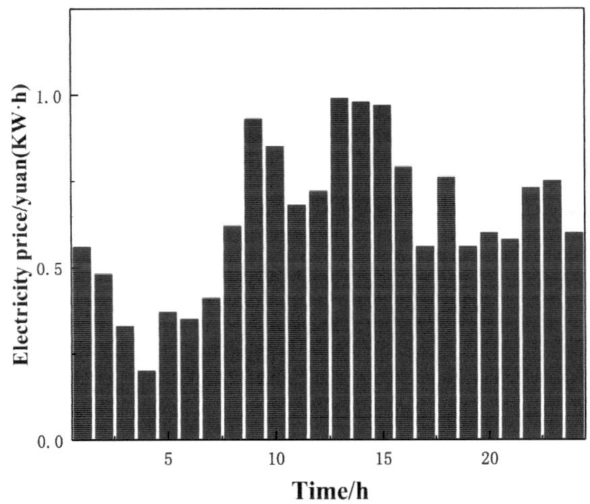

Figure 2. Real-time electricity price within a certain day

Figure 2 shows the real-time electricity price of the power company in a certain day. The price of natural gas is converted into 0.380 yuan/(kW·h) in this article. Photovoltaic, wind power system, outdoor temperature parameters are shown in Figure 3.

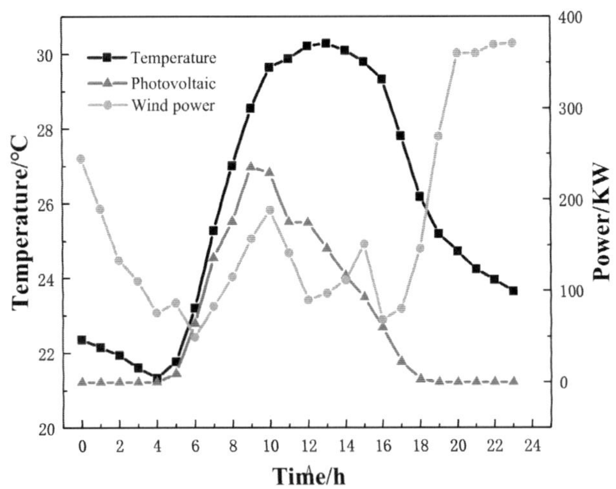

Figure 3. PV and wind power forecast output and outdoor temperature

Figure 4 shows the electricity, cooling, and heating load forecasts of the industrial park with the energy storage characteristics of the cooling area of the building. The equal load composition is a fixed predicted value, and the thermal load is smaller than the electric load.

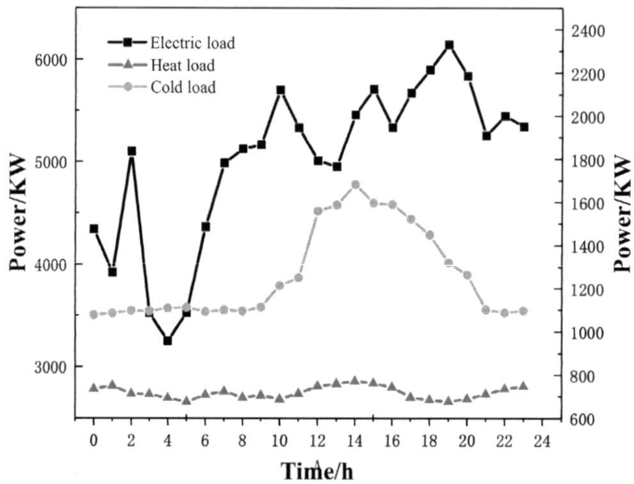

Figure 4. Electricity, heating and cooling load

In the optimization model, due to the consideration of the storage characteristics of the cooling area of the building, it is affected by the outdoor ambient temperature, the cooling load is not fixed to take the predicted value, but the result of the optimization of the park scheduling model based on the temperature constraints in the plant, thus through the multi-energy complementary characteristics of the industrial park, the output of the relevant components of electricity, cooling, and heat in the integrated energy system is affected. It can be seen from figure 4 that when the outdoor ambient temperature is relatively high, the cooling load required is also more. The cooling load power starts to rise from 11:00 and drops around 15:00 to meet the environmental temperature changes in the park. n the early hours of the morning, based on the lower ambient temperature, the required cooling load is relatively low. At the same time, due to the certain energy storage characteristics of the building's cooling area, the cooling load power remains unchanged for a period of time, but it does not affect the temperature inside the park. Require. In addition, because the selected building parameters are literature data, they are not the actual building parameters of the cooling zone in the park. Therefore, the energy storage characteristics represented by thermal inertia are not fully utilized, resulting in inertia between the outdoor temperature and the load value provided by the actual park.

4.2. Comparative analysis

This article proposes two different schemes based on whether to consider the energy storage characteristics of the building cooling area:

1). The energy storage characteristics of the cooling area of the building are not considered. Assuming that without considering the energy storage characteristics of the building's cooling area, all loads in the park are fixed forecast values at each forecast moment, and there is no adjustable capacity.

2). Consider the energy storage characteristics of the building's cooling area. Assuming that considering the energy storage characteristics of the building's cooling area, the internal room temperature range of the building's cooling area in the park is set to 20-25°C, which has a certain demand response capability.

Figure 5 shows the comparison between the total operating cost of the industrial park and the power purchased if the energy storage characteristics of the cooling area of the building are taken into account. It can be seen from the figure 5 that in case 1, the park does not take into account the energy storage characteristics of the building cooling area because the cooling load is basically fixed, and the various types of resources in the park cannot be effectively adjusted during each time period, but the supply is provided according to the fixed forecast value. The total operating cost of the integrated energy system of electricity, cooling and heating in the park is relatively high, which is 85585.2 yuan.

After taking into account the energy storage characteristics of the building's cooling area, the goal is also to minimize the operating cost, and the energy supply equipment in the park at each time period is optimally dispatched according to the electricity price signal. Compared with case 1, in case 2, the power purchased in the industrial park has been significantly reduced during most of the time period, and the total operating cost in the park has dropped to 8,3749.7 yuan, making full use of the energy storage characteristics of the cooling area of the building, which can delay cooling. The loss time of cold energy produced by the equipment is reduced by reducing electricity purchases at relatively high electricity prices to effectively reduce operating costs.

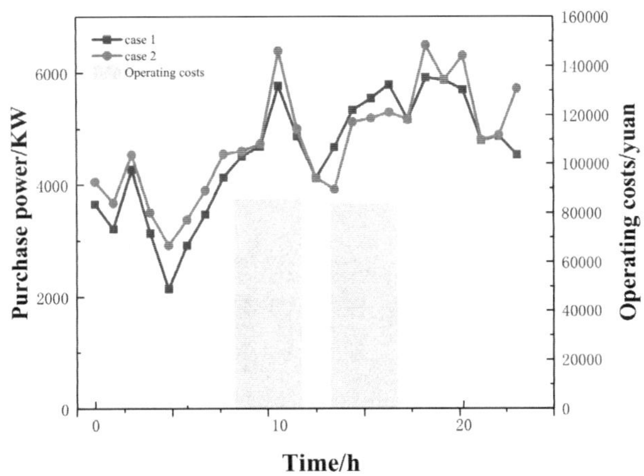

Figure 5. The total operating cost and power purchase of the park in case 1 and 2

It can be seen from figure 5 that from some relatively low electricity prices, such as 16:00, the power purchase in case 2 has increased compared to case 1. It is mainly based on the environmental temperature of the park and the energy storage characteristics of the cooling area of the building to provide a certain response ability to maintain the indoor temperature of the industrial park plant within the required range.

Figure 6 shows the energy consumption of the related equipment in the park configuration in case 1 and case 2. It can be concluded from figure 6 that after considering the energy storage characteristics of the building's cooling area, the total power consumption of electric boilers, ice storage devices, and industrial air conditioners in case 2 is significantly lower than that in case 1 at most of the time. Similarly, in case 2, the cooling power of the building cooling area, that is, the cooling load, and the heat power of the absorption chiller are also reduced.

However, it can be seen from figure 6(b) and (c) that both the cooling power and the heat power consumption of case 2 will increase compared with case 1 for a period of time after 13:00. It is mainly caused by ensuring that the indoor temperature in the cooling area of the building does not exceed the upper limit under the condition of considering the outdoor temperature. Therefore, the power consumption of the corresponding equipment is increased at the same time, but it is mainly distributed at the time when the electricity price is relatively low. This fully demonstrates that by using the energy storage characteristics of the building's cooling area, the cooling power of the refrigeration equipment is flexibly reduced, and the relevant electrical and thermal energy equipment are affected while meeting the requirements of the park to improve the total operating cost.

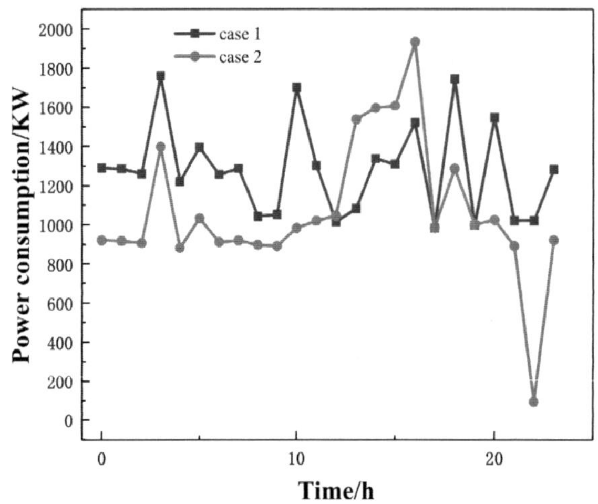

(a) Power consumption of electric boilers, ice storage, industrial air conditioners

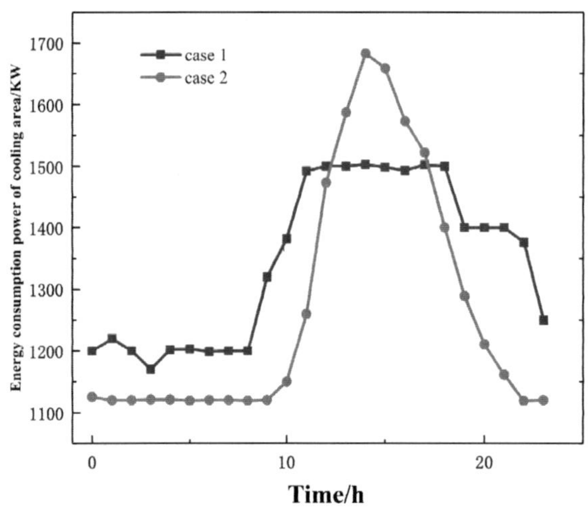

(b) Cooling power

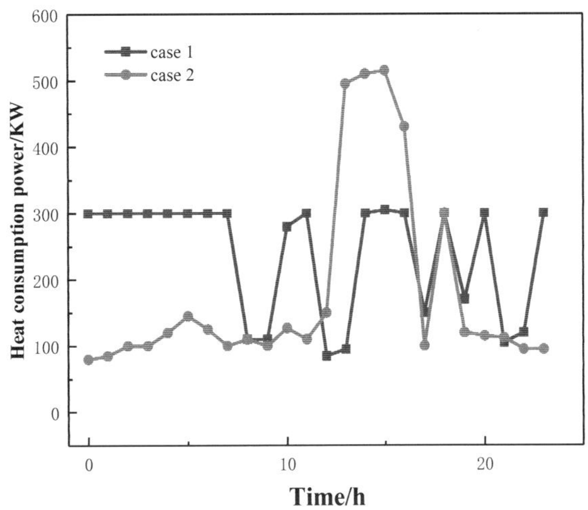

(c) Heat consumption power of absorption chiller

Figure 6. Power comparison of related equipment in the park in case 1 and case 2

5. Conclusion

This chapter first establishes the thermal inertia model of the building cluster, and fully considers the energy storage characteristics based on the detailed building thermal inertia model in the building cooling area to optimize the power purchase strategy of the industrial park. Then, component modeling is carried out for the energy supply and energy storage equipment of the industrial park. With the goal of minimizing the total operating cost of the park's integrated energy system, the components and energy storage models are integrated into the park's integrated energy system scheduling model, the scheduling framework is analyzed, and the optimal scheduling model is constructed. Finally, the validity of the model is verified through the analysis of a numerical example.

The main conclusions include: the energy storage characteristics of the cooling area of the building can help reduce the interaction power with the grid and the operating cost. Ambient temperature will affect the value of the cooling load and further affect the scheduling results; the thermal inertia model of the building cluster can provide additional operational flexibility for the integrated energy system of the park, and provide a direction for improving the economics and reliability of operation.

References

[1] Cong D, Liang L, Jing S, et al. Energy supply efficiency evaluation of integrated energy systems using novel SBM-DEA integrating Monte Carlo[J]. Energy, 2021, 231(1):120834.

[2] D Yang, Wang M, Yang R, et al. Optimal Dispatching of an Energy System with Integrated Compressed Air Energy Storage and Demand Response[J]. Energy, 2021(19):121232.

[3] Electric/thermal hybrid energy storage planning for park-level integrated energy systems with second-life battery utilization[J]. Advances in Applied Energy.

[4] Edtmayer H, Nageler P, Heimrath R, et al. Investigation on Sector coupling Potentials of a 5th Generation District Heating and Cooling Network[J]. Energy, 2021, 230:120836.

[5] Wirtz M, Neumaier L, Remmen P, et al. Temperature control in 5th generation district heating and cooling networks: An MILP-based operation optimization[J]. Applied Energy, 2021, 288(24).

[6] Buffa S, Cozzini M, D'Antoni M, et al. 5th generation district heating and cooling systems: A review of existing cases in Europe[J]. Renewable and Sustainable Energy Reviews, 2019, 104(APR.):504-522.

[7] Li Z, Wu W, Shahidehpour M, Wang, J, Zhang B. Combined heat and power dispatch considering pipeline energy storage of district heating network[J]. IEEE Transactions on Sustainable Energy, 2015, 7(1): 12-22.
[8] Kensby J, Truschel A, Dalenback J O. Potential of residential buildings as thermal energy storage in district heating systems – Results from a pilot test[J]. Applied Energy, 2015, 137(jan.1):773-781.
[9] Wei G, Wang J, Shuai L, et al. Optimal operation for integrated energy system considering thermal inertia of district heating network and buildings[J]. Applied Energy, 2017, 199:234-246.
[10] Yuanhang, Dai, Lei, et al. Dispatch Model for CHP With Pipeline and Building Thermal Energy Storage Considering Heat Transfer Process[J]. IEEE Transactions on Sustainable Energy, 2019, 10(1):192-203.
[11] Pan Z, Guo Q, Sun H. Feasible region method based integrated heat and electricity dispatch considering building thermal inertia[J]. Applied Energy, 2017, 192(APR.15):395-407.
[12] Wu C, Wei G, Ping J, et al. Combined Economic Dispatch Considering the Time-Delay of District Heating Network and Multi-Regional Indoor Temperature Control[J]. IEEE Transactions on Sustainable Energy, 2017, P (99):1-1.
[13] Zheng J, Zhou Z, Zhao J, et al. Integrated heat and power dispatch truly utilizing thermal inertia of district heating network for wind power integration[J]. APPLIED ENERGY -BARKING THEN OXFORD-, 2018, 211(FEB.1):865-874.

Journal of Physics: Conference Series 2195 (2022) 012020 doi:10.1088/1742-6596/2195/1/012020

Research on Standards for Information Supportive System in District Energy Internet Construction

Decheng Wang[1], Yan Li[2], Qun Zhang[3], Qian Chen[4,a]*, Mengshuang Feng[5], Xiaotong He[6]

[1,2,3]Economic and Technological Research Institute, State Grid Jiangsu Electric Power Co., Ltd, Nanjing, China

[4,5,6]State Grid (Suzhou) City & Energy Research Institute, Suzhou, China

[a]2279462317@qq.com

Abstract—At present, the concept of Energy Internet has attracted great attention around the world, and is generally considered to be an important means to realize carbon neutrality. In terms of the overall structure, the Energy Internet is composed by physical network system, information supportive system, and value creation system. As is known to all, the information supportive system is the nerve center of the Energy Internet. While to establish the information supportive system of the Energy Internet, standards and specifications are in urgent need. Therefore, this paper mainly focuses on standards for information supportive system in the Energy Internet construction. First, the overall structure of the information supportive system is proposed. Then, the functional characteristics of each layer in information supportive system are analyzed respectively. In addition, the existing standards for information supportive system are summarized. Meanwhile, standards that need to be formulated in the future are analyzed as well. Finally, suggestions are given for the formulation of construction of information supportive system in the Energy Internet.

1. Introduction

Energy Internet is an inevitable trend for the coordinated development of energy, environment, and economy [1,2]. The production, transmission, conversion, storage of multiple energy systems can be flexibly interacted with users through the advanced Internet of Things (IoT) technology in the Energy Internet, which can enhance the complementarity efficiency of different energy sources, and smooth the fluctuations caused by renewable energy with high penetration rate [3].

Energy Internet is composed by three important parts, including physical network system, information supportive system, and value creation system. Physical network system mainly refers to the traditional physical infrastructures of multiple energy systems. Information supportive system is a digital intelligent system covering all aspects of energy development, energy utilization and related social activities for information collection, transmission, processing, storage, and control. Value creation system is not an actual system. It is the value extension of Energy Internet. Among all the above-mentioned systems, information supportive system is the nerve center of Energy Internet [4]. At present, researches on the information supportive system of Energy Internet are mainly focused on key digital technologies [5], system reliability analysis [6], application scenarios [7], etc. While the norms and standards for information support system, which will support the construction of Energy Internet, are still relatively defective. Due to the fact that district government managers have greater autonomy and initiative in the construction of actual Energy

Internet projects. Therefore, this article focuses on the standards for the information supportive system construction of Energy Internet at the district level.

In this paper, an overall structure of information supportive system is put forward in Section II, and the functional characteristics of each layer in information supportive system are analyzed respectively as well. Then, the existing standards for information supportive system are summarized in Section III, as well as standards that need to be formulated in the future. Finally, suggestions are given for the formulation of construction standards of information supportive system in the Energy Internet in Section IV.

2. OVERALL STRUCTUR OF INFORMATION SUPPORTIVE SYSTEM

In this paper, information supportive system is divided into four layers, namely data collection layer, communication network layer, development platform layer and system application layer. Overall structure of information supportive system is shown in Fig.1.

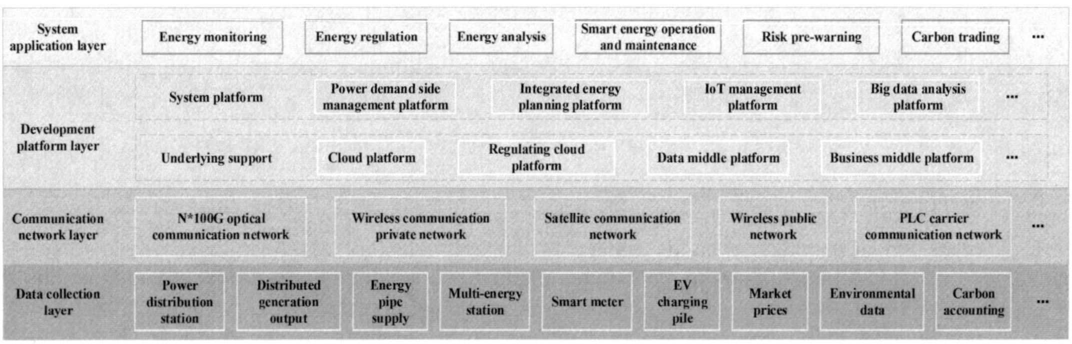

Fig.1 Overall structure of information supportive system

2.1.Data Collection Layer

The essence of data collection is to use ubiquitous sensing technology to realize real-time collection and aggregation of multi-source devices, heterogeneous energy systems, and other environmental information. The core of this layer is to build a set of accurate, real-time, and efficient data collection systems. In this layer, some data is processed locally through protocol conversion and edge computing, while the other data is uploaded to the cloud for cloud computing processing.

The collected data type is various. From the perspective of energy categories, it can be divided into electricity data, thermal data, natural gas data, etc. From the perspective of energy supply forms, it can be classified into distributed energy data, centralized energy station data, energy network data, etc. Apart from direct energy-related data, indirect data that affects the operation of energy systems is also within the collection range, such as market price data, meteorological data, and environmental data. Generally speaking, the data collection frequency of direct energy-related data from physical network system is not less than 15 minutes per time, and data accuracy depends on the sensitivity of the sensors. While the indirect energy-related data like the GDP, corporate information, and district population can be read automatically through the third-party platform.

2.2.Communication Network Layer

Communication network layer of Energy Internet includes two parts: the backbone communication network and the terminal access network. The former mainly covers the construction of regional and municipal backbone transmission networks, and data communication networks, which strives to improve the network's reliable and efficient capacity. The latter mainly focuses on realizing the integrated energy service access through remote access network and local access network. As for Energy Internet at district level, the latter should be paid more attention on.

Remote access network mainly meets the high-reliability, low-latency, and differentiated communication requirements between the cloud master station and the terminal aggregation node. The communication methods include optical fiber technology, power wireless private network, wireless public network like 4G/5G, etc. Under normal circumstance, optical fiber communication method is the first choice. In areas that cannot be reached by optical fiber communication networks, carrier and wireless communication are used as supplements. Local communication network mainly satisfies communication demands between terminal aggregation nodes and the terminal equipment. Power line carrier communication, RS485 bus mode, as well as some short-distance wireless communication methods are adopted in local communication network.

2.3. Development Platform Layer

Development platform layer of the information supportive system in Energy Internet, which is similar to the PaaS layer of industrial internet, provides an entrance to the platform basic functions for developers. Developers can access the basic service resources of the platform and build personalized industrial system applications through the interface provided by the development platform layer. There are three platform development modes, which are summarized as follows.

The first development mode is oriented to system application software developers. In this mode, the platform provides software developers with software development kits (SDKs) that support different development languages and application program interfaces (APIs) that comply with http or https specifications. The SDK packs the basic computing functions of the platform. Developers can build personalized applications locally through the SDK. The second mode is ordinary user-oriented. In this mode, the platform provides users with easy-to-use system development services, such as a drag-and-drop software development method. The drawback of this approach is that it cannot build a highly personalized system application for users, due to the limited basic services provided by the platform. The third mode is hybrid development mode. This mode is for the developers who have a basic software development technology reserve and expect to build personalized applications based on the platform deployment operation and maintenance environment.

2.4. System Application Layer

System application layer is concentrated on the interaction between system users and Energy Internet. In this paper, system users are divided into four major categories, including governments, energy consumers, energy operators, and energy producers. Energy Internet needs to provide functions or services such as energy monitoring, carbon measurement, energy management, energy operation and maintenance, energy optimization, transaction settlement. All the above-mentioned applications should be running throughout the whole process of energy industry services. In this layer, big data, artificial intelligence and other technical intelligent analysis are adopted to improve the efficiency of multi-energy management, reduce energy costs, and build a win-win energy ecosystem for Energy Internet.

3. Standards Analysis of Construction for Each Layer of Information System

So far, there have already been some standards regarding communications and information technology in the field of industry internet. However, when it comes to the construction of information supportive system in Energy Internet, standards are still in a deficient state.

3.1. Data Collection Layer

In data collection layer, there have already been several national standards or industry standards, including basic general standards, sensor standards, measuring terminal standards, etc. [8,9]. Details are summarized in Tab. 1. However, the existing standards for data collection layer of Energy Internet are not yet complete, and some intelligent terminals still remain a low degree of standardization. For example, technical specifications for smart energy meter of Energy Internet, related standards for energy controllers, and technical specifications for energy routers have yet to be worked out.

TABLE 1. ANALYSIS OF STANDARDS FOR DATA COLLECTION LAYER

Fields	Standard Title	Level	Standard Number
Basic	Basic environmental testing procedures for electric and electronic products	National standard	GB/T 2423
	Classifications and codes for sensors	National standard	GB/T 36378
	Technical guidelines for the perception layer of the Energy Internet	/	To be formulated
	Layout Guidelines for Energy Internet Sensing Applications	/	To be formulated
Sensor	Specification for sensors information model of power internet of things	Industry standard	DL/T 1732—2017
	Requirements for smart sensor technology	/	To be formulated
	Network access detection method for sensors	/	To be formulated
	Value traceability method for partial discharge sensors (e.g. high frequency, ultrahigh frequency, ultrasonic and transient ground wave)	/	To be formulated
Measuring terminal	Instrument transformers-Part 1: General requirements	National standard	GB 20840.1—2010
	Testing equipment for digital input electricity meters	National standard	GB/T 37006—2018
	Technical specifications for smart energy meter of Energy Internet	/	To be formulated
	Related standards for energy controllers	/	To be formulated
	Technical specifications for energy routers	/	To be formulated
Local data transmission	Information technology-Sensor networks	National standard	GB/T 30269
	Information security technology-Security technical requirements for application of sensing terminals in internet of things	National standard	GB/T 36951—2018
	Multi-function watt-hour meter communication protocol	Industry standard	DL/T 645—2007
	Data acquisition and management system for electrical energy	Industry standard	DL/T 698
	Micropower wireless network communication protocol for power internet of things	/	To be formulated
	Wireless networking protocol for node devices in power internet of things	/	To be formulated
	Security protection specification for edge IoT agent in power internet of things	/	To be formulated
	Safety certification technical requirements for perception layer devices of power internet of things	/	To be formulated
Edge IoT agent	Technical requirements for the internet of things at the edge of Energy Internet	/	To be formulated
	Technical specification for IoT agent terminal of power plant and stations	/	To be formulated
	Interaction protocol specification for smart terminals and IoT management platform	/	To be formulated

3.2. Communication Network Layer

In communication network layer, there have already been several international standards, industry standards, and enterprise standard. Details are summarized in Tab. 2. However, the existing standards for access network communication layer of Energy Internet are not yet complete. For example, technology standards for ultra-wideband short-range wireless communication, standards for ubiquitous network interconnection, technology standards for visible light communication, technology standards for low-power wide area network have yet to be formulated.

TABLE 2. ANALYSIS OF STANDARDS FOR COMMUNICATION NETWORK LAYER

Fields	Standard Title	Level	Standard Number
Transmission network communication	Types and characteristics of SDH network protection architectures	International standard	ITU-T G.841
	Characteristics of a single-mode optical fiber and cable	International standard	ITU-T G.652:2009
	Interfaces for the optical transport network (OTN)	International standard	ITU-T G.709/Y.1331
	Technical specification of integrated intradyne coherent receiver	Industry standard	YD/T 2799
	Code for engineering design of wavelength division multiplexing (WDM) optical fiber transmission systems	Industry standard	YD 5092-2014
Access network communication	Technical Requirements for Access Network-Multi-service Delivery over EPON/GPON Systems	Industry standard	YD/T 1953-2009
	Power ethernet passive optical network system part 3: Technical requirements and testing methods for interoperability	Enterprise standard	Q/GDW1553.3-2014
	Technical guidelines for conception planning of electric wireless private network	Enterprise standard	Q/GDW 11664-2017
	Technology standards for ultra-wideband short-range wireless communication	/	To be formulated
	Standards for ubiquitous network interconnection	/	To be formulated
	Technology standards for visible light communication	/	To be formulated
	Technology standards for low-power wide area network	/	To be formulated

3.3.Development Platform Layer

In development platform layer, there have already been several international standards and national standards, which are summarized in Tab. 3. The existing standards for this layer are concentrated on two fields, one is data sharing and management, the other is cloud computing and analysis platform. As for data sharing and management, standards for data governance and analysis algorithm, and standards for management and analysis based on digital technology have yet to be formulated. When it comes to standards for cloud computing and analysis platform, standards for cloud computing network, and standards for cloud computing services are still in need.

TABLE 3. ANALYSIS OF STANDARDS FOR DEVELOPMENT PLATFORM LAYER

Fields	Standard Title	Level	Standard Number
Data sharing and management	International technology-metadata registries standard	International standard	ISO/IEC 11179
	Information technology-procedures for achieving metadata registry	International standard	ISO/IEC TR 20943
	Standard series for data quality	International standard	ISO TS 8000
	Information technology-database languages	International standard	ISO/IEC 9075-1:2008
	Technical requirements of relational database management system	International standard	GB/T 28821-2012
	Information technology-reference model of data management	International standard	GB/Z 18219-2008
	Standards for data governance and analysis algorithm	/	To be formulated
	Standards for management and analysis based on digital technology	/	To be formulated
Cloud computing and analysis platform	Information technology-cloud computing-reference architecture	International standard	ISO/IEC 17789：2014
	Information technology-open virtualization format specification	International standard	ISO/IEC 17203:2011
	Information technology-cloud data management interface	International standard	ISO/IEC 17826:2012
	Graphic technology-extensible metadata platform specification	International standard	ISO 16684-1:2012
	Information technology-Elastic computing application interface	National standard	GB/T 31915-2015
	Information technology-cloud data storage and management	National standard	GB/T 31916.2-2015
	Information security technology-security capability requirements of cloud computing services	National standard	GB/T 31168-2014
	Standards for cloud computing network	/	To be formulated
	Standards for cloud computing service	/	To be formulated

3.4. System Application Layer

In system application layer, there have already been several national standards and industry standards, which are summarized in Tab. 4. The existing standards for this layer are concentrated on mobile internet and human-computer interaction. Similarly, standards for this layer of Energy Internet are not yet complete. For example, technology standards for mobile application development and evaluation, third-party service interaction standards for mobile internet applications, Security protection standards for mobile applications, and New type of human-computer interaction instant communication technology, three-dimensional space human interaction system standards are relatively lacking.

TABLE 4. ANALYSIS OF STANDARDS FOR SYSTEM APPLICATION LAYER

Fields	Standard Title	Level	Standard Number
Mobile internet and human-computer interaction	Technical requirements of security for operating system in smart mobile terminal	National standard	GB/T 30284-2013
	Technical requirements of authorization framework for mobile internet API	Industry standard	YD/T 2912-2015
	Security protection requirements for networked application over mobile internet	Industry standard	YD/T 2694-2014
	Technical requirement of internet-based service deployment and operation platform	Industry standard	YD/T 3016-2016
	Technology standards for mobile application development and evaluation	/	To be formulated
	Third-party service interaction standards for mobile internet applications	/	To be formulated
	Security protection standards for mobile applications	/	To be formulated
	New type of human-computer interaction instant communication technology, three-dimensional space human interaction system standards	/	To be formulated

4. CONCLUSION

This paper focuses on the construction standards of information supportive system in district Energy Internet. First, the overall structure of the information supportive system is put forward, and the functional characteristics of each layer of information system are analyzed as well. Then, the existing standards are summarized, and standards that need to be formulated in the future are proposed. Last but not least, suggestions for standard establishment are given for the construction of information system in the Energy Internet.

In data collection layer, standards like technical specifications for smart energy meter of Energy Internet, related standards for energy controllers, and technical specifications for energy routers have yet to be worked out. In communication network layer, technology standards for ultra-wideband short-range wireless communication, standards for ubiquitous network interconnection, technology standards for visible light communication, technology standards for low-power wide area network are in need of establishment. In development platform layer, standards for data governance and analysis algorithm, standards for management and analysis based on digital technology, standards for cloud computing network, and standards for cloud computing services are still in need. In system application layer, technology standards for mobile application development and evaluation, third-party service interaction standards for mobile internet applications, Security protection standards for mobile applications, and New type of human-computer interaction instant communication technology, three-dimensional space human interaction system standards are relatively lacking.

ACKNOWLEDGMENT

This work is supported by State Grid Jiangsu Electric Power Co., Ltd. Economic and Technological Research Institute under project named “Research on the Standards for Typical District Energy Internet Construction”.

References

[1] C. Y. Dong, J. H. Zhao, F. S. Wen, and Y. S. Xue, "From smart grid to energy internet: basic concept and research framework," Automation of Electric Power Systems, vol. 38, pp. 1-11, 2014.

[2] S. M. Tian, W. P. Luan, D. X. Zhang, C. H. Liang, and Y. J. Sun, "Technical forms and key technologies on energy internet," Proceedings of the CSEE, vol. 35, pp. 3482-3494, 2015.

[3] J. Yang, N. Zhang, and Y. Wang, "Multi-energy System Towards Renewable Energy Accommodation: Review and Prospect," Automation of Electric Power Systems, vol. 42, pp. 11-24, 2018.

[4] Q. Liu, "Information and Communication Technology in Energy Internet," Computer knowledge and technology, vol. 15, pp. 50-51, 2019.

[5] J. Y. Wnag, J. H. Guo, J. W. Cao, L. C. Gao, Z. W. Hu, J. Zhou, Y. Y. Ming, and Z. W. Fang, "Review on information and communication key technologies of energy internet," Smart grid, vol. 3, pp. 473-485, 2015.

[6] X. L, and H. H. Li, "Study on information and Communication technologies and reliability of energy internet," North china electric power, vol. 2, pp. 59-64, 2017.

[7] Z. Wang, H. D. Zhao, M. X. Zhu, X. Zhou, X. P. Xie, H. F. Xie, and B. P. Zang, "Application of power wireless private network in ubiquitous power internet of things," Electric power, vol. 52, pp. 27-38, 2019.

[8] S. L. Lv, J. Ding, H. B. Zhang, D. D. Mei, and Y. Hou, "Research and thinking on the standard system of new power system smart IoT perception technology," Electric power ICT, vol. 19, pp. 39-46, 2021.

[9] Y. Q. Lei, J. Tong, S. H. Zhang, L. R. Wang, and W. Q. Lei, "Research on technical standard system of energy internet sensing layer," Distribution and utilization, vol. 38, pp. 14-20+33, 2021.

Optimization of minimum crossing distance of transmission line based on multi objective ant colony algorithm

Qiaoyun Xu,[1] Ji Tian[2]*, Zhenpeng Tang[3], Ronghuan Mai[4], Yuancheng Qin[5]

[1,3,4]Jiangmen Electric Power Bureau, Guangdong Power Grid Company, Jiangmen, China

[2,5]School of Automation, Wuhan University of Technology, Wuhan, China

* Corresponding author: tianji@whut.edu.cn

Abstract—In the test of cross span distance of transmission line, the estimation of cross span distance is not accurate, which leads to the increase of subsequent test error. Therefore, a test method of transmission line cross span distance based on a multi-objective ant colony algorithm is proposed. Statistics of the actual information of different crossovers including trees and forest areas. Considering the transmission line as a flexible cable, the objective function for estimating the cross span distance of the transmission line is constructed according to the catenary model of the flexible cable model. On this basis, the multi-objective ant colony algorithm is used to solve the function, and the optimization results of transmission line crossing distance are output. The experimental results show that the method has high accuracy, small error, short time and reliability.

1. INTRODUCTION

With the continuous expansion of the construction scale of long-distance, large-capacity and ultra-high-voltage transmission lines, transmission corridors are becoming more and more tense, leading to the phenomenon of crossing over many overhead lines, and also causing line discharge accidents due to too small crossing distances. .In order to reduce the occurrence of accidents and reduce the cost of construction, the crossing distance of transmission lines has become the most effective means to solve these problems. Therefore, in order to achieve more precise results, a cross-over distance test is required [1]. Cross-crossing distance test of transmission lines refers to the test of the intersection distance between transmission lines and other lines such as roads, railways, cableways, pipelines, power lines and communication cables. Alternatively, the distances of transmission lines through rivers, mountains, forests and buildings are tested to avoid the occurrence of flashover or interference from the electromagnetic network, resulting in safety incidents [2].

Great attention has been paid to the research of transmission line cross the over distance tests at home and abroad, and fruitful research results have been obtained. The research of transmission line cross over distance test in foreign countries focuses on the selection of measuring points and estimation of sag variation. In our country, the research on transmission line crossover distance test is relatively late, and the emphasis is put on the crossover distance test in special environments. Based on the research at home and abroad, some scholars have put forward a method of measuring the crossover distance of transmission lines based on stereo vision measurement. Based on the principle of binocular ranging, the geometric measurement model of transmission line crossing is established to realize the test of transmission line crossing distance [3]. Some scholars have also proposed a lidar -based transmission line cross distance test method, this method is mainly based on lidar cross distance automatic detection system. In the process of system design, the working principle and composition of the system are described, the test results are pretreated and optimized, and the final test results are obtained [4]. When these two methods are used to test the cross over distance of transmission lines, the objective function of transmission lines can not be constructed because of the complex environment, so the test error is large. So this paper introduces a multi-objective ant colony algorithm, which has many advantages, such as fast searching speed and high precision, to solve the objective function of transmission line crossover distance estimation, so as to obtain the optimal solution of transmission line crossover distance test.

2. Test Method Design of Transmission Line Crossing Distance

2.1. Statistics of actual information about crossover objects

First of all, the actual information of different crossover objects is counted, including eight aspects, as shown in Table 1.

TABLE 1. STATISTICAL CONTENT OF DIFFERENT CROSSOVERS

Serial number	Crossover runner	Statistical content
1	Trees and forest area	The spatial coordinates of tree age, tree height, tree species and distance between poles and towers with low poles were calculated.
2	Buildings (houses, etc.)	The spatial coordinates of building materials, building quantity and distance of low hanging point pole and tower are calculated.
3	Aqueduct, ropeway and other facilities	Statistics facilities materials, access, name, quantity, low suspension point pole and tower near the space coordinates of facilities.
4	Underground pipelines (oil and gas pipelines, etc.)	Statistics transportation materials, pipeline materials, pipeline access, pipeline name, low suspension point near the space coordinates of underground pipelines.
5	Railway and highway	Calculate the actual coordinates of the intersection, the railway and highway mileage, the actual route names of the railway or highway at the intersection.
6	Lakes and rivers	Calculate horizontal elevation, distances across, and names of lakes or river courses.
7	Weak wire	Statistical intersection point corresponding space coordinates, wire material, line name.
8	Transmission high voltage line	The actual coordinates of the intersection point, the actual coordinates of the hanging point, the corresponding coordinates of the four poles and towers, the pole and tower model, the projection angle of the line, the wire material, the voltage class of power transmission, and the name of the line.

In statistical table content, when the transmission line suspensions of equal height phenomenon, any one of the towers on the coordinates to calculate [5].

2.2. Construct objective function

The transmission line is regarded as a flexible cable and the catenary model in the conductor flexible cable model is used as the objective function of the transmission line [6]. Firstly, the transmission line is analyzed, and the transmission line is separated, as shown in Figure 1.

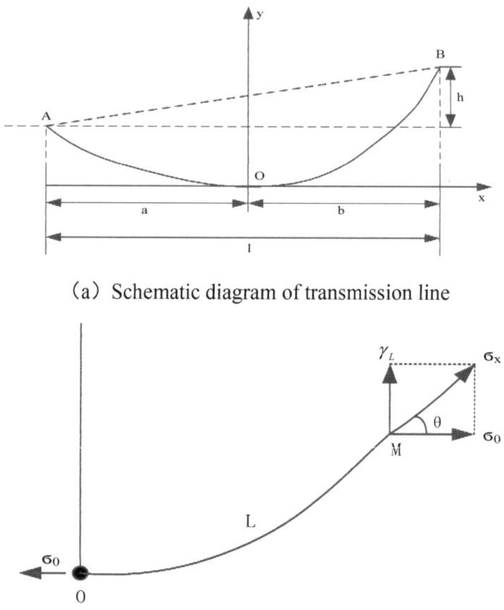

(a) Schematic diagram of transmission line

(b) Schematic diagram of transmission line separation

Figure 1. Schematic diagram of transmission lines and their separation

In Figure 1 (a), overhead wires are suspended between points A and B, where A is the lower end and B is the upper end, the wire gap is h, O is the lowest, A is a, and B is b [7].

Assuming that the load γ acting on the overhead conductor is uniformly distributed, the corresponding horizontal component σ_0 of the actual stress at each point on the transmission line is equal on the basis of the force balance condition [8]. In order to facilitate the analysis of the forces on transmission lines, one of the conductors needs to be separated, as shown in Figure 1 (b). The length of separated conductor is expressed by L, the corresponding load of transmission line is expressed by γ_L, the corresponding axial stress of point M is expressed by σ_x, and the angle formed by horizontal is expressed by θ. Based on the actual conditions of the transmission line, the following equations exist:

$$\tan\theta = \frac{dy}{dx} = \frac{\gamma}{\sigma_0}L \qquad (1)$$

In equation (1): x and y represent the horizontal and vertical axes of the coordinate system with O as the origin. Therefore, the equation is also the corresponding differential equation of the catenary, which is treated differentially again as shown in the following equations:

$$d\left(\tan\theta\right) = d\left(\frac{\gamma}{\sigma_0}L\right) = \frac{\gamma}{\sigma_0}\left(\sqrt{1+\tan^2\theta}\right) \qquad (2)$$

When the equation (2) is shifted and integrated, the following equations are obtained:

$$\frac{dy}{dx} = sh\frac{\gamma}{\sigma_0}\left(x+C_1\right) \qquad (3)$$

In equation (3): s for the number of shifts, C_1 for the integral threshold [9].

The corresponding catenary point equation of transmission line is obtained according to equation (3) and boundary conditions. Boundary conditions such as equations (4). The corresponding catenary point equation of transmission line is concrete as the equation (5).

$$\begin{cases} x=0 \\ y=0 \end{cases} \qquad (4)$$

$$y=\frac{\sigma_0}{\gamma}\left(ch\frac{\gamma x}{\sigma_0}-1\right) \quad (5)$$

In equations (4) and (5): c represents the boundary threshold.

According to equation (5), the relationship between a and b can be analyzed as follows:

$$\begin{cases} a+b=1 \\ y(b)-y(-a)=h \end{cases} \quad (6)$$

$$\begin{cases} a=\dfrac{1}{2}-\dfrac{\sigma_0}{\gamma}sh^{-1}\dfrac{\gamma h}{sh\dfrac{\gamma L}{2\sigma_0}} \\ b=\dfrac{1}{2}+\dfrac{\sigma_0}{\gamma}sh^{-1}\dfrac{\gamma h}{sh\dfrac{\gamma L}{2\sigma_0}} \end{cases} \quad (7)$$

Based on equation (6) ~ (7), an objective function for estimating the crossover distance of transmission lines is constructed, as shown in equation (8):

$$H=y(a)=\frac{\sigma_0}{\gamma}\left\{ch\frac{\gamma L}{2\sigma_0}\left[1+\left(\frac{\gamma h}{sh\dfrac{\gamma L}{2\sigma_0}}\right)^2\right]^{\frac{1}{2}}-1\right\}-\frac{h}{2} \quad (8)$$

In equation (8): H represents the height from point A to the bottom of the wire [10]. The optimization criteria and constraints of this function are all the optimization of transmission line crossover distance.

2.3. Transmission line crossover distance test

Based on the statistical information of different crossover objects and the constructed objective function [11], the multi-objective ant colony algorithm is used to solve the function iteratively, and the optimal solution of the function is output to obtain the optimized results of the crossover distance test of transmission lines.

At the initial time, n nodes shall be set up on the lines where the power transmission lines intersect with highways, railways, cableways, pipelines, power lines and communication cables, as well as the lines formed by the power transmission lines crossing rivers, mountains, forests and buildings. And put m ants on the nodes randomly.

Each ant chooses the next mobile node according to the state transition rule. During this process, the ant chooses a higher pheromone concentration and a shorter route [12]. Each ant is equipped with a search tabu list, will not go to the node they have been to. In each cycle, local pheromone updates are performed on the ant's path. In each round, all the ants looped once, then recorded the shortest path, and enhanced the left pheromone on the shortest path according to the global pheromone updating rules. Iterations are then repeated until the transmission line crossover distance is obtained [13].

The specific test procedures are as follows:

Step 1: Ant Colony $A(t)$ Initialization: Set the relevant parameters as shown in Table 2.

TABLE 2. PARAMETER SETTINGS AND THEIR MEANINGS

Serial number	Parameter	Meaning
1	m	Number of ants
2	n	Number of nodes
3	nc	Cycles
4	$nc\_\max$	Maximum number of cycles
5	$Alpha$	Heuristic pheromone factor
6	$Beta$	Heuristic expectation factor
7	ρ	Volatile Pheromone Factor
8	Q	The global pheromone update rule enhances the strength of the pheromone left on the shortest path
9	ξ	Local pheromone volatilization factor
10	q	Pseudorandom factor

Step 2: Random Distribution: Randomly place m ants on n nodes [14].

Step 3: Status update

(1) Keep updating the local pheromone until all the ants have constructed a complete path.

(2) Global update pheromone.

The status update rules to be followed are as follows:

For each ant k, the path R^k memory vector records the specific serial numbers of all the passing nodes of the ant according to the visit order [15]. Given that it is currently on the i node, the next node it will visit is j, as shown in equation (9):

$$j=\begin{cases}\arg\max j\in_{J_k(i)}\left\{\left[\tau(i,j)\right]^{Alpha}\left[\eta(i,j)\right]^{Beta}\right\} & \begin{matrix} if \\ q\le q0\end{matrix} \\ S \quad others & \end{cases} \quad (9)$$

In equation (9): $J_k(i)$ represents the set of path R^k memory vectors; $\tau(i,j)$ represents the amount of pheromone existing on edge (i,j); $\eta(i,j)$ represents heuristic information; and S represents random variable generated by probability distribution.

The formula for updating pheromones locally is as follows:

$$\tau(i,j)=Q(1-\xi)\cdot\tau(i,j)+\xi\cdot\tau_0 \quad (10)$$

In equation (10): τ_0 represents the actual initial value of the pheromone.

The equations used to update pheromones globally are as follows:

$$\tau(i,j)=(1-\rho)\cdot\tau(i,j)+\rho\cdot\Delta\tau_k(i,j),\forall(i,j)\in R^k \quad (11)$$

In equation (11): $\Delta\tau_k(i,j)$ represents the global range of pheromone values [18].

Step 3: Terminate: When the termination condition is met, the test is terminated. When the termination condition is not met, then $nc=nc+1$, go to the third step to loop again. The termination condition is the maximum number of times $nc\_\max$ in the loop to be satisfied.

Step 4: The result is output.

The process of multi-objective ant colony algorithm is shown in Figure 2.

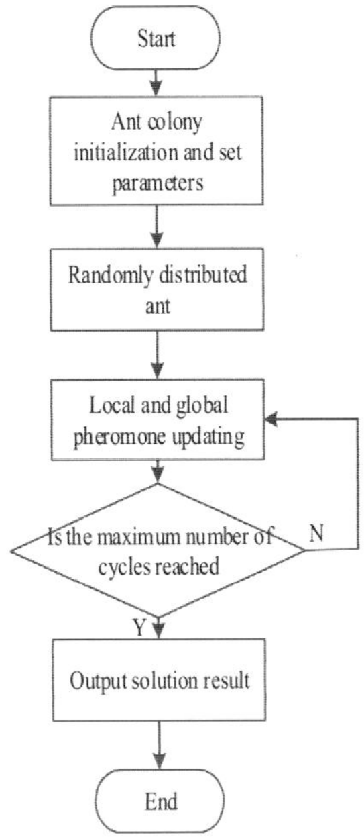

Figure 2. Multiobjective ant colony algorithm process

3. EXPERIMENTAL VERIFICATION

3.1.Experimental method

In order to verify the performance of the multi-objective ant colony algorithm based transmission line crossover distance test method designed in this paper, experimental verification is carried out. The transmission lines in the experiment are two parallel single-loop transmission lines, ZB401 and ZMC1-3A1 respectively, which are shown in Figure 3.

(a)ZB401 (b)ZMC1-3A1

Figure 3. Tower type of two transmission lines

Among them, ZB401 is the use of steel-cored aluminum wire six split conductor, the specific model for the LGJ X6-50/400, ground for a 120-OPGW cable, the other is 100-GJ stranded steel wire. ZMC1-3A1 uses 2×425-JLHA3 type two split aluminium alloy stranded wire, one of which is 120-OPGW type optical fiber cable, the other is 100-GJ type steel stranded wire.

The mechanical characteristic parameters of the experimental transmission lines are shown in Table 3.

TABLE 3. MECHANICAL CHARACTERISTIC PARAMETERS OF EXPERIMENTAL TRANSMISSION LINES

Project	Earth wire: 120-OPGW type	Earth wire: 100-GJ type	Wire: LGJ X6 - 50/400 type	Wire: 2×425-JLHA3 type	Unit
Outside diameter	16.0	13.0	30.0	27.3	mm
Cross section	145.67	101.98	530.86	450.44	mm^2
Subconductor actual spacing	-	-	400	400	mm
Mass per unit length	0.948	0.810	1.677	1.522	kg/m^{-1}
Tensile Strength	174.3	116.5	122.9	118.3	kN
Coefficient of elasticity	$1.54X10^5$	$1.82X10^5$	$6.6X10^4$	$7.0X10^4$	N/mm^2
Expansivity	$1.3X10^{-5}$	$1.2X10^{-5}$	$2.1X10^{-5}$	$2.0X10^{-5}$	$℃^{-1}$
Actual strength safety factor	3.6	3.1	2.4	2.4	-

Since the baseline distance is the range within 15m-35m where the test errors are most likely to occur, experimental tests are performed within a range.

The parameters of multi-objective ant colony algorithm used in this paper are $m = 32$, $n = 14$, $Q = 100$, and the number of iterations is 500.

In order to avoid the experimental results being too single and lack of comparison, the above two methods are used as the comparison method. The two methods are based on stereo vision measurement and based on lidar, which are used in the test of transmission line crossover distance.

(1) Distance estimation accuracy

In order to test the performance of the three methods, the accuracy of the distance estimation was tested. The formula for calculating the accuracy of the distance estimation is as follows:

$$W = \frac{H}{H_{actual}} \times 100\% \qquad (12)$$

In equation (12): H_{actual} refers to the measured value of the crossing distance of transmission lines.

The comparison of the distance estimation accuracy of the three methods is shown in Figure 4.

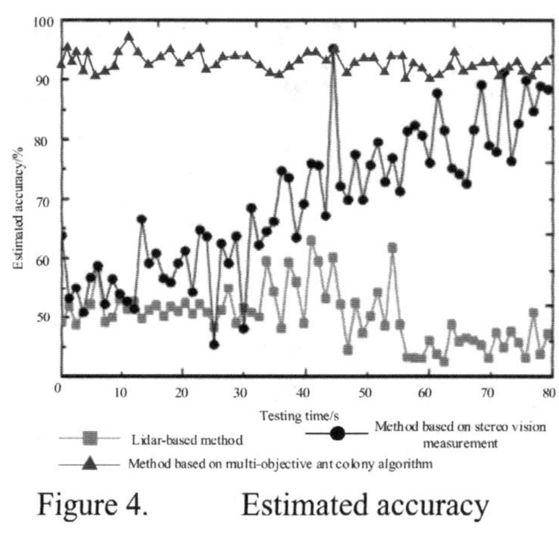

Figure 4. Estimated accuracy

Analysis of Fig. 4 shows that the accuracy of range estimation based on stereo vision method is between 45% and 95%, and that based on lidar method is between 43% and 63%. Compared with the experimental results, the estimation accuracy of the multi-objective ant colony algorithm for transmission line crossover distance test is maintained at more than 89%. It is shown that the proposed method can obtain accurate estimation results of cross over distance of transmission lines, which lays a solid foundation for future research.

(2) Measurement error

In the 15m-35m baseline distance range, the measurement errors of the three transmission line cross-distance test methods are shown in Table 4.

TABLE 4. TEST ERROR COMPARISON

Baseline distance (m)	Test error (m)		
	Method based on multi-objective ant colony algorithm	Method based on stereo vision measurement	Lidar-based method
15	0.21	0.85	0.46
16	0.15	0.56	0.42
17	0.18	0.55	0.47
18	0.17	0.57	0.51
19	0.18	0.58	0.39
20	0.37	0.51	0.48
21	0.14	0.50	0.54
22	0.17	0.60	0.54
23	0.09	0.64	0.56
24	0.12	0.64	0.48
25	0.11	0.56	0.48
26	0.10	0.57	0.50
27	0.14	0.58	0.60
28	0.17	0.58	0.48
29	0.09	0.51	0.54
30	0.05	0.50	0.48
31	0.18	0.54	0.52
32	0.37	0.55	0.44
33	0.16	0.55	0.50
34	0.11	0.57	0.54
35	0.17	0.55	0.46

According to the test error comparison data in Table 4, within the baseline distance of 15M-35M, the maximum error based on stereo vision measurement method is 0.85m, and the minimum error is 0.5m. The maximum error of lidar method is 0.6m and the minimum error is 0.42m. The maximum test error of the method based on multi-objective ant colony algorithm is 0.37m, and the minimum is 0.05m. The test error of this method is lower than that of experimental comparison method, which verifies the superiority of this method.

(3) Testing time

On the basis of the above experiments, the test time of three methods is compared, and the results are shown in Figure 5.

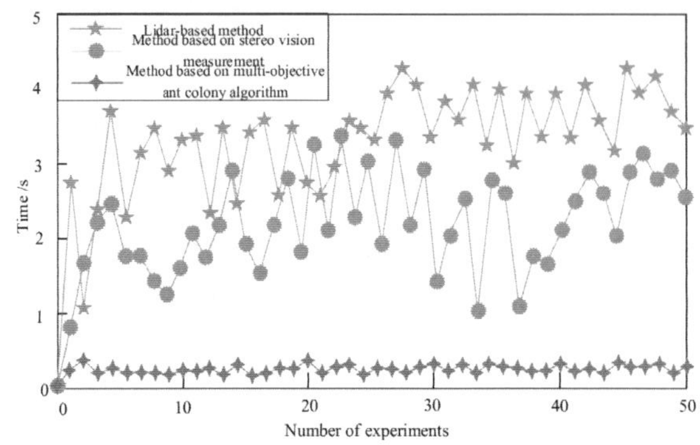

Figure 5. Test time comparison

Analysis of Figure 5 shows that the test time based on stereo vision measurement varies between 0 and 4.2s. The test time based on the lidar method varies from 0-3.3s. However, the test time of the method based on multi-objective ant colony algorithm is always less than 0.5 s, which shows that the method is short and can complete the test in a short time.

4. Conclusion

The transmission line is the channel of electric energy transmission, is an essential part of the grid, its reliability is directly related to the smooth development of national economy. However, the safe operation of transmission line is affected by crossing for a long time, and the distance between transmission line and crossover will change when strong wind, subsidence, tree growth, temperature change, wire aging and other factors. When the spacing is too small, a discharge will occur between the conductor and the crossover, which will lead to the accident. With the development of the power industry, it is more and more important for line operators to choose a reasonable method to estimate the crossing distance of overhead transmission lines. In order to improve the security of transmission lines, a cross-over distance testing method based on multi-objective ant colony algorithm is designed. The multi-objective ant colony algorithm is used to solve the model for several times, and the results of transmission line crossover distance test are given. Experimental results show that the proposed method has high precision, small error and short test time, and can effectively avoid the occurrence of line discharge accidents.

Acknowledgment

The study was supported by "Guangdong Power Grid Co., Ltd. (Grant Number: 030700KK52180142)".

References

[1] Zhang G X, Mu J L, Ma H, et al. Design of cross-span distance online monitoring system for power transmission lines[J]. Modern Electronics Technique, 2018, 41(24):14-17.
[2] Huo F, Lu W, Huang D C, et al. Study on Electric Field Distribution Characteristics of Live-Working and Protection for Crossing Project of 1000kV-Over-500kV AC Transmission Lines[J]. Power System Technology, 2019, 43(1):386-392.
[3] Sun S N, Lv Y X, Wang S, et al. Measurement of Space Crossing Transmission Line Based on Binocular Ranging[J]. High Voltage Engineering, 2018, 44(11):3657-3663.
[4] Tang C J, Xu J, Liu T Y, et al. The automatic lidar-based detecting system for crossing distances between overhead transmission lines[J]. Machine Design And Manufacturing Engineering, 2018, 47(11):68-72.
[5] Gong G S. Design of Automatic Detection System for Crossover Distance of Overhead Transmission Line[J]. Journal of Chifeng University, 2019, 35(05):71-73.

[6] Lv J S, Li J, Luo C. Influence of Uneven Ice on Distance to Ground and Crossed Crossing Distance of UHV Transmission Lines in Heavy Icing Area[J]. Shaanxi Electric Power, 2017, 45(03):35-38.
[7] Xiong J. Construction Technology for Crossover and Live Crossover of 220kV Transmission Line[J]. Science & Technology Information, 2018, 16(34):110-111.
[8] Zhang Y C, He Z H. Analysis of velocity field of high-speed railway transmission line based on Fluent[J]. Electrical Engineering, 2019, 20(05):5-9.
[9] Wang D J, Zhu X L, Yin M, et al. Research on measurement method of space electric field in cross transmission area of AC transmission lines[J]. Electronic Design Engineering, 2019, 27(20):79-83.
[10] Song F G, Lin H. Analysis of space electric field intensity of transmission lines in the crossing area[J]. Electric Machines and Control, 2017, 21(08):72-77.
[11] Yao Y, Fu W Z, He W L, et al. Study of Radio Interference Characteristics of Multi-turn EHV/UHV AC Transmission Lines in Crossing Erection[J]. Water Resources and Power, 2018, 36(06):201-204,170.
[12] Wang D L, Lu T B, Cui X, et al. Simulation of Total Electric Field under the Crossing of Two Circuit HVDC Transmission Lines[J]. Transactions of China Electrotechnical Society, 2017, 32(2):77-84.
[13] Fu W Z, Jin M, Bao H, et al. Study on Power Frequency Electrical Field Characteristics of Multi-turn UHV/EHV Transmission Lines in Parallel and Crossing Erection[J].
[14] Water Resources and Power, 2018, 36(05):201-204,162.
[15] Tang X, Liang Y S, Du W L, et al. Short Circuit Current Calculation Method for Crossing Fault of Two Transmission Lines with Mutual Inductance[J]. Southern Power System Technology, 2017, 11(09):41-47.
[16] Yao W Q, Yi B S, Hu F. Monitoring System for Cross Frame State and Its Wireless Data Transmission[J]. Southern Power System Technology, 2019, 13(1):48-52.

A Short Term Forecasting Method for Regional Power Consumption Considering Related Factors

Chang Liu*, Yuanliang Zhang, Weisong Chen, Haitong Gu, Hui Li, Shaoliang Chen

Metrology Centre of Guangzhou Power Supply Bureau, Guangdong Power Grid Co.,Ltd. Guangzhou, China

* Corresponding author: 18202710433@139.com

Abstract—Analysis and prediction of power consumption law is the basis of power grid planning and construction, and is also an effective guide for energy demand side management. With the rapid development of economy and the complex change of industrial structure in recent years, the internal structure of power demand is changing to some extent. Therefore, a short-term forecasting method of regional electricity consumption considering the related factors is proposed. Based on the analysis results, a short-term prediction model of regional electricity consumption considering the related factors is established, and the short-term prediction is realized by the calculation of the model. Through the example analysis, it is verified that the forecasting deviation of the short-term forecasting method is low and meets the basic requirements of electric quantity forecasting.

1. Introduction

The medium and long term load forecasting is mainly aimed at the load demand of monthly time scale and above. In recent years, China's reform and opening up into the deep-water zone, the domestic economic structure is undergoing profound adjustment, load demand changes and economic and social development. At the same time, with the advent of the information age, the statistical accuracy of data, the frequency of publication, and the degree of informatization continue to increase [1-3]. Detailed industry and business climate indexes provide sufficient data sources and data foundations for medium and long-term load forecasting.

In fact, the development of load demand can be divided into trend part, cycle part and other parts. Based on the monthly electricity consumption of the whole society, the different components are extracted by the method of seasonal decomposition, combined with a variety of economic data, and targeted models are used to forecast the electricity consumption[4~6]. However, the development trend of electricity demand of the whole society is relatively clear, and the macro external environment is relatively stable, and the change of influencing factors is relatively small. But the factors affecting the maximum load are more various, among which the relationship between regional load and regional meteorological factors is close [7]. At the same time, there is a complex nonlinear relationship between the related factors and load in the medium and long term prediction. Most of the above literatures use local nonlinear models to fit. Firstly, aiming at the problem that it is difficult to distinguish the dominant factors of the maximum load [8]. This paper adopts the method of constructing the correlation coefficient matrix to select the influential factors with strong correlation with the maximum load as the input variables; secondly, aiming at the problem that the regression model may overfit, this paper adopts principal component analysis (PCA) to reduce the dimension of the sample data, so as to avoid the overfitting effect caused by the contradiction between the sample size and the type of influencing factors. Finally, aiming at the limitation of the local nonlinear model, this paper uses several nonlinear models to forecast, and analyzes the applicability of the model and the ability of data feature extraction.

2. INFLUENCING FACTORS AND ANALYSIS OF RELATED FACTORS

The definition of the Granger causal test in this case can be shown in Figure 1. Granger causality test was initially used to analyze the causality between economic variables. Later, with the convergence of various research fields, scholars gradually applied this method to the study of causality between economic variables and variables in other fields, such as the Granger causality test for electricity consumption and macroeconomics, as shown in Figure 1:

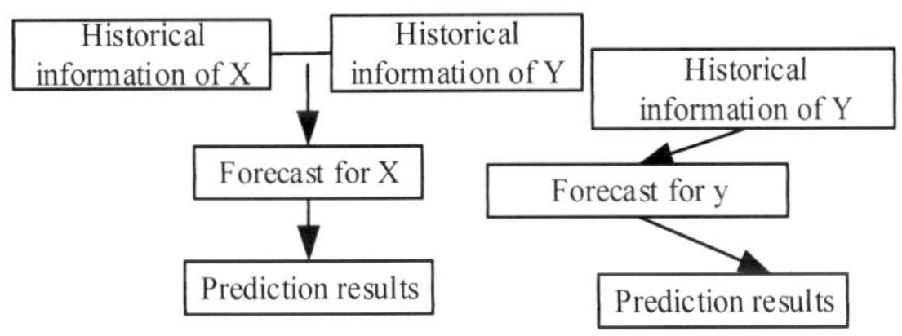

Figure 1. Granger causality test of electricity and macroeconomics

One of the prerequisites of Granger causality test for time series is that the series must be stationary, or false regression may occur. Therefore, the unit root test for the stationarity of each time series should be performed before performing the Granger causality test. The augmented Dicky-Fuller test (ADF unit root test) is a commonly used unit root test. The general steps of the ADF unit root test are as follows:

(1) Verifying the original time series;

(2) If the original time series fails to pass the test, that is, the original series is not stable, the smoothness shall be tested after the first order difference of the series;

(3) If the first-order difference sequence has not passed the test, the second-order difference transformation shall be carried out continuously, and the time sequence after the second-order difference is generally stationary.

3. SHORT-TERM FORECASTING MODEL OF REGIONAL ELECTRICITY CONSUMPTION CONSIDERING RELATED FACTORS

In this paper, the Lagrangian interpolation method is used to process the data. The shortest number of edges on the path connecting the two nodes v_i and v_j in the network represents the distance between the two nodes d, and d_{ij} represents the efficiency between the nodes v_i and v_j, and represents the speed of information transmission, and has $\varepsilon_{ij} = 1/d_{ij}$. $d_{ij} = 0$, $\varepsilon_{ij} = 0$ denotes a pathless connection between two nodes. The diameter of the network D is expressed in terms of the maximum distance between any two nodes of the network, namely:

$$D = \max d_{ij} \quad (1)$$

In formula (1) $1 < i < N$, N is the total number of network nodes, where $N = 340$. The diameter of the dynamic Bayesian network is calculated to be 14. The average distance between any two nodes represents the average path length L of the network:

$$L = \frac{1}{C_N^2} \sum_{j=1}^{n} d_{ij} \quad (2)$$

The diameter of the network can be understood as the maximum relational chain in the network. The average path length of the dynamic Bayesian network is 4.8439. The neighbor of a node in the network is defined as the node to which it is directly connected by an edge. The cluster coefficient C_i of node

v_i is defined as the number of E_i edges that actually exist between neighbouring nodes of κ_i in v_j divided by the total number of possible edges· In order to reduce the influence of each component forecast error on the overall forecast results, the relationship between each component is expressed in addition model:

$$C_i = \frac{L}{C^2_{\kappa i}} E_i \quad (3)$$

Among them, the triple connected with node i refers to the three nodes including node i, and there are at least two edges from node i to the other two nodes. As shown in Figure 2:

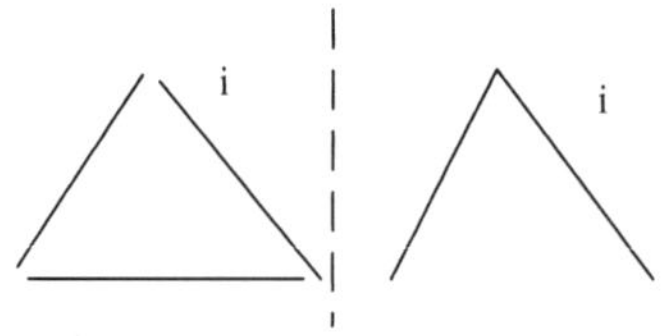

Figure 2. Two possible forms of triples with node i as one of the vertices

Spearman correlation coefficients indicate the correlation direction of data sets X and Y, where X is called an independent variable and Y is a dependent variable. If the value of Y tends to increase with the increase of X value, the correlation coefficient of Spearman is positive. if the value of Y tends to decrease with the increase of X value, the correlation coefficient of Spearman is negative. If the Spearman correlation coefficient is zero, it indicates that there is no tendency for the value of Y to change as the value of X increases. If a node of the network v is passed by the shortest path of many other non-adjacent nodes, then the node's role in the network is irreplaceable. The influence and importance of a node is expressed by its intersection number B:

$$B_i = \sum_{j=1}^{n} \left[n_{j,i}(i) / n_{j,i} \right] \quad (4)$$

In formula (4), the shortest number of paths between nodes v_i and v_j shall be expressed by n; the shortest number of paths between v_i and v_j passing through nodes v_i shall be expressed by n and (i), and the total number of nodes shall be expressed by N. In order to avoid the influence of overfitting caused by various data types and complex regression models on the prediction accuracy, the dimensionality of each component data should be reduced. The main cause analysis (PCA) method is used to deal with the factors that have passed the cointegration test. Firstly, Z-score is used to standardize the various factors, and then the covariance matrix is constructed. Secondly, the covariance matrix is solved by using the eigenvalue, and the degradation dimension is determined by using Cumulative Partial Variance (CPV). Finally, the new dimension reduction matrix is reconstructed by using the PCA dimensionality reduction algorithm.

4. XAMPLE TEST

In order to verify the reliability and validity of the annual electricity consumption forecasting model based on weighted partial least squares regression, this paper uses the annual social and economic development index of a certain area and the data of the whole society to model and forecast. Verify the practicability and reliability of the short-term prediction method of regional electricity consumption which is designed in this paper.

4.1.Case analysis data sources

The data source still uses the monthly data on electricity consumption, weather, economy and transportation in XX city. After processing the missing values of the input data using the hybrid information patching algorithm, the model is constructed using the improved DBN method introduced in 3.3. First, in the first stage of the reasoning process, a first-order dependent scoring matrix, Si (340 ×

340), is obtained. The element in S1 is the score of each edge of a directed acyclic graph, which is the inference value. The lower the score of DBN, the more significant the first order dependency is. The first five rows and the first five columns of the first step scoring matrix, Si, are listed here. V plus subscripts represent variable names, as shown in Table 1:

TABLE 1. ESTABLISHMENT OF SCORING MATRIX

	V1	V2	V3	V4	V5
V1	0.730247	0.999279	0.011232	0.007797	0.197536
V2	0.089954	0.000061	0.000134	0.254780	0.245780
V3	0.745219	0.384123	0.854216	0.001245	0.236980
V4	0.145863	0.256874	0.315522	0.245879	0.236587
V5	0.124580	0.264781	0.214587	0.236985	0.214586

Enter the time series data of the first step scoring matrix S and 340 variables, the given time window width = 0.1, if the number of time slices in the time window is still less than the time window width, the process is repeated after the evidence information of the next time slice is obtained. When the number of time slices is equal to the width of the time window, and the reasoning results are updated, then the reasoning process of the second stage is entered. The first five rows and the first five columns of the Sz scoring matrix are listed here. Since a lower score indicates a more significant causal relationship, an inference value greater than 0.1 is set to NA. As shown in Table 2:

TABLE 2. ESTABLISHMENT OF SCORING MATRIX2

	V1	V2	V3	V4	V5
V1	NA	NA	0.011232	NA	NA
V2	0.089954	0.000061	0.000134	NA	0.245780
V3	NA	NA	NA	NA	0.236980
V4	NA	NA	NA	NA	NA
V5	0.124580	NA	NA	NA	NA

Visualize the DBN complex network using the sna package in R l50-51]. From the scoring matrix S, we can get the directed edges of the network, which are sorted according to their scores from small to large. The first column is the predictive variable, the second is the explanatory variable, and the third is the corresponding edge score. There are 650 dependencies. The adjacency matrix A (340 × 340) can be constructed from the directed edge case of a network, and the element a in the matrix is 0 or 1. A, = 1 means that the j variable is linearly dependent on the i variable; a = 0 means that the j variable is not linearly dependent on the i variable. Only the first five rows and the first five columns of the network adjacency matrix are listed here, as shown in Table 3:

TABLE 3. ADJACENCY MATRIX OF NETWORK

	V1	V2	V3	V4	V5
V1	0	0	1	0	0
V2	1	1	1	0	1
V3	0	0	0	0	1
V4	0	0	0	0	0
V5	1	0	0	0	0

The proportion of all the shortest paths that pass through the node in the network is the number of v_i intervals. List the top 10 variables in number of mediations, as shown in Table 4:

TABLE 4. TOP TEN INTERMEDIARIES

Serial number	Variable name	Medium number
1	Domestic electricity consumption of urban and rural residents	2635.1
2	Electricity consumption of secondary industry	2041.78
3	Mean atmospheric pressure	1925.41
4	Electricity consumption of the whole society	1360.24
5	Electricity consumption in textile	1266.32

	industry and light textile industry	
6	Electricity consumption in construction industry.	1118.56
7	Electricity consumption in smelting nonferrous metals	1059.42
8	Total electricity consumption of physical stores	941.23
9	Total electricity consumption of gas industry	841.23
10	Electricity for animal husbandry.	814.52

For the data in Table 4, there is a complex non-linear relationship between load density and load density because of the close relationship between load density and load density. If the input sample data of the prediction model is not properly selected, the prediction accuracy of the model will be greatly reduced. Therefore, it is necessary to introduce the appropriate sample at the beginning of the prediction model. In this section, based on the spatial load forecasting of LSSVM load density method, grey relational analysis is introduced to improve the forecasting precision and speed of LSSVM.

4.2.Application results

The steps of selecting LSSVM model training samples with grey incidence degree are: 1) nondimensionality of the sample data of original influencing factors; 2) difference sequence; 3) maximum and minimum difference between two poles; 4) correlation coefficient; 5) calculation of incidence degree. Taking the load density prediction of a certain residential area as an example, the validity of the model proposed in this paper is analyzed and verified. A large sample of residential load and its influencing factors is collected as shown in Table 4.1. Among them: A, population density; A, per capita income; A, per capita electricity; As, per capita coal-electricity price ratio growth rate; D, load density. As shown in Table 5:

TABLE 5. SAMPLE DATA

Serial number	A_1/(person/km^2)	A_2/（yuan）	A_3/（Kw h）	A_4/（%）	D/(Kw h)
1	30651	1325	519	1.2547	11.2364
2	16150	4837.42	1033	1.0635	27.62
3	29410	1122	537	1.677	10.3
4	25413	1000.3	834	-1.313	11.58
5	16600	1222.7	1019	2.365	2.54
6	15855	2608.4	628	1.6587	1.117

Select these four variables as explanatory variables to forecast the 12 months' power consumption of Shanghai in 2020. The gap between the actual and predicted values of the method designed in this article is shown in Figure 3:

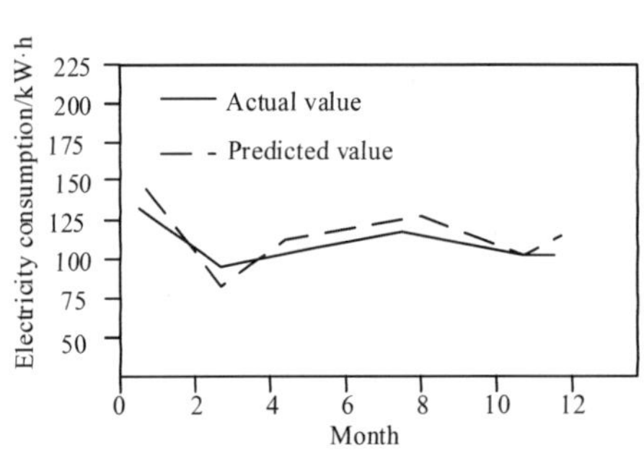

Figure 3. Prediction results

From the Figure 3, we can draw the conclusion that the improved DBN model proposed in this paper is more effective than the other models. The classical model VAR is easy to calculate, but the error is also the largest among all the models. It shows that the prediction of 12 time points has exceeded the range of its accuracy. BP network does not have memory, at the same time only consider the current time point data, and do not consider the characteristics of the time series before and after the connection, so the error is second only to the VAR model, is also very large. The traditional DBN combines the causal relationship and the Markov model to describe the probability transfer of time series, so the error is obviously reduced. But it is a Bayesian network based on steady-state assumption. Elastic network and Granger causality test model analyze time sequence before causality, but DBN model can not analyze time sequence and causality at the same time. First, the output data is processed by using mixed information patching algorithm, then the improved DBN model is built by using data analysis software R. By relaxing the traditional DBN's assumption of time homogeneity, the disadvantage that the traditional DBN can not model the unsteady process is solved. Finally, the monthly data of electricity consumption, weather, economy and transportation in Shanghai from 2011 to 2017 are used to forecast the electricity consumption of the whole society for 12 months in 2020. Compared with VAR (15.1%), BP (10.8%), traditional DBN (7.37%) and elastic network model (3.07%), the improved DBN model proposed in this paper is superior to other algorithms in the prediction accuracy of monthly electricity consumption (1.63%).

5. Conclusion

In this paper, a short-term forecasting method of regional electricity consumption considering related factors is proposed. The main conclusions are as follows: (1) In this paper, a monthly data set of 340 variables including electricity consumption, meteorology, economy and transportation, covering 96 time points from 2011 to 2020 is established. (2) A prediction method for monthly electricity consumption of elastic network model is proposed. After processing the missing value of the input data by multiple interpolation method, the paper uses elastic network and Granger to carry out two-step factor screening so as to achieve the goal of dimension reduction. (3) An improved dynamic Bayesian network forecasting method for monthly electricity consumption is proposed to further improve the forecasting accuracy. By relaxing the traditional assumption of time homogeneity in dynamic Bayesian networks, the unsteady process can be modeled, and the algorithm flow is designed. At the same time, the complex network method is used to analyze the dynamic Bayesian network. Compared with VAR (15.1%), BP (10.8%), traditional DBN (7.37%) and elastic network model (3.07%), the improved DBN model is superior to other algorithms in prediction accuracy (1.63%), and the advanced property of the improved DBN is verified by actual data.

Acknowledgment

The study was supported by “Science and Technology Project of Guangzhou Power Supply Bureau, China (Grant No. GZHKJXM20200036)”.

References

[1] Aladin S , Tran A , Allogba S , et al. （2020） Quality of Transmission Estimation and Short-Term Performance Forecast of Lightpaths. Journal of Lightwave Technology, , PP(99):1-1.
[2] Hong D Y , Ji T Y , Li M S , et al. （2019） Ultra-short-term forecast of wind speed and wind power based on morphological high frequency filter and double similarity search algorithm. International Journal of Electrical Power & Energy Systems, 2019, 104(JAN.):868-879.
[3] Kang H , Lee K , Shin D H . （2019） Short-term Forecast Model of Apartment Jeonse Prices using Search Frequencies of News Article Keywords. KSCE Journal of Civil Engineering, 23(12):4984-4991.
[4] Roushangar K , Alizadeh F . （2019） Scenario-based prediction of short-term river stage-discharge process using wavelet-EEMD-based relevance vector machine. Journal of Hydroinformatics, 21(1-2):56-76.
[5] Allogba S , Aladin S , Tremblay C . （2021） Multivariate Machine Learning Models for Short-Term Forecast of Lightpath Performance. Journal of Lightwave Technology, PP(99):1-1.

[6] Krylov S S , Perepelkin V V , Soe W Y . (2020) Short-Term Forecast of the Earth's Pole Motion, Taking Into Account Lunar Disturbances. Mechanics of Solids, 55(6):892-897.
[7] Tang Q , Yang M , Yang Y . (2019) ST-LSTM: A Deep Learning Approach Combined Spatio-Temporal Features for Short-Term Forecast in Rail Transit. Journal of Advanced Transportation, 2019:1-8.
[8] Kka C , Aba C , Mgb C . (2021) Analysis of short-term operational forecast deviations and controllability of utility-scale photovoltaic plants. Renewable Energy, 167:343-358.

Forecasting Method of Energy Demand of Integrated Energy System Considering Seasonal Catastrophe

Qingsong Hua[1], Qiang Li[2]*, Shengyu Gao[3], Yongqing Liu[4], Hong Zhu[5], Zhengyi Zhu[6], Qilin Shuai[7]

[1,7]Beijing Normal University, Beijing, China

[2,4]State Grid Information and Telecommunication Group Co.,Ltd. Beijing, China

[3,5,6]State Grid Nanjing Power Supply Company, Nanjing, China

* Corresponding author: qiang-li@sgitg.sgcc.com.cn

Abstract—The change of season will cause a variety of factors affecting energy demand to change, resulting in severe fluctuations in energy demand. Accurate prediction is of great value for energy management. Therefore, a prediction method of energy demand of integrated energy system considering seasonal mutation is proposed. Based on the analysis of the basic concepts, attributes and influencing factors of energy and energy demand, a seasonal energy demand impact decomposition model is constructed by lmdi-i to decompose the impact of season on energy demand. By dividing the energy indicators of the integrated energy system, a fuzzy neural network with error output and correction mechanism is established to predict energy demand. The test results show that the maximum relative error of the prediction results of the design method is 5.89%, the minimum relative error is 1.03%, the average absolute error is 3.21%, the root mean square error is 0.019, and the hill inequality coefficient is 0.020. Are better than the comparison method.

1. Introduction

One of the necessary supporting conditions for rapid economic growth is the rapid growth of energy use. Therefore, whether China's energy industry can support the rapid growth of China's economy in the future has become a matter of concern at home and abroad. Energy problem has become one of the main factors restricting economic and social development. According to the analysis, the national economic loss caused by energy shortage is about 20-60 times the value of energy itself. It can be seen that energy plays a special and important role in the process of national economic development [1-3]. Therefore, doing a good job in the future energy strategy and energy demand prediction to provide a scientific basis for energy planning and policy-making is of great practical significance for China to fully realize the strategic goal of building a well-off society and the grand task of building a conservation and harmonious society. Due to the limited data, the research object of this paper is the total energy demand in China, but the model established is also applicable to the energy demand prediction of energy users at all levels. Future energy competition will be an important aspect of enterprise competition [4-6]. Therefore, for energy users at all levels, the improvement of energy demand prediction and prediction accuracy can bring direct economic benefits. The joint study of 16 U.S. household energy units shows that the power demand forecast can help to arrange the production plan, and for every 1% increase in the relative error of the average absolute value of the forecast results, it can save us $143 per megawatt peak every year. Therefore, it is of direct economic significance to establish an energy demand forecasting model with higher accuracy than the previous methods. Theoretically, energy demand forecasting is an eternal research topic. Because the energy demand system is a nonlinear complex system, social progress, economic development, environmental change and macro policies will directly affect the operation of the energy system [7-8]. On the contrary, energy demand also directly affects economic structure, social development and environmental change. A lot of work has been done on energy demand forecasting at home and abroad. There are many commonly used energy demand forecasting methods, mainly including scenario analysis method, operations research planning optimization technology, elastic coefficient method, static or dynamic input-output method, linear time series method, nonlinear / chaotic time series method, grey theory, artificial neural network method and genetic algorithm, However, in terms of prediction accuracy, no generally applicable method has been accepted [9-11]. The traditional linear prediction method has clear economic significance, but it is difficult to accurately describe the nonlinear relationship in the energy system; Among the emerging nonlinear methods, chaotic time series method can only do short-term prediction of energy demand, because the premise of applying chaotic time series method is that the research object has chaotic characteristics. According to the sensitive characteristics of initial conditions of chaotic system, its long-term prediction is not feasible [12].

In order to realize the accurate prediction of energy demand, this paper designs an energy demand prediction method of integrated energy system from the perspective of seasonal change, analyzes the impact of season on energy demand, realizes its prediction, and verifies the effectiveness of the design method through testing. Through this study, in order to provide help for the practical development of related fields.

2. BASIC CONCEPTS OF ENERGY AND ENERGY DEMAND

2.1. Concept and attribute of energy

Energy is a natural resource that can be used by human beings to obtain energy. It mainly includes solar energy, wind energy, hydro energy, biomass energy, geothermal energy, marine energy, nuclear energy, fossil fuels, natural gas, etc. The energy referred to in this paper includes China's energy products, including raw coal and its products, crude oil and its products, natural gas and electric power, including thermal power (coal and gas power), hydropower and nuclear power. Strictly speaking, energy demand and energy consumption are two different concepts [13-15]. Energy consumption refers to the utilization or use of energy by households or manufacturers, including the domestic energy consumption of individuals and families; Energy input in industry, agriculture, service industry, transportation, etc; Conversion between various energy forms and energy loss. It belongs to statistical significance, not economic significance. Energy demand refers to the quantity that families or manufacturers are willing and able to buy energy products at a certain price level, which belongs to the category of economics. When there is energy shortage, energy demand will be greater than energy consumption. When energy supply is sufficient or oversupply, energy demand will be equal to energy consumption. However, in the study of actual energy demand, it is very difficult to accurately count or estimate the actual energy demand. Therefore, energy consumption is often used when discussing the energy demand in the past. This paper will also follow this practice. Energy demand is factor demand, because the energy demand in the production process does not directly obtain utility and income from the energy itself, but because the energy can be converted into the light, heat and power required by the production process, so the energy demand is a derived demand. The so-called derivative demand is people's demand for production factors, which is rooted in people's demand for products themselves. In order to produce products for people's consumption, people have the demand for production factors, which is derived from the demand for products [16-18].

2.2.Analysis of factors affecting energy demand

The development of energy demand is an extremely complex system, which is closely related to many factors. The energy demand is mainly determined by the economic, social and political environment. The energy demand of a region will be restricted by various internal factors and external environment in the process of development [19-20]. To sum up, the demand for energy is mainly determined by various factors from various aspects, such as the development level of the national economy, industrial structure, energy price, energy demand, technical level and the government's energy policy guidance. The specific explanation of the impact of each factor is as follows:

(1) Economic development level

The development level of energy demand is directly restricted by the level of economic development, so the level of economic development is a very important factor affecting energy demand. Since entering the 21st century, the development trend of China's economic growth has shown a very positive correlation with the growth of heavy industry. As the development of heavy industry is mainly based on energy and mineral products, China's economic growth has rapidly increased the demand for energy such as oil, coal and electricity. At present, the promotion of China's industrialization still needs a lot of energy to support. To sum up, for the influencing factors of economic development, indicators such as GDP and energy consumption elasticity coefficient can be selected as the economic influencing factors of energy demand.

(2) Industrial structure

With the development of economy, the industrial structure is constantly changing, and the change of industrial structure will also lead to the change of energy demand. The energy consumption level of different industries is different. If high-energy consuming industries account for a large part of the national economy, it will stimulate the overall energy consumption, and vice versa, it will reduce the energy consumption. The energy consumption levels of the three industries are different, Among them, the energy consumption of the secondary industry is relatively high, and in China's industrial structure, the secondary industry is especially a high energy consuming industry (the energy consumed accounts for 70% of the total energy consumption). Compared with western developed countries, the proportion of China's secondary industry is very high. Therefore, if the proportion of industry increases in the development of national economy, it will inevitably lead to a significant increase in energy demand.

(3) Energy price

Consumer demand theory points out that the change of commodity price will produce income effect and substitution effect, which will affect consumers' purchase behavior. From the perspective of energy demand, energy can also be regarded as a commodity in the market, so energy price will inevitably affect the development of energy demand. However, energy is also a special commodity, which is different from general commodities The impact of source price on energy demand is not fully consistent with the demand theory. Energy is a strategic asset competed by countries all over the world. Energy price is closely related to the national economic lifeline, especially the change of crude oil price. However, generally speaking, energy price is still closely related to energy demand.

(4) Energy supply

At present, the energy consumed worldwide is mainly non renewable energy, such as oil, coal, natural gas, etc. This determines that the supply of energy such as oil, coal and natural gas is limited in a certain period of time. Energy is an important material basis for national economic development and national political and economic security. In a sense, energy supply determines energy consumption, and energy supply is not limited In case of sufficient, energy demanders will also look for corresponding energy substitutes, so as to reduce energy dependence or improve energy efficiency by improving technology, which will also reduce energy demand to a certain extent.

(5) Technical level

Under the condition that the total economic output remains unchanged, the energy intensity directly determines the total energy consumption. The progress of technological level can not only greatly reduce the production cost of various production activities, but also reduce the energy consumption rate of various production activities, improve the energy utilization rate, so as to reduce the demand for energy. On the other hand, the progress of technological level can improve energy consumption Therefore, the progress of technological level is also a factor affecting energy demand.

3. Seasonal Energy Demand Impact Decomposition

3.1. Building LMDI model

In the field of energy research, there are two main factor decomposition methods: one is structural decomposition analysis (SDA) based on input-output; the other is index decomposition analysis (IDA) based on depolymerization Among them, IDA method can be divided into linear growth Laspeyre method and exponential growth Divisia method. By comparing the factor decomposition results of these two methods, it is found that logarithmic mean Divisia index (LMDI) It has better decomposition effect and a wide range of application. LMDI decomposition method can not be used only when the data is negative or zero. It effectively solves the problem of residual term and zero value in decomposition and breaks through the limitations of LMDI. In addition, LMDI method is divided into I method and II method. I method can ensure the consistency of decomposition results, but II method is not It can achieve this effect, and it is found that the decomposition results obtained by the two algorithms are similar, but the calculation process of I method is simpler.

To sum up, according to the research conclusions of existing IDA methods, this paper uses LMDI-I method to analyze the impact of seasonal energy demand. LMDI has two modes: addition and multiplication. In this paper, the addition mode is used to realize the decomposition process, and its principle is as follows.

Set E as the total amount related to energy, there are n factors $(x_{1i}x_{2i}...x_{ni})$ that have different degrees of impact on e, then these factors can be quantified through factor quantitative decomposition, where $E_i = x_{1i}x_{2i}...x_{ni}$ subcategory of. Then the general seasonal index factor decomposition (IDA) formula is

$$E=\sum_{i=1}^{T} E_i \tag{1}$$

When the total amount changes from E_0 in the base year to E_T in the T year, the addition decomposition formula is

$$\Delta E = E_T - E_0 = \Delta E_{x1} + \Delta E_{x1} + ... + \Delta E_{xn} \tag{2}$$

Among them,

$$\Delta E_{xi} = \sum_{i=1}^{n} L(E_{iT}, E_{i0}) \tag{3}$$

It is not difficult to see that this paper can use it to analyze the impact on seasonal energy demand.

3.2. Construction of LMDI model of seasonal energy demand impact

Although the urbanization rate is an important factor affecting energy demand, due to the difficulties in obtaining the data of municipal cities, there is no further decomposition of residents' domestic energy consumption and urbanization level. In this paper, the total energy demand is divided into production and domestic energy consumption and residents' domestic energy consumption. On the basis of Kaya identity, the constructed seasonal energy demand LMDI model is as follows:

$$E=E' + E_l = P(\frac{GDP}{P} + \frac{GDP_i}{GDP} + \frac{E_i}{GDP}) + E_l \tag{4}$$

Among them, E' is production and domestic energy, E_l is residents' domestic energy, i is the industrial type, T is the seasonal period, P is the total population at the end of the year, GDP is the regional GDP, GDP_i is the GDP of the i industry, and E_i is the energy consumption of the i industry. Each factor is defined as. Among them,

$A = \frac{GDP}{P}$ represents per capita GDP, that is, economic development

$S_i = \frac{GDP_i}{GDP}$ refers to the proportion of the GDP of the i industry in the regional GDP, that is, the industrial structure

$I_i = \frac{E_i}{GDP}$ represents the energy consumption per unit GDP of the i industry, that is, the energy consumption intensity

Therefore, the decomposition formula of influencing factors of seasonal energy demand can be further expressed as

$$E=P(A+S_i+I_i)+E_l \tag{5}$$

Correspondingly, the change of total energy consumption in season T relative to the base period can be expressed as

On this basis, the contribution rate of each decomposition factor is calculated:

$$\Delta E_p = \frac{(E_T - E_0)}{\ln(E_T - E_0)} \ln(\frac{P_T}{P_0}) \tag{6}$$

$$\Delta E_{sca} = \frac{(E_T - E_0)}{\ln(E_T - E_0)} \ln(\frac{A_T}{A_0}) \tag{7}$$

$$\Delta E_{stc} = \frac{(E_T - E_0)}{\ln(E_T - E_0)} \ln(\frac{S_T}{S_0}) \tag{8}$$

$$\Delta E_{\text{int}} = \frac{(E_T - E_0)}{\ln(E_T - E_0)} \ln(\frac{I_T}{I_0}) \tag{9}$$

Among them, ΔE_t is the comprehensive effect, E_0 is the total energy consumption in the base period, E_T is the total energy consumption in season T, ΔE_p is the population scale effect, ΔE_{sca} is the economies of scale effect, ΔE_{str} is the industrial structure effect, ΔE_{int} is the technological progress effect, ΔE_t is the change of the total domestic energy consumption of residents.

4. ENERGY DEMAND FORECAST OF INTEGRATED ENERGY SYSTEM

Before optimizing energy demand, we must first clarify the total energy demand in China, which is one of the constraints of optimization. In this paper, the dynamic weight combination based on fuzzy neural network algorithm and support vector machine algorithm is selected to reasonably predict China's energy demand during the 13th Five Year Plan period, so as to facilitate the later research on the optimization of energy demand.

4.1. Data processing of energy demand forecast

In the multi index comprehensive evaluation, some are the evaluation indicators that the larger the index quantity value, the better the evaluation, which are called positive evaluation indicators (also known as benefit indicators or expected large indicators); some are the indicators that the smaller the index value, the better the evaluation, which are called reverse evaluation indicators (also known as cost indicators or expected small indicators) , others are indicators that perform better when the indicator value is closer to a certain value, which are called appropriate evaluation indicators. In the comprehensive evaluation, the indicators must first be trended, generally converting the reverse indicators and appropriate indicators into positive indicators, so they are also called positive evaluation indicators. Different evaluation indicators often have different dimensions and dimensional units, so they can be synthesized directly Combination is inappropriate and has no practical significance. Therefore, the index value must be transformed into a dimensionless relative number. This process of removing the dimension of the index is called dimensionless of the index (also known as the same degree quantization) , which is the premise of index synthesis. Combined with the multi index comprehensive evaluation system, the index attributes of the influencing factors of China's energy demand analyzed above are classified, as shown in Table 1.

TABLE 1. ENERGY DEMAND OPTIMIZATION INDEX SYSTEM

Composition of integrated energy system	Index	Attribute
economic system	Gross National Product	Benefit type
	Disposable income of urban residents	Benefit type
energy system	Total energy consumption	Moderate
	Total coal consumption	Moderate
	Total oil consumption	Moderate
	Total natural gas consumption	Moderate
	Total renewable energy consumption	Moderate
environmental system	Total Population	Moderate
	Urbanization rate	Moderate
	industrial structure	Moderate
	Energy industry conversion rate	Benefit type

It can be seen from table 1 that the indicator units of each factor of energy demand are different, so there are differences in the dimensions of indicators, which makes it impossible to compare and contradictory between indicators. In this paper, the impact can be eliminated by dimensionless treatment of benefit indicators. In this paper, the appropriate indicators are standardized to quantify different specifications. The specific implementation methods are as follows:

When the index r_{ij} is benefit index

$$r_{ij} = \frac{r_{ij} - \min(r_{ij})}{\max(r_{ij}) - \min(r_{ij})} \quad (10)$$

When the index r_{ij} is moderate index

$$r_{ij} = 1 - \frac{\left|\max(r_{ij}) - r_{ij}\right|}{\max(r_{ij})} \quad (11)$$

In the formula, $\max(r_{ij})$ is the maximum value of data in the influencing factors of seasonal energy demand, and $\min(r_{ij})$ is the minimum value of data in the influencing factors of seasonal energy demand. Each factor is dimensionless through the above formula to prepare for subsequent prediction.

4.2. Energy demand forecasting based on Fuzzy Neural Network

Based on the above, this paper uses fuzzy neural network to predict the energy demand under the condition of seasonal mutation. Fuzzy mathematics, also known as fuzzy mathematics, is a mathematical method to study and deal with fuzzy phenomena. It uses "subordinate function" to describe the intermediate transition process of phenomenon differences, thus breaking the subordinate relationship of classical set theory that absolutely belongs to or does not belong to. The basic idea of fuzzy mathematics is to use accurate mathematical expressions to describe and simulate a large number of fuzzy concepts and phenomena in the real world, so as to achieve the purpose of correct processing. Fuzzy mathematics is based on the uncertainty of things. The emergence of fuzzy sets is the requirement of mathematics to adapt to the description of complex things. T-S fuzzy model is a nonlinear fuzzy reasoning model. Because the conclusion of T-S model is based on the rules of various linear equations, its mathematical expression characteristics are also effective. In addition, the global output of the model has a good multivariable system, and can reduce the number of fuzzy rules in the processing of multivariable systems, Therefore, it has very good advantages.

In this paper, T-S is defined as "*If -then*". Under the condition that the rule is the correlation of the influencing factors of seasonal energy demand analyzed above, the fuzzy derivation method is

$$if\ x_{ij}\ is\ r_{ij}\ then\ y=\sum_{i=1}^{n} p_{ij}x_{ij} \tag{12}$$

Among them, x_{ij} is the fuzzy set of fuzzy system, p_{ij} is the parameter of fuzzy system, y is the output derived from the fuzzy rule. The fuzzy part is the system input part (i.e. *if* part), and the determined part is the output part (i.e. *then* part). Such a linear combination is used to represent the input and output of the fuzzy rule.

Suppose that for input x_{ij}, firstly, the membership degree of each input variable is calculated based on fuzzy rules, which changes with the change of season. At this time, the determination center point of energy demand membership function and function operation width variables will directly determine the accuracy of the reddest prediction result. Therefore, this paper makes fuzzy calculation for each membership function, and uses fuzzy operator as the continuous multiplication operator of fuzzy function. The fuzzy neural network is divided into four layers: input layer, fuzzy layer, fuzzy rule calculation layer and output layer. There are two levels connected: the input layer is connected with the input vector x_{ij}, in which the number of two nodes is the same, that is, the number of connected nodes is the same as the dimension of the input vector. The second layer uses the membership function to fuzzify the input value to obtain the fuzzy membership value. The calculation method of error e in the learning algorithm of fuzzy neural network is

$$e=\frac{(z_d-z_c)^2}{2} \tag{13}$$

Where, z_d is the expected output of the network, and z_c is the actual output.

According to the calculation results, the output prediction results are corrected, and the correction coefficient at this time is ϑ

$$\vartheta=p_{ij}(t-1)-\alpha\frac{\partial e}{\partial p_{ij}} \tag{14}$$

Where, α is the learning speed of the network.

In this way, the prediction result of the final output is ensured to have high accuracy.

5. Empirical analysis

5.1. Overview of regional load

This chapter selects a new park in a certain area of a city to analyze the application mode of distributed power generation. According to its geographical location and cooling / heating / electric load demand, the comprehensive energy demand of cooling, heating and power in the park is as follows: the design load installation power in the power supply area is 667 kW. According to the functions and characteristics of the building, the cold index of the building area is 85W/m^2 and the heat index of the building area is 59W/m^2. The park is an office area, so the cooling, heating and power load during non working hours is low. Since the cooling, heating and power loads need to be supplied at the same time, the cooling, heating and power cogeneration equipment is the best choice. Moreover, due to the large difference between the peak load in the daytime and the valley load at night, the minimum operating condition of the gas turbine can meet the nighttime load, and the cold and heat storage device and energy storage device can be used to store energy, Therefore, the gas turbine and photovoltaic whose output characteristics are basically consistent with the power load trend are preliminarily selected for the renewable distributed power generation, and the fan is not considered for the time being. Adhering to the principle that the distributed power generation in the park can meet the park load as much as possible and reduce the connecting line with the distribution network, it is known that the month with the highest power load output at night is January, with an average of about 420 kW, and the lowest operation condition of gas turbine is generally about 30%, so the initial configuration of gas turbine is 1400kW. The peak value of daytime electric load is in June, and the average load is about 1700 kW, so the remaining electric load is supplied by photovoltaic. The results of Tianjin distributed photovoltaic actual power tracking and installed capacity show that the average actual output of photovoltaic in summer is about 75% of the theoretical output. Therefore, considering the load difference under the configuration capacity margin, the required distributed photovoltaic installed capacity is about 400k WP. The distributed power actually included in the system mainly includes gas triple supply 1.5MW, parking lot photovoltaic 400 kWp, ground source heat pump unit 1MW, battery energy storage 300kWh and water cold storage device.

It is proposed to predict the cooling / heating load of buildings 1-5 in the regional plan, and the annual energy load of buildings 1-4 is shown in Table 2.

TABLE 2. STATISTICS OF ANNUAL ENERGY LOAD IN THE PARK

Region	Built-up area /m^2	Cooling load /kw	Thermal load /kw
Building 1	15880	1366	7712
Building 2	15690	1405	7563
Building 3	15630	1440	7625
Building 4	22400	2026	13169

5.2. Prediction results and analysis

Based on the above, the annual energy demand of Building 5 is predicted by using SVM, BPNN and the methods proposed in this paper, and compared with the actual data. The results are shown in Figure 1 and Figure 2.

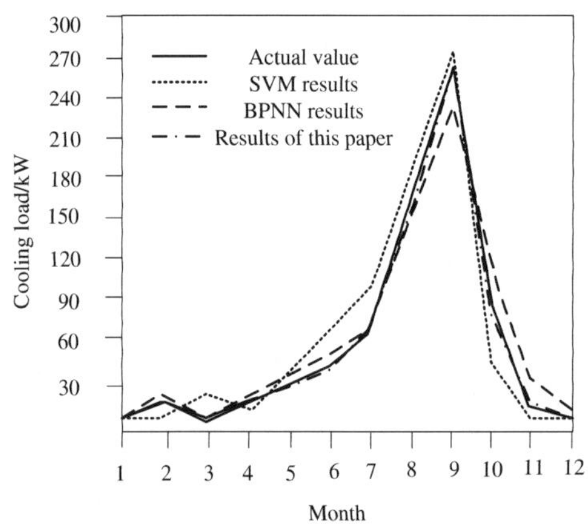

Figure 1. Cooling load forecast results

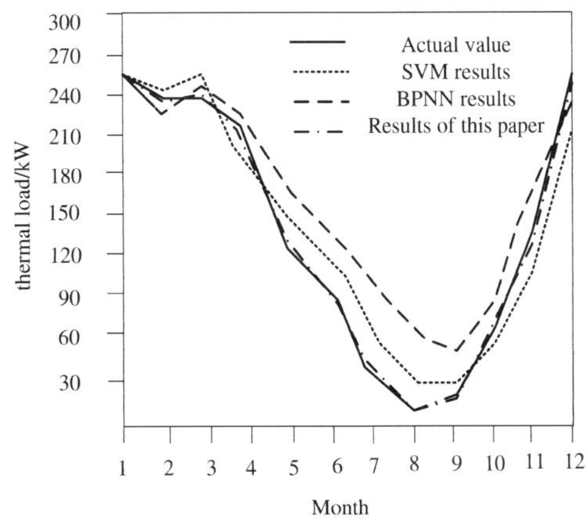

Figure 2. Thermal load prediction results

It can be seen from figures 1 and 2 that the error between the predicted data and the actual data of various types of energy loads by this method is small, the trend of the predicted curve is basically consistent with the actual curve, and the closeness of the two curves is high, indicating that the prediction error is relatively small. In contrast, the prediction results of SVM and BPNN are less stable, and there is a large floating trend in the goodness of fit with the actual results.

In order to verify the effectiveness and feasibility of the prediction methods designed in this paper, support vector machine (SVM) and BP neural network (BPNN) are also selected for comparative analysis. The calculation results of the prediction results and evaluation indexes of various methods for various types of energy are shown in Table 3, where $MaxR_E$ is the maximum relative error, $MinR_E$ is the minimum relative error, MAPE is the average absolute percentage error, R_{MSE} is the root mean square error, T_{IC} is hill inequality coefficient.

TABLE 3. EVALUATION STATISTICS OF PREDICTION RESULTS INDICATORS OF DIFFERENT METHODS

Index		$MaxR_E$/%	$MinR_E$/%	MAPE/%	R_{MSE}	T_{IC}
Prediction results of	SVM	8.96	3.79	5.24	0.049	0.065
power consumption	BPNN	8.55	4.29	6.33	0.064	0.092
related indicators	Paper method	5.66	1.02	3.21	0.015	0.020
Prediction results of	SVM	7.72	3.56	4.49	0.039	0.062
gas consumption	BPNN	7.87	3.50	4.89	0.059	0.092
related indexes	Paper method	5.44	1.01	2.96	0.014	0.019
Prediction results	SVM	7.59	3.22	4.56	0.085	0.077
with cold related	BPNN	7.88	4.44	5.70	0.094	0.085
indexes	Paper method	5.89	1.03	2.74	0.019	0.014

It can be seen from table 3 that in all kinds of energy load demand forecasting, the five evaluation index values of the design method in this paper are less than SVM, BPNN and RBF, indicating that the prediction accuracy of the design method in this paper is higher, and the learning ability and generalization ability of the method are strengthened through data preprocessing and information feature extraction. However, SVM and BPNN methods can not get the correlation between multiple features of data information, so there is a large error in prediction accuracy.

6. Conclusion

In order to effectively control energy, it is very necessary to accurately predict its demand. Based on the change of energy demand caused by seasonal change, this paper puts forward an energy demand prediction method of comprehensive energy system considering seasonal mutation. Although this paper has made some research results, there are many factors affecting power demand, and there is a considerable degree of randomness. Moreover, power demand also changes with the process of power marketization, which will lead to more and more complex research on energy demand. Therefore, this paper believes that on the basis of this research, we can further study the content.

(1) Comprehensively consider various factors affecting energy demand in season

This paper focuses on the study of energy demand from the perspective of economy. But in fact, the influencing factors of energy demand exist in various angles, such as climate conditions, population size, residents' living standards, major policies and so on. As mentioned above, if we comprehensively consider various factors affecting energy demand and want to establish an energy demand model, we need to collect a lot of data and information and make greater efforts. There are many prediction methods for energy demand, but many are relatively rigid, because they only consider the data itself and ignore the characteristics of energy demand itself. Therefore, the actual choice of prediction methods needs to be closely combined with the actual situation and pay attention to the scope of application of means. Energy demand is related to many factors. It may be inaccurate to predict only from the energy data itself. When selecting the relevant variables for predicting energy demand, we must pay attention to two conditions: first, we must use scientific and reasonable methods to measure, analyze and determine the relevant variables to ensure the perfection and accuracy of the index system; Second, the variables selected as indicators should be able to predict accurately. Once the independent variable data in the model is inaccurate, there will be a certain error in the model itself, and the error of the prediction result will be magnified infinitely. Therefore, choosing reasonable and certain factors to carry out prediction is the top priority of energy demand prediction.

(2) We should do more and more in-depth research on promoting energy market-oriented reform

There is no doubt that we can foresee that the reform direction of energy marketization will not change again. However, the current organizational structure, operation system and business philosophy can not fully match the marketization. In particular, distributed energy and direct power purchase by large users are connected to the power grid, which will undoubtedly lead to significant changes in the pattern of the energy market, which will make the competition more intense and the resource allocation more and more reasonable. This is not only an opportunity but also a challenge. Therefore, academic research should pay more attention to, do a good job in-depth discussion, strive to make relevant achievements, and provide guidance and development suggestions for energy market-oriented reform.

Acknowledgment

The study was supported by "the National Key Research and Development Program of China: Research and Application Demonstration of Intelligent IoT and Control Technology for Urban Integrated Enerty, China (grant number: 2020YFB2104500)".

References

[1] Sanz-Hernandez A . Social engagement and socio-genesis of energy poverty as a problem in Spain[J]. Energy Policy, 2019, 124(JAN.):286-296.

[2] Ciller P , Lumbreras S . Electricity for all: The contribution of large-scale planning tools to the energy-access problem[J]. Renewable and Sustainable Energy Reviews, 2020, 120:109624.

[3] Coelho R L . On the Energy Concept Problem: Experiments and Interpretations[J]. Foundations of Science, 2020(368–397).

[4] Gazafroudi A S , Shafie-Khah M , Heydarian-Forushani E , et al. Two-stage stochastic model for the price-based domestic energy management problem[J]. International Journal of Electrical Power & Energy Systems, 2019, 112(NOV.):404-416.

[5] Xb A , Xiang L B , Rj C , et al. A distributionally robust credibilistic optimization method for the economic-environmental-energy-social sustainability problem[J]. Information Sciences, 2019, 501:1-18.

[6] Shariatmadar K , Arrigo A , Hallez H , et al. Day-Ahead Energy and Reserve Dispatch Problem under Non-Probabilistic Uncertainty[J]. Energies, 2021, 14(1016):1016.

[7] Irawan C A , Hofman P S , Chan H K , et al. A stochastic programming model for an energy planning problem: formulation, solution method and application[J]. Annals of Operations Research, 2021:1-36.
[8] Thollander P , Palm J , Hedbrant J . Energy Efficiency as a Wicked Problem[J]. Sustainability, 2019, 11.
[9] Wakui T , Hashiguchi M , Yokoyama R . Structural design of distributed energy networks by a hierarchical combination of variable- and constraint-based decomposition methods[J]. Energy, 2021, 224(6):120099.
[10] Casteleiro-Roca J L , JF Gómez-González, Calvo-Rolle J L , et al. Short-Term Energy Demand Forecast in Hotels Using Hybrid Intelligent Modeling[J]. Sensors, 2019, 19(11):2485-2503.
[11] Jamil R . Hydroelectricity consumption forecast for Pakistan using ARIMA modeling and supply-demand analysis for the year 2030[J]. Renewable Energy, 2020, 154.
[12] Bedi Jatin,Toshniwal Durga . Deep learning framework to forecast electricity demand[J]. Applied Energy,2019,238(MARa15):1312-1326.
[13] Lin J , Zhu K , Liu Z , et al. Study on A Simple Model to Forecast the Electricity Demand under China's New Normal Situation[J]. Energies, 2019, 12(11):2220-2248.
[14] L Rivera-González, D Bolonio, Mazadiego L F , et al. Long-Term Electricity Supply and Demand Forecast (2018-2040): A LEAP Model Application towards a Sustainable Power Generation System in Ecuador[J]. Sustainability, 2019, 11(19):1-19.
[15] Gharabaghi S , Stahl E , Bonakdari H . Integrated nonlinear daily water demand forecast model (case study: City of Guelph, Canada)[J]. Journal of Hydrology, 2019, 579:124182.
[16] Kl A , Zm A , Dr A , et al. A data-driven strategy to forecast next-day electricity usage and peak electricity demand of a building portfolio using cluster analysis, Cubist regression models and Particle Swarm Optimization[J]. Journal of Cleaner Production, 2020, 273.
[17] Jiang P , Li R , Lu H , et al. Modeling of electricity demand forecast for power system[J]. Neural Computing and Applications, 2019, 32:6857–6875.
[18] JM Jiménez, Stokes L , Moss C , et al. Modelling energy demand response using long short-term memory neural networks[J]. Energy Efficiency, 2020(3):1263-1280.
[19] Jamali A , Aghaei J , Esmaili M , et al. Self-Scheduling Approach to Coordinating Wind Power Producers With Energy Storage and Demand Response[J]. IEEE Transactions on Sustainable Energy, 2020, 11(3):1210-1219.
[20] Semieniuk G , Taylor L , Rezai A , et al. Plausible energy demand patterns in a growing global economy with climate policy[J]. Nature Climate Change, 2021, 11:313-318..

Research status of partial discharge detection of power transformer based on pulse current method

Jiahao Sun [1a*], Shuyan Pan [1b*], JinZhao Deng [1c]

[1]State Grid Electric Power Research Institute, Nanjing 211106, Jiangsu Province, China

[a*]1303372729@qq.com,[b*]panshuyan@sgepri.sgcc.com.cn,[c]dengjinzhao@sgepri.sgcc.com.cn

Abstract—There are many important equipment running in the power grid, and the power transformer is one of them. It plays an extremely important role in the stable operation of the power system. Therefore, whether the power transformer can operate safely and stably becomes particularly important. When the power transformer is running online, partial discharge will occur if there are insulation defects, which will further develop into breakdown without timely detection and treatment. The insulation inside the transformer often has certain defects and produces certain partial discharge, and the partial discharge signal is often a high-frequency pulse signal. In this paper, the definition of partial discharge, the generation process of pulse current and the detection method of transformer partial discharge based on pulse current method will be introduced in detail. Several common methods are combed and summarized, the current development status of pulse current method is introduced, and the possible research direction in the future is prospected.

1. Introduction

These faults often exist when electrical equipment is running: conductor fault, insulation fault and mechanical fault. In practice, insulation is the fault that accounts for the largest proportion of all faults, so it is the most common. In recent years, the AC voltage level of power grid equipment has been continuously improved, and the power system is also constantly improving the measures to limit overvoltage. In this case, the electrical equipment will be directly affected by the higher working voltage, and the probability of insulation circuit fault will increase accordingly. Insulation failure is often accompanied by partial discharge, which further promotes the aging of insulation and finally develops into breakdown [1-3].

At the Gottingen Royal Society meeting in 1777, Lichtenberg showed the world his new progress in partial discharge test. In 1873, Maxwell deduced four equations describing electromagnetic field. In 1896, the basic hypothesis of the existence of electromagnetic wave and its propagation characteristics in space and time was verified by Hertz's experimental results for the first time, which provided a new theoretical and experimental basis for the detection of a large number of local discharge and the research and development of instruments and equipment [5]. Xilin bridge was first proposed in 1919. This research method has laid a solid foundation for the integral bridge technology developed by Dakin and Malinaric in 1960. It can be clearly considered that the partial discharge integral measurement method used on two parallelogram is the original ancestor of modern partial discharge detection technology using integral bridge. Up to now, the integral bridge method is still widely used in many experiments to measure partial discharge [4-5].

At present, the insulation fault of power equipment has attracted great attention at home and abroad. It is committed to looking for the partial discharge and its related characteristics that develop rapidly and seriously threaten the insulation of transformer. Canada, Japan and other countries first studied the occurrence of partial discharge in the operation of power equipment, and put the developed detection system into online operation. The occurrence of partial discharge in transformers has attracted the attention of power operation departments and major power equipment manufacturers in the 1950s. Since the 1960s, many scientific research institutions have begun to study the partial discharge characteristics of transformers. Up to now, the detection methods of transformer partial discharge can be roughly divided into two main categories: using the electrical physical quantity of transformer for measurement and using non electrical physical quantity for measurement. The pulse current measurement method belongs to the method of using electrical quantity for measurement. Pulse current method has mature technology and wide application. At present, the relevant standard (IEC-270) formulated by IEC (International Electrotechnical Commission) is widely used to detect partial discharge by pulse current method [6-7].

This paper is divided into three parts. The first section summarizes the definition, causes and process of partial discharge in power transformer; The second section summarizes the partial discharge detection methods of transformer based on pulse current, and the third section looks forward to the possible research directions in the future.

2. Generation mechanism of partial discharge and pulse current

When the power transformer is running online, partial discharge will occur if there are insulation defects, which will further develop into breakdown without timely detection and treatment. Therefore, it is of great significance to deeply study the relevant theories of partial discharge in transformer and explore various parameter characteristics when partial discharge occurs for the application of pulse current method to identify and protect the internal discharge defects of transformer [8].

2.1. Definition and causes of partial discharge

Partial pulse discharge is usually a kind of pulse discharge between two local electrodes that have not been filled with insulating gas medium and have not completely formed conductive communication. A partial pulse discharge of the equipment is generally considered to be caused by the partial concentration of the electric field in an insulator or on the surface of other insulating media, resulting in short-term partial discharge, that is, it is caused by some defects that may be caused by some equipment in the actual production and use process or some weaknesses of the insulating media in the equipment. When a partial pulse discharge occurs in the equipment, it will directly produce a pulse discharge with a duration of usually less than $\mathbf{1\mu m}$, often accompanied by mutual reactions such as light, heat, sound and chemistry. In case of partial discharge, the energy released is very small, so the damage to the equipment can be ignored. However, if there is partial discharge in electrical equipment for a long time, the weak effect caused by partial discharge will gradually accumulate and eventually develop into insulation breakdown [9].

When partial discharge occurs in the insulation of power transformer, in addition to pulse current due to uneven distribution of insulation electric field, heating and luminescence of gas in corresponding dielectric, high-frequency ultrasound and decomposition of insulating grease may also occur, resulting in changes in the composition of insulating gas [10-11].

2.2. Generation process of pulse current

There is an air gap in the insulation medium of transformer winding due to insulation defects, the equivalent circuit of the whole system is the capacitance C_g of the air gap and capacitance C_b of the part of the medium connected in series with the air gap in series, and then with the dielectric capacitance C_a of the other intact parts in parallel. As shown in Fig.1, the total capacitance of the whole system is

$$C = C_a + \frac{C_b C_g}{C_b + C_g} \quad (1)$$

Under the action of power supply voltage

$$u = U_m \sin(\omega t) \quad (2)$$

The voltage on C_g is

$$u_g = \frac{C_b}{C_b + C_g} U_m \sin(\omega t) \quad (3)$$

When air gap capacitance C_g voltage at both endsu_g reach discharge voltage u_s, partial spark discharge will occur in the air gap C_g along circuit$C_g \to C_a \to C_b \to C_g$discharge when u_g drop to extinction voltage u_r, the spark is extinguished, a partial discharge is completed, and a high-frequency partial discharge pulse current is generated in the circuit [12].

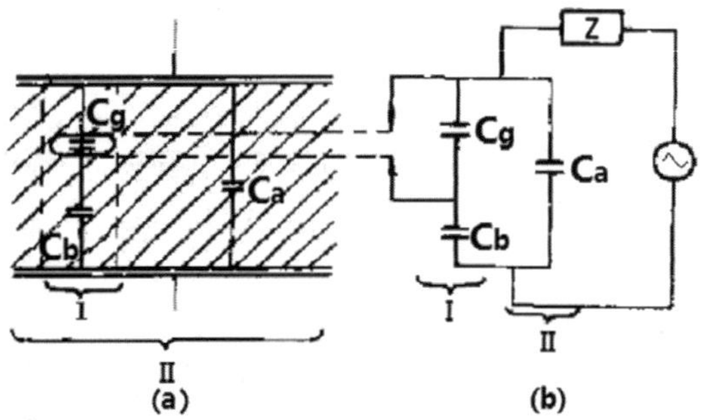

Fig.1. Pulse current equivalent circuit.

3. Partial discharge detection method of transformer based on pulse current

The pulse current method explores the situation of partial discharge by detecting the megahertz pulse current generated when partial discharge occurs. In the actual detection test, the pulse current method is to detect the amplitude and frequency of pulse current and discharge amount when partial discharge occurs in power transformer winding and other transformer structures [13]. The experiment analyzes and discriminates the collected information by detecting the propagation characteristics of pulse signal in the winding and the high-frequency pulse response of the winding, and cuts off the fault in time [14-15]. This method is sensitive and easy to operate. However, in the pulse current amplification detection circuit, the impedance of the current amplification circuit and the measurement circuit has a bad impact on the measurement of pulse current data, and the detection accuracy of the measurement instrument is reduced [16-23]. In recent years, with the increasing maturity of relevant detection theory and the continuous improvement of relevant measurement technology test quantization methods, the basic measurement methods and relevant measurement test quantization methods of partial AC discharge in transformer circuit have a more solid theoretical foundation [24]. At present, there are two types of transformer detection methods: on-line and off-line. The advantages of on-line detection are: it can detect the local discharge signal in real time. The disadvantages are: the equipment is complex, and it is easy to affect the normal operation of the power grid, the detection accuracy is low, it is easy to be disturbed by various signals, resulting in misjudgment and high investment cost. The advantages of off-line detection are: simple equipment, high measurement accuracy, low interference and low investment cost. The disadvantages are: the pulse signal of local discharge can not be detected in time and corresponding protective measures can be taken. The time interval between two tests is long. Some insulation defects can not be found immediately, resulting in further deterioration of the defects. The pulse current detection circuit is shown in Fig.2.

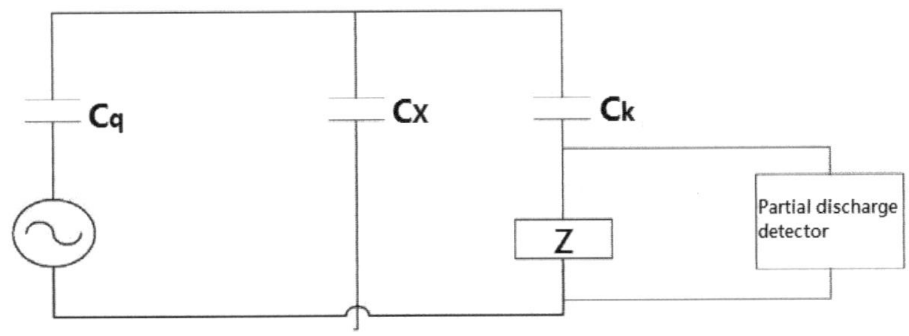

Fig. 2. Pulse current detection circuit

3.1. UHF detection method

UHF detection method is an off-line detection method. Through the real-time collection of various characteristics and variable information of pulse current, and the real-time analysis and judgment of the real-time collected pulse data information. We can obtain the partial discharge information and determine the fault point. This detection method usually collects the partial discharge signal on the grounding wire of the end screen of the power transformer bushing, the grounding wire of the iron core or the grounding wire of the winding neutral point by using the high-frequency detection instrument composed of the high-frequency detection CT connected by the high-frequency detection discharge impedance and the Roche detection coil. After completing the calibration, measure the apparent discharge, determine the fault position of the partial discharge, and take corresponding protective measures according to the specific situation.

At present, the computer-aided automatic test system is widely used in the practical work of transformer partial discharge detection in power grid. Reference [25] points out that this computer-aided measurement system is widely used in various UHF signal detection experiments. Firstly, various forms of partial discharge signals measured in the laboratory are successively amplified and filtered to complete various real or A / D signal conversion, digitize the signal, send the measured data to the computer processing system in turn, and process the high-frequency signal data to produce various statistical spectrum diagrams and ray diagrams, which can be used to analyze various forms of partial discharge. However, in the high-frequency circuit, the impedance of transformer winding is very large, and the pulse signal attenuates rapidly in the propagation process, which is very unfavorable to the measurement of pulse signal, and the difficulty of UHF detection increases greatly.

Reference [26] proposed an on-line detection and control system for partial discharge. It uses the broadband oscilloscope in the detection circuit of partial discharge pulse current to process the data of partial discharge. In addition, reference [27] also proposes a partial discharge detector composed of a high-frequency CT designed by Roche coil and a spectral frequency analyzer with high-frequency amplifier which can detect transformer partial discharge signals in different frequency bands of 100kHz-200MHz. It mainly places the high-frequency CT made of Roche discharge coil on the transformer bushing, so as to detect the partial discharge signals of the primary and secondary windings of the whole transformer.

3.2. Differential balance method

In the complex electromagnetic environment, the pulse current generated by partial discharge of power transformer is easily affected by electromagnetic interference, so it becomes very difficult to obtain useful information. In the whole online detection system, eliminating external electromagnetic interference has always been a difficult problem [28-32]. The experimental schematic diagram and the comparison between output and original waveform are shown in Fig. 3 and Fig. 4 respectively.

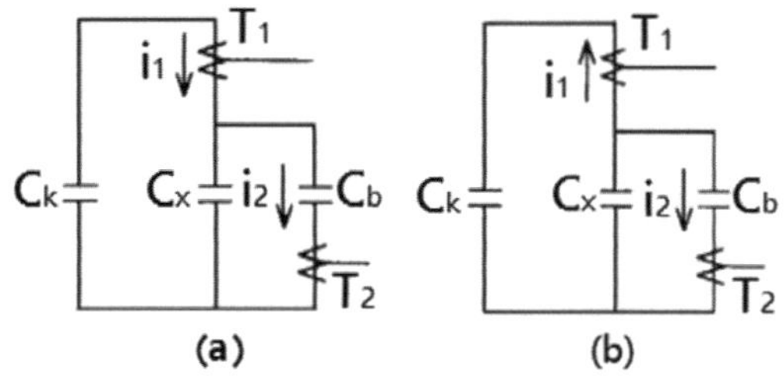

Fig. 3. Experimental principle circuit diagram of differential balance method

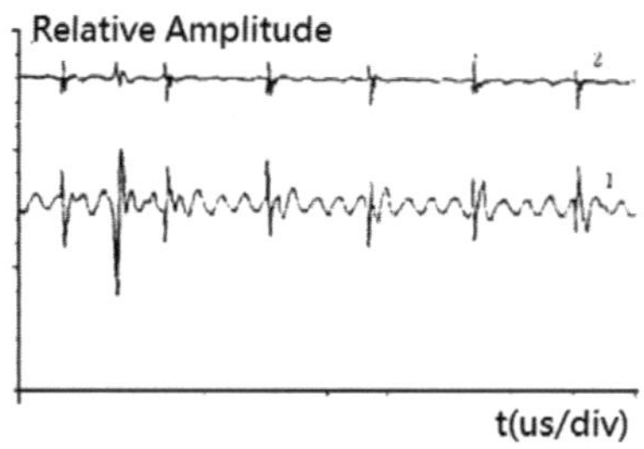

Fig. 4 . Comparison between differential output signal and original signa

Differential balance method is an on-line detection method. Its working principle is to use high-frequency CT to detect current signals from high-voltage lead, winding neutral point grounding wire and bushing end screen grounding wire respectively. When electromagnetic interference enters a phase, high-frequency pulses in the opposite direction will be formed on the end screen grounding wire and high-voltage lead, The partial discharge signal of this phase winding forms a high-frequency pulse in the same direction on the end screen grounding wire and the high-voltage lead. Using the characteristics of this interference signal and partial discharge signal, the signal extracted by high-frequency CT can be input into the common mode differential circuit, which greatly weakens the external interference signal, enhances the internal partial discharge signal, well suppresses the common mode interference, improves the signal-to-noise ratio of the detection signal, and lays a foundation for further determining the fault location and taking corresponding protection measures.

Based on the above principle, reference [31] developed a direction sensor for on-line monitoring of transformer partial discharge. Based on the analysis of the relationship between sensor damping resistance and coil turns and sensitivity and frequency characteristics, reference [31] studies the sensor transmission characteristics based on closed-loop and open-loop forms, and then tests the anti-interference characteristics by using the differential balance method. The test shows that this method can effectively suppress external interference and help to identify internal partial discharge signals.

3.3. Pulse polarity identification method

Pulse polarity identification method is usually used in off-line detection experiments. Its working principle is that when there is insulation defect in the transformer, partial discharge will occur between two electrodes with insulation defect with the increase of test voltage level. The partial discharge signal will be transmitted to each outgoing terminal, and the signal detected by each detection terminal will have a certain polarity. Use these characteristics to locate the fault point [33-34]. The experimental schematic diagram and the waveform characteristics of pulse signals at the output and each outlet are shown in Fig. 5 and Fig. 6 respectively.

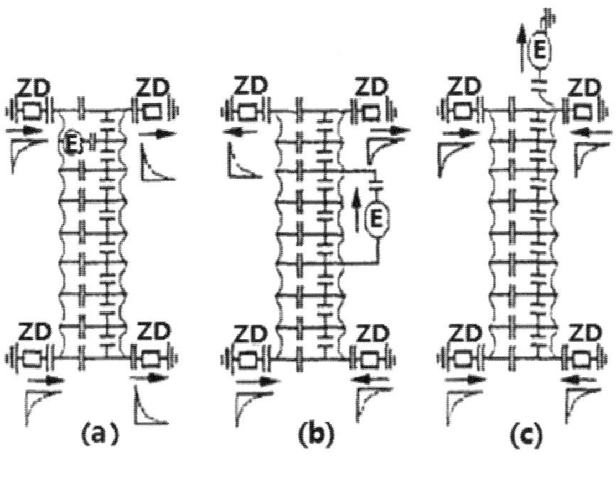

Fig. 5. Schematic diagram of pulse polarity identification

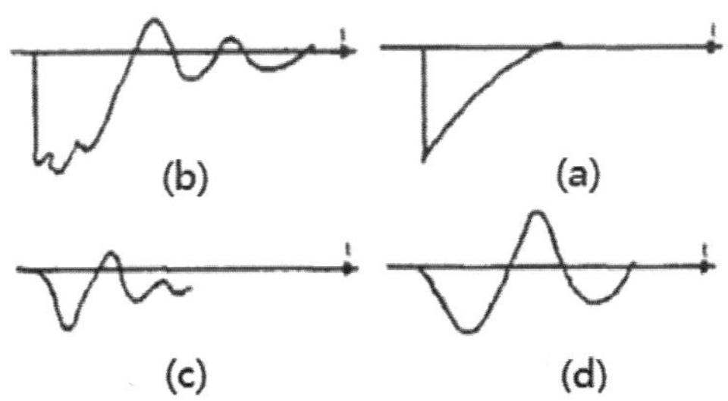

Fig. 6. Waveform of pulse signal at each outgoing terminal

Reference [34] attempts to apply the pulse polarity discrimination method to the on-line detection of transformer partial discharge. It installs high-frequency CT at the base flange of transformer high-voltage outlet bushing to collect the partial discharge signal of high-voltage outlet. Considering that the sensitivity of the series circuit is greater than that of the parallel circuit and has a good filtering effect on the low-frequency signal, the filter capacitor and acquisition resistance are connected in series at both ends of the magnetic core high-frequency CT coil. The enclosed CT shell is made of metal, which can effectively shield external interference and collect high-voltage lead and casing end screen partial discharge signals. It can also prevent the influence of rain and dust on the sensor and improve the vibration resistance of the coil. Partial discharge and external electromagnetic interference are identified by collecting and comparing the polarity of pulses collected by two high-frequency CT.

3.4. Multi terminal adjustment method

References [35] and [36] mentioned a partial discharge signal detection and control method using multi terminal signal regulation. It can be used to effectively suppress high-frequency interference. Similar to the differential balance method, the control method is extracted from the four points of high-voltage bushing and its end screen grounding wire, transformer core and winding neutral grounding wire. The external interference is suppressed through the common mode differential circuit, and then the multi terminal operation adjustment method can be used to make the circuit effectively suppress various interphase interference signals, It includes suppressing the interphase interference of adjacent two phases and the interference of adjacent two windings of this phase. The multi terminal operation adjustment method has strong anti-interference ability in the experiment.

4. Conclusion

The improvement of voltage level and the continuous expansion of Wang have brought more requirements and challenges to the normal operation of power transformer, and the transformer insulation fault, which accounts for a high proportion in many fault types of transformer, has high research value. Through the study of transformer partial discharge based on pulse current method, it is of great significance to distinguish and protect the defects of transformer internal insulation.

This paper summarizes the definition of partial discharge in power transformer, the formation mechanism of pulse current and various protection and discrimination methods based on pulse current method. The following are some personal prospects for the possible research directions of partial discharge discrimination and protection in power system in the future.

1) Further strengthen the research on internal partial discharge of power transformer, summarize the relationship between insulation aging of transformer winding and various types of partial discharge, and further work is needed on how to improve the detection accuracy of pulse signal generated by partial discharge;

2) In the research on the detection method of partial discharge of transformer, although some achievements have been made in eliminating electromagnetic interference, the external electromagnetic interference is very large in the actual on-line operation, and further research is needed on how to identify and suppress the interference more accurately;

3) The high-frequency impedance of transformer winding is large and the partial discharge signal attenuates rapidly. How to improve the partial discharge detection circuit so that the high-frequency pulse can be well coupled to the detection impedance still needs further research.

Acknowledgments
First of all, I thank the project supported by Joint Funds integration project of the National Natural Science Foundation of China (U1866603).The group of Fundamental Research on Multivariable-Based Adaptive Protection and Safe Operation for Power Transformers provided its technical and suggestion to this paper.

Finally, I sincerely thank my mentor, Pan Shuyan, for her guidance in writing this article. It is her support that has provided me with great help in completing this article.

References

[1] Jin Zhuorei. Research on Fractal Antenna and Optimal Wavelet Denoising and Signal Secognition for UHF Monitoring of Transformer Partial Discharge [D]. Chongqing University, 2008.
[2] Bai Shaozuo. Study on Propagation Characteristics of Partial Discharge Signal in Dry-type Transformer Winding [D]. Harbin University of Technology, 2008.
[3] He Peng. Partial Discharge Test of Transformer in Substation Construction [D]. Shandong University, 2011.
[4] Dakin T W , Malinaric P J . A Capacitance Bridge Method for Measuring Integrated Corona-Charge Transfer and Power Loss per Cycle [J]. Transactions of American Institute of Electrical Engineers. Part III: Power Apparatus and Systems, 1960, 79(3):648-652
[5] Izeki Tatsuta F. Three Types of Partial Discharge in Electrical Insulation . Conf.Recordon IEEE Intern. Symp. in El. Ins Montreal 1984: 327～331.
[6] Shanghai Electric Power Company. Code for Handover and Preventive Test of Power Equipment [M]. China Electric Power Press, 2008.
[7] Zhou Zecun. High Voltage Technology. Beijing: Water Resources And Electric Power Press, 1987:103-105.
[8] Zhou Yuchuan. Research on Partial Discharge Diagnosis Method of Transformer [J]. Electrical Switch, 2012, 050 (002): 47-49,53.
[9] Issued by State Grid Corporation of China. Power Safety Regulations of State Grid Corporation of China. Power Transformation Part [M]. China Electric Power Press, 2009.
[10] Wang Hui. Study on Typical Partial Discharge Development Process of Transformer Oil Paper Insulation [D]. North China Electric Power University (Beijing), 2010.
[11] Li Jinming. Research on Partial Discharge Detection Technology of 110kV Transformer [D]. North China Electric Power University, 2013.

[12] Zhao Zhida. High Voltage Technology. Beijing: China Electric Power Press, 2013:85-90.
[13] Sun Xiaofeng, Chen Weiyu. Current Situation and Development of Partial Discharge Detection Technology for Power Transformers [J]. Scientific Chinese, 2017 (14).
[14] Wensheng G , Kexiong T . Study on The Influence of Pulse Propagation in Transformer Winding on Partial Discharge Pattern Recognition[C]. International Conference on Properties & Applications of Dielectric Materials. IEEE, 2000.
[15] R Malewslei， etal. Insulation Diagnostic System for HV Power Transformer in Service[J]. IREQ.CIGRE, 1986,12-01:13-15.
[16] C. Boisseau. Instrument Transformers on Line Monitoring by Means of Partial Discharge Measurement [C], 7th ISH.August 1991.
[17] Song Keren, Feng Yuquan. On Line Partial Discharge Measurement of High Voltage Transformer [J]. High Voltage Technology, 1992, 000 (001): 40-44.
[18] Zhu Deheng, Tan Kexiong. Microcomputer System for On-line Detection of Transformer Partial Discharge [J]. High Voltage Technology, 1992, 000 (001): 45-49.
[19] Guo Heng, Wang Changchang, Zhu Deheng, et al. Study on The Method of Charged Correction of Partial Discharge [J]. Journal of Electrotechnics, 1990 (03): 53-57.
[20] Guo Heng. Study on On-line Detection of Partial Discharge of Power Equipment by Pulse Current Method [D]. 1988.
[21] W. R. Rutgers. Y. H. Fu. UHF PD-detection in Power Transformer[C], 10th ISH, Montreal, August 25-29, 1997, (4):2109-2112.
[22] M Pompili. et al. Early Stage of Negative PD in Dielectric Liquids[J]. IEEE Trans. On Diel. and Elect. Insul. , 1995. 2(4):602-613.
[23] Ge jingpang. Partial Discharge Measurement [M]. China Machine Press, 1984.
[24] Xing Weina. Research and Design of Partial Discharge Detection Technology for Power Transformer [D]. Hebei Agricultural University, 2014.
[25] Krivda, A. Automated Recognition of Partial Discharges[J]. IEEE Trans. On Diel. And Elect. Insul. 1995, 2(5):796-821.
[26] X.Ma, C.Zhou, I.J.Kemp. Novel Computer-based Processing System for Partial Discharge Detection and Diagnosis[J]. IMTC2004, Como, Italy, 2004, 684-688.
[27] N. H. Ahmed, N. N. Srinivas. On-line partial Discharge Detection in Transformer[J]. The 1998 IEEE International Symposium on Electrical Insulation (ISEI), Arlington, Virginia, USA, 1998, 39-42.
[28] Gao Ning, Zhu Deheng, Tan Kexiong. Electromagnetic Interference and Suppression in On-line Partial Discharge Monitoring Signal of Transformer [J]. New technology of electrical energy, 1999 (04): 25-29.
[29] Su Pengsheng, Wang Huan. Analysis of Power System Equipment Condition Monitoring and Fault Diagnosis Technology [J]. Power System Automation, 2003, 27 (1): 61-65.
[30] Guo Jun, Wu Guangning, Zhang Xueqin, Shu Wen. Current Situation and Development of Partial Discharge DetectionTechnology [J]. Journal of Electrotechnics, 2005 (02): 29-35.
[31] Wang Xian. Experimental Study on Anti-interference Method in Transformer Partial Discharge Measurement [D]. North China Electric Power University (Beijing), 2003.
[32] Sun Caixin, Luo Bing. Study on Directional Coupling Differential Balance Method Against Electromagnetic Interference in On-line Partial Discharge Monitoring of Transformer [J]. Chinese Journal of Electrical Engineering, 1998, 18 (5): 340-344.
[33] Tang Zhiguo, Jiang Tongtong, Liu Changbiao, Wang Qian. Study on Anti-interference of Partial Discharge Detection of Capacitor by High Frequency Current Method [J]. Power Capacitor and Reactive Power Compensation, 2016,37 (05): 67-72 + 77.
[34] Xu Yang, Qiu Changrong, Fang Qiong, Cao Xiaolong. Anti Interference Method for Large-scale magnetic Core Sensor and Directional Coupling Pulse Identification [P]. Shanxi: cn13978062003-02-19.

[35] Li Jian, Tang Ju, Sun Caixin, et al. Study on Multi Terminal Regulation Directional Coupling Differential Balance Method for Anti-interference of Transformer Partial Discharge Monitoring [J]. Journal of Electrotechnics, 1998 (5): 47-51.

[36] Jing Qiang. Research on Noise Suppression in On-line Partial Discharge Monitoring of Transformer [D]. Xi'an University of technology, 2009.

Current Status And Prospects Of Research On New Urban User-side Energy Interconnection System Planning

Peng YE[1], Zheng bin ZHANG[2*], Huan WANG[3], Duo jiao GUAN[4], Ming li ZHANG[5], Na ZHANG[6]

[1,2,3,4]Shenyang Institute Of engineering, Shenyang, China

[5,6]Economic and Technological Research Institute of State Grid Liaoning Electric Power Co., Ltd, Shenyang, China

[*] Corresponding author: zhangzhengbin1998@163.com

Abstract—Constructing a new-type urban user-level energy interconnection system, comprehensively considering user-side energy characteristics to optimize the "source-network-load" structure, which is conducive to awakening the potential of users to delay load peak shaving, and realizing the local consumption of renewable energy, which is a solution for small towns. An important means to balance supply and demand. Based on the initiative of users to use energy, this paper divides the user energy interconnection system into residential users, industrial and commercial users, and electric vehicle users. First, it outlines the connotation of the energy Internet and the architecture of the user-side energy interconnection system. Secondly, it summarizes the planning model of user energy interconnection system. Subsequently, according to the "source-network-load" structure, the configuration of the user-side energy interconnection system is analyzed from the perspective of planning. Finally, this article briefly prospects and discusses the factors that need to be considered in the future user energy interconnection system planning.

1. INTRODUCTION

With the continuous improvement of the level of social productivity in my country, the development of science and technology, and the optimization of the industrial structure, the level of urbanization is getting higher and higher. The prerequisite for urbanization and urban development is urban energy planning. The town-level energy interconnection system can be divided into three levels: hub, region, and user according to the size of the interconnected area and the connection method[1]. The user energy system is the main end consumer of urban energy, and reasonable planning is of great significance for improving the utilization rate of renewable energy and promoting the local consumption of renewable energy. At present, the urban user energy system is mainly the building cooling, heating and power cogeneration system, with the cooling, heating and power cogeneration system as the core. Space heating, space refrigeration, hot water, etc., realize the cascade utilization of energy by centrally providing users with electric energy, cold energy and heat energy[2]. However, the energy gradient utilization method cannot meet the "diversified needs" of urban energy users. At the same time, in the context of the Energy Internet, users no longer exist only as energy consumption links. The development of new energy power generation technology has led to the participation of users in power generation methods such as photovoltaic user groups. Users can also act as energy producers to further influence the urban energy interconnection system. The planning brings challenges.

Therefore, this article uses the urban energy Internet as the background to divide urban users into residential users, industrial and commercial users, and electric vehicle users. Firstly, it sorts out the connotation of the energy Internet and proposes the basic architecture of the new urban user-side energy interconnection system; secondly, it summarizes the modeling methods of the user-side energy interconnection system; thirdly, according to the "source-network-load" structure, the user-side energy interconnection The system configuration is analyzed; finally, the future direction of planning the user-side energy interconnection system that needs further research is prospected.

2. Connotation of Energy Internet and System Architecture

2.1. Connotation Of Energy Internet

The essence of the energy Internet is the large power grid as the "backbone network". Each distributed energy network forms an energy local area network (LAN), and multiple LANs are connected through electricity, gas, and heat networks to form an energy wide area network (WAN). The Energy Internet is based on the power grid and integrates multiple renewable energy power generation technologies, distributed energy technologies, and information and communication technologies. It combines energy flow and information flow to establish energy in which multiple entities participate in supply, multiple entities participate in regulation, and multiple entities participate in consumption. Shared network.

2.2. New Urban User-side Energy Internet Architecture

The new urban users are mainly residential users, industrial and commercial users, and their load is limited by location constraints. Among them, industrial and commercial users are composed of multi-energy conversion equipment and energy storage equipment, which are the core of the energy coupling of the user-side energy system; residential users are composed of many It is composed of energy consumption equipment and is the main energy consumption terminal in cities and towns. In addition, with the increasing development of new energy power generation technology, a large number of electric vehicle users have emerged, which is characterized by the random distribution of load positions in the transportation network. The architecture of the user-side energy interconnection system is shown in Figure 1.

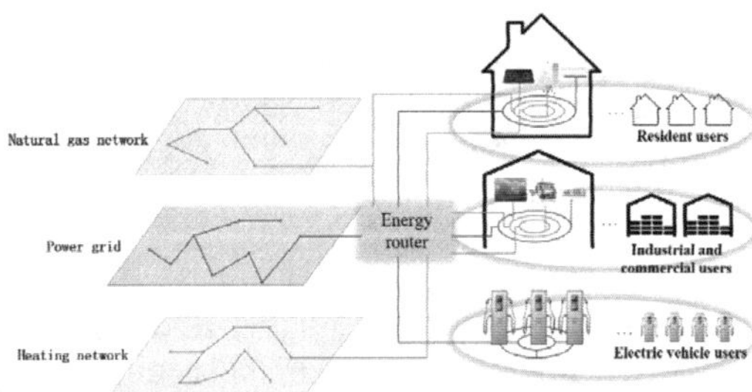

Figure 1. Basic architecture diagram of the new urban user-side energy interconnection system

3. User-side Energy Interconnection System Modeling

3.1. Energy Conversion Device

Industrial user energy interconnection systems are usually equipped with multiple energy conversion devices. Among them, combined heating and power (CHP) and power to gas (P2G) are the core of industrial users' energy conversion, which promotes a variety of Energy coupling.

1) Combined heat and power plant

The proposal of CHP deepens the coupling between the power grid and the heating network. Its typical mathematical model is[3]:

$$P_{H,MTE}(t) = \frac{P_{MT}(t)(1-\eta_{MT}-\eta_L)}{\eta_{MT}} \tag{1}$$

$$P_{H,MT}(t) = P_{H,MTE}(t)\eta_r C_o \quad (2)$$

$$H_{MT}(t) = P_{H,MT}(t)\Delta t \quad (3)$$

In the formula, $P_{H,MTE}(t)$, $P_{MT}(t)$, η_{MT}, η_L Respectively, the output waste heat power, output electric power, power generation efficiency and heat loss rate of the micro-gas turbine at time t; $P_{H,MT}(t)$, η_r, C_o Respectively, the heating power, heating coefficient and flue gas recovery rate at time t; $H_{MT}(t)$ Heating capacity at time t.

2) Electric to gas device

The P2G concept first proposed by Germany is mainly divided into two steps: electricity-to-hydrogen and electricity-to-natural gas. The proposal of P2G promotes the coupling of power grid and gas grid, and its basic principle is shown in Figure 2.

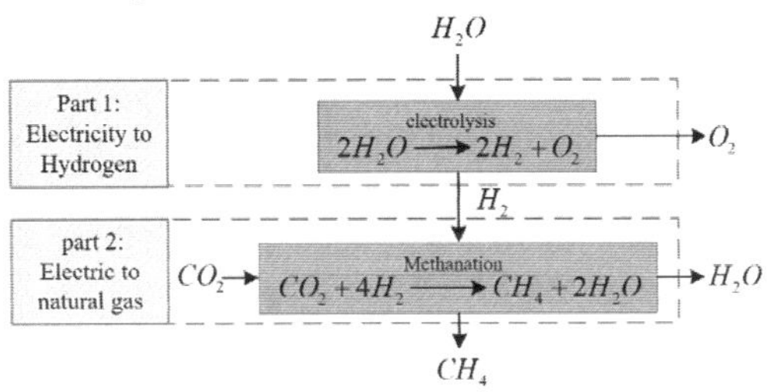

Figure 2. P2G device flow chart

3.2. Household Appliances Model

In the consumer segment of the urban energy Internet, the traditional household power supply method can no longer meet the development needs of the energy Internet. At present, the main solution to such problems is to divide the household electricity load into delayable load and non-delayable load, and build a family microgrid to flexibly dispatch energy to meet the electricity demand of household users[4].

1) Electrical load can be delayed

The load can be adjusted by delaying the load of electrical appliances, such as air conditioners, water heaters, etc. It has the characteristics of flexible scheduling, and its mathematical model is:

$$\sum_{h=\alpha_a}^{\beta_a} s_a(h) = d_a \quad (4)$$

$$s_a(h) = \begin{cases} 0, & \text{if } h \in [1,\mathrm{H}] \text{ and } h \notin [\alpha_a, \beta_a] \\ 0 \ or\ 1, & \text{if } h \in [\alpha_a, \beta_a] \end{cases} \quad (5)$$

$$P_{def}(h) = \sum_{a=1}^{A} S_a(h) P_a, \forall h \in [1,2,\cdots H] \quad (6)$$

In the formula, $[\alpha_a, \beta_a]$ Indicates the allowable working range of the electrical load that can be delayed; d_a Indicates the period of time during which the electrical load can be delayed; 0 and 1 respectively indicate that the electrical load can be delayed in the standby state and the working state; H indicates that the time of day is divided into several equal time periods, period $h \in [1,2,\cdots H]$; A indicates the number of electrical loads that a resident user can delay, $a \in [1,2,\cdots A]$; $P_{def}(h)$ Indicates the total electrical power used by the electrical load that can be delayed.

2) Do not delay electrical load

Non-delayable electrical load, that is, non-adjustable load, refers to electrical equipment that cannot participate in optimal dispatch, such as induction cookers, lights, etc. If controlled to participate in dispatch optimization, it will cause extreme discomfort to residents. Therefore, the electrical load that cannot be delayed can be considered as a constant power.

4. RESEARCH ON THE "SOURCE-NETWORK-LOAD" PLANNING OF USER ENERGY INTERCONNECTION SYSTEM

4.1.Multi-source Collaborative Planning Based On User-side Energy Routers

At present, the construction of an energy Internet with energy routers as the core coordinated control is considered to be an important means to achieve a balance between supply and demand after a high proportion of renewable energy is connected to the grid[5]. The energy transmission model of the energy router is shown in Figure 3.

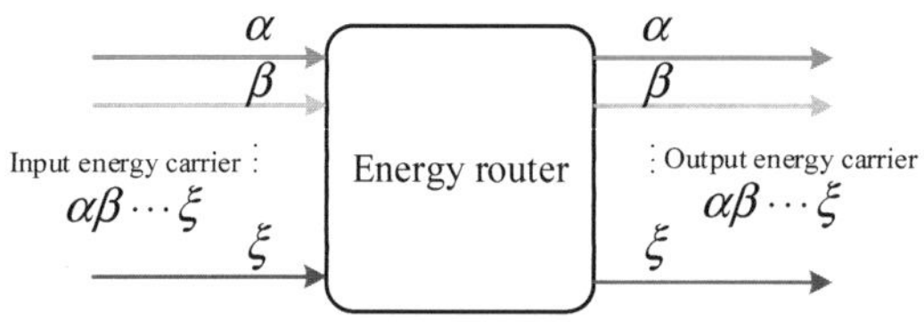

Figure 3. Energy transmission model of energy router

In view of the functions and characteristics of user-side energy routers, [6] proposes that energy routers should meet multiple energy access ports, have a plug-and-play interface, and include multiple communication technologies to meet the needs of two-way intelligent interconnection of energy. For the user-side energy LAN planning, [7] proposes to use the energy router as the core control device of the energy LAN, divide the user-side energy Internet into multiple energy LANs, and establish a two-layer planning model for the user-side energy interconnection system. The routers coordinate and complement each other to realize the complementarity of multiple energy LANs. On this basis, [8] proposed the concept of virtual routers, that is, through information and communication technology, multiple distributed energy routers are centralized, coordinated and optimized, and a virtual router multi-energy coupling correlation model is established to realize the connection between user-side energy LAN and energy WAN. Energy interaction.

4.2.Cooperative Planning Based On Multiple "networks" Of User-side Energy Network-transportation Network Coupling

At present, the research on urban user-side energy network planning is mainly divided into: grid-gas network coordinated planning[9], grid-heating network coordinated planning[10], grid-gas network-heating network coordinated planning[11], and energy network The resulting traffic electrification is also an important part of energy Internet planning.

The current research on user-side energy grid-transportation collaborative planning is mainly aimed at rationally planning new energy charging stations to promote grid-transportation coupling. Reference [12] proposes an integrated photovoltaic storage and charging electric vehicle charging station that comprehensively considers the intermittentity of photovoltaic power generation and the economics of charging and discharging. The capacity and number of charging piles in the station are decision variables to establish a multi-objective function planning model. The calculation example verifies that the integrated optical storage and charging station has better economic efficiency, and at the same time promotes the coupling of the energy network and the transportation network.

With the continuous development of user energy interconnection systems, electric vehicle users, as small and irregular distributed energy storage devices in the energy interconnection system, that is, distributed power sources, how to establish a distributed power source model, simulate their driving and charging behavior, and promote electrical -The transportation network is highly coupled, which is worthy of further in-depth study.

4.3.Multi-"load" Collaborative Planning Based On User-side Comprehensive Demand Response

The traditional demand response (DR) response model is not enough to respond to the response demand of the energy interconnection system. The integrated demand response (IDR) has effectively solved this problem. IDR is the foundation of DR In the above, considering the high proportion of new energy access and the complementary characteristics of each energy, that is, the user-side demand is no longer unilaterally defined as electricity demand, and the electricity users are expanded to electricity-heat-gas and other multi-energy users, which effectively promotes renewable energy Of consumption.

In the residential, industrial and commercial user-side energy interconnection system planning that takes into account IDR, for industrial users, [13] proposes to minimize cost as the goal, and establishes a planning model that takes IDR into account. Finally, heat-electricity coupled industrial users As an example, it is verified that the planning model that takes IDR into account can better utilize the complementary characteristics of industrial parks. For residential users, [14] considers the electricity consumption habits of residential users and establishes an IDR optimization model for urban residents, which verifies that considering the IDR of residential users can improve the response potential.

Different from traditional industrial, commercial and residential users, electric vehicle users are affected by traffic congestion, charging price, charging waiting time, remaining battery life, etc., and the location of load power is not fixed and random, as shown in Figure 4. How to make reasonable use of the response potential of electric vehicle users while ensuring the satisfaction of electric vehicle users is an important topic for user-side energy Internet planning. For the energy interconnection system planning that considers the IDR of electric vehicle users, the charging behavior of users must first be considered.

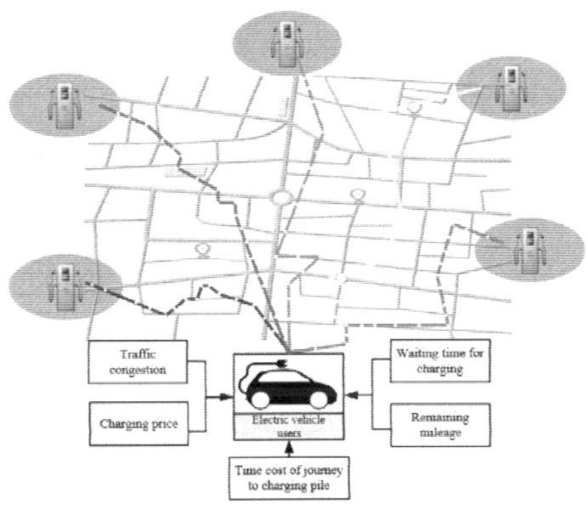

Figure 4. Factors to be considered when planning electric vehicle users

As mentioned above, considering the user-side IDR can improve the grid's ability to absorb clean energy to a certain extent and reduce the pressure on the grid. However, most of the current research on user-side IDR studies its impact on system reliability from a macro perspective, taking the user-side as a whole. However, the user's subjective willingness, user comfort and other behaviors are also issues that cannot be ignored.

5. PROSPECTS FOR IN-DEPTH RESEARCH

5.1.Coordinated Planning Of DC Conversion Of Electrical Equipment Considering User Load

At present, most of the research related to user-side energy Internet planning focuses on the coupling of multiple energy sources. However, with the continuous development of power electronics technology, electrical equipment has gradually moved towards the "direct current" road. Joint planning of a multi-energy coupling interconnection system and a low-voltage DC system in the user-side energy interconnection system to reduce unnecessary AC/DC conversion links is of great significance to reducing system losses and improving system efficiency.

5.2. Considering Users' participation In The Coordinated Planning Of Electricity Spot Market Transactions

With the increasing access to a high proportion of new energy sources, user needs are gradually showing diversified characteristics. The traditional mid- and long-term power market transaction model can no longer take into account the volatility and intermittent characteristics of new energy. Electricity spot market transactions adopt the trading mechanism of day-to-day real-time energy trading, which is conducive to promoting new energy consumption, realizing short-term real-time energy trading, and having better economic benefits. How to realize a transaction mode in which residential users participate in bidding with households as a unit and industrial and commercial users with parks as a unit, and build an energy interconnection system that considers users' participation in power spot market transactions, will help to further optimize the configuration of the energy interconnection system for urban users.

6. Summarize

In the context of the continuous increase in the proportion of urbanization and the continuous development of energy interconnection technology, research on user-side energy interconnection systems has long-term significance for the realization of the global energy interconnection. This article starts from the current research status of the urban user energy interconnection system, and summarizes the system architecture, system modeling, system planning, and future issues to be considered. It is of great significance to the development of the new urban energy Internet.

References

[1] Wang D, Meng Z J, Jia H J, Li J R, Yu J C, Zhang Z Y. Research and prospects on key technologies of energy interconnection system planning for new towns[J]. Automation of Electric Power Systems, 2019, 43(14): 16-28.

[2] Zhang X, Zhang Y M, Ji X Q, Han X S, Yang M, Xu B. Considering the coordinated optimal dispatch of transmission and distribution of the electricity-gas-heat integrated energy system[J/OL]. Power grid technology: 1-14[2021-11-24].https://doi.org/10.13335/j.1000-3673.pst.2021.2065.

[3] Liu H, Zhu T, Zhang T. Evaluation and analysis of CCHP-ORC system for different types of buildings in Shanghai area[J]. Proceedings of the Chinese Society for Electrical Engineering, 2016, 36(12): 3198-3206. DOI: 10.13334/j. 0258-8013.pcsee.152217.

[4] Li S L, Yang J J. Real-time optimization strategy of household renewable energy based on dual battery packs [J]. Renewable Energy, 2019, 37(07): 1007-1014.

[5] Wang J Y, Li Y, Lu Z M, Sheng W X, Cao J W. Research on local energy internet based on energy switches and routers[J]. Proceedings of the Chinese Society of Electrical Engineering, 2016, 36(13): 3433-3439+3362.

[6] Liu X Z, Zheng Z D, Li Y D. Multi-port energy router using power electronic transformer[J]. Electric Drive, 2016, 46(04): 80-83.

[7] Wu C, Tang W, Bai M K, Zhang L, Cong P W. User-side energy internet planning based on two-tier planning[J]. Journal of Electrotechnical Technology, 2017, 32(21): 122-131.

[8] Zhang L Q, Chen A K, Gu J, Xie D, Gu C H. Regional energy wide area network collaborative planning based on multi-energy network routing algorithm[J]. Proceedings of the Chinese Society of Electrical Engineering, 2020, 40(23): 7499-7511.

[9] Yang Z J, Gao C W, Zhao M. A review of research on power-natural gas network coupling system[J]. Automation of Electric Power Systems, 2018, 42(16): 21-31+56.

[10] Yu D L, Cao J, Tu C W, Wang Z L, Lv C, Wang H F. Optimal energy flow analysis of electric-thermal multi-energy system considering linearized network constraints[J]. Proceedings of the Chinese Society of Electrical Engineering, 2019, 39(07):1933-1944.

[11] Liu X R, Li Y, Sun Q Y, Pan Y L. Dynamic state estimation of electricity-gas-heat coupling network based on multiple time scales[J]. Power System Technology, 2021, 45(02): 479-490.

[12] Zhang Y, Han W, Song C, Yang S Y. Joint optimization of facility planning and operation of integrated solar storage and charging power stations with electric vehicles [J/OL]. Energy Storage Science and Technology: 1-12 [2021-11- 18].https://doi.org/10.19799/j.cnki.2095-4239.2021.0481.

[13] Gao Q, Liu C, Jin D J, Dong Y B, Chen C M, Lin Z Z. Optimal configuration of the park's integrated energy system considering comprehensive demand response[J]. High Voltage Electrical Apparatus, 2021, 57(08): 159-168.

[14] Luo J M, Wen Z C, Dong W J, Li Z W, Liu Z X, Wu H T. Quantitative model of comprehensive energy demand response potential based on residential user portraits[J]. Renewable Energy, 2020, 38(10): 1407-1414.

Optimal Control Measures For Dynamic Stability Level Of Northeast Power Grid

Shi xuan LU[1], Yang di SHAO[2*], Peng YE[3], Tian he SUN[4], Xue jie WANG[5], Hong peng ZHANG[6], Chun shan LI[7]

[1,2,3,4,5]Shenyang Institute Of engineering, Shenyang, China

[6,7]Northeast Branch of State Grid Corporation of China, Shenyang, China

[*] Corresponding author: 1619198316@qq.com

Abstract—Due to the rapid development of new energy such as nuclear power at the transmission end of Northeast power grid, the improvement of transmission capacity of Lugu UHVDC and the large-scale development of power base, the problem of dynamic stability of Northeast Power Grid has surfaced. Based on the reality of Northeast power grid, this paper deeply and carefully carries out the dynamic stability level analysis of Northeast power grid from three aspects of source, network and load, studies the effective methods and measures to improve the damping ratio and robustness of main vibration modes between power grid regions, takes into account the scientificity, systematicness and scientificity of the measures, and provides reference for domestic and foreign related grid operation.

1. Introduction

With the rapid and vigorous development of new energy in the whole network, the continuous expansion of the interconnection scale of regional power grids and the increasingly tense relationship between energy supply and demand, the installed capacity of Liaoning Hongyanhe Nuclear Power Plant is also increasing, and the optimization of large-scale weak transmission end energy base in Northeast Power Grid continues to appear.

The dynamic stability of power grid is generally characterized by low-frequency oscillation with weak damping or even negative damping. Low frequency oscillation not only threatens the dynamic stability of large regional power grid, but also restricts the transmission capacity of long-distance transmission[1]. The research methods of low-frequency oscillation mainly include eigenvalue analysis method, time-domain simulation method and signal analysis method. Reference [2] uses eigenvalue analysis to linearize the system near the working point, form a state matrix, and analyze the sensitivity and oscillation mode of the system. If the real parts of the eigenvalues obtained by solving the equation of state are all negative, the system is stable; If there is a positive real part in the characteristic root, the system is unstable. Reference [3] uses the time-domain simulation method to calculate the complete time response of system variables by using the nonlinear equation generated by system disturbance, and obtains the damping characteristics and frequency of system oscillation mode through curve simulation. Reference [4] uses the signal analysis method to identify the oscillation mode, frequency and other information of the system mainly through the measured or simulated data, and quantitatively analyze the oscillation damping problem. Commonly used signal analysis methods include Fourier transform analysis method, wavelet analysis method, Prony method, HHT, etc.

Based on the actual operating conditions of the Northeast Power Grid and starting from the basic principles of dynamic stability, this paper mainly analyzes the impact of changes in the internal parameters of the power grid on the dynamic stability of the Hongyanhe Nuclear Power Plant after the grid-connected operation of the Hongyanhe Nuclear Power Plant, and proposes relevant measures to improve the dynamic stability of the power grid.

2. DYNAMIC STABILITY ANALYSIS MECHANISM OF POWER SYSTEM

2.1.Classification And Significance Of Power System Dynamic Stability

The dynamic stability of power system belongs to the power angle stability of power angle, voltage and frequency. It refers to that the power system can recover to the operation state before disturbance or close to the acceptable stable operation state under the action of automatic regulation and control device after small disturbance and large disturbance[5]. The main form of dynamic stability is low-frequency oscillation. There are two forms of low-frequency oscillation: one is interregional oscillation, which is the oscillation of one part of the system cluster relative to another part of the cluster. It involves a wide range and does great harm, and will be transmitted to the whole system through the tie line[6]; The second is local oscillation, which is the mutual oscillation between several generators with close electrical distance or the oscillation between other units in the system as a whole. Local oscillation is limited to the region and belongs to small-scale oscillation. It is usually manifested as the low-frequency oscillation of a single power plant through a long-distance transmission line[7].

The essential reason of low frequency oscillation in power system is that some inherent oscillation frequencies of the system lack sufficient damping. The dynamic stability of the system depends on the inherent characteristics of the system and is independent of the size of the disturbance[8].

2.2.Dynamic Stability Model Analysis

In the power system, taking the conventional thermal power unit as an example, the generator excitation system parameters of thermal power unit will affect the damping ratio of the system, which is affected by the generator coefficient, grid structure, installed capacity and startup mode. For a single machine infinite bus system, a fifth-order model can be used for data expression[9]:

$$K_D = K_1 K_2 K_3 K_4 \tag{1}$$

In the formula, K_D is the same damping torque coefficient; K_1 is the influence of the generator's own parameters on the damping coefficient; K_2 reflects the influence of the equivalent reactance of the tie line; K_3 is the terminal load of the generator, which reflects the influence of the equivalent system on the local load; K_4 is the influence of tie-line power on the damping coefficient.

K_1, K_2, K_3 and K_4 can be expressed as:

$$\begin{cases} K_1 = \dfrac{C_D}{2T_J} \\ K_2 = \dfrac{X_G}{X_G + X_E} \\ K_3 = \dfrac{1}{U_2^4 + \dfrac{X_G^2 X_E^2}{(X_G + X_E)^2 P_L^2}} \\ K_4 = U_2^4 \sqrt{1 - \dfrac{P_S^2 X_E^4}{U_2^4} - \dfrac{X_G X_E^2 P_S P_L}{U_2^4 (X_G + X_E)}} \end{cases} \tag{2}$$

In the formula, T_J is the inertia time constant of the generator; X_G is the internal impedance of the isolator; X_E is the equivalent reactance between the system tie lines; P_L is the local load equivalent to the machine-end system; P_S is the active power delivered by the tie line to the receiving end system; U_2 is the bus voltage of the receiving end system. Among them:

$$C_D = U_2^2[(\frac{1}{X_d^{'}} - \frac{1}{X_d})T_d^{'} + (\frac{1}{X_d^{"}} - \frac{1}{X_d^{'}})T_d^{"} + (\frac{1}{X_q^{"}} - \frac{1}{X_q})T_q^{"}]\omega_0 \quad (3)$$

In the formula, X_d, $X_d^{'}$ and $X_d^{"}$ are generator d-axis synchronous reactance, transient reactance and sub-transient reactance respectively; X_q and $X_q^{"}$ are synchronous reactance and subtransient reactance of generator q-axis respectively; ω_0 is the rated angular velocity of the system.

The relationship between power grid dynamic stability level and system damping ratio can be found:

The damping level of the system is closely related to the reactance of the tie line. As the reactance of the tie line increases, the damping torque coefficient of the system will decrease, and the damping level of the system will also decrease. Therefore, serious low-frequency oscillation problems will occur in long-chain, weakly interconnected systems.

Along with the influence of the operation mode of the system, whenever the complexity of the system tie line increases, the damping level of the system will decrease accordingly. Therefore, the phenomenon of low frequency oscillation problems in large power grids and large systems will be more frequent.

2.3.Dynamic Stability Of Northeast Power Grid

Northeast power grid is located in Northeast China, including three provinces and one Mongolia (Heilongjiang Province, Jilin Province, Liaoning Province and Eastern Inner Mongolia).

Under the current grid structure, the main vibration mode of Northeast power grid is the oscillation mode of Southern northern cluster, that is, the oscillation mode of Liaoning and Mengdong Chifeng cluster to Heilongjiang and Mengdong Hulunbuir cluster, as shown in Figure 1.

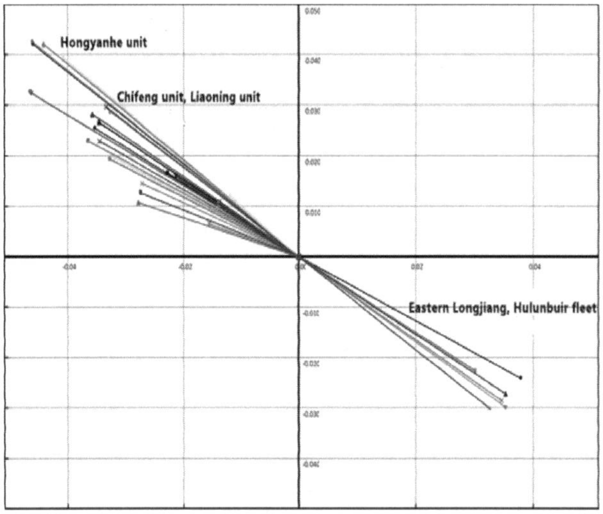

Figure 1. Northeast Power Grid Group Oscillation Mode

If all 6 units of Hongyanhe Nuclear Power Plant are connected to the grid and put into operation, and single-phase instantaneous fault occurs in Chifeng outgoing 500 kV transmission line under full power generation mode, Chifeng unit and Liaoning unit will produce equal amplitude oscillation. Figure 3 shows the low frequency range oscillation curve of Northeast power grid.

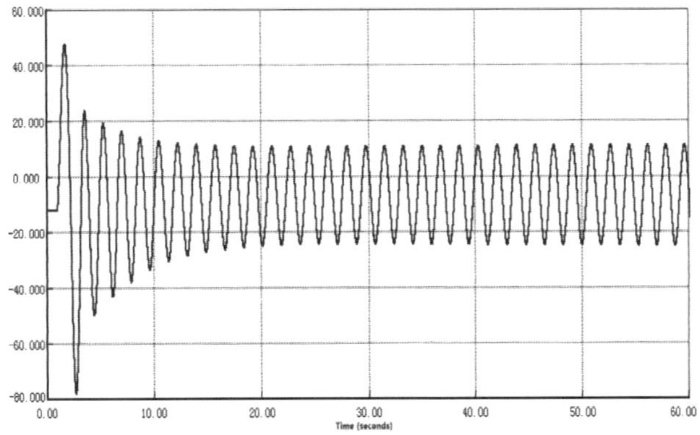

Figure 2. Equal amplitude oscillation of Northeast Power Grid units

3. Analysis of Influencing Factors of Dynamic Stability Level

The dynamic stability level of the power grid is affected by multiple factors such as power structure, power distribution, excitation system and debugging system of power regulation, grid structure, load distribution, load type, etc. The detailed analysis is given below.

3.1.The Influence Of Different Load Types On Dynamic Stability

The load model of electromechanical simulation analysis in China is mainly based on "constant impedance and motor" mode, and the load model of Northeast Power Grid adopts 50% constant impedance + 50% motor model.

TABLE 1. Main vibration mode of Northeast Power Grid under different constant impedance / static characteristic load modes

Constant Impedance Load/Static Characteristics Load/%	Oscillation Frequency/Hz	Damping Ratio/ %
40	0.632	8.160
50	0.632	8.011
60	0.632	7.876

Through the analysis, it can be concluded that the lower the proportion of constant impedance and the higher the proportion of motor, the higher the system damping ratio, which has a favorable impact on the dynamic stability of the system, but the impact is small. There is no dynamic instability problem under small disturbance under this oscillation mode.

3.2.Influence Of Different Load Scale On Dynamic Stability

Through the analysis of three modes of large load, waist load and small load in Northeast power grid at the end of the 14th five year plan, it is concluded that the larger the load scale, the smaller the oscillation frequency of the system and the higher the damping ratio.

TABLE 2. Main vibration modes of Northeast Power Grid under different load levels

Load Level /MW	Oscillation Frequency /Hz	Damping Ratio /%
Heavy load（69000）	0.613	8.748
Lumbar load（58000）	0.632	8.011
Small load（54000）	0.641	7.988

3.3.Influence Of Different Transmission Capacity Of Lugu DC On Dynamic Stability

Through the analysis, it can be concluded that the larger the Lugu DC transmission capacity, the lower the main vibration frequency and the higher the damping ratio of the Northeast power grid, and the oscillation center is close to the Jilin power grid side of the Liaoji section. Under this oscillation mode, there is no dynamic instability problem under small disturbance.

TABLE 3. MAIN VIBRATION MODES OF NORTHEAST POWER GRID UNDER DIFFERENT LUGU DC TRANSMISSION MODES

Lugu DC Transmission Level /MW	Oscillation Frequency /Hz	Damping Ratio /%
6100	0.632	8.011
8000	0.625	7.945
10000	0.613	7.763

3.4.Influence Of Grid Reinforcement On Dynamic Stability

Grid structure is the material basis of power transmission and the most important guarantee for system security and stability. It can be seen that the grid strengthening measures can not only improve the dynamic and stable system damping ratio, but also increase the voltage stability and frequency stability of the power grid, which is of great engineering significance.

TABLE 4. INFLUENCE OF STRENGTHENING GRID STRUCTURE ON MAIN VIBRATION FREQUENCY AND DAMPING RATIO OF NORTHEAST POWER GRID

Grid Reinforcement	Main Vibration Frequency /Hz	Damping Ratio /%
No reinforcement	0.564	5.071
Strengthen Chifeng outgoing grid	0.566	6.016
Strengthen the incoming grid of load center	0.576	6.217
Strengthen the North-South transmission grid	0.632	8.011

3.5.Influence Of Synchronous Power Distribution On Dynamic Stability

Based on the analysis of the different distribution positions of water, fire, and nuclear power using synchronous generators in the current Northeast power grid, it can be seen that when the synchronous generator is connected to the power base at the end of the power grid, the corresponding system damping ratio shows a downward trend; When the machine is connected to the center of the grid, the damping ratio of the system is on the rise.

TABLE 5. INFLUENCE OF SYNCHRONOUS POWER DISTRIBUTION ON DYNAMIC STABILITY

Project	Oscillation Frequency /Hz	Damping Ratio /%
Current grid structure	0.564	5.071
Hongyanhe Nuclear Power Plant adds 2 units	0.543	4.673
Add 6 units of synchronous power supply to the grid center	0.549	5.287

3.6.Influence Of Excitation And Speed Regulation System Parameters On Dynamic Stability

Hongyanhe Nuclear Power Plant of southern Liaoning power grid will add No. 5 and No. 6 units. At that time, the installed capacity of a single power plant will exceed 6000mw, aggravating the North-South main vibration mode of Northeast power grid.

If the current PSS and governor models and parameters are still adopted, the system damping ratio will be reduced to 2.58%, which is lower than the grid requirement of 3%.

If the active power measurement delay time of type 9 speed governing system of units 2 to 6 of Hongyanhe Nuclear Power Plant is adjusted from 0.545 s to 2.038 s, the output power range P_{max} of prime mover is changed from 1.05 to 1.0; Increase the gain K_p of PSS from 4 to 8; After the amplitude limit is increased from 5% to 10%; The damping ratio is increased from 2.58% to 7.02, which is the best economic and practical measure.

4. Conclusions

Based on the actual situation of Northeast power grid, this paper focuses on the research on the dynamic stability of power grid from the aspects of load model, load scale, Lugu DC transmission capacity, grid strengthening measures, power distribution, proportion of new energy, excitation of synchronous power supply and influence of excitation system, and puts forward effective measures to improve the dynamic stability performance of power grid. The following conclusions are drawn:

4.1. Give Priority To PSS And Speed Regulation System Model Parameter Optimization Measures To Improve The Dynamic Stability Of Power Grid.

The important inducement of dynamic stability of modern power system is fast excitation and speed regulation system. Therefore, optimizing and adjusting the models and parameters of Power System PSS and speed regulation system is the most economical, fastest and effective measure to improve power grid damping and dynamic stability.

4.2. Strengthening The Grid Structure Is The Most Stable Measure To Deal With The Problem Of Power Grid Dynamic Stability.

The other important inducement to the dynamic stability of modern power system is the long-distance and weak connection transmission of power. Strengthening the grid structure can significantly shorten the electrical distance of power transmission, which can not only improve the dynamic stability of power grid, but also improve the transient stability and static stability of power grid. It is the cornerstone of long-term stability of power grid development.

4.3. Strengthening The Flexibility Of Active And Reactive Power Resource Allocation Of Large-scale Weak Transmission End Energy Base Can Improve The Dynamic Stability Of Power Grid.

Large scale weak transmission end energy base will inevitably aggravate the dynamic stability of power grid. In addition to relying on the long-term network architecture, it is necessary to improve the dynamic voltage stability and frequency stability of the power base transmission system and improve the damping characteristics of the system from the power side.

References

[1] Xue A C, Wang J W, Liu X B. Ultra low frequency oscillation of power system Mechanism analysis and suppression research status and prospect [J]. China Electrical Engineering Report, 2020, 41 (2): 553 – 568.
[2] Zhao X W, LV S X, Xie H. Overview of low frequency oscillation in power system [J]. Heilongjiang electric power technology, 2019, 41 (5): 392 – 395.
[3] Li X, Gong Q W. Review of analysis methods for dynamic characteristics of power system low frequency oscillation modes [J]. Shaanxi electric power, 2013, 41 (12): 24 – 29.
[4] Wang C, Li J H, Li Y Y. Analysis of low frequency oscillation characteristics of East China Power Grid after huaihu UHV is put into operation [J]. Power system automation, 2013, 37 (18): 120 – 125.
[5] Deng B W. Voltage stability analysis of AC / DC hybrid transmission system based on continuous power flow [J]. Journal of Shenyang Institute of Engineering (NATURAL SCIENCE EDITION). 2018, 14 (4): 334 – 338.

[6] Zhu F, Zhao H G, Liu Z H. Influence of regional power grid interconnection on power system dynamic stability [J]. Chinese Journal of electrical engineering, 2007, 27 (1): 1 – 7.
[7] Liang Z F, Xiao M, Zhang K. Discussion on low frequency oscillation control strategy of China Southern Power Grid [J]. Power system automation, 2011, 35 (16): 54 – 58.
[8] Wang Z Y, Chang X Q, Long C. Two stage low frequency oscillation modal identification algorithm based on dynamic system [J]. Journal of power system and automation, 2020, 32 (6): 108 – 114.
[9] Wang B, Xu Y P, Dang J, Li W F, Ma S J, Wang Y. Evolution law of dynamic stability characteristics and construction of prevention and control system of central China Power Grid [J]. Power grid technology, 2021,45 (08): 3238-3246.

Overview Of Measurement And Evaluation Methods Of Wind Farm And System Inertia Under Large-scale Wind Power Grid Connection

Peng YE[1], Hong yu YANG[2*], Shi cheng WANG[3], Xiao yi QIAN[4], Xue jie WANG[5], Feng SUN[6], Qiang ZHANG[7]

[1]Shenyang Institute Of engineering Shenyang, China

[2]Shenyang Institute Of engineering Shenyang, China

[3]Liaoning Siji technology service Co., Ltd Shenyang, China

[4]Shenyang Institute Of engineering Shenyang, China

[5]Shenyang Institute Of engineering Shenyang, China

[6]Electric Power Research Institute of State Grid Liaoning Electric Power Co .,Ltd Shenyang, China

[7]Electric Power Research Institute of State Grid Liaoning Electric Power Co .,Ltd Shenyang, China

[*] Corresponding author: yanghyu1995@163.com

Abstract: With the large-scale development of new energy, large-scale wind power is incorporated into the power grid, the inertia level of the power system continues to decrease, the system regulation ability continues to weaken, and the system stability is seriously challenged. The evaluation of inertia has become a hot research issue for scholars. Firstly, this paper introduces the basic concept and components of power system inertia. Then, the different methods of inertia measurement and evaluation of wind farm and power system are reviewed. Finally, the future research directions in the field of power system inertia evaluation are summarized.

1.INTRODUCTION

In order to solve the current energy crisis and meet the requirements of sustainable development, the state has strongly supported the development of new energy power generation in recent years . The traditional thermal power generation mode is gradually being replaced and supplemented by new energy power generation such as wind power and photovoltaic . The proportion of new energy power generation in many regions has reached 50% of the total[1].

In order to meet the future energy demand and make wind power better used. In recent years, scholars at home and abroad have done a lot of research on the measurement and evaluation of wind farm inertia and power system inertia. The research on power system inertia mainly includes system identification, inertia evaluation methods based on measured data processing and so on. Reference [2] calculates the inertia of the system by measuring the frequency change of the system and combining the motor rotor motion equation. However, it is difficult to obtain the initial value of power disturbance and disturbance time accurately. Reference [3] based on the measured data of Iceland, it is proposed that the inertia can be estimated from the environmental measurement data. A method for estimating the effective inertia of power system from ambient frequency and active power is proposed. Reference [4] adopts the sliding window processing method for the frequency and power measurement data, sets the appropriate window length and sliding step, and realizes the online estimation of inertia by using the system equivalent rotor motion equation. For the inertia evaluation of wind farm stations, the main research direction is quantitative evaluation. Reference [5] studies the quantitative analytical characterization method of equivalent virtual inertia of wind farm. Reference [6] analyzes the evaluation method of available virtual inertia of doubly fed fan under all wind conditions by modeling the wind speed of wind farm. Reference [7] establishes a doubly fed wind turbine model with virtual inertia control to calculate the equivalent virtual inertia time constant after wind farm aggregation. Reference [8] established CAR identification model by measuring kilometer increment and frequency deviation of wind farm, and verified the accuracy of the model by comparing the coincidence degree of predicted frequency disturbance and measured frequency disturbance curve of the model.

In view of the above problems. Firstly, this study introduces the definition and composition of power system inertia. Then, the research on the measurement and evaluation of inertia of wind farm and power system at this stage is reviewed, and the existing problems of inertia evaluation at this stage are discussed. This paper summarizes the measurement of wind power grid inertia in China.

2.INERTIA OF POWER SYSTEM

2.1.Definition And Composition Of Inertia

In power system, inertia is defined as the ability of the system to resist energy fluctuation and provide the fastest and most direct response to frequency change. Therefore, maintaining sufficient inertia is of great significance to frequency stability [6]. The traditional inertia is usually composed of grid connected synchronous generator, asynchronous motor and load. In recent years, due to the impact of carbon emission requirements, the power system has gradually turned to new energy such as wind and light. Therefore, at this stage, the inertia of power system is mainly composed of synchronous generator and asynchronous motor.

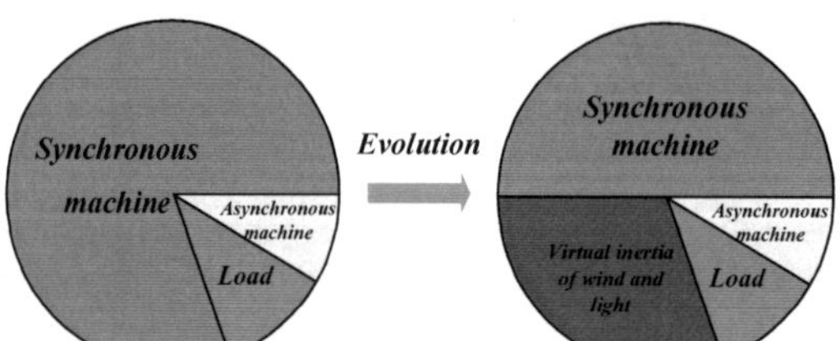

Figure 1. Evolution of inertia in modern power system

2.2.Traditional Inertia

In traditional inertia, synchronous generator is the main body of inertia. The rotational inertia of the synchronous motor is stored in the rotating rotor and expressed as formula (1):

$$E_{gen} = \frac{J\Omega^2}{2} \tag{1}$$

Where: E_{gen} is the inertia of the synchronizer; J is the moment of inertia; Ω is the mechanical angular speed of the synchronizer. It can be seen from equation (1) that the rotational inertia provided by the synchronous machine is determined by its own parameters and has nothing to do with its output.

The inertia constant of the synchronous machine is defined as the ratio of the rotating kinetic energy of the synchronous machine rotor to the rated capacity of the synchronous machine at the rated speed, which is formula (2):

$$H_{gen} = \frac{E_{nom}}{S_{gen}} = \frac{J\Omega_N^2}{2S_{gen}} \tag{2}$$

Where: H_{gen} is the inertia constant of the synchronizer; E_{nom} is the rotational kinetic energy stored at the rated speed of the synchronous machine; S_{gen} is the rated capacity of the synchronizer; Ω_N is the rated mechanical speed of the synchronizer.

2.3. Virtual Inertia

The representative of virtual inertia is wind power, photovoltaic and other new energy power generation technologies. According to the control mode of inertia source, it is divided into current source virtual inertia and voltage source virtual inertia

The current source type virtual inertia is to introduce the frequency change of the system into the active power control link of the converter, change the active power reference of the fan, and make the fan provide the active power proportional to the change rate to the power grid, and the expression is formula (3):

$$\Delta P = K\frac{df}{dt} \tag{3}$$

In the formula: K is the virtual inertia coefficient, which is directly proportional to the rotating inertia of the synchronous machine. The virtual inertia coefficient is not subject to physical constraints and can be set flexibly, but the system requirements and fan working conditions need to be considered. The current source virtual inertia control changes the output power in a short time, improves the unbalanced power of the system, and then reduces the frequency change rate. It does not increase the inertia of the system. It is the power response.

Voltage source virtual inertia mainly refers to virtual synchronous machine technology. Virtual synchronous machine refers to the technology of introducing synchronous machine rotor motion and electromagnetic transient equation into the converter control link to make it have the external characteristics of synchronous machine grid connected operation[9], which can simulate the characteristics of synchronous electromechanical voltage source.

3. MEASUREMENT AND EVALUATION OF WIND FARM EQUIVALENT INERTIA

3.1. Equivalent Virtual Inertia Estimation Principle

The wind turbine can provide dynamic active power support to the power grid by releasing its own kinetic energy or absorbing the system kinetic energy. Because the wind turbine applies virtual inertia control and the generator speed is coupled with the grid angular speed, according to the definition of virtual inertia of inertia, the virtual inertia of wind turbine can be expressed as formula (4):

$$H_{equ} = \frac{\Delta\omega_r \omega_{s0} \omega_{r0}}{\Delta\omega_s \omega_{nom}^2} H_{DFIG} \tag{4}$$

In the formula: H_{equ} is the virtual inertia of the wind turbine, ω_{r0} is the initial speed of the generator, $\Delta\omega_r$ is the speed increment of the generator, ω_{s0} is the initial angular frequency of the power grid, $\Delta\omega_s$ is the angular frequency increment, ω_{nom} is the rated angular frequency of the wind turbine unit, and H_{DFIG} is the natural inertia of the doubly fed wind turbine unit.

When the virtual inertia control strategy is applied to each unit, the whole wind farm will also show the equivalent virtual inertia characteristics, and the equivalent virtual inertia H_{WF} of the wind farm can be used to reflect its inertia response strength. The electromechanical transient characteristics can be expressed by the equivalent rotor motion equation in incremental form, and the time-domain expression can be obtained by Laplace transform (5):

$$g(t) = -\frac{1}{2H_{WF}}\exp(-\frac{D}{2H_{WF}}t) \tag{5}$$

In the formula: D is the damping coefficient. The equivalent frequency response process of the wind farm can be shown in Figure 2.

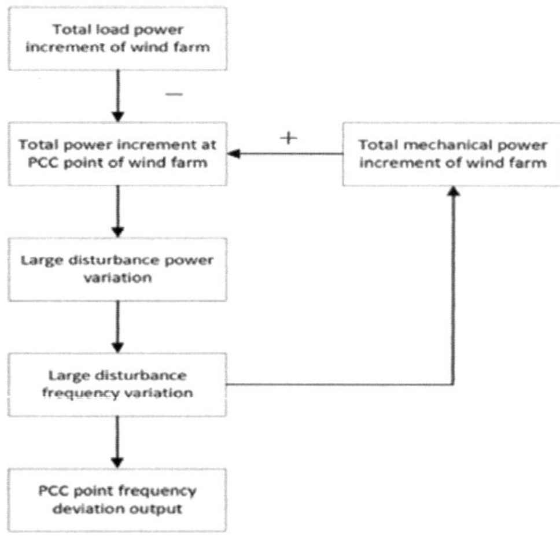

Figure 2. Equivalent frequency response process of wind farm

3.2.Measurement And Evaluation Method Of Equivalent Inertia Of Wind Farm

Reference [8] describes the wind farm system by using the controlled autoregressive model by estimating the equivalent virtual inertia of the wind farm as a whole. Because the car model contains unknown parameters, the author uses the time-varying forgetting factor recursive least squares algorithm to further refine the car model. The fast extraction process of virtual inertia needs to determine the model with the lowest order and the highest fitting degree. The final prediction error criterion is selected to carry out the two processes at the same time, which simplifies the complex problem. Finally, the identification parameters are brought into the identification model to compare the coincidence degree between the predicted frequency disturbance curve and the measured frequency disturbance curve. The higher the coincidence degree of the two curves, the higher the accuracy of the identification model. Reference [7] studies the relationship between the equivalent virtual inertia time constant and the dynamic frequency characteristics of power grid. The generator electromechanical transient model, rotor side converter model, virtual inertia controller model and speed controller model are established. Combined with the above models, the doubly fed fan model with virtual inertia control is obtained. The equivalent virtual inertia time constant is calculated by a fan, and the virtual inertia time constant of the whole wind farm is aggregated. Reference [6] combines the air propagation principle, obtains the spatial distribution and time variation characteristics of wind energy in the wind farm by analyzing the wake effect, wind shear effect and time delay effect, constructs the instantaneous wind speed condition probability distribution model under the influence of atmospheric turbulence by using the mixed copula function, and analyzes the operation condition of doubly fed fan under all wind conditions. The rotor speed of the wind turbine is obtained, and the available inertia and the confidence interval of the available inertia of the wind farm are calculated.

4.SYSTEM INERTIA MEASUREMENT AND EVALUATION

4.1.Power System Inertia Response

provides inertia for the system will spontaneously respond to the transformation of system frequency and absorb or release active power to the system. This process is called inertia response. For example, when the system power is unbalanced, the synchronous machine will spontaneously absorb or release the rotational kinetic energy in the rotor, so as to absorb or inject electromagnetic power into the power grid. So as to hinder the frequency fluctuation of the system. The formula can be expressed as (6):

$$p_e(t) \approx -\frac{2H_{gen}}{f_0}\frac{df(t)}{t}P_N \tag{6}$$

In the formula: $P_e(t)$ is inertia support power; P_N is the rated active power of the synchronizer; f_0 is the rated frequency of synchronizer and $f(t)$ is the frequency of synchronizer. Virtual inertia control and energy storage release or absorb energy to the system for a short time according to the control command to provide power support, so as to have the inertia support ability similar to the synchronous machine.

4.2.Measurement And Evaluation Method Of System Inertia

Reference [10] considers the physical meaning of inertia in the form of nodes. The relationship among inertia, system inertia and node inertia is considered. It is concluded that in the dynamic process, the support of inertia to frequency is embodied in the blocking effect on the frequency change of each node, which constitutes the blocking effect on the frequency change of the system as a whole. On this basis, it is obtained that the power system inertia has space-time characteristics, that is, the inertia of the same node changes with time, and the inertia of different nodes is different. Reference [11] also considers the frequency distribution characteristics. Firstly, the concept of square deviation of center frequency is defined. Considering the influence of frequency distribution characteristics on inertia evaluation, the concept of inertia graph center is proposed to determine the appropriate frequency acquisition location. The influence of primary frequency modulation on the accuracy of inertia evaluation is reduced. Reference [12], the wind farm and load response are considered, and the virtual inertia control is considered for the wind farm, which is equivalent to a fan. The zip model and zip plus asynchronous motor comprehensive model are compared and analyzed. The change rate of disturbance is obtained. The objective function is established by considering four parameters at the same time. However, the calculation is limited because the power loss of the power grid needs to be considered. Reference [13] studies the method of continuous estimation of grid inertia using active frequency noise disturbance information. By dividing the inertia estimation region, the overall inertia is divided into different parts. Equivalent modeling of power grid. For the small noise disturbance and large noise disturbance of the system, the inertia is estimated respectively.

5.CURRENT DEVELOPMENT AND PROSPECT

At present, scholars at home and abroad mostly focus on optimizing the frequency modulation control strategy of wind turbine. In contrast, there are relatively few studies on the participation of wind turbines or wind farms in the evaluation of power system frequency modulation service capacity. Whether it is the wind farm evaluation based on the virtual inertia control principle or the evaluation of the available inertia kinetic energy of wind turbine, the virtual inertia of the whole power plant can be estimated. The impact of specific environment on wind power generation is not considered. The impact of shutdown and fault caused by frequency fluctuation on the overall inertia of the wind farm is not considered. An accurate estimate that cannot be made. Therefore, the following research on the inertia of the wind farm should be specific to how to measure the inertia of the current wind farm, how to make different disturbances in the wind farm response system and the evaluation of the wind farm's ability to respond to disturbances. These are specific issues that need to be considered in reality. It is of great help to the current new energy power generation industry.

The proportion of new energy power generation in China is rising, and the current policy still supports new energy power generation. Thirdly, the test of power system becomes extremely arduous. Combined with the current research status, the following suggestions are put forward for the development of inertia control technology in China and the application of inertia demand assessment method in the system. (1) Develop virtual inertia technology for new energy power generation. The frequency security of the system is closely related to the permeability of renewable energy. Under the same power disturbance, the higher the permeability of renewable energy, the greater the frequency fluctuation. Therefore, developing virtual inertia technology for new energy power generation is an effective means to solve this problem.

(2) Put forward the inertia demand evaluation index and standard. At present, there is no definite standard and index for inertia in power system. As a result, the new energy power plant cannot correctly deal with inertia. Corresponding indicators and standards should be proposed from different scales of planning and operation, and the evaluation indicators should be simple and easy to use. At present, there is no clear standard and index for inertia in power system. (3) improve the on-line monitoring ability of system inertia. On the one hand, improving the inertia monitoring ability can support the formulation of more accurate inertia demand assessment indicators, on the other hand, it can provide guidance for the stable operation of the system. (4) Inertia service marketization. At present, the market operation of inertial service in China is not perfect. In the synchronous machine based power system, inertia has always been used as an "accessory" to provide services free of charge. With the increasing proportion of new energy power generation, the structural contradiction of power supply has become increasingly prominent. The provision of inertial services presents technical and economic differences, that is, it has become a tradable commodity with value differences.

6.CONCLUSION

As the dominant position of asynchronous power supply becomes more and more obvious, the characteristic of "low inertia" of power system will be further highlighted. Starting from the definition and composition of power system inertia, this paper discusses the basic principle of measuring inertia of wind farm. The current methods of measuring wind farm inertia are reviewed. After that, the change of system measurement inertia after large-scale wind power is connected to the power grid is discussed. Finally, the wind farm and power system are prospected and suggestions are put forward. I hope this paper can provide reference for carrying out relevant work.

REFERENCES

[1] Jian S , Li M , Zhang Z. Renewable energy transmission by HVDC across the continent: system challenges and opportunities[J]. CSEE Journal of Power and Energy Systems, 2017, 3(4):353-364.

[2] Noue T , Taniguchi H , Ikeguchi Y. Estimation of Power System Inertia Constant and Capacity of Spinning-Reserve Support Generators Using Measured Frequency Transients[J]. IEEE POWER ENGINEERING REVIEW, 1997.

[3] Tuttelberg K , Kilter J , Wilson D. Estimation of Power System Inertia from Ambient Wide Area Measurements[J]. IEEE Transactions on Power Systems, 2018, PP(6):1-1.K. Elissa, “Title of paper if known,” unpublished.

[4] Wall P , Terzija V . Simultaneous Estimation of the Time of Disturbance and Inertia in Power Systems[J]. IEEE Transactions on Power Delivery, 2014, 29(4):2018-2031.

[5] Liu H M, Ren Q Y, Zhang Z K. Calculation of equivalent inertia time constant and slip rate feedback inertia control strategy of doubly fed fan [J]. Power system automation, 2018, 42 (17): 9.

[6] Han S, Zhang F, Ding L. Evaluation method of available inertia of wind farm based on hybrid copula function [J]. Power automation equipment, 2021, 41 (3): 8.

[7] Li S C, Deng C H, long Z J. Calculation of equivalent virtual inertia time constant of wind farm [J]. Power system automation, 2016, 40 (7): 8.

[8] Li S C, Xu S L, Li H Z, Shu Z Y, Huang S Y, Tian B J. Fast estimation method of equivalent virtual inertia of wind farm [J / OL]. Power grid technology: 1-10 [2021-11-25].

[9] Zhong Q C. Virtual synchronous machine and autonomous power system [J]. Chinese Journal of electrical engineering, 2017,37 (02): 336-349. Doi: 10.13334/j.0258-8013. Pcsee. 162325.

[10] Zeng F H, Zhang J B. Temporal and spatial characteristics and analysis method of power system inertia [J]. Chinese Journal of electrical engineering, 2020, 40 (1): 9.

[11] Li D D, Zhang J L, Xu B. Evaluation of equivalent inertia of new energy power system considering frequency distribution characteristics [J]. Power grid technology, 2020, 44 (8): 9.

[12] Xu B, Peng Y H, Zhang L W. Measurement of power system inertia time constant considering wind farm and load inertia response [J]. Hydropower and energy science, 2021, 39 (1): 5.

[13] Li S C. Normalized continuous estimation method of power grid inertia based on noise like disturbance [J]. Chinese Journal of electrical engineering, 2020, 40 (14): 10.

Power load forecasting and interpretable models based on GS_XGBoost and SHAP

Mingming Li[1], Yulu Wang[1*]

[1] School of Business, Shanghai DianJi University, Shanghai, 201306, China

*Corresponding author's e-mail: wangyl@sdju.edu.cn

Abstract: At present, the prediction accuracy of the power load prediction model is not high and the black box model prediction results are not explanatory, so this paper proposes an improved XGBoost based grid search for power load forecasting model (GS-XGBoost), and uses SHAP for model analysis and interpretation to show the impact of the feature set on the model in a visual way according to the marginal contribution of power load feature samples The impact of the prediction results is shown in a visual way to improve the interpretability of the model, and the key factors affecting the electricity load are identified based on the ranking results of the feature importance to provide decision support for the power system equipment maintenance plan, guide the formulation of strategies related to electricity production, resource procurement and pricing, and avoid market risks. In this paper, the electricity consumption dataset of an enterprise in Jiangsu is used as the research object, and the dataset is pre-processed with unique thermal coding, outlier analysis, null filling, and standardization. Through experimental comparison, the GS-XGBoost load forecasting model has the best prediction accuracy compared to SVR, random forest, decision tree, and XGBoost machine learning models, and then SHAP is applied to interpret the model and improve the model interpretability.

1. Introduction

Power load forecasting plays a crucial role in power system scheduling, maintenance and capacity planning, and is of great significance in guiding the planning and construction of power grids and improving the optimal allocation of power system resources[1]. All along, short- and medium- and long-term power load forecasting has been an important task for power supply enterprises, and various load forecasting methods and researches have emerged, including classical forecasting methods such as time series, neural network, SVM (Support Vector Machine) and gray forecasting method. The mainstream electric load forecasting mainly focuses on regional and industry macro load data. The load data of individual enterprises lack cyclical patterns and are highly volatile due to the influence of uncertain factors such as enterprise production and macro policies, and there are large model prediction errors. In the current context of power reform, grid enterprises can use their technical and data advantages to provide monthly load forecasting results for large industrial users, help enterprises to reasonably choose billing methods, reduce electricity costs and provide value-added services to customers[2].

In this paper, we improve the power load forecasting accuracy and model interpretability based on improved XGBoost model and SHAP model analysis. Using an enterprise electricity consumption dataset in Jiangsu as the research object, preprocessing, feature selection, model training, hyperparameter optimization, performance metrics analysis[3], and SHAP model interpretation are performed to construct a prediction model with excellent performance and high interpretability. The main work focuses on the following two aspects.

An electric load forecasting model based on the GS-XGBoost model is developed and compared with existing load forecasting models (in terms of relevant performance metrics) to verify the superior performance of the model in the paper.

The SHAP model is used to analyze the XGBoost load forecasting model, transform the black-box model into a visual chart, enhance the interpretability of the model, provide decision support for the power system equipment maintenance plan, and guide the formulation of strategies related to electrical energy production, resource procurement, and pricing, thereby effectively avoiding market risks.

2. Analysis of model construction

2.1. Model building process

Process Analysis. After the original training samples are processed for feature engineering such as unique heat coding, outlier analysis, null filling, and data normalization, the data are input into XGBoost to train the model, find the optimal parameters by grid search, and compare with other machine learning models to find the optimal load prediction model, and then analyze by SHAP model interpretation to visually demonstrate the feature set on the model prediction results Impact[4].

2.2. Flow chart

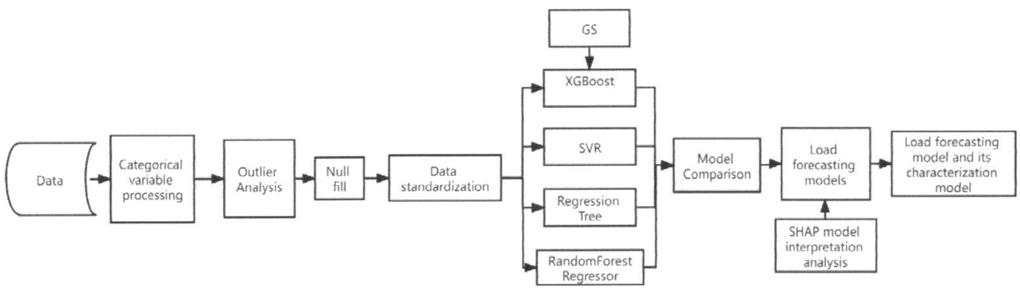

Figure 1. Flow chart of model construction

3. analysis of real-life examples

3.1. Evaluation indicators

Based on the direction of power load forecasting in this paper, the evaluation indicators use root mean square error (RMSE), mean square error (MSE), mean absolute value error (MAE), and coefficient of determination (R^2) where the smaller the three evaluation indicators RMSE, MSE, and MAE represent the smaller the error, and the closer R^2 is to 1 represents the higher the degree of explanation of the target variable on the characteristic input and the better the fitting effect [5], the above evaluation The specific formulas of the above evaluation indicators are as follows.

Root Mean Square Error (RMSE)

$$RMSE = \sqrt{\frac{1}{m}\sum_{i=1}^{m}\left(y_i - \hat{y}_i\right)^2} \tag{1}$$

Expected Variance (EVS)

$$EVS(y, \hat{y}) = 1 - \frac{Var\{y - \hat{y}\}}{Var\{y\}} \tag{2}$$

Mean absolute percentage error (MAPE)

$$MAPE = \frac{100\%}{\mathrm{m}}\sum_{i=1}^{m}\left|\frac{\hat{y}_i - y_i}{y_i}\right| \tag{3}$$

Coefficient of determination (R 2)

$$R^2 = \sum_{i=1}^{m}(y_i - \hat{y}_i)^2 / \sum_{i=1}^{m}(y_i - \overline{y}_i)^2 \tag{4}$$

where m is the number of samples predicted

3.2. Optimal feature selection

The values of the prediction evaluation indicators for the optimal set of features selected for the prediction model at different number of features are shown in Figure 2, and it can be seen that: different combinations of features lead to different biases in the prediction model, and when the number of feature combinations is 6, all four error evaluation indicators are better than other combinations.

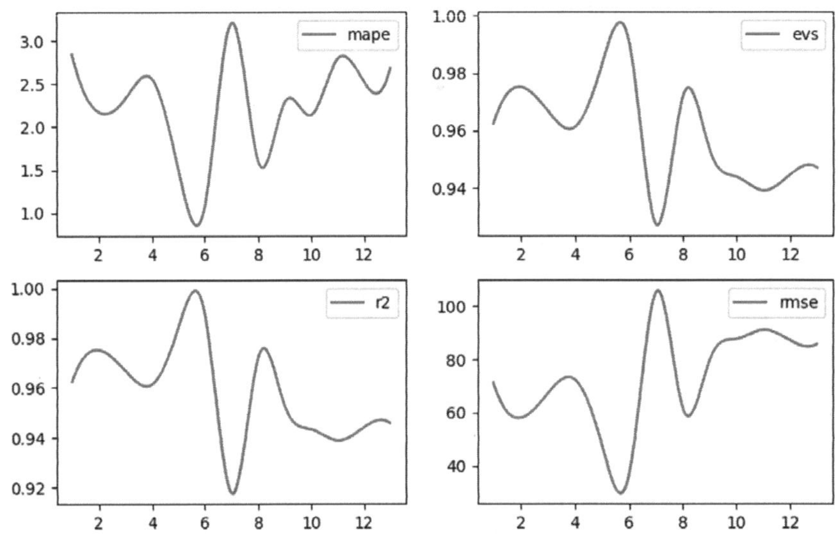

Figure 2. Comparison of errors under different combinations of feature numbers

3.3. Predictive model performance comparison

Table 1. Performance comparison

Model	RMSE	MAPE	R^2	EVS	Prediction time/s
SVR	174.533	6.519%	0.775	0.780	0.04
RF	86.575	2.784%	0.945	0.946	0.26
DTR	253.860	8.437%	0.525	0.529	0.01
XGBoost	90.408	2.963%	0.940	0.950	0.02
GS_XGBoost	36.928	1.069%	0.989	0.989	0.02

From Table 1, we can see that: compared with the XGBoost prediction model with direct input features, the RMSE of the GS-XGBoost optimal feature combination method decreases by 59.2%, and the three error evaluation metrics of MAPE, R 2, and EVS are better than the comparison model. Overall, the GS_XGBoost model has the best performance with short prediction time, highest prediction accuracy and best fitting ability.

The comparison of the prediction curve of this method with the real curve in different prediction models is shown in Fig. 3, which can visualize the fitting of each prediction model to the real load curve, and the figure can clearly reflect the load prediction of each model, in which GS_XGBoost load model fits the real load curve the best.

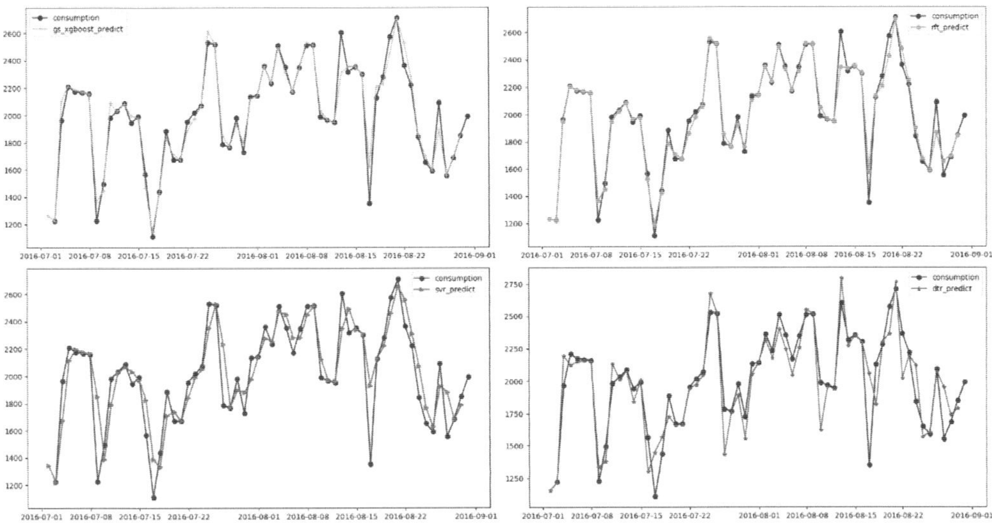

Figure 3. Comparison of each model for load forecasting

4. SHAP interpretive analysis

SHAP is a method used to interpret predictions that are based not only on the importance of features based on a complex training model, but also on the predictions of the current test sample [6].

The overall SHAP feature summary analysis was conducted in the SHAP framework, and the GS-XGBoost model variable importance rankings and the positive and negative effects affecting the results are shown in the SHAP summary graph. As can be seen in Figure 4, the y-axis ranks the previous day's electricity consumption, current day's holiday condition, previous two days' holiday condition, current day's minimum temperature, previous day's holiday condition, and current day's temperature difference according to the importance of the input features, from largest to smallest. x-axis indicates the SHAP values corresponding to the features. The point of each feature in the graph represents a sample of features in the corresponding dataset, and the change in color from Low to High indicates the sample feature array from small to large, and the positive and negative SHAP values indicate the positive and negative correlation between the features and the prediction results [7].

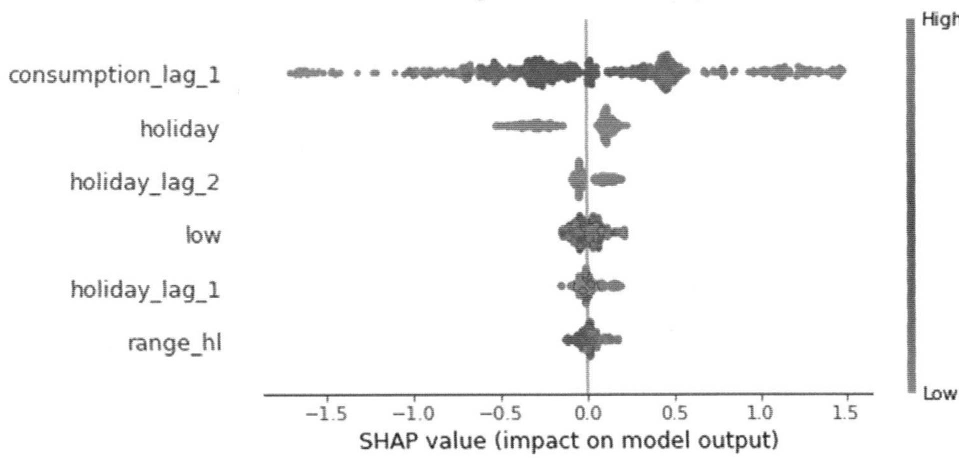

Figure 4. Summary analysis of overall SHAP features

5. Conclude

In this paper, GS-XGBoost method is adopted to study power load forecast. The main conclusions can be summarized as follows: (1) the GS-XGBoost model has a great improvement in the accuracy of load prediction compared with other models. (2) There are many factors affecting load prediction, the input of different feature combinations has a great influence on the accuracy of the prediction model and the generalization ability of the model, and the optimal combination of features can be obtained without losing the amount of data information by the quantitative selection of the input features, which can further improve the precision and generalization of the machine learning load prediction model. (3) The SHAP analysis framework is introduced to explain the GS-XGBoost prediction black box model, to explain the importance of input features in the prediction process, and to qualitatively analyse the value size of feature samples and load prediction worthy of positive and negative correlation. (4) To explains the reasons for the load forecasting model to make corresponding predictions of load values, increases the credibility of the forecast results, and provides an analytical basis for power dispatchers to formulate strategies. In terms of the future work, Add other associated factors as features should be carried out to enhance power load prediction accuracy.

References

[1] Yang Y,Gu Zhenhao. Medium-term power load forecasting based on XGBOOST-DNN [J]. Computer System Applications,2021,30(09):186-191.

[2] Qian Zhongwen,Chen Hao,Ji Deliang. A monthly load forecasting method based on XGBoost algorithm[J]. Zhejiang Electric Power,2019,38(05):77-82.

[3] Cao Rui, Liao B, Li M, Sun Rina. XGBoost-based price prediction and feature analysis model for online short term rental market[J]. Data Analysis and Knowledge Discovery,2021,5(06):51-65.

[4] Chen, Zhuang, Zhou, Buyong. Residential water consumption data prediction based on MIC-XGBoost algorithm[J]. Computer Applications and Software,2021,38(10):125-130.

[5] Chen, W.-H. XGBoost-based pre-lending overdue identification and model expression for Internet finance [D]. Harbin Institute of Technology,2019.

[6] Yang Chao,Chen Mingyang,Yuan Quan. the application of XGBoost and SHAP to examining the factors in freight truck-related crashes: an exploratory analysis[J]. Accident Analysis and Prevention,2021,158:

[7] Zhao Ya. Research on consumer finance default prediction based on LightGBM algorithm [D]. Dalian University of Technology,2021.

[8] Chen SQ, Hu T, Shen YH, Cao Y, Li J. Power generation forecasting for thermal power plants based on MIC and XGBoost[J]. Qinghai Electric Power,2021,40(02):41-45+72.

Power system solution based on root matching method in FPGA environments

Ke Xue [1], Le Cao [2], GuoChu Chen [1a*]

[1]School of Electrical Engineering, Shanghai DianJi University, Shanghai 201306, China

[2]Shanghai Zhenhua Heavy Industries Co,LTD

xueke_bimmer@163.com, caole@zpmc.com, chengc@sdju.edu.cn

Abstract—Digital electromagnetic transient simulation is an important means to analyze the operation, planning and control of power systems. Numerical integration algorithm is the core of electromagnetic transient simulation analysis. By analyzing the existing electromagnetic transient simulation methods of power systems, it is difficult to solve the numerical oscillation and decoupling problems, and the root matching method is proposed to solve such problems. From the perspectives of basic component modeling methods and basic simulation algorithms, the disadvantages of mainstream algorithms in solving electromagnetic oscillations are analyzed. Combining the calculation and simulation features in FPGA, the three-phase two-level inverter model of the power system built is verified.

1. Introduction

Real-time simulation of the power system needs to reproduce the output (voltage/current) waveforms with the required accuracy, and these waveforms represent the behavior of the actual power system being modeled. In order to achieve this goal, the digital real-time simulator needs to solve a model equation of a time step at the same time in the actual clock[1]. Therefore, it produces output at discrete time intervals, where the system state is calculated using a fixed time step at a specific discrete time. The system is represented by the component models available in the software tool library by using a graphical interface, and simulation is performed on a hardware platform using parallel computing. Since RTDS Technologies Inc. used DSP to demonstrate the first commercial real-time digital simulator (RTDS) in 1991, more and more real-time simulators such as RT-LAB, dSPACE, starsim, etc. appeared on the market[2]. The real-time simulator needs to solve the mesh model with a time step of about 50μs or less to faithfully reproduce the transient. Computing power can be defined as the product of the number of nodes/buses in the simulated power grid and the number of time steps taken per second. It requires more computing power and the ability to simulate in a small time step. However, obtaining these hardware emulation platforms often requires high costs[3]. And these commercial simulation platforms often use FPGAs to build their own company's products. Compared with traditional application specific integrated circuits, FPGA has the advantages of fast hardware speed and flexible software. In the past ten years, the application field of FPGA has been expanding continuously[4]. With the rapid development of the current power electronic technology and the introduction of a large number of power electronic devices in the power grid, many calculation and control problems in the power system have become increasingly complex[6]. At present, the modeling and analysis methods used in the more mature power system simulation platform for power system simulation can be roughly divided into: nodal analysis method, state-space techniques and state space nodal method. SSN)[7]. These three analysis methods use different numerical calculation methods to solve the circuit. Each algorithm has its own advantages and disadvantages. Different algorithms show significant differences in response to model switching events. It is very necessary to choose a suitable algorithm. And the modeling is subject to three constraints of space scale, time scale, and component complexity. The core part of power system simulation is the calculation part with matrix calculation as the core. This paper considers the different modeling analysis methods and matrix multiplication architectures used in power system solving, and is used in FPGA hardware implementation. The structure of the paper is as follows. Section 2 briefly summarizes the basic simulation algorithm and component modeling analysis method; section 3 introduces the difference between the root matching method and the above method and its advantages in component decoupling;section 4 introduces the application of this method in matrix operation of the electrical systems; section 5 introduces the design and implementation on FPGA.

2. Power system basic simulation algorithm and component modeling methods

For different types of applications, electromagnetic transient simulation can be divided into offline simulation tools and real-time simulators[8]. Offline simulation tools include various common software packages, such as EMTP-RV, ATP, EMTDC, MicroTran, etc. In the analysis and calculation of the transient stability of the power system, the computer simulation of the dynamic process of the power grid including the components is very valuable, and the basic method is the numerical integration method[9]. At this stage, the solution of the electrical system is essentially to use different methods for the solution of the system in the time domain. The main analysis methods are divided into three types: 1. Nodal analysis method, 2. State variable method, 3. State space node method. Since it is possible to observe the amount of electricity that cannot be measured in a live power system due to technical requirements, the simulation results are helpful to analyze the impact of the transient (abnormal) state of the power system operation and also provide valuable data[10]. If you want to use a computer to realize system simulation, it is necessary to convert a continuous model to a discrete model. There are many different ways of expressing differentiation and integration, and transformations are not unique. Therefore, the choice of numerical method directly determines the nature of the discrete model. The numerical value of the discrete model is uncertain, which is specifically manifested as undamped oscillation, even if the corresponding continuous model is stable. Arithmetic summaries that affect the accuracy of numerical calculations may also lead to instability of the discrete model. In the power grid, the basic differential equation of the branch with linear components (R, L, C) is in the form:

$$\frac{dy(t)}{dt} + \lambda y(t) = bw(t) \tag{1}$$

Among them, w(t) is voltage, y(t) is current; λ, b are network parameters respectively.

2.1. Nodal analysis

The use of nodal analysis methods such as RTDS, offline simulation software PSCAD/EMTDC and so on. This type of software is called EMTP[11]. The root of this type of software is that the transient process of the circuit is solved by trapezoidal integration.

$$y(k) = y(k-1) + \frac{Tb(w(k) + w(k-1))}{2} \tag{2}$$

The core idea of the implicit ladder method is to replace all component models in the network with constant equivalent resistances and equivalent current sources that reflect historical records. Then establish the nodal equation according to this equivalent circuit, form the admittance matrix, and perform the inversion operation to find the equation solution. The implicit trapezoidal integration method has the advantages of small error, good numerical stability, local truncation error of o(h(3)), and large step size simulation. However, when there is a circuit breaker operating in the electrical system, numerical oscillations may occur.

2.2. State variable analysis

State variable (or state space) analysis uses a set of first-order differential equations to characterize the power system, which is then solved by numerical integration. The solution of the state equation is based on the number of state variables and the number of independent energy storage elements. The identification of state variables can be obtained by establishing a path correlation matrix[11,12]. The ladder column of the branch using Gaussian elimination calculation indicates the state variables, and then the state variables are specified in a special order (such as current source, voltage source, inductance, capacitance, resistance) . But there are also certain problems. The system must re-operate every time the system topology changes. Therefore, this method is not suitable when the circuit contains similar switching components frequently breaking. One possible way to reduce the amount of calculation is to divide the power system into two parts: the steady part and the frequent topology change part. Then use the voltage source and the current source to link the two parts of the interface. At present, the simulation software using state variable analysis method includes Simulink, PLECS, etc. There are two main methods for the formation of the main state equations: transformation method and graphic method.

Based on what type of branch is connected to the node, it can be roughly divided into three types: (1) node α is connected to at least one capacitive branch; (2) node β is connected to at least one resistive branch, without capacitance; (3) γ The node is only connected to the inductor. You can use A_{RN}^T、 A_{LN}^T、 A_{CN}^T to represent the branch nodes of R, L, and C respectively. The basic branch equation can be derived:

Resistance branch
$$I_r = R_r^{-1}(A_{Ra}^t V_a + A_{Rb}^t V_b \ ) \tag{3}$$

Inductive branch
$$E_l\text{-p}(L_l * I_l) - R_l * I_l + A_{la}^t V_a + A_{lb}^t V_b + A_{lc}^t V_c)\text{=}0 \tag{4}$$

Capacitor branch
$$A_{nc} I_c + A_{nr} I_r + A_{nl} I_l + A_{ns} I_s = 0 \tag{5}$$

For calculations within one step, if the integral calculation converges, subsequent calculations need to calculate the dependent variable after calculating the state variable, and then calculate the derivative of the state variable from the state equation.

2.3. State space method

For power systems with short transmission lines, complex networks, and numerous power equipment, it is difficult to ensure that modeling and analysis are completed under the premise of a certain step size and accuracy by using traditional power system model decoupling methods, and the circuits are processed in parallel to achieve real-time The purpose of transformation. The basic principle of the state space method is to divide the circuit model into several sub-networks of arbitrary size, and use the state space method to model and calculate each sub-network, and solve the equivalent circuit

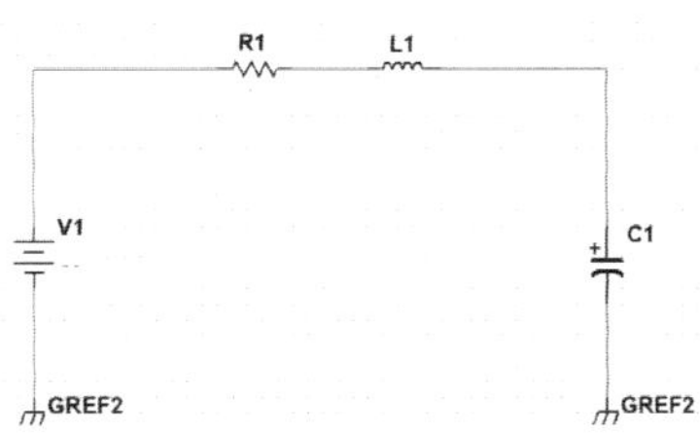

Fig.1 RLC series resonance circuit

corresponding to the network. By dividing the state space group, reducing the number of equivalent electrical nodes in the system, reducing the size of the sub-network, and reducing the number of switches connected to the Zhouwei network in each small grid network, high-level algorithms can be used inside the sub-network . The node method is used between each sub-network to establish an instant solution of the node network equation, and within the same simulation step, the equivalent circuit calculated by each network is combined to obtain the global result of the entire circuit. At the same time, avoid the space complexity caused by decoupling and the corresponding artificial delay, and improve the simulation accuracy.

According to the known circuit, the differential equation of the circuit can be obtained. The exact solution obtained after the state equation is transformed into discretization is:

$$x_{k+1} = e^{A\Delta t} x_k + \int_t^{\Delta t} e^{A(t-\tau)} Bu(\tau)d\tau \quad (6)$$

2.4. Component modeling method

In a power grid with lumped parameters, the basic differential equation describing the dynamic relationship between the physical quantities observed in the branch with linear elements (R, L, C) takes the following form. Resistor is the simplest circuit element. It is shaped like a resistor connected between nodes k and m. It is described by the following equation based on Ohm's law:

$$i_{km}(t) = \frac{1}{R}[v_k(t) - v_m(t)] \quad (7)$$

Inductance:

$$v_L = v_k - v_m = \frac{di_{km}}{dt} \quad (8)$$

Obtained by using the trapezoidal method:

$$i_{km}(t) = I_{history}(t - \Delta t) + \frac{1}{R_{eff}}[v_k(t) - v_m(t)] \quad (9)$$

Similarly, the capacitor can also use the trapezoidal integral to obtain the equivalent current history item:

$$I_{history}(t - \Delta t) = -i_{km}(t - \Delta t) - \frac{2C}{\Delta t}[v_k(t - \Delta t) - v_m(t - \Delta t)] \quad (10)$$

Any form of implicit integral formula can be brought into the differential equation to form different differential equations and at the same time form the corresponding Norton equivalent circuit. Numerical integration substitution technology can be used to solve electromagnetic transient problems as a universal method. Its main characteristics are simplicity, comprehensiveness of application and high efficiency of calculation. Simple circuit components such as resistors, inductors, capacitors, and various transmission lines can be simplified to simply solve the Norton equivalent circuit with nodal equations. By choosing an appropriate integration step size, numerical integration substitution can be applied to all transient phenomena and any modeling system. However, in some cases, the truncation error of the trapezoidal method based on the truncated Taylor series will cause numerical oscillations. When simulating a power system that includes switches, the network will contain a large number of step responses, causing numerical oscillations.

3. The advantages of root matching method component decoupling

In recent years, as the development of electromagnetic transient programs has gradually deepened, it is inevitable that several major companies have invested a lot of time and energy to maintain their simplicity and efficiency. One of the major characteristics of the power system is that the power devices frequently complete the switching action under the action of the control system, which is also one of the difficulties that needs to be overcome in the development of electromagnetic transient simulation programs. The fundamental cause of numerical oscillations is that when some form of change occurs in the network, the non-state variables have abrupt changes. Methods to eliminate numerical oscillations: 1. Correctly obtain the non-state variables after the mutation; 2. Choose an integral format to avoid non-state variables. The former method is difficult to achieve for large-scale power system calculations. We can only find a way to avoid non-state variables and use integration methods with higher accuracy instead. The integral method of truncated Taylor series has numerical oscillation problems during simulation, and the alternative method of numerical integration is to use numerical integration.

The purpose of root matching method in eliminating numerical oscillations is to accurately transfer the zeros and poles of the s-plane to the z-plane. Ensure that the continuous time of the simulation is lacking. For the RL series branch, the numerical integration substitution method and the trapezoidal rule can be used to obtain the following branch difference equation:

$$i_k = \frac{1-\Delta tR/(2L)}{1+\Delta tR/(2L)} i_{k-1} + \frac{\frac{\Delta t}{2L}}{1+\frac{\Delta tR}{2L}}(v_k + v_{k-1}) \tag{11}$$

From the above formula, we can find that the first coefficient term is a first-order approximation of e^x, where x=ΔR/L. The second coefficient term is the first-order approximation of (1-e^x)/2, and we can see that the elimination of truncation error can be in the form of exponential expression. Then (9) can be expressed as:

$$i_k = e^{-\Delta tR/L} i_{k-1} + \left(1 - e^{-\frac{\Delta tR}{L}}\right) v_k/R \tag{12}$$

Although the difference equation can be obtained through numerical integration, and then the corresponding exponential form can be obtained from the difference equation, it is difficult to identify the original data form from the truncated formula. The root matching method completes the conversion of the continuous time system to the discrete time system mapping, and equivalently converts the s domain to the z domain. The root matching method changes the mapping process, thereby changing the resulting error.

Digital simulation needs to be carried out in the z-domain, or expressed in the form of a transfer function, or based on an equivalent difference equation. When the transfer function is used for research, the bilateral recursive forms are:

$$\begin{aligned}(a_0 + a_1 z^{-1} + a_2 z_{-2} + \cdots + a_m z_{-m})U(z) \\ = (1 + b_1 z^{-1} + b_2 z^{-2} + \cdots b_m z^{-m})Y(z)\end{aligned} \tag{13}$$

The corresponding difference equation is:

$$\begin{aligned}y(k\Delta t) = (a_0 u + a_1 u_{-1} + a_2 u_{-2} + \cdots + a_m u_{-m}) \\ - (b_1 y_{-1} + b_2 y_{-2} + \cdots + b_m y_{-m})\end{aligned} \tag{14}$$

The poles remain stable in the s-domain on the left half plane, and the corresponding z-domain poles fall inside the unit circle. The purpose of digital simulation for frequency domain conversion is for the correct conversion of zeros and poles. In order to ensure the continuity of the simulation, the final value of the poles and zeros must match the actual system. Therefore, as long as the system is stable, regardless of the step size change, the difference equation will be stable, so $z^{-1} = e^{-s\Delta t}$ transformation equation.

4.Acknowledgments

In power system simulation, electromagnetic transient simulation is one of the main system simulation applications, and it is also a key component of power system safety analysis and operation. The core algorithm of this application is the solution of large-scale linear equations $Ax = b$. In the third section, through the root matching method, we get a matrix expression that avoids oscillations. Field Programmable Gate Array has the advantages of large on-chip buffer, customizable transmission bandwidth, flexible scheduling, etc.[13], which can maximize the use of computing power of the computing system. The design of FPGA mainly involves several processes of design input, functional simulation, synthesis optimization, placement and routing, and online debugging. When it comes to matrix operations, the first problem to be solved is to perform matrix inversion operations on FPGA. Convert $Ax = b$ into the form of $x = A^{-1}b$, and then solve the value of x through a multiplier[14].

Matrix decomposition is to decompose a complex or unidentified matrix into several matrix products with simple forms or obvious features. The decomposed matrix can still effectively reflect some of the numerical features before the tapping matrix, such as the rank and singularity of the determinant. Value, characteristic value, etc. Most of the process of matrix inversion is implemented according to different matrix decomposition methods. A single matrix triangulation decomposition mainly includes QR decomposition, LU decomposition and Cholesky decomposition. Design three IP cores respectively corresponding to the inverse of the matrix L, the lower triangle multiplying the diagonal matrix IP core and the upper triangle multiplying the lower triangle IP core.

In order to avoid the square root operation, we update the inversion algorithm of the matrix to:

$$A^{-1} = (LDL^H)^{-1} = (L^H)^{-1}D^{-1}L^{-1} = B^H DB \tag{15}$$

This conversion brings certain advantages. The main performance is that the operation is easy to implement by FPGA, and it can take advantage of the parallel operation of FPGA to only perform multiplication and addition and reciprocal operations, and the results generated in the continuous storage and reading required during the operation can be obtained by Realization of RAM resources in FPGA.

5. Example verification and conclusion

Our design uses a three-phase inverter system simulation to verify. The circuit is divided into three parts: a 3-phase inverter part, the power electronic switch converts the input direct current under the action of the PWM control signal; a filter part; a 3-phase load output part References

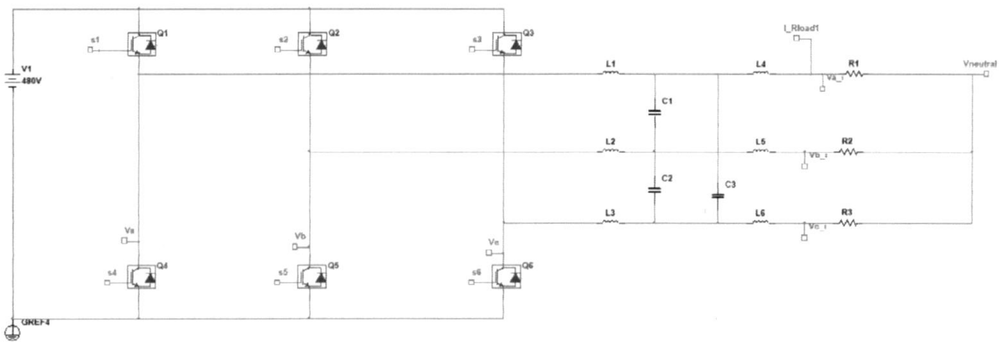

Fig.2 Three-phase inverter

For the FPGA board, we choose the ARTIX-7 XC7A35T FGG484ABX2017 (black gold series) board. After comparing the same model simulation we built in Matlab, we can see that except for the inevitable glitch, the rest are basically similar and meet the simulation requirements.

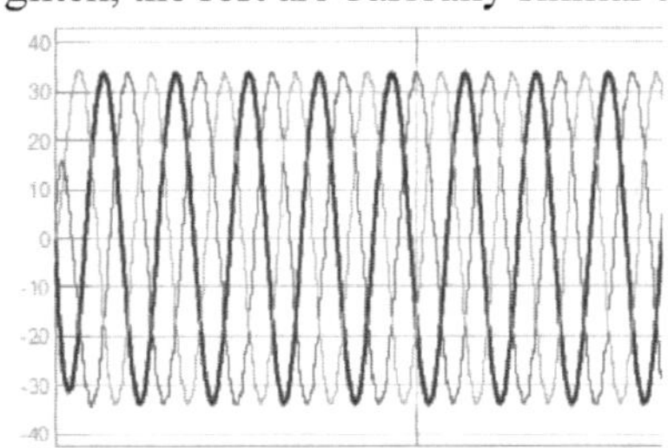

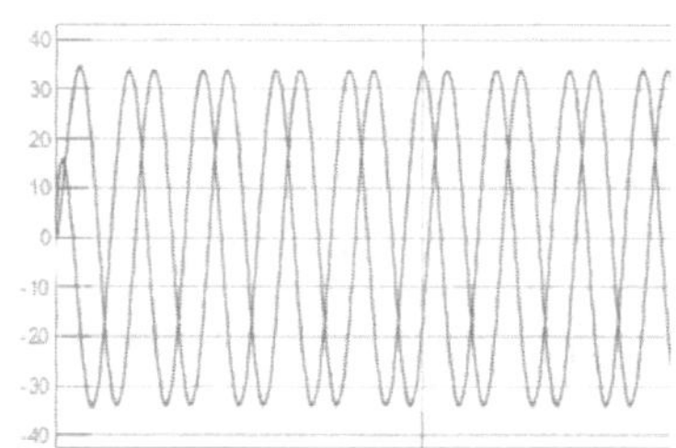

Fig.3 FPGA on-chip simulation Fig.4 matlab simulation

This section introduces the simulation test verification using the root matching method in the FPGA environment, and initially verifies the feasibility, correctness and effectiveness of the root matching method through the steady-state simulation test of the three-phase inverter circuit.

Acknowledgments

This work was financially supported by:

Scientific research project of Shanghai Municipal Science and Technology Commission, "Research on real-time simulation and grid connection test platform for high-power wind turbines and certification test technology" (17DZ1201200);

Scientific research project of Shanghai Zhenhua Heavy Industry (Group) Co., Ltd., "Container Lock

Pin Visual Recognition and Positioning System"(Z-19B306);

A project funded by the Shanghai Graduate Education Association, "Research on the professional degree graduate training model based on the whole school-enterprise cooperation"(ShsgeY201711).

References

[1] A. M. Gole, S. Filizadeh, R. W. Menzies and P. L. Wilson, "Optimization-enabled electromagnetic transient simulation," in *IEEE Transactions on Power Delivery*, vol. 20, no. 1, pp. 512-518, Jan. 2005, doi: 10.1109/TPWRD.2004.835385.

[2] Wang Chengshan,Li Peng,Wang Liwei.Research progress of electromagnetic transient simulation algorithms for power systems[J].Automation of Electric Power Systems,2009,33(07):97-103.

[3] Lin Jikeng, Yan Yipeng, Liu Tao, Zheng Weihong.Summary of external system equivalence methods for electromagnetic transient simulation of power system[J].Automation of Electric Power Systems,2012,36(11):108-115. 281-304.

[4] Dong Yifeng,Wang Yanliang,Han Ji,Li Yaowang,Miao Shihong,Hou Junxian.Summary of high-efficiency electromagnetic transient simulation technology for power system[J].Proceedings of the Chinese Society of Electrical Engineering,2018,38(08):2213-2231+2532.

[5] J. Xie, D. Henneberg, M. Suberski, U. Rädel and J. Petzoldt, "Analysis and Implementation of a Novel four-step Commutation Method for Direct Matrix Converters based on FPGA and DSP," 2021 23rd European Conference on Power Electronics and Applications (EPE'21 ECCE Europe), 2021, pp. P.1-P.9.

[6] Song Yankan,Chen Ying,Huang Shaowei,Zhang Jietan,Yang Libin.Parallel simulation algorithm and implementation of electromagnetic transients in large-scale power systems[J].Power Construction,2015,36(12):9-15.

[7]Liu Dong,Tang Shaopu,Hu Xiangnan,Zhang Shuqing,Shu Dewu,Hu Lu,Lu Junguang. Comparison and analysis of basic simulation algorithms for power systems[J].Global Energy Internet,2018,1(02):137-143.

[8] J. Gao, P. Zhan, S. Wang, N. Wang, J. Wang and Y. Liu, "Hardware-in-the-loop Real-time Simulation of High-power Inverter Based on Electro-thermal Coupling Effect," 2020 IEEE 9th International Power Electronics and Motion Control Conference (IPEMC2020-ECCE Asia), 2020, pp. 3531-3535, doi: 10.1109/IPEMC-ECCEAsia48364.2020.9368239.

[9] Gao Shilin,Song Yankan,Chen Ying,Huang Shaowei,Yu Zhitong,Tan Zhendong.Summary of the principle and application of frequency shift electromagnetic transient simulation in power system[J].Automation of Electric Power System,2021,45(14):173-183.

[10] P. Le-Huy, S. Guerette, L. A. Dessaint and H. Le-Huy, "Real-Time Simulation of Power Electronics in Power Systems using an FPGA," *2006 Canadian Conference on Electrical and Computer Engineering*, 2006, pp. 873-877, doi: 10.1109/CCECE.2006.277356.

[11] K. Meddah, T. Ould-Bachir, M. Becherif and A. Benmouna, "An FPGA-based Real-Time Simulation of a Hybrid Fuel Cell/Battery Source," 2021 IEEE Canadian Conference on Electrical and Computer Engineering (CCECE), 2021, pp. 1-5, doi: 10.1109/CCECE53047.2021.9569167.

[12] X. Dai, C. Ke, Q. Quan and K. -Y. Cai, "Simulation Credibility Assessment Methodology With FPGA-based Hardware-in-the-Loop Platform," in IEEE Transactions on Industrial Electronics, vol. 68, no. 4, pp. 3282-3291, April 2021, doi: 10.1109/TIE.2020.2982122.

[13] F. Xu, Z. Guo, H. Chen, D. Ji and T. Qu, "A Custom Parallel Hardware Architecture of Nonlinear Model Predictive Control on FPGA," in IEEE Transactions on Industrial Electronics, doi: 10.1109/TIE.2021.3118427.

[14] Y. Wang and P. Li, "Algorithm and Hardware Co-Design for FPGA Acceleration of Hamiltonian Monte Carlo Based No-U-Turn Sampler," 2021 IEEE 32nd International Conference on Application-specific Systems, Architectures and Processors (ASAP), 2021, pp. 9-16, doi: 10.1109/ASAP52443.2021.00009.

The Auxiliary System of Video Surveillance in Smart Substation

Yu Liu*, Chao Zhang

College of Electrical Engineering and Automation, Shandong University of Science and Technology, Qingdao 266590, China

*curentliuyu@163.com

Abstract—With the continuous development of artificial intelligence technology, the current substation operation and maintenance personnel have more and more urgent needs for the visualization, remoteness and intelligence of the operation and maintenance system. In order to solve the existing problems and present situation, key technologies such as image acquisition, image recognition, and visual display are studied. A smart substation video monitoring auxiliary system was developed. The system is used advanced cameras and acquisition equipment to monitor and detect related electrical equipment such as hard pressure plate switches in the substation in real time. The goal of intelligent inspection is achieved. In the smart substation project, it has achieved good results and has a wide range of application prospects.

1. Introduction

In the field of the power industry, substations as an intermediate link connecting power plants, grids, and power users, perform the role of centralizing, transmitting and distributing electrical energy, and are an indispensable basic facility of the power system. In recent years, with the continuous improvement of the grid structure, the continuous development of substations, the continuous increase in the number of commissioning, and the gradual expansion of the infrastructure in the station, resulting in an increasingly significant contradiction between enterprise development and operation and maintenance management costs. With the development and application of information, Internet of Things, Robotics, Artificial intelligence and other technologies, substation operation and maintenance personnel have increasingly prominent requirements for operation and maintenance intelligence, visualization, and remoteness. How to combine various new technologies with the actual needs of operation and maintenance personnel to realize a practical intelligent operation and maintenance system is the challenge facing the current intelligent operation and maintenance of substations[1]. With the introduction and improvement of the concept of smart substations, the traditional "four remote" model no longer meets the needs of the continued development of today's power system industry. Under the continuous development of intelligent technology, a variety of operation and maintenance systems have appeared in substations in recent years[2], such as: outdoor laser navigation robot inspection system, three-position inspection system, etc. But these systems are passive operation and maintenance methods. The lack of interactivity between information leads to serious information islands.

Equipment inspection is a basic and important task for substation staff. Manual on-site inspection under the traditional mode has the inevitable shortcomings of inaccurate inspection data and low inspection efficiency, and cannot meet the requirements of informatization and intelligent inspection of electric power enterprises[3]. In the resident operation and maintenance class of operation and maintenance personnel, there is a lack of effective monitoring methods for substation equipment[4]. Therefore, the substation intelligent auxiliary system based on video surveillance technology was born in response to the requirements of the times, relying on visual equipment and its technology to provide reliable support for the management and operation of smart substations. Therefore, the research and practical application of smart substation video monitoring auxiliary system is of great significance in the future development of smart substation and power system industry!

2. Key Technology and Analysis of Video Surveillance System in Substation

2.1. Technology and Analysis on Image Acquisition

In the initial stage of the application of video surveillance technology in substations, it was mainly aimed at the monitoring and operation management of electrical switchgear such as hard press plates in the substation. Image acquisition technology consists of three parts: optics, imaging, and processing[5]. Collecting images is the basis of image recognition, and high-definition and high-quality images will make image recognition more effective. Use cameras as the main means of video monitoring to monitor the real-time environment in the substation, monitor and detect the operation of the hard plate switch equipment in the station in real time, and then collect the monitoring information through the video transmission host and transmit it to the main control terminal. The staff of the substation realizes the panoramic monitoring of the station in the relevant monitoring room. The quality of the image collection is not only affected by the hardware of the camera quality, but also by multiple environmental factors such as the installation position of the camera, the sunlight and the mirror reflection, the camera shooting angle, etc, making the quality of the captured image not high. In order to improve the picture quality and make the picture clear, it needs to be processed in advance.[6]. Therefore, in the image acquisition stage, multiple technologies need to be combined to ensure the accuracy and stability of the acquired images. For the mechanical errors caused by the rotation of the camera pan/tilt, it will cause inaccurate tracking and acquisition of images during the image capture process[7], using image correction technology to ensure that the shooting angle is the same every time, and the captured images are basically the same; for the mechanical error generated by the camera during the focusing process, the image blur retrieval technology is used to ensure the captured image Clear and reliable; for the overexposure or underexposure of the image caused by light and specular reflection, the use of high dynamic lighting rendering technology, also known as HDR technology, ensures that the highlights and dark details of the image are clearly visible.

2.2. Technology and Analysis on Image Recognition

Image recognition is a technology that uses machine learning, pattern recognition and other methods to process, analysis and understand images to automatically identify targets and objects in various patterns in the image[8]. It is an important branch of artificial intelligence. Traditional image recognition technologies include: image acquisition, pre-treatment, feature extraction, classification, etc.[9]. Compared with traditional methods, image recognition technology based on deep learning can provide feature representation based on learning, and has a good performance in automatic feature extraction and segmentation recognition accuracy[10]. Therefore, the basic core of image recognition can be simply summarized as eight characters, namely: feature extraction and step-by-step approximation. After acquiring the original image, it needs to be identified. The first is the pre-treatment process. The image pre-treatment technology includes: image transformation and correction technology, ROI (interest) region creation, image enhancement technology, image smoothing and denoising technology, Image morphology processing, etc.; the feature extraction process generally includes: regional shape feature extraction, feature extraction based on grey-scale value, feature value extraction based on image texture, etc. However, there is no clear dividing line between pre-treatment and feature extraction. In general circumstances, pre-treatment and feature extraction are performed and used at the same time to gradually approach the final region of interest to obtain target tasks and data.

2.2.1. Three-Channel Image Split

The original image is generally an RGB image composed of three channels of red, green and blue. R is Red, G is Green, and B is Blue. Each pixel block in the RGB image is shown by the brightness changes of these three colours and their superposition. A pixel block needs to be saved by multiple matrix elements, and the columns in the matrix will contain multiple sub-columns, and the number of sub-columns is equal to the number of channels. The matrix of the saved colour image is generally as shown in Figure 1:

	column 0			column 1			column ...			column m		
Row 0	0,0	0,0	0,0	0,1	0,1	0,1	...	...	...	0,m	0,m	0,m
Row 1	1,0	1,0	1,0	1,1	1,1	1,1	...	...	...	1,m	1,m	1,m
Row ...	...,0	...,0	...,0	...,1	...,1	...,1	...	...	...	...,m	...,m	...,m
Row n	n,0	n,0	n,0	n,1	n,1	n,1	n,...	n,...	n,...	n,m	n,m	n,m

Fig.1 Matrix form of colour image

After splitting, it is split into 3 single-channel images. Each pixel block in each single-channel image only needs one matrix element to save, and the pixel value is generally 0~255, as shown in Figure 2, it is a single matrix form of the channel image. Finally, select the corresponding single-channel image according to the needs.

	column 0	column 1	column ...	column m
Row 0	0,0	0,1	...	0,m
Row 1	1,0	1,1	...	1,m
Row ...	...,0	...,1	...	...,m
Row n	n,0	n,1	n,...	n,m

Fig.2 Matrix form of colour image

2.2.2. Image Transformation and Correction Technology

For two-dimensional images, in order to correct the distortion of the image during shooting, some simple geometric transformations can be performed on the image, such as translation, rotation, and scaling, can be performed on the image. Suppose that the position of a point P_0 in the image can be represented by two coordinates (x_p, y_p). Before performing simple geometric transformation, it needs to be transformed into *P(x, y, 1)* by homogeneous transformation. Equation 1 is translation transformation, equation 2 is rotation transformation, and equation 3 is scaling transformation.

$$P_t = T * \mathrm{P} = \begin{bmatrix} 1 & 0 & x_t \\ 0 & 1 & y_t \\ 0 & 0 & 1 \end{bmatrix} * \mathrm{P} \tag{1}$$

where T is the translation matrix, P_t is the coordinate after translation, x_t is the pixel distance moved by the original point along the *X*-axis, and y_t is the pixel distance moved by the original point along the *Y*-axis.

$$P_r = R * \mathrm{P} = \begin{bmatrix} \cos\alpha & -\sin\alpha & 0 \\ \sin\alpha & \cos\alpha & 0 \\ 0 & 0 & 1 \end{bmatrix} * \mathrm{P} \tag{2}$$

where R is the rotation matrix, α is the rotation angle of the original point around the origin of the coordinate, and P_r is the coordinate after rotation.

$$P_s = S * \mathrm{P} = \begin{bmatrix} s_x & 0 & 0 \\ 0 & s_y & 0 \\ 0 & 0 & 1 \end{bmatrix} * \mathrm{P} \tag{3}$$

where S is the zoom matrix, s_x is the magnification along the *X*-axis, s_y is the magnification along the *Y*-axis, and P_s is the scaled coordinate.

2.2.3. Image Smoothing and Denoising Technology

There are three main methods of smoothing filtering, which are mean filtering, median filtering, and Gaussian filtering.

Mean filtering is also called linear filtering. The main method used is neighborhood averaging, that is, its principle is: surrounding square pixels centered on the target pixel constitute a filtering template, and the average value of all pixels in the selected template is calculated instead the original target pixel value.

$$X = \frac{1}{n}\sum_{j=1}^{n} x_j \tag{4}$$

where n is the number of pixels in the filter template, x_j is the pixel value of each pixel area in the filter template, and X is the pixel average value, that is, the latest pixel value of the central target pixel after average filtering. The side length of the selected square pixel filter template is generally an odd number of unit pixel blocks, because an odd number can ensure that the target pixel is in the middle of the filter. From the point of view of the frequency domain, mean filtering is a low-pass filter, high-frequency signals will be removed, so it can help eliminate sharp noise in the image, and achieve image smoothing, blurring and other functions.

Median filtering is a non-linear smoothing technique. Median filtering is a non-linear signal processing technique that can effectively suppress noise based on ranking statistical theory. Its basic principle is to use the value of a point in a digital image or a digital sequence. The median value of each point in a neighborhood of is replaced, so that the surrounding pixel values are close to the true value, thereby eliminating isolated noise points.

$$Y = Med[x(z-N), \ \ldots, \ x(z), \ \ldots, \ x(z+N)] \tag{5}$$

where $x(z\text{-}N),\ldots,x(z),\ldots,x(z+N)$ are the $2N+1$ pixel values in the neighbourhood of the target pixel block, and Y is the value of these $2N+1$ pixels The median value is the latest pixel value of the center target pixel after median filtering. This filtering method is very effective for removing some isolated noise points, and it can also retain most of the edge information. However, it is necessary to pay attention to the size of the selected square pixel filter template during use. If the size is too large, it is easy to cause blurring of the image.

Gaussian filtering is a linear smoothing filter, which can be simply understood as Gaussian filtering denoising is a weighted average of image pixel values. The area of the target pixel area is selected as a square pixel filter template, then the new pixel value of the target pixel area is It is the weighted average of its own value and other pixel values in the neighborhood.

$$G(x) = \frac{1}{\sqrt{2\pi}\sigma} e^{-\frac{x^2}{2\sigma^2}} \tag{6}$$

$$G(x, \ y) = \frac{1}{\sqrt{2\pi}\sigma^2} e^{-\frac{x^2+y^2}{2\sigma^2}} \tag{7}$$

Equation 6 is a one-dimensional Gaussian distribution, and equation 7 is a two-dimensional Gaussian distribution, where x is the coordinate point in the X-axis direction, y is the coordinate point in the Y-axis direction, σ is the standard deviation, and $G(x)$ is the one-dimensional coordinate the new value after Gaussian filtering, $G(x,y)$ is the new value after Gaussian filtering in two-dimensional coordinates. The Gaussian smoothing filter is very effective in suppressing the noise that obeys the normal distribution.

In image recognition and processing technology, dynamic threshold segmentation technology is often combined with filtering technology. The filtered image often becomes blurred, and it does not perform a further precise search for the target area, and the dynamic threshold segmentation is to further confirm the search for the target area under the premise of filtering. Dynamic threshold segmentation is generally divided into: brightness area selection, darkness area selection, median area selection, non-median area selection, respectively, as equation 8-11:

$$g_0 \geq g_t + Offset \tag{8}$$

$$g_0 \leq g_t - Offset \tag{9}$$

$$g_t - Offset \leq g_0 \leq g_t + Offset \tag{10}$$

$$g_t - Offset > g_0 \&\& g_0 > g_t + Offset \tag{11}$$

where g_o is the pixel value after filtering; g_t is the pixel value before filtering, and *Offset* is the set parameter value. Choose different dynamic threshold segmentation methods to select different image regions.

3. Test and Analysis

The research in this article mainly involves three-channel image conversion, ROI region creation, smoothing filter processing, dynamic threshold segmentation, image morphology processing, target task feature extraction and other technologies.

The image acquisition is shown in Figure 3, and the switch state of the hard plate is judged according to the target task.

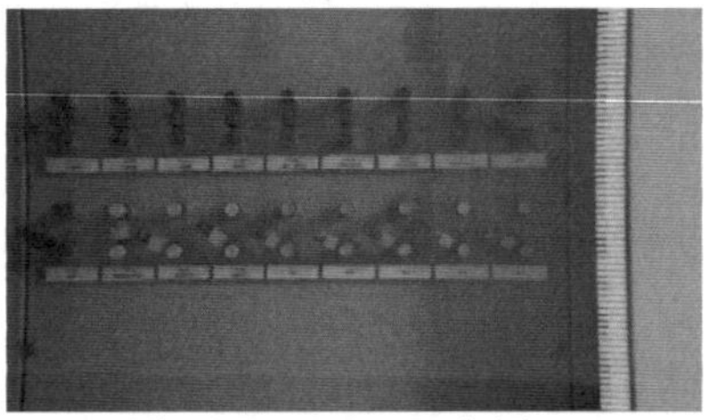

Fig.3 Image acquisition

First, perform grayscale processing on the color image. Since the target task contains red and yellow, the channel separation method can be used to convert the three-channel image to convert the obtained color image into three single-channel images of red, green, and blue. For the red and yellow switches, the red channel image is selected for processing. The grey value difference between the two is large, and it is easy to extract separately. Figure 4 shows the red channel image.

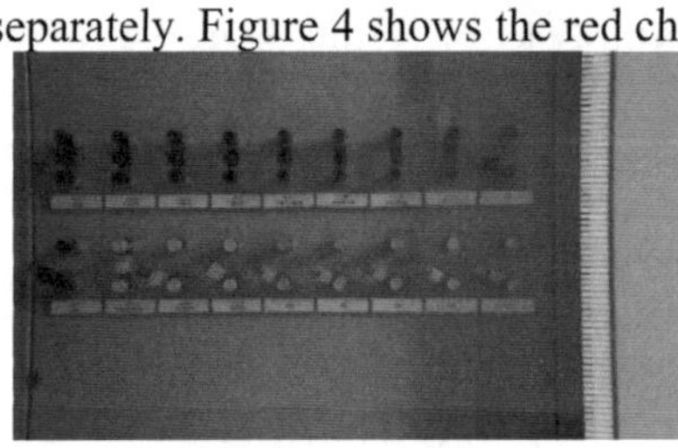

Fig.4 Red channel image

In the process of machine vision image processing, the research area that needs to be processed is outlined in a geometric shape from the processed image, which is called the region of interest, abbreviated as ROI (region of interest). ROI technology can ignore the images that do not need to be processed, and greatly improves the efficiency of image processing[11].Due to the complex environment and other conditions, it is necessary to create an ROI area for it. Due to the large difference in gray value, it is necessary to create an ROI area for the red switch and the yellow switch separately. Figure 5 shows the red switch ROI area, and Figure 6 shows the yellow switch ROI area.

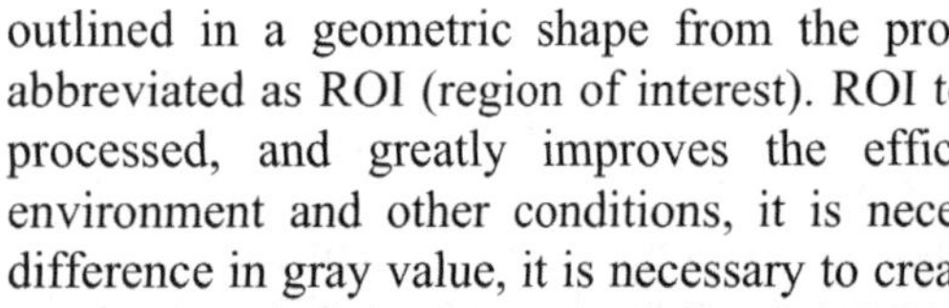

Fig.5 Red switch ROI area

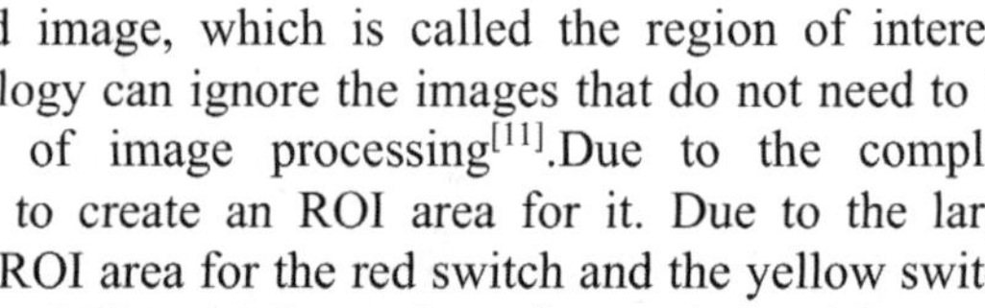

Fig.6 Yellow switch ROI area

Aiming at the fine particle noise in the image, this thesis selects the mean filter. This thesis selects n as 49 to form a pixel area filter template with a length and a width of 7 pixels. Figure 7 shows the red switch ROI area after averaging filtering, and Figure 8 shows the yellow switch ROI area after averaging filtering.

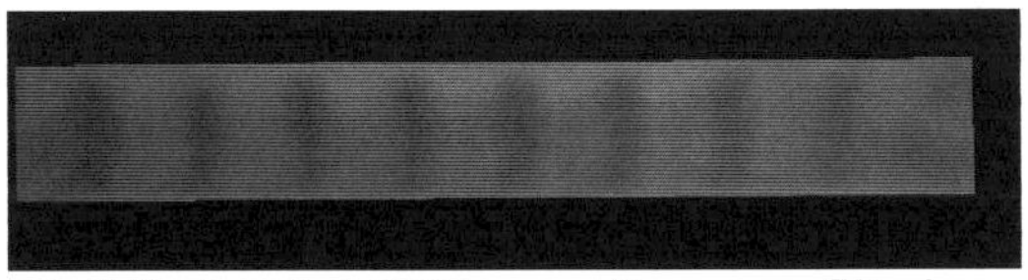

Fig.7 Red switch ROI area after mean filtering

Fig.8 Yellow switch ROI area after mean filtering

Mean filtering also has its inevitable shortcomings, that is, its fuzzification. Therefore, the mean filter is often used in conjunction with the dynamic threshold local segmentation algorithm to further find the approximate target area. The *Offset* setting parameter value in this article is *11~12*. Figure 9 shows the red switch area segmented image, and Figure 10 shows the yellow switch area segmented image.

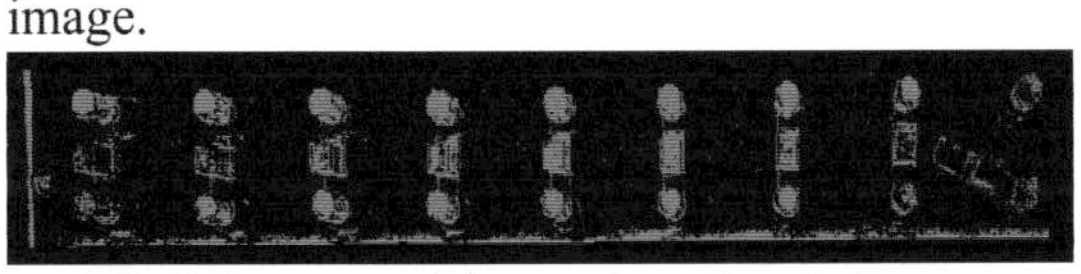

Fig.9 Segmented image in red switch area

Fig.10 Yellow switch area segmentation image

Because the local dynamic segmentation image will produce more particles and impurities, the region division operation is first performed on it, that is, the connected parts are divided into the same area, and the unconnected parts are divided into different areas. Then select the area based on the area characteristic value to exclude particulate impurities. Finally, join the regions.

The main methods of morphological processing are: corrosion and expansion, opening and closing operations, top hat operations and bottom hat operations, and so on. In this thesis, the method of combining erosion and expansion and opening operation and closing operation is used to enhance the target of the dynamic threshold local segmentation image, and prepare for the subsequent target task feature extraction. Figure 11 shows the processed image of the red ROI region, and Figure 12 shows the processed image of the yellow ROI region.

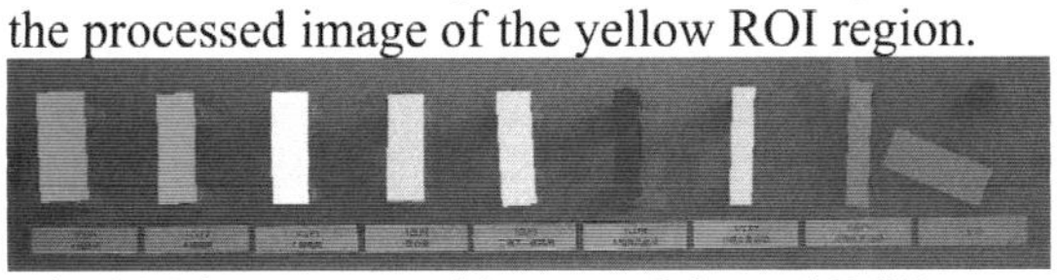

Fig.11 The processed image of the red ROI area

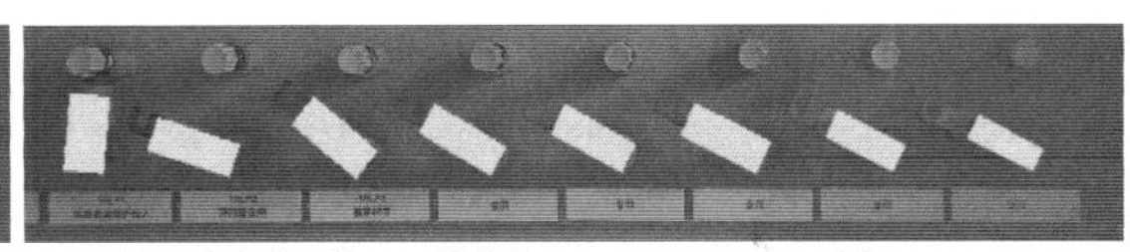

Fig.12 Yellow ROI area processed image

4. Test Results

The target task feature extraction is selected according to the task requirements. The main feature extraction includes: target area rotation angle extraction, target area extraction, target area central point extraction, and so on. The task of this article is to accurately determine the switch state of the hard plate. Due to the influence of some factors such as the shooting angle, the area selection is often inaccurate. Therefore, this article uses the rotation angle to determine the switch state. When the switch is closed, the rotation angle is about 90°, and when the switch is open, the rotation angle is about 135°:

$$\begin{cases} 75° \leq phi \leq 105°, \text{ Switch closed} \\ 120° \leq phi \leq 150°, \text{ Switch on} \\ else, \text{ Wrong judgment, Alarm in time} \end{cases} \quad (12)$$

where *phi* is the rotation angle of the target area. Due to some inevitable minor errors such as mechanical errors, there should be a margin of 15° when judging the switch state based on the angle. Figure 13 shows the red switch test results, and Figure 14 shows the yellow switch test results.

Fig.13 Red switch test result

Fig.14 Yellow switch test result

Before the switch state is judged, the recognition of the number of regions is also added. As shown in Figure 11 and Figure 12, the number of red switches is 9 and the number of yellow switches is 8. If it is detected that the number of red switches is not equal to 9, and the number of yellow switches is not equal to 8, the system will alarm in time to remind the relevant staff to check whether the system is abnormal or the state of the switch is abnormal. This further improves the detection accuracy and reduces the occurrence of errors. It can be seen from Figures 13 and 14 of the test results that the test results are correct and accurate.

5. Conclusion

The intelligent substation video monitoring auxiliary system adopts advanced cameras and acquisition equipment to achieve stable capture of the environment, so that the angle of multiple shots is consistent, and the captured images are consistent. Then, through a series of image recognition technologies such as pre-treatment technology and feature extraction, stable monitoring and Judge whether the switch status of the hard pressure plate is correct, and display it on the main control terminal, which reduces the intensity of inspection work of related staff, and realizes unmanned inspection, automated operation and less humanized maintenance of the smart substation operation and maintenance system.

References

[1] Chunxiao Zhang, Zhihao Lu, Xiangcai Liu. Smart Substation Joint Inspection Technology and Its Application[J]. Power System Protection and Control, 2021, 49(09): 158-164.
[2] Zhengqin Zhou, Zhenbo Du, Wenrui Wang, etc. Intelligent technology and application of substation live operation and maintenance based on hierarchical distribution[J]. Power system protection and control, 2019, 47(1): 150-157.
[3] Zhiqiang Peng, Hang Zhou, Yu Han. Construction and application of transparent operation and maintenance system for intelligent substation automation equipment[J]. Power System Protection and Control, 2020, 48(13): 156-163.
[4] Guohua Chen, Weinan Chen. Research on Technical Scheme of Remote Centralized Operation and Maintenance System for Substation Automation Equipment[J]. Electrical Engineering Technology, 2017(9): 121-122.
[5] Jianguo Sun, Zhipeng Cai, Chi Zhang, Peng Wang. Research on the Application of Machine Vision Recognition and Inspection Technology in Engine Manufacturing[C]. Proceedings of China Society of Automotive Engineering 2021 (5), 2021: 357-362.
[6] Yinghu Lu, Qingchun Zhang, Jinyao Wang, Yang Yang. Research on image acquisition and processing technology of patrol robots[J]. Computer and Digital Engineering, 2021, 49(04): 813-817.
[7] Yingdong Wu, Qingping Mu, Fei Dong, Beiping Hou, Huang Jun. Study on cloud image acquisition and segmentation for photovoltaic power forecasting[J]. Journal of Zhejiang University of Science and Technology,2018,30(05):391-397.
[8] Mei Pan.The Application of Image Recognition in the Classification and Recognition of Fruits and Vegetables[J]. Modern Agricultural Science and Technology,2021(16):257-259.
[9] Rong Xu, Xiaohui Qiu. An improved YOLO V3 target detection method[J]. Computer Technology and Development, 2020, 30(7): 30-33.
[10] Yuanpan Zheng, Guangyang Li, Ye Li. Summary of Research on Application of Deep Learning in Image Recognition[J]. Computer Engineering and Applications, 2019, 55(12): 20-36.
[11] Lijuan Zhang, Dongliang Li. The research and application of image recognition technology in electrolytic aluminum production[J]. Special Casting & Nonferrous Alloys, 2021, 41(08): 968-973.

A Method of Integrated Energy Metering Simulation Data Generation Algorithm Based on Variational Autoencoder WGAN

Penghe Zhang, Yang Xue, Runan Song*, Yining Yang, Cong Wang, Liu Yang

China Electric Power Research Institute Co., Ltd.

Haidian District, Beijing, 100192

*Corresponding author's e-mail: songrunan@epri.sgcc.com.cn

Abstract—With the development of integrated energy systems, the penetration rate of all types of distributed power sources continues to increase. In a grid oriented to integrated energy metering, the metering business of different energy sources needs to fully consider the dynamic characteristics of different user loads. This paper aims at the generation of large-scale integrated energy metering data in large-scale integrated energy digital simulation technology, starting from supporting the construction of an integrated energy metering simulation system, combining various business applications and artificial intelligence technology under the background of the energy Internet. For the requirements of energy metering simulation, a method of integrated energy metering simulation data generation based on variational autoencoder WGAN is proposed, and the effect of the method proposed in this paper is verified through actual metering scene data.

1. INTRODUCTION

In recent years, with the continuous improvement of the level of clean and flexible energy production in our country, the demand for diversified and interactive energy consumption has become increasingly significant, and energy business development, technological innovation, and policy promotion are all facing new challenges. Integrated energy metering for multiple business application scenarios such as electric vehicle charging and replacement, customer-side energy storage, distributed energy access, and multi-meter integration collection is an important foundation for evaluating various energy utilization conditions and implementing energy conservation supervision and management in accordance with the law. Among them, the integrated energy metering system is inevitably a complex and diverse system, and energy companies have gradually increased their requirements for reliability, safety, and stability of the system. However, there is no systematic and large-scale research on simulation technology for energy metering equipment. The metering system simulations carried out by various energy metering industries are mostly physical simulations and field tests[1], large-scale verification has not yet been achieved, the simulation method is single, and the results of laboratory test, field test, and the effect of large-scale application are quite different, which cannot meet the verification requirements for the feasibility of various new services, new methods, and new technologies, so it seriously affects the fairness and justice of measurement , and has severely restricted the development and innovation of our country's energy measurement industry. Therefore, it is urgent to carry out the research of data generation technology in the digital simulation of large-scale integrated energy metering system to provide simulation technology support for the construction of the system and the development of key equipment.

In order to further promote the construction of a smart energy service system and meet the reliability, safety, and stability requirements of various energy metering businesses, it is urgent to build a comprehensive efficient, accurate and reliable v energy metering digital and true hybrid simulation system to achieve the true restoration of sundry scenes, and effectively support the exploration and research of various new energy metering technologies. Aiming at the problem of generating large-scale integrated energy metering data in large-scale integrated energy digital simulation technology, this paper proposes a integrated energy metering simulation data generation method based on variational autoencoder WGAN, starting from supporting the construction of an integrated energy metering simulation system, combining with the needs of various business applications, artificial intelligence technology and other key energy metering simulations under the background of the energy Internet, it provides complete and sufficient data for integrated energy metering simulation in real time.

2. Related Work

The generative confrontation network uses the generator to generate generated data with the same distribution as the input data, and uses the discriminator to determine the gap between the generated data and the real data, during the confrontation training process of the generator and the discriminator, it gradually approaches the Nash equilibrium, so that the generated data is as similar as possible to the original data[2]. The generative confrontation network improves the generation ability and robustness of the model through the idea of game theory, and has a good data generation effect, but the generative confrontation network does not have the function of specifically responsible for feature extraction, and it is difficult to train[3].

Therefore, in order to solve the problem of training stability of the generative confrontation network, this paper chooses WGAN to replace the GAN generative model. The variational autoencoding network VAE can encode data into low-dimensional factors with original data characteristics[4]. Therefore, this paper adds the coding network of the VAE encoder to the WGAN generator part for data feature extraction, and proposes a method of data generation based on the variational autoencoder WGAN (VAE-WGAN).

3. Mathematical Model

3.1 Objective Function

The simulation data generation model based on variational autoencoder WGAN is composed of three parts: VAE encoder, WGAN generator and discriminator.

Assuming that the output of the hidden layer of the network is different in Gaussian way[5]:

$$p(Dis_l(x) \mid z) = N(Dis_l(x) \mid Dis_l(\tilde{x}), I) \tag{1}$$

In the formula, $Dis_l(x)$ is the l-th hidden layer of the network. Maximizing $E_q(z|x)[\log p(x|z)]$ is equivalent to maximum likelihood estimation, we can use $E_q(z|x)[\log p(Dis_l(x)|z)]$.

Therefore, the objective function of the VAE encoder in the integrated energy simulation data generation model is:

$$\max[\gamma_1 \cdot E_{q(z|x)}[\log p(Dis_l(x) \mid z)] - D_{KL}(q(z \mid x) \mid p(z))] \tag{2}$$

The objective function of the WGAN generator in the integrated energy simulation data generation model is:

$$\max[\gamma_2 \cdot E_{q(z|x)}[\log p(Dis_l(x) \mid z)] + E[\log(D(\hat{x}))] + E[\log(D(\tilde{x}))]] \tag{3}$$

The objective function of the WGAN discriminator in the integrated energy simulation data generation model is:

$$\max[E_{x \sim P_{\text{data}}}[\log D(x)] + E[\log(1 - D(\hat{x}))] + E[\log(1 - D(\tilde{x}))]] \tag{4}$$

Journal of Physics: Conference Series **2195** (2022) 012031 doi:10.1088/1742-6596/2195/1/012031

3.2 *Restrictions*

In order to ensure that the generated measurement data can fit the actual value, it is necessary to meet the authenticity constraint and the similarity constraint.

The authenticity constraint is to ensure that the generated data can be close to the real scene, and the authenticity loss Lr is defined as:

$$L_r = W\left(G\left(z;\theta^{(G)}\right);\theta^{(D)}\right) \tag{5}$$

In the formula, G(z;θ) is the generated data of the generator; W(~;θ(D)) represents the Wasserstein distance between the generated sample and the real sample.

$$Ls = \left\| G\left(z;\theta^{(G)}\right) - I \right\|_2 \tag{6}$$

In the formula, I represents the real sample, and two-norm are used to measure the similarity of two matrices.

In summary, the final constraint for the generation of integrated energy metering data is:

$$\min_{Z \sim p_z(z)} L_r + L_s \tag{7}$$

Use Adam as the optimizer to optimize the latent variable z, so that the generated measured value is as close as possible to the real value. The final total sample is:

$$\hat{I} = I + G\left(\hat{z};\theta^{(G)}\right) \tag{8}$$

3.3 *Network Structure*

According to the parameters set for the network structure and the activation function in literature[6], the network structure of VAE-WGAN in this paper is shown in Fig.1, and the hyperparameter selection is determined by experimental tests. The detailed network parameters of the VAE-WGAN generator in this paper are shown in Table 1, and the detailed network parameters of the discriminator are shown in Table 2. In order to improve the training speed of the generator network, the ReLU activation function is used. The discriminator network is basically symmetrical with the generator network. In order to improve the recognition performance of the discriminator, LeakyReLU is selected as the activation function of the convolutional layer[7].

The calculation formula of the activation function is as follows[8]:

$$\mathrm{ReLU}(x) = \begin{cases} x, x > 0 \\ 0, x \le 0 \end{cases} \tag{9}$$

$$\mathrm{LeakyReLU}(x) = \begin{cases} x, x > 0 \\ 0.2x, x \le 0 \end{cases} \tag{10}$$

$$Sigmoid(x) = 1/(1+e^{(-x)}) \tag{11}$$

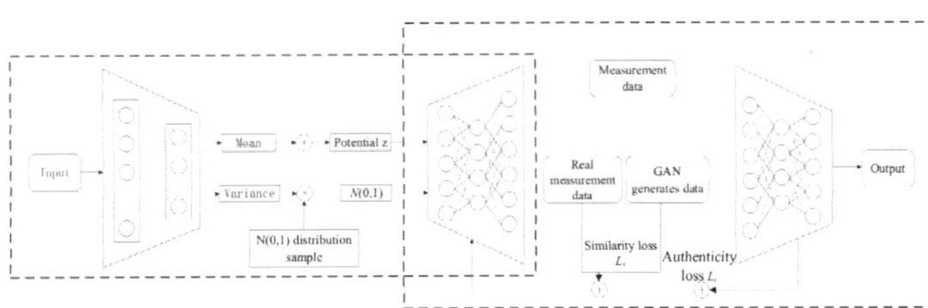

Figure 1. Data generation framework based on VAE-WGAN

TABLE 1. GENERATOR STRUCTURE PARAMETERS

Name of network layer	*Parameters*	*Numerical value*
Convolutional layer 1	Size of convolution kernel	3
	Number of filters	64
	Step length	1
	Activation function	ReLU
Convolutional layer 2	Size of convolution kernel	3
	Number of filters	128
	Step length	1
	Activation function	ReLU
Convolutional layer 3	Size of convolution kernel	3
	Number of filters	256
	Step length	1
	Activation function	ReLU
Fully collected layer	Number of neurons	1
	Activation function	sigmoid

TABLE 2. DISCRIMINATOR STRUCTURE PARAMETERS

Name of network layer	*Parameters*	*Numerical value*
Convolutional layer 1	Size of convolution kernel	3
	Number of filters	64
	Step length	1
	Activation function	LeakyReLU
Convolutional layer 2	Size of convolution kernel	3
	Number of filters	128
	Step length	1
	Activation function	LeakyReLU
Convolutional layer 3	Size of convolution kernel	3
	Number of filters	256
	Step length	1
	Activation function	LeakyReLU
Fully collected layer	Number of neurons	1
	Activation function	sigmoid

4. CALCULATION EXAMPLE

The calculation example in this paper takes the user power data in the actual integrated energy metering scenario as an example to verify the effect of the VAE-WGAN data generation method proposed in this paper.

4.1 Data generation process

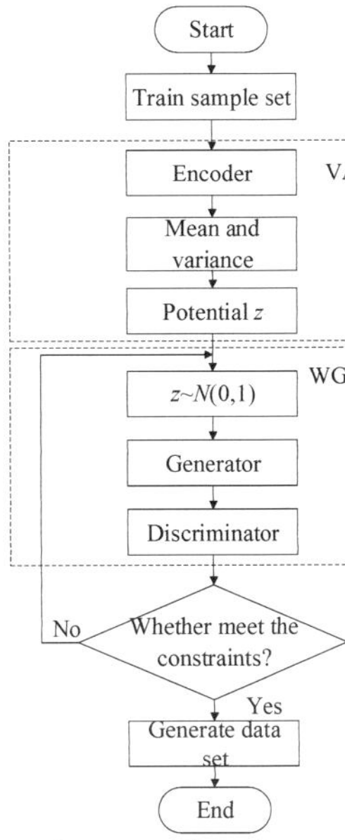

Figure 2. Flowchart for generating of comprehensive energy metering simulation data based on variational autoencoder WGAN

The data generation process is shown in Fig. 2. The integrated energy metering simulation data generation method based on variational autoencoder WGAN includes the following process steps:

Step 1: Build a integrated energy measurement sample data set based on the existing integrated energy metering data;

Step 2: Use the VAE coding network of the variational autoencoder to extract the features of the integrated energy metering data in the sample library in step 1 to obtain the encoded data features;

Step 3: Input the data features encoded in step 2 into the WGAN network, and conduct confrontation training between the WGAN generator and the WGAN discriminator;

Step 4: Determine whether the data generated by the WGAN network in Step 3 meets the authenticity constraint and the similarity constraint;

Step 5: Obtain the generated integrated energy metering simulation data set.

4.2 Training process

Process 1： A sample set is generated through the generation network, which is a fake label, and is used as the training set of the discriminator A together with the input real sample.

Process 2： Migrate the trained discriminator A parameters to the opposite discriminator B, use the generated sample set and the real samples in the training set to construct the mixed sample to train the discriminator B, update the network parameters, so that the discriminator A cannot distinguish the real samples from the generated samples.

Repeat the above two processes until the classifier A and classifier B cannot distinguish the authenticity of the sample, and the generated integrated energy metering simulation data is obtained.

The change process of the loss function of the generator and the discriminator during the training process of GAN and the VAE-WGAN network proposed in this paper is shown in Fig. 3 and Fig. 4.

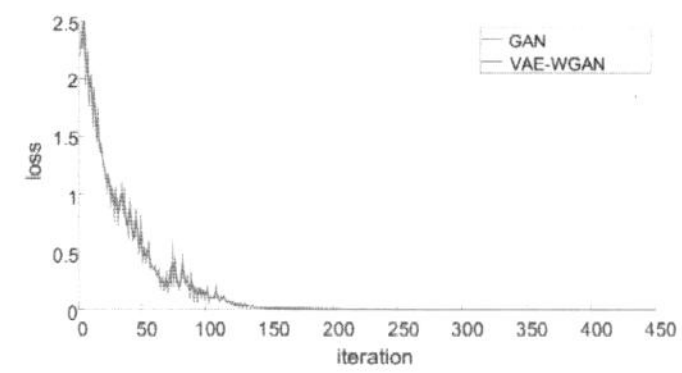

Figure 3. Change of loss function during the training of the generator

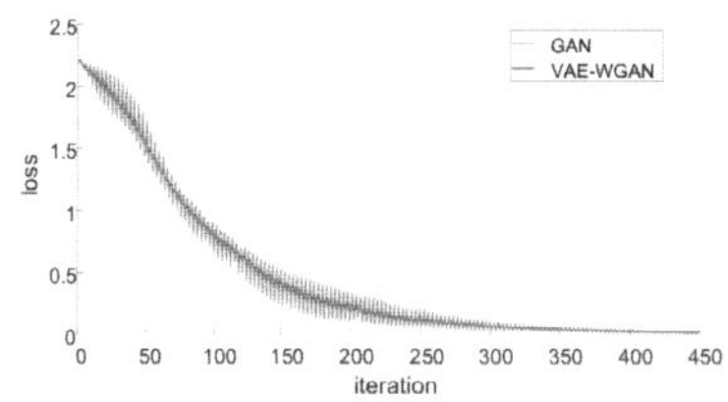

Figure 4. Change of loss function during the training of the discriminator

Comparing the training process of GAN and VAE-WGAN proposed in this article, it can be seen that the generators of the two basically reach convergence after 150 iterations, and the discriminator basically converges after 300 iterations, but the loss functions of GAN generator and discriminator are oscillating, the convergence effect is not good.

In summary, through the change of the loss function curve of the generator and the discriminator, the training process of GAN is unstable, while VAE-WGAN has better stability.

4.3 Result of generated data

Fig. 5 shows the comparison between the generated data of the VAE-WGAN network after 30 rounds of training and the real samples. It can be seen from the figure that the gap between the generated data after 30 rounds of training and the real data is not too big, which is close to the distribution of the real samples, and can reflect the data characteristics of the real measurement scene.

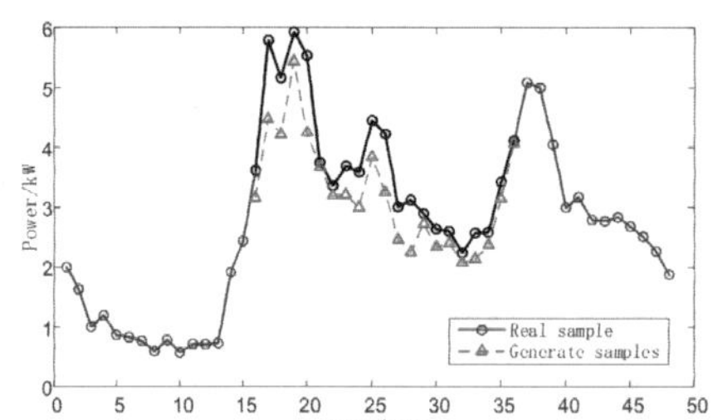

Figure 5. Result of generated data

5. Conclusion

The integrated energy metering simulation data generation method based on variational autoencoder WGAN proposed in this paper can effectively generate simulation data in integrated energy metering scenarios, realize the reasonable generation of basic data in large-scale integrated energy metering simulation, and solve user's data problems in the process of large-scale integrated energy metering online simulation. It can provide complete and sufficient data for large-scale integrated energy metering simulation in real time, and improve the rationality of data generation in the simulation process, so further improve the efficiency and effect of integrated energy simulation.

ACKNOWLEDGMENTS
This work is supported by General Science and Technology Project of State Grid Corporation of China-Research on Key Technologies of Digital and True Hybrid Simulation of Large-scale Integrated Energy Metering System.

REFERENCES
[1] CUI Yingbin, ZHANG Weixin, SHAN Zengli, et al. Systematic Test Method Based on Low-voltage Power Line Carrier Communication [J]. Proceedings of the CSU-EPSA, 2018, 30(01): 81-87.
[2] WANG Shouxiang, CHEN Haiwen, PAN Zhixin, et al. A Reconstruction Method for Missing Data in Power System Measurement Using an Improved Generative Adversarial Network [J]. Proceedings of the CSEE, 2019, 39(01): 56-64+320.
[3] YANG Yulian, QI Linhai, WANG Hong, et al. Reconstruction of Missing Measurement Data in Distribution Network Based on Generative Confrontation and Dual Semantic Perception [J]. Automation of Electric Power System, 2020, 44(18): 46–56..
[4] ZHANG Wenqiang, CHEN Jing, YANG Wei, et al. Photovoltaic Uncertainty Modeling Based on VAE-CGAN [J]. Power System Technology, 2021, 45(04): 1273–1280.
[5] YANG C, WANG Z. An Ensemble Wasserstein Generative Adversarial Network Method for Road Extraction From High Resolution Remote Sensing Images in Rural Areas[J]. IEEE Access, 2020, 8: 174317–174324.
[6] XIAO Rui. Research on Data Generation Model Based on Generative Adversarial Network[D]. University of Electronic Science and Technology, 2019.
[7] Bao Ping, Liu Yunjie. Research on Fault Identification Based on Improved Deep Model in Combination of Generative Adversarial Networks Under Unbalanced Data Sets [J]. Journal of Electronic Measurement and Instruentation, 2019, 33(03): 176-183.
[8] PATHAK D, KRAHENBUHL P, DONAHUE J, et al. Context Encoders: Feature Learning by Inpainting[C]//2016: 2536–2544.

System Design of Automatic Parking Assist Based on ISO26262

Xiaojun Kuang[1a], Yafei Zhang[1b], Xiyang Wang[2c*]

[1]SAIC-GM-Wuling Automobile Co., Ltd.., china

[2]China Automotive Technology and Research Center Co.,Ltd.,china

[a]xiaojun.kuang@sgmw.com.cn, [b]yafei.zhang@sgmw.com.cn, [c*]wangxiyang@catarc.ac.cn

Abstract—The technology of intelligent driving cars is one of the key technologies of current automobile development, and the automatic parking assist system (APA, Auto Parking Assist) is a representative function of low-speed automatic driving. In order to improve the safety and reliability of the APA system, a hazard analysis and risk assessment of the APA system was carried out based on ISO 26262, safety objectives and functional safety requirements were derived, and then technical safety requirements were put forward and the system architecture design was carried out. And from the software point of view, the algorithmic design of the horizontal parking function is carried out, which further illustrates the feasibility of the technology.

1. Introduction

Auto Parking Assist (APA, Auto Parking Assist) is an intelligent parking assistance system that can quickly and safely drive the vehicle into a parking space automatically[1-3]. It recognizes the parking space by sensing the surrounding environment information of the vehicle through ultrasonic and image sensors, and according to the relative position information of the vehicle and the parking space, generate the corresponding parking trajectory and control the speed of the vehicle and the steering wheel to complete the automatic parking. A schematic diagram of the parallel parking process of the automatic parking system is shown in Figure 1[4].

For the automatic parking system, ultrasonic radar is used to monitor environmental obstacles, parking spaces and other information, and the APA controller makes corresponding judgments based on the received information, and controls the steering and braking of the vehicle[5]. However, with the upgrade of electronic control systems, electrical and electronic systems have become more complex[6]. The safety risks caused by systemic failures and random hardware failures will further rise, so functional safety issues are very important, so this article focuses on intelligent driving cars The key technology of automatic parking system is designed for system functional safety and algorithm design based on horizontal parking function.

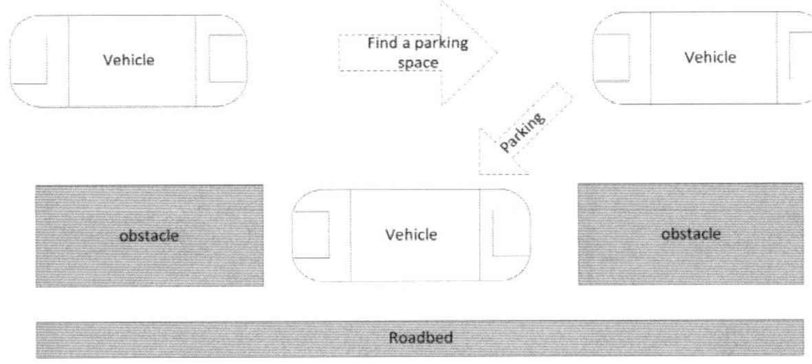

Figure 1 Schematic diagram of parallel parking of automatic parking system

2. Function definition of APA system

By analyzing the functions of the parking system and the parking process, the parking system mainly constructs five control systems: ultrasonic control system, image processing system, electric power steering control system, vehicle speed control system and vehicle control system. The above system constitutes the control system of the automatic parking system. There is a unified CAN (Controller Area Network) communication protocol between each system. The vehicle control system can collect the communication information of each module and coordinate the cooperation between the modules to realize the parking Control of the entire vehicle's actuators during the vehicle process, so as to complete the parking operation safely and quickly. The boundary block diagram of the automatic parking system is shown in Figure 2, and the external element function list is shown in Table 1.

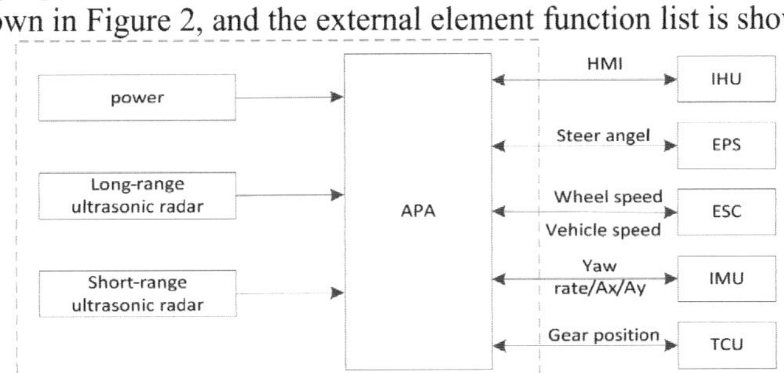

Figure 2 Boundary block diagram of APA system

Table 1 Description of APA system boundaries and external elements

ID	Element	Function description
1	Long-range ultrasonic radar	Original information about obstacles
2	Short-range ultrasonic radar	Original information about obstacles
3	APA controller	Receive and process the original data of the ultrasonic radar and the information of related parts, and realize the functions of parking space detection, path planning, follow-up line prediction and alarm prompt
4	IHU	Complete the human-computer interaction during the parking process
5	EPS	Provide lateral movement information
6	ESC	Provide longitudinal movement information
7	IMU	Provide body posture information
8	TCU	Provide vehicle gear information
9	Power	Provide power supply information

3. Analysis of the top-level requirements for the functional safety of the APA system

The combination of functional failure and driving scenario is called a hazard event. After the hazard event is determined, the safety integrity (ASIL) of the entire vehicle of the hazard event is evaluated based on three factors: severity (Severity), exposure rate (Exposure) and controllability (Controllability). Therefore, the Hazard Analysis and Risk Assessment (HARA) method can be used to identify the hazards caused by faults in related items, and divide them into categories and formulate corresponding safety goals, so as to avoid unreasonable risks.

For the function of "detecting obstacles or people during parking", to assess the risks and hazards of related items, first analyze its potential hazard events "no obstacles or pedestrians detected" and "provide wrong information". Obstacles or pedestrian locations". The safety goals and their corresponding ASIL levels are determined by systematic assessment of the hazardous event itself. The following HAZOP analysis, hazard analysis and risk assessment are carried out on the function of "detecting obstacles or people during parking", as shown in Table 2 and Table 3:

Table 2. HAZOP analysis of APA system

ID	Function	malfunction	Vehicle-level hazards	scene
1	Detect obstacles or people during parking	No obstacles or pedestrians detected	No obstacles or pedestrians detected	Automatic parking, pedestrians on the track, the speed is less than 10Km/h, city streets, normal weather conditions
2		Provide wrong obstacle or pedestrian location	Can't stop in time, hit obstacles or pedestrians	Automatic parking, pedestrians on the track, the speed is less than 10Km/h, city streets, normal weather conditions

Table 3. Hazard analysis and risk assessment of APA system

ID	malfunction	S	E	C	ASIL
1	No obstacles or pedestrians detected Provide wrong obstacle or pedestrian location	Parking at low speed, hitting people, causing injuries, S=2	Normal parking, E=3	There is a blind spot when parking, and the controllability is low, C=2	ASIL A
2	Provide wrong obstacle or pedestrian location	Parking at low speed, hitting people, causing injuries, S=2	Normal parking, E=3	There is a blind spot when parking, and the controllability is low, C=2	ASIL A

The safety goal is the top-level goal derived from the development of functional safety technology for the entire system. The above-mentioned function of "detecting obstacles or people during parking" has carried out a hazard analysis and risk assessment. According to the analysis result, the safety target "SG_01 should be able to detect pedestrians on the parking track when parking automatically".

After obtaining the safety goal, the corresponding functional safety requirements (FSR) need to be proposed. The following functional safety requirements are proposed for "SG_01", as shown in Table 4.

Table 4. Functional safety requirements of APA system

ID	Description of FSR	ASIL
1	Short-range ultrasonic radar should provide real-time radar environment sensing signals to APA	ASIL A
2	Short-range ultrasonic radar should avoid providing false radar environment sensing signals to APA	ASIL A
3	Long-range ultrasonic radar should provide real-time radar environment sensing signals to APA	ASIL A
4	Long-range ultrasonic radar should avoid providing false radar environment sensing signals to APA	ASIL A
5	The APA power supply system should ensure that the ECU receives the correct power of the vehicle	ASIL A
6	APA should perform power-on and runtime diagnosis for short-range ultrasonic radar and long-range ultrasonic radar. When one of the above sensors fails, APA should be able to detect and exit the function, and send a system failure alarm signal to IHU	ASIL A
7	APA should protect the signal received from ECU (IHU/EPS/ESC/IMU/TCU)	ASIL A
8	APA should calculate the parking position spatial information data based on the received radar environment sensing signals, determine the expected driving trajectory, and detect pedestrians and two-wheeled vehicles on the driving trajectory	ASIL A
9	When APA judges that there are pedestrians or obstacles on the driving track, it should send an obstacle warning signal to IHU	ASIL A

4. Technical safety requirements and system design of APA system

The functional safety concept specifies the safety goals of the system and the safety functions required by the system to achieve these safety goals. The technical safety concept needs to implement the following two parts.

(1) Further refine the safety functions proposed by the safety concept, and obtain the technical implementation plan of the safety functions.

(2) Analyze the realization path of the safety function, find the single point and latent fault that causes the safety function failure in the system or technical scheme, and propose safety measures or safety mechanisms to cover these faults.

Specifically, the steps to derive technical safety requirements and technical safety concepts from functional safety requirements are as follows.

(1) For each FSR, formulate the technical plan of the functional path in the system's initial architecture elements in detail.

(2) FTA or FMEA analysis is carried out with the FSR implementation technical solution as the object, and the single-point failure and double-point failure that violate the FSR in the implementation technical solution are identified.

(3) Develop corresponding safety mechanisms for single point of failure.

(4) Develop diagnostic mechanisms or safety measures to avoid latent faults for double-point faults.

(5) The safety mechanisms in (3) and (4) form the technical safety requirements TSR.

(6) Allocate the exported TSR to the hardware and software of the specific realization elements, and optimize the system architecture design.

The preliminary architecture of the automatic parking system after being assigned to the safety-related elements is shown in Figure 3.

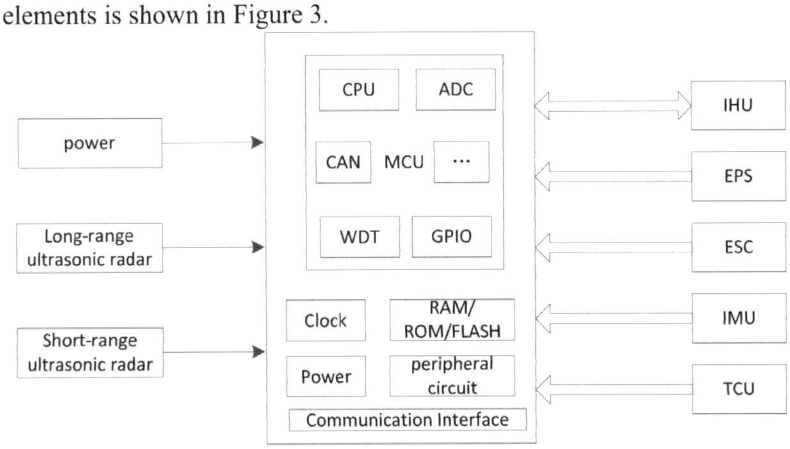

Figure 3. Block diagram of the APA system architecture

In order to achieve the above FSR, the TSR in Table 5 is obtained according to the FMEA analysis.

Table 5. Technical safety requirements of APA system

ID	Description of TSR	ASIL	Remark
1	The LIN communication diagnostic module should monitor the LIN messages input by the short-distance ultrasonic radar and the long-distance ultrasonic radar, and perform communication protection verification. When the verification fails, the MCU exits the function response, records the DTC, and sends the system fault signal.	ASIL A	FSR1 FSR2 FSR3 FSR4
2	The power module should monitor the voltage input from the KL30. When the voltage is too high or too low, the MCU stops the function response, but does not stop the CAN message transmission, and records the voltage status in the DTC	ASIL A	FSR5
3	APA should perform power-on and runtime diagnosis for short-range ultrasonic radar and long-range ultrasonic radar. When one of the above sensors fails, APA should exit the function, record DTC, and send a system failure signal	ASIL A	FSR6

4	The CAN communication diagnostic module should monitor the CAN messages input by IHU/EPS/ESC/IMU/TCU and do communication protection verification. When the verification fails, the MCU exits the function response, records the DTC, and sends a system failure signal	ASIL A	FSR7
5	APA should jointly calculate the parking position spatial information data based on the radar environment sensing signals of the short-range ultrasonic radar/long-range ultrasonic radar, determine the expected driving trajectory, and detect pedestrians and two-wheeled vehicles on the driving trajectory	ASIL A	FSR8
6	When APA judges that there are pedestrians or two-wheeled vehicles on the trajectory according to the radar environment sensing signal, APA sends an obstacle warning signal to IHU	ASIL A	FSR9
7	APA should perform CRC check on the signal sent to IHU, frame counter, overtime monitoring	ASIL A	FSR9

5. Software implementation strategy for horizontal parking

The horizontal parking module contains a signal input module, including parking information acquisition after finding a parking space, vehicle self-positioning information, vehicle coordinate update, front and rear radar signals, and side radar signals. The real-time status update of the vehicle is mainly used for the initial calculation of the parking starting point information of the first parking after the parking space is found, and the subsequent real-time posture and position information of the vehicle. The horizontal parking module is divided into an algorithm module, a logic module, an anti-collision module, and an adaptive angle control module. The horizontal parking algorithm module calculates the parking starting point during horizontal parking according to the initial parking position information after the vehicle searches for a parking space, and realizes path planning according to two arcs.

The algorithm flow chart of horizontal parking is shown in Figure 4.

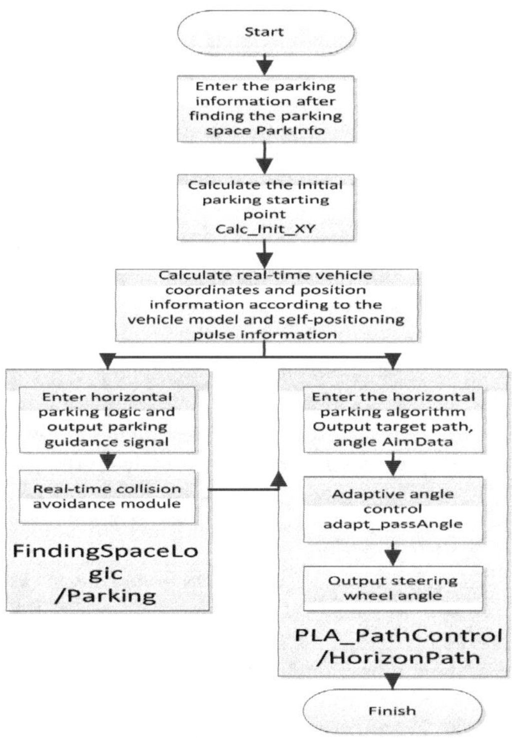

Figure 4. Algorithm flow chart of horizontal parking

(1) Find a parking space. After the car finds a parking space and parks, it outputs the parking position information to the path planning module.

(2) The path planning module starts to calculate the initial parking starting point.

(3) The vehicle module calculates real-time vehicle position and coordinates based on parking information and positioning pulses.

(4) Then enter the algorithm module, update the algorithm information in real time, output the target path, the vehicle wheel angle, and output the steering wheel angle through adaptive angle control.

(5) At the same time, the horizontal parking logic module outputs the parking guidance signal of the vehicle according to the real-time coordinate information of the vehicle.

(6) The anti-collision module receives front and rear radar information and outputs anti-collision prompt information.

6. Conclusion

In this article, we start from the general and very high-level consideration of the related hazards of pedestrians that cannot be detected in the automatic parking process. Through the process described in ISO26262, we gradually prove the needs and usefulness of specific technologies. Through the process of functional safety concepts and technical safety requirements, we first proved the requirements for the key functions provided by automatic parking, and then covered how to use them correctly to achieve safe operation in our automatic parking process. It also describes the function of horizontal parking from the perspective of algorithm, which shows the feasibility of this technology.

References

[1] LI Hong, GUO Kong-hui, SONG Xiao-lin, et al. Trajectory planning of automotive parallel parking with multi-constraints based on MATLAB [J]. Journal of Central South University: Science and Technology, 2013,44(1):101-107.

[2] Zhang Jiaxu, Guo Chong, Wang Chen, Zhao Jian, Wang Xinzhi. Performance evaluation of automatic parking system based on hardware-in-the-loop simulation platform [J/OL]. Journal of Jilin University (Engineering Edition): 1-8 [2021-09 -29].

[3] Zhang Jiaxu, Guo Chong, Wang Chen, Zhao Jian, Wang Xinzhi. Performance evaluation of automatic parking system based on hardware-in-the-loop simulation platform [J/OL]. Journal of Jilin University (Engineering Edition): 1-8 [2021-09 -29].

[4] You Zuwang. Research on the key technology of automobile automatic parking assist system[J]. China New Technology and New Products, 2020(20):11-14.

[5] Meng Haolan, Chen Junyi, Zuo Renjing, Yu Zhuoping, Chen Lei. Test method for autonomous parking system of smart car[J]. China Journal of Highway and Transport, 2019, 32(06): 158-168.

[6] Guo Xin, Han Yanming, Bai Yong, Cheng Bin. Research on Intelligent Parking Assist System (APA)[J]. Automotive Practical Technology, 2016(01): 28-31.

Research on the Semantic Composite Network Recognition Method of Power Distribution DTU Automation Acceptance Virtual Seat System

Hongwei Liu[1,a], Yuwei He[2,b], Jian Jiang[3,c], Xingye Liu[4,d], Haoyang Cui[5,e*]

[1,2,3,4]State Grid Zhejiang Electric Power Co, Ltd. Hangzhou Power Supply Company, Hangzhou, China

[5]Shanghai University of Electric Power, School of electronic and information engineering, Shanghai, China

[a]liuhw20212021@163.com, [b]yvwei_he@126.com, [c]jiangjianhangzhou@163.com, [d]liuxxingye2021@163.com, [e]cuihy@shiep.edu.cn

Abstract—Aiming at semantic inaccurate recognition problem caused by the lack of professional vocabulary in the existing corpus, a composite network recognition method for the semantics of the power distribution DTU acceptance virtual agent system is proposed. This method is used to extract and merge the entity, attribute and relationship information of power equipment from multiple data sources to construct a power distribution DTU acceptance knowledge graph. In addition, the graph database is the main body and the SQL database is the extension, and the corpus is stored in a distributed manner. The acceptance test results show that the overall recognition rate of the system platform for the instructions issued by the acceptance personnel in the complex environment is 92.8%, which reduces the acceptance time by more than 70%.

1. INTRODUCTION

DTU (Data Transfer Unit) is the key equipment of the distribution network, and its automated acceptance efficiency determines the delivery cycle of the distribution network upgrade [1]. At present, the automatic acceptance of distribution terminals is still generally based on the mode of communication and cooperation among field operation, maintenance personnel and the staff of the main station to test the trigger signal of electronic control. Statistics show that the acceptance of column opening and 6-interval ring network, etc. , takes 20 and 90 minutes, respectively, and opening and closing the cell change acceptance takes longer [2]. In the stage of extensive construction of intelligent distribution network, the manual intervention method of the existing distribution terminal acceptance has been difficult to meet the speed requirements of rapid distribution network construction. Intelligent and efficient configuration and automatic acceptance of DTU construction has become a key technical problem that needs to be solved urgently in the development of intelligent distribution networks. In view of the current situation that the installation process of DTU is still generally used by operation and maintenance personnel, it is more important to realize the intellectualization of acceptance links that the main station acceptance is intellectualized. Through intelligent extraction of communication semantic information and automatic comparison with DTU telemetry information, the construction quality and efficiency assessment can be effectively completed [3].

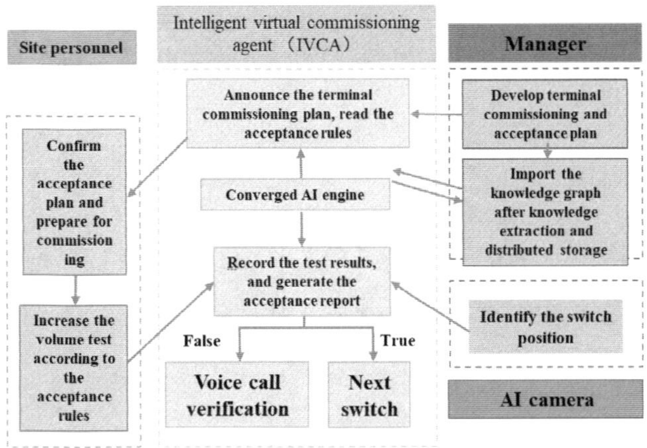

Figure 1. Virtual agent assistant acceptance mode

The intelligent virtual agent technology, which integrates intelligent speech recognition, text recognition, and multi-round human-machine dialogue, has the advantages of repetitive tasks and automatic execution of massive transactions, and is widely used in customer service, call centers, sales transactions and other fields. If the potential of virtual agent technology can be fully tapped to form an automated acceptance mode for virtual agents as shown in Figure 1. to assist on-site personnel, it will greatly promote the application and verification of data-driven concepts in the construction of intelligent terminals in distribution networks. Scholars in domestic and foreign fields have begun to use virtual agent technology in the field of electric power customer service. For example, literature [4]~[6] proposes to build a special map for electric power communication services for power communication network management, operation inspection, risk monitoring, business scheduling, and knowledge accumulation Provide human intelligence services. In reference to the problem of repetitive command issuing in power grid operation and dispatching, literature [7] constructed a power operation dispatching map and proposed a smart assistant program for power grid regulation and control operations. Literature [8] performed historical operation information based on the BP model data mining, realized the power system of professional dictionary library of structures. General corpus not included in the electric power business involves specialized vocabulary of semantic recognition accuracy is difficult to meet the needs of engineering applications, so the use of virtual agent technology to achieve the most important issue of the acceptance and distribution automation DTU is to build a dedicated language materials library.

Reasonable semantic extraction network and corpus storage structure have great influence on the accuracy and speed of speech recognition. In reference [9], hidden Markov model is used to segment the existing power equipment defect records, and cosine similarity of discriminant vector is used to eliminate co referential words, which improves the accuracy of semantic retrieval. In reference [10], semantic enhancement method based on topic comparison is used to reduce the redundancy of entity recognition results of LSTM network and improve the speed of speech recognition of virtual power customer service system.

The above research provides important theoretical support and reference for the construction of data transfer unit acceptance knowledge graph. However, current research on knowledge graphs in the power field mainly focuses on the construction of customer service and corpus and virtual customer service system, while the special corpus and its construction method for DTU acceptance of distribution network have rarely been reported. Based on this, this paper proposes a semantic composite network identification method for distribution DTU automatic acceptance virtual agent system according to the command joint triggering scenario and requirements of power DTU automatic acceptance business. The purpose of this paper is to identify and extract the professional words in the operation procedures, alarm information and historical operation logs of distribution DTU automatic acceptance business, construct the special corpus of distribution DTU acceptance, and form the operation mode of semantic search and joint trigger based on virtual agent.

2. Model Building

In the process of implementing DTU automatic acceptance business, the first problem to be solved by virtual agent system is to build a knowledge graph of distribution terminal automatic acceptance domain (as shown in Figure 2.) which includes all acceptance professional vocabulary and has higher retrieval efficiency, so as to improve the accuracy and recognition speed of key words in acceptance speech by virtual agent system.

In this paper, the composite network and ESIM model are used to complete the recognition, extraction and fusion of professional words from multiple data sources. The distributed database model is used to store the corpus for the nodes with different entity correlation degrees. While improving the recognition rate of key words, the redundancy of nodes in graph database is reduced, and the retrieval speed is improved.

2.1. Compound network knowledge extraction model

The objects of knowledge extraction are the semi-formatted and non-formatted sentences in the distribution network regulations, historical alarms, and operation logs. The difficulty of extraction is that it is difficult to align the data by using graph mapping to obtain the relationship between power equipment from the circuit diagram topology (semi-formatted data) [11]. And the other difficulty of extraction is that when using traditional machine learning methods such as conditional random fields, support vector machines, and decision trees extracts information from texts (unformatted data) such as distribution network regulations and historical operation information, due to the high degree of discretization of differences part in co-referential vocabulary and the sparse distribution of differences part in co-referential vocabulary in the NLP data set, the accuracy and coverage of knowledge extraction are difficult to meet the requirements of engineering applications [12].

The deep neural network has powerful vector distance analysis capabilities, which can effectively identify the same vocabulary in unformatted text. However, there are obvious differences in the extraction effects of different parts of speech texts on various networks. The CNN network determines the length of the sliding extraction window according to the number of convolution kernels, which can effectively extract device attribute information with a compact input structure and a certain length. The RNN network composed of the information backward linear conduction structure has a better effect on the information extraction of the acceptance business entity with greater uncertainty in the input length. The BiLSTM network introduces a direction control factor to extract the relationship information that has a strong correlation with the context [13]. The F1 scores of these three neural network sub-modules on the NLP general data sets of three different websites are 91.7%, 99%, and 87%. The accuracy and coverage of keyword extraction from text data sources are relatively high.

The core idea of knowledge extraction is using deep neural networks to format the unformatted data in the text in order to distinguish the entities, relationships, attributes and other knowledge elements in the unformatted data [11]. This part is divided into three sub-tasks including named entity recognition, attribute extraction and relationship extraction. The three sub-task modules are implemented by CNN, RNN and BiLSTM.

The data obtained by knowledge extraction contains a lot of repetitive information. For example, 3U0, ABC-phase zero-sequence voltage, and three-phase zero-sequence voltage refer to the same objects, and distribution DTU and distribution DTU devices have the same meaning. The repetitive co-referencing text has a greater impact on the uniqueness of the keyword reference in the automatically generated acceptance report and the standardization of the report. In this paper, the relationship extens-

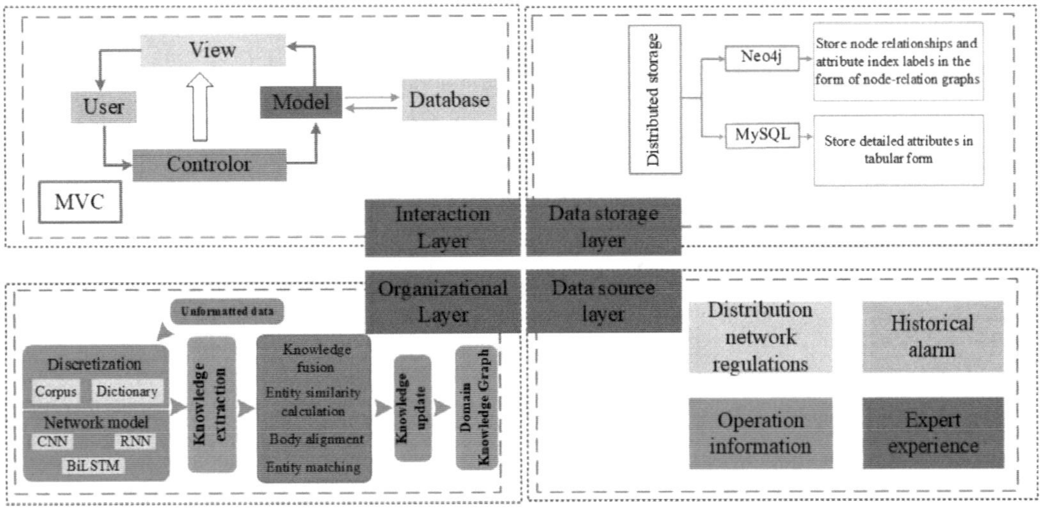

Figure 2. Knowledge Graph Architecture in the Field of Automated Acceptance

ion calculation is used to achieve common reference resolution. The filtered data is not stored in the map, which can be stored in a relational database and then linked with entities in the knowledge graph [14]~[17]. The following takes BiLSTM network output result processing as an example to show the process of coreference resolution. The processing methods of CNN and RNN network output results are similar.

Using ESIM co-referential resolution model

$$\alpha_t = \omega_\alpha \cdot FFNN_\alpha(x^*) \tag{1}$$

$$\alpha_{i,t} = \frac{\exp(\alpha_t)}{\sum_{k=START(i)}^{END(i)} \exp(\alpha_k)} \tag{2}$$

$$\hat{x}_i = \sum_{k=START(i)}^{END(i)} \alpha_{i,t} \cdot x_t \tag{3}$$

$x^*=[h_{t,1}-h_{t,-1}]]$is the output of the BiLSTM network. 1 and -1 represent the direction in which the LSTM recognizes the context. $\hat{x}_i$ represents the attention coefficient, which is the accumulation of the weights of the co-referential words in a sequence *span i*. The ESIM co-referential resolution model is shown in Figure 3.

After the weighted summation of the output x^* of the BiLSTM network, the sequence integrated into the attention mechanism can be obtained, denoted by $\hat{x}$. The sequence is comprehensively evaluated on its accuracy, recall, and harmonic F value through a three-layer scoring structure. The sequence with the highest core is considered to be the target sequence that needs to be stored in the map. The data source information is classified and modularized according to the length of the input text and the contextual relevance. The repeated information in the extraction results is resolved by the ESIM model, which improves the accuracy and coverage of keyword extraction, and eliminates the repetitive common reference. The influence of the text on the standardization of the acceptance report saves the time cost of node traversal in the graph database and improves the keyword retrieval speed. The data source information is classified and mod-

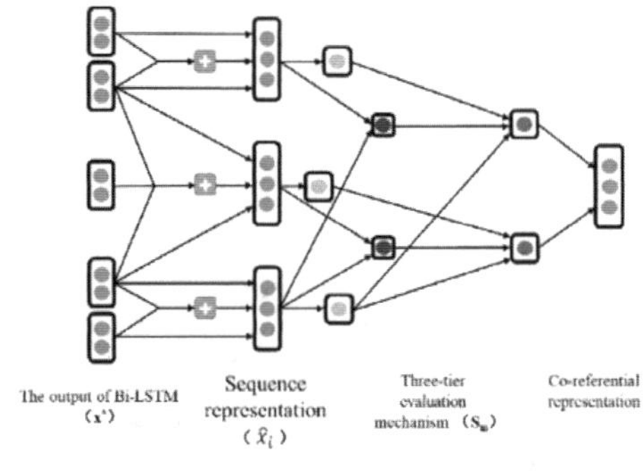

Figure 3. ESIM co-referential resolution model

ularized according to the length of the input text and the contextual relevance.

2.2.Distributed storage model of joint related database and relational database

The power equipment of the distribution network is deeply coupled and has a high degree of data correlation. When the SQL-type database is used to store the target sequence obtained from knowledge extraction, it is necessary to establish a complex relationship between different rows of tables and multiple tables to characterize the coupling relationship between multiple power devices. The index relationship of the Entity Relations is difficult to construct, and the retrieval time is long, which cannot meet the requirements of real-time extraction of keywords in the on-site acceptance speech.

Graph database is a kind of NoSQL database, which is different from the storage form of two-dimensional tables in relational database. The data in graph database is stored in the form of node-relation (N-R) graph [18]~[20]. The topological relationship among the power equipment participating in the joint debugging is accustomed to be represented by circuit diagrams in engineering. It has a natural similarity with the data storage form of graph database. This can effectively reduce the complexity of relationship representation. An instruction in the automatic acceptance of a power distribution DTU usually contains multiple interrelated entities. Compared with relational databases that store data in tabular form, native graph databases represented by Neo4j have built-in more efficient traversal algorithms. When retrieving a group of highly related entity nodes, there is no need to establish links between multiple data tables. Just specify the starting node and the traversal direction. The time and complexity of data retrieval operations are greatly reduced. Therefore, it is more convenient and quicker to "add, delete, modify, and check" data.

However, after repeated access tests on the data with different attribute richness in the two databases, it is found that for entity classes with rich entity attribute information and some frequently updated data (such as events, logs, fault records, etc.), use Relational database storage can update the attribute information of an entity multiple times without traversing all associated entities in the database. For example, a group of entities has a relevance of 10, and each entity contains data of 3 attributes as experimental samples. One attribute of any entity is updated 100 times. The result shows that the data is updated when the same memory is occupied. The speed is increased by about 25%.

In order to ensure the real-time performance of keyword retrieval during on-site speech recognition, and take into account the data update speed when the corpus is used for case recording, reference, reasoning and other in-depth applications, a distributed storage technology of joint graph database and relational database is proposed. The pivot nodes selected by the extended calculation of entities and relationships with an entity relevance greater than 10 can be found in the knowledge graph by creating a formatted index label in the attribute column of the pivot node of the graph database to point it to the corresponding table in the relational database. Link to data. The distributed storage structure with entity association as the extension calculation trigger condition simplifies the relationship construction process. While retaining all the results of knowledge extraction, it solves the problem of low voice recognition efficiency caused by high coupling of equipment and improves the data update speed.

2.3.Semantic search model of aggregation engine based on sentence pattern matching and knowledge graph

The situation of the distribution DTU automation acceptance site is complex, and the question and answer sentences are changeable. Therefore, this paper proposes a dual-engine semantic search technology based on sentence matching and the distribution DTU automation acceptance knowledge graph. First, through text analysis, clarify the relationship between the node to be identified and the object to be searched. For "Hello, I am XXX", "Dial XXX", "Correct, please start", the more fixed command responses are based on ESIM. The model is matched with the sentence patterns in the predefined corpus, and the predetermined operation in the template library is executed when the results are consistent. For a long sentence scenario that contains multiple equipment operating parameters, the knowledge graph is directly called to form the answer. In the process of virtual agent-assisted acceptance of DTU, the search method of aggregation engine reduces the return time of speech recognition results and improves the efficiency of acceptance.

3. EXPERIMENTAL RESULTS AND ANALYSIS

Based on the Distributed DTU automation acceptance business scenario, with the help of the Open5200 distribution automation management platform of a State Grid electric power company, the IVCA system is used to test the accuracy of the semantic recognition of the distribution DTU acceptance corpus constructed by this method and the rapidity of the virtual agent-assisted acceptance mode.

3.1.Accuracy and analysis

In an environment of 40dB background noise, using a sampling frequency of 11.025KHz, 16-bit quantized numbers, and mono wav format audio, the IVCA system was subjected to an acceptance rule increase test. The acceptance rule is a data set composed of four voice data sets: Mandarin data, accent Mandarin data, Mandarin natural conversation data and real network customer service voice data. The result is shown in Figure 4.

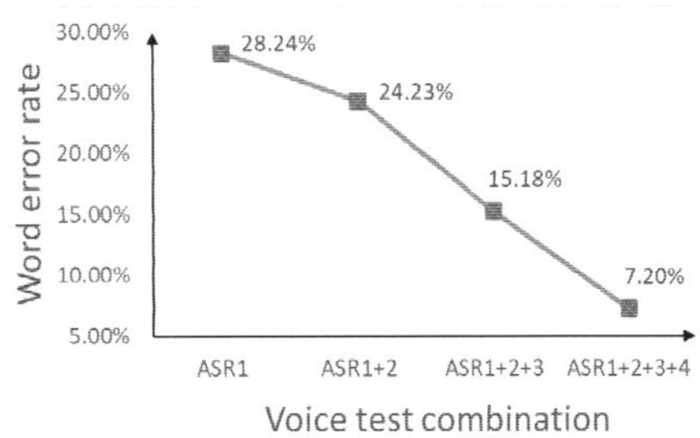

Figure 4. Multi-language acceptance rules plus test results

When only using Mandarin speaking data for training, the word error rate of the speech recognition model is 28.42%, after superimposing the accent in Putonghua, Putonghua natural conversation, and live online customer service voice in the data hall, the word error rate WER was reduced to 7.2%. The overall system recognition accuracy rate ACR can be expressed as

$$ACR = 1 - WER\% \tag{4}$$

The word error rate WER is defined as follows

$$WER = 100 \cdot \frac{S + D + I}{N} \% \tag{5}$$

Among them, *S*, *D*, *I* are the number of words added, replaced, and deleted in the recognition result, and *N* is the total number of words. Applying aggregate AI scheduling engine technology driven by knowledge, the overall recognition rate ACR (calculated according to the word error rate) of the IVCA system is 92.8%. In the multi-bay ring network DTU acceptance service, the working hours of the main station operation and maintenance personnel are reduced from h-level to m-level, which can improve the problem of low efficiency of traditional mode acceptance under the background of massive power DTU access in the process of distribution network upgrade and transformation.

3.2.Rapidity and analysis

3.2.1 Rapidity when the acceptance result is normal: On September 15, 2020, the acceptance personnel used the IVCA system to conduct an automated acceptance testing on the distribution DTUs of 10 switch stations (the green part in Figure 5) between the two main power distribution stations of Fuqian and Xixing. Take the acceptance process of DTU (equipment name new1Tq1) as an example, The on-site acceptance personnel

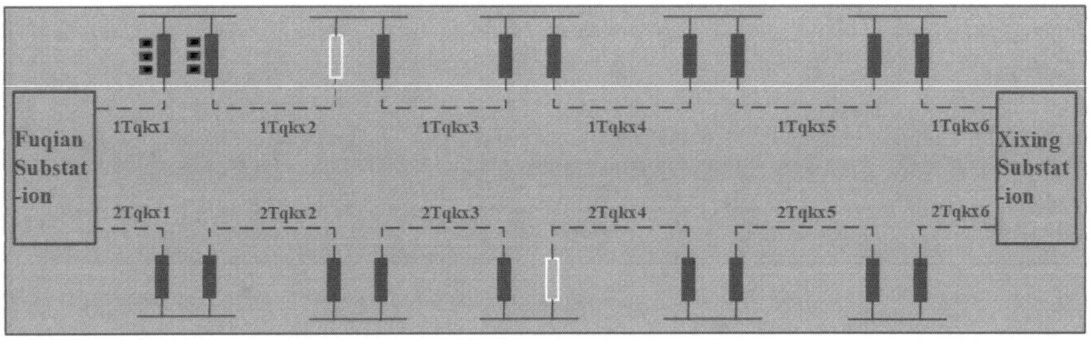

Figure 5. 10 DTUs between Fuqian substation and Xixing substation

<TESTRERURNINFO::DMS>

Measuring Point ID	Equipment		Feeder	Switch Station	Test Content	On-site Result	Actual Result	Debug Result	Date
3800475307413492143	1Tqkg1		1Tqkx	new1Tql	close	close	close	success	2020-9-15_20:00:00
3800475307413492143	1Tqkg2		1Tqkx	new1Tql	open	open	open	success	2020-9-15_20:00:00
3808637909939014265	1Tqkg1	A phase current	1Tqkx	new1Tql	26	26	26	success	2020-9-15_20:00:00
3808637909939014285	1Tqkg1	B phase current	1Tqkx	new1Tql	27	27	27	success	2020-9-15_20:00:00
3808637909939014287	1Tqkg1	C phase current	1Tqkx	new1Tql	26	26	26	success	2020-9-15_20:00:00

<TESTRERURNINFO::DMS>

Figure 6. Acceptance report

measured the opening and closing status of interval 1 switch (equipment name 1Tqkg1) and interval 2 switch (equipment name 1Tqkg2). After the three-phase current of the network bus is distributed, the IVCA is input by voice After confirming that the voice recognition result is correct, the system will report it to the power distribution master station. The system automatically generates an acceptance report as shown in Figure 6. The aggregate AI confirms that the return value in the acceptance report is consistent with the acceptance rules (the remote signal is correct, and the remote measurement accuracy is within the operating range). After that, make a prompt for the DTU acceptance of the next switch station.

After the system is put into operation, under the condition that the on-site collection and telemetry results are consistent, the work time of the main station operation and maintenance personnel the acceptance of a single switch station DTU is reduced from the original 20 minutes to 0 minutes, which greatly improves the automatic acceptance of the power distribution DTU Efficiency, significantly eases the work pressure of the main station operation and maintenance personnel.

3.2.2 Rapidity when the acceptance result is abnormal: When the return value in the acceptance report is inconsistent with the acceptance rules, the aggregate AI dispatch engine sends alarm information to the main stations' operation and maintenance personnel, according to the acceptance rules, the distribution master station observes the main station graphics, alarm window information, and check the actual position of the interval switch, the actual value of the three-phase current of the distribution bus, the voltage of the DTU battery of the switch station and other information with the on-site acceptance personnel. Then modify the terminal debugging plan and execute it in accordance with the processing flow of the DTU terminal data abnormality. During this process, the operation and maintenance staff of the main station reduced the working time for acceptance from the original 20 minutes to 5 minutes. Alleviate the problem of a large gap in the joint commissioning of distribution network automation effectively. In addition, the handling of power distribution DTU acceptance abnormalities is more refined and standardized.

4. Conclusion

This paper proposes a composite network recognition method based on the semantics of the virtual agent system for automatic acceptance of power distribution DTU. For professional vocabulary information in unformatted texts such as distribution network regulations, a composite network extraction and distributed storage method is adopted to construct a special corpus for the field of power distribution DTU acceptance. The virtual agent assists the field staff in the power distribution DTU acceptance test, and the recognition rate of the field staff's acceptance speech in the complex system environment is extremely high, reducing the main station operation and maintenance staff's acceptance work time by more than 70%. The joint debugging test results show that the method can significantly improve the efficiency of the automatic acceptance of distribution DTUs. During the upgrading and transforming of the distribution network, the method also can effectively cope with the problems of the access of a large number of newly added power distribution terminals, the cumbersome maintenance of old terminals which leads to insufficient use of the traditional acceptance mode, and the increase of the shortage of personnel in the automation and joint debugging of the distribution network.

The research on the construction method of special knowledge graph for distribution network automation management is still in its infancy, and effective mining methods for historical operation logs, alarm information, fault records and other internal laws have not yet been formed. Manual intervention is still necessary to change the acceptance plan after the acceptance report is abnormal. Therefore, how to fully mine the inherent laws contained in the data source and guide the adjustment of the acceptance plan will be the future research direction.

Acknowledgment

This study was supported by the Science and Technology Project of State Grid Zhejiang Electric Power Co., Ltd(5211HZ19014R). The authors gratefully acknowledge the anonymous reviewers for their valuable comments.

References

[1] R. Zeng, Y. Xu and D. Wu, "Normalization Performance Evaluation Method for Distribution Automation Terminal," 2019 IEEE 28th International Symposium on Industrial Electronics (ISIE), Vancouver, BC, Canada, 2019, pp. 1833-1838, doi: 10.1109/ISIE.2019.8781296.
[2] Loken R S J, Apostolov A. Testing of Protection in a Digital Substation Protection Automation and Control System (SPACS)[J]. 2020.
[3] Song X, Yuan H, Xiaoxuan Z, et al. Evaluation Model for World-Class Distribution Network Management[C]//2018 China International Conference on Electricity Distribution (CICED). IEEE, 2018: 59-62.
[4] Yang H, Luo L, Chueng L P, et al. Deep Learning and Its Applications to Natural Language Processing.
[5] Shen G , Wang W , Mu Q , et al. Data-Driven Cybersecurity Knowledge Graph Construction for Industrial Control System Security[J]. Wireless Communications and Mobile Computing, 2020, 2020(6):1-13.
[6] ZHANG C, Zhiyong Y U, Zhen Z, et al. User image construction framework based on big data fusion background for power grid business travel[J]. Telecommunications Science, 35(12): 148.

[7] Tang, Xisheng (Institute of Electrical Engineering, Chinese Academy of Sciences, Beijing; 100190, China) Source: Dianli Xitong Zidonghua/Automation of Electric Power Systems, v44, n1, p23-28, January 10, 2020
[8] Chen, Zhengping (State Grid Fujian Electric Power Dispatching and Control Center, Fuzhou; 350001, China); Mi, Weimin; Lin, Jinghuai; Wang, Heng; Wang, Hao; Dong, Genyuan Source: Dianli Xitong Zidonghua/Automation of Electric Power Systems, v43, n22, p173-178 and 186, November 25, 2019
[9] Liu, Ziquan (College of Electrical Engineering, Zhejiang University, Hangzhou; 310027, China); Wang, Huifang Source: Dianli Xitong Zidonghua/Automation of Electric Power Systems, v 42, n14, p158-164, July 25, 2018
[10] Huang H, Hong Z, Zhou H, et al. Knowledge Graph Construction and Application of Power Grid Equipment[J]. Mathematical Problems in Engineering, 2020.
[11] Wang J, Wang X, Ma C, et al. A survey on the development status and application prospects of knowledge graph in smart grids[J]. IET Generation, Transmission & Distribution, 2021, 15(3): 383-407.
[12] Qiao, Ji (China Electric Power Research Institute, Haidian District, Beijing; 100192, China); Wang, Xinying; Min, Rui; Bai, Shuhua; Yao, Dong; Pu, Tianjiao Source: Zhongguo Dianji Gongcheng Xuebao/Proceedings of the Chinese Society of Electrical Engineering, v40, n18, p5837-5848, September 20, 2020
[13] Li Z , Li Q , Zou X , et al. Causality extraction based on self-attentive BiLSTM-CRF with transferred embeddings[J]. Neurocomputing, 2021, 423:207-219.
[14] Li Z , Ding X , Liu T . Constructing Narrative Event Evolutionary Graph for Script Event Prediction[J]. 2018.
[15] Sun, Hongbin (Department of Electrical Engineering, Tsinghua University, Haidian District, Beijing; 100084, China); Huang, Tianen; Guo, Qinglai; Zhang, Boming; Guo, Wenxin; Liu, Wentao; Xu, Taishan; Xu, Tao Source: Dianwang Jishu/Power System Technology, v 44, n 1, p 1-8, January 5, 2020
[16] V Yawahare H R , Karde P P , Thakare V M . A Hybrid Database Approach Using Graph and Relational Database[C]// 2018:1-4.
[17] Zhao, M.; Wang, H.; Guo, J.; Liu, D.; Xie, C.; Liu, Q.; Cheng, Z. Construction of an Industrial Knowledge Graph for Unstructured Chinese Text Learning. Appl. Sci. 2019, 9, 2720.
[18] Gao Y , Li Y F , Lin Y , et al. Deep Learning on Knowledge Graph for Recommender System: A Survey[J]. arXiv, 2020.
[19] Wang Y, Zhang K, Dai Q, et al. Construction and Application of Knowledge Graph in Full-service Unified Data Center of Electric Power System[J]. 2018.
[20] Wikidata as a knowledge graph for the life sciences[J]. eLife Sciences, 2020, 9.

A Design of fast-setting on-chip Charge Pump Circuit

Yannan ZHAI[1,a], Ling Gao[1], Jingquan Li, Qangli Qiu[1]

[1]Aviation University Air Force ,130022 Changchun Jilin China

Abstract : A circuit of fast-setting charge pump on-chip is designed based on the Dickson circuit. The charge pump circuit, which is improved, increases the initial node voltage. With the consideration of the current mismatch, the accurate clock of the duty circle below 50% is proposed. The HSPICE simulation result indicates that the setting time from 0V to 20V only needs 51.650μs for the charge pump, and it is faster than the traditional Dickson charge pump 26.03μs. In summary, the settling time of the output voltage of the charge pump is prominently decreased and the performance of the charge pump is obviously improved.

1.Introduction

With the rapid development of integrated circuits, the requirements of various devices for power supply are also increasing. As an internal power supply, on-chip charge pump is also widely used. It is used in the fields that need high voltage, such as serial communication circuit, EEPROM, dynamic random access memory and so on. In EEPROM, the charge pump is used to write or erase the suspended gate device. If the charge pump pumping speed is slow, EEPROM can not realize high-speed data rewriting, and the data may be lost and mistransmitted, affecting the device performance. Therefore, reducing the voltage rise time has become the focus of the design of on-chip charge pump.

Now most charge pumps are based on Dickson charge pump circuit [1], which uses charge to accumulate in capacitance to produce high voltage. When the capacitor is charged and discharged, the current mismatch leads to the low voltage of the first stage of the charge pump, which will seriously affect the boost speed of the charge pump. Therefore, starting with the study of reducing the initial voltage of the charge pump, a 16 stage Dickson charge pump boost circuit is designed in this paper. The circuit improves the traditional Dickson charge pump and is driven by a clock signal with a duty cycle of less than 50% to achieve the purpose of fast boost.

2. Dickson charge pump analysis

Fig. 1 is a circuit of Dickson charge pump and its clock signal. In Fig. 1, the input signal of the charge pump is the power supply voltage VDD, Vout is the output voltage, φ and φ/ Is a two-phase non overlapping clock signal. In microelectronic technology, the diode is difficult to manufacture, so the gate drain short circuited NMOS tube is used instead of the diode.

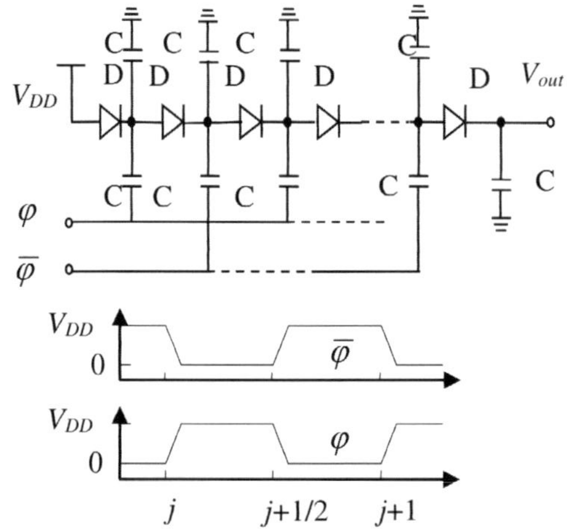

Figure 1The Dickson charge pump circuit and driving clocks

2.1 static analysis

The operation process of Dickson charge pump is the process of charging and discharging the capacitor with the help of the driving signal with opposite polarity. It continuously pushes the power supply and the charge of the previous stage into the next stage, so that the potential of the latter stage is higher than that of the previous stage, so as to obtain high voltage at the last stage. Clock phase φ At low level, diode D1 is turned on and V_{IN} charges C_1 until the voltage of node 1 becomes V_{in}-V_d; When φ Jump to V_φ , the voltage of node 1 becomes Vin + (V_φ-V_d), causing diode D_2 to turn on and charge C_2 until the voltage of node 2 becomes V_{in}+(V_φ-V_d)-Vd; When φ becomes low level Once again, and the voltage of node 2 jumps to V_{in}+2 (V_φ-V_d). In this way, the charge is transferred from the left to the right. Due to the unidirectional conductivity of the diode, the charge cannot return from the right to the left. With the increase of the number of charge pump stages, the charge is continuously transferred from the power supply to the output, so as to obtain the required high voltage. After stage n, the output voltage is:

$$V_{out} = V_{in} + N \cdot (V_\varphi - V_d) - V_d \tag{1}$$

2.2 dynamic analysis

However, when Dickson charge pump is working, there is a loss of charge in the process of transfer, and the output voltage is not constant. The Dickson charge pump is dynamically analyzed [29-31], and the output voltage value of each clock cycle is obtained:

$$V_{out}(j) = (N+1)V_g - NV_g\beta^j \tag{2}$$

$$\beta = \frac{1}{1 + \frac{C}{NC_{load}}} \tag{3}$$

If the rise time of the output voltage Vout (J) from VG to VPP is represented by TR, TR satisfies $V_{out}(T_r) = V_{pp}$,So tr can be expressed as:

$$T_r = \ln\lfloor 1 - (V_{pp} - V_g)/NV_g \rfloor / \ln\beta \tag{4}$$

Where n is the number of stages of the charge pump and VG is the maximum voltage gain of each stage. Since equation (4) is represented by the number of cycles of the clock, in practice, equation (4) should be multiplied by the cycle of the driving clock. If used α Representing the influence of the body effect on the transmission transistor, it can be deduced that the boost time of the output voltage rising from VG to VPP is:

$$T_r = \ln\left[1-\left(V_{pp}-V_g\right)\Big/\left(\sum_{i=1}^{N+1}\alpha^i-1\right)V_g\right]\Big/\ln\beta \tag{5}$$

According to the above analysis and equation (2), equation (4), equation (5) and equation (5), whether the Dickson charge pump connected with diode or the Dickson charge pump connected with MOS tube, the rise time of their output voltage is related to the power supply voltage, transmission tube threshold voltage, charge pump stage, coupling capacitance, load capacitance The period of the driving clock and the initial voltage of the node are closely related.

Therefore, reasonable setting of power supply voltage, transmission tube threshold voltage, charge pump stage, coupling capacitance, load capacitance and other parameters is very important to reduce the voltage rise time. Designing a high-performance clock generation circuit to output a clock signal with high frequency will also reduce the voltage rise time. The initial voltage of the node is very low during the initial boost cycles of the charge pump. Only by increasing the initial voltage of the node and reducing the number of clock cycles required for voltage rise can the voltage rise time be reduced. In this paper, the above factors are considered in order to reduce the boost time.

According to the previous theoretical analysis of Dickson charge pump, the initial voltage of the node has a great impact on the voltage rise time. According to equation (2), without considering the volume effect, the voltage of each node can reach the following ideal value only when the number J of clock cycles is large

$$V_{out}(\infty) = (N+1)V_g \tag{6}$$

In the first clock cycle, i.e. J = 1, the voltage Vout (1) at each node can only reach

$$V_{out}(1) = (N+1)V_g - NV_g\beta \tag{7}$$

It can be seen that the voltage of each node needs a certain clock cycle to reach the theoretical value. If the initial voltage value of each node is increased, the voltage rise time can be significantly reduced.

3. circuit design

3.1 improved charge pump model

NMOS precharge tube [2-12] is added to each stage of MOS charge pump to precharge the voltage of each node to a certain value, which can reduce the time when the node voltage reaches the theoretical value. However, the circuit with precharge tube also has some defects, that is, it increases the area of the circuit too much.

Simulation results show that adding an even number of precharges can improve the boost speed. In this paper, the 16 stage MOS Dickson charge pump is simulated by adding different even number of precharges, and the effect of the number of precharges on the boost time is studied. The voltage rise time of each charge pump from 0V to 20V is shown in table 1.

Table 1 The relationship between the number of precharging FET and rise time

Number of precharge tubes	Rise time(μm)	Number of precharge tubes	Rise time(μm)
0	77.680	10	67.402
2	73.610	12	66.643
4	70.346	14	65.905
6	69.472	16	65.385
8	68.514		

Journal of Physics: Conference Series **2195** (2022) 012034 doi:10.1088/1742-6596/2195/1/012034

Therefore, this paper improves the Dickson charge pump and proposes a charge pump model that only adds a precharge tube in the first four stages of the circuit, which not only improves the boost speed, but also saves the circuit area. The improved Dickson charge pump model is shown in Figure 4-20

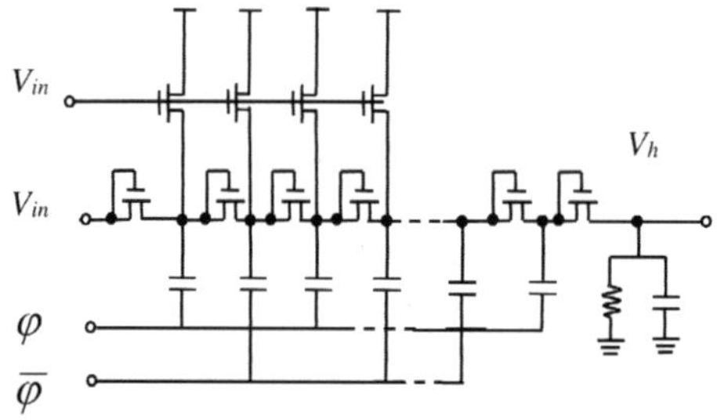

Figure 2 Optimized Dickson charge

When the input signal Vin is high, the precharge tube is turned on and the node voltage is charged to a certain value. The working principle of the charge pump core is exactly the same as that of the traditional Dickson charge pump.

3.2 improve the drive clock signal

The traditional Dickson charge pump uses a two-phase non overlapping clock driving signal with a duty cycle equal to 50%. Each time the signal changes, the charging current and discharge current of the capacitor turn on at the same time, which will undoubtedly cause adverse factors such as charge leakage and charge discharge current mismatch.

Therefore, this paper proposes to use a clock signal with a duty cycle equal to 30% as the drive of the charge pump, as shown in Fig. 3.

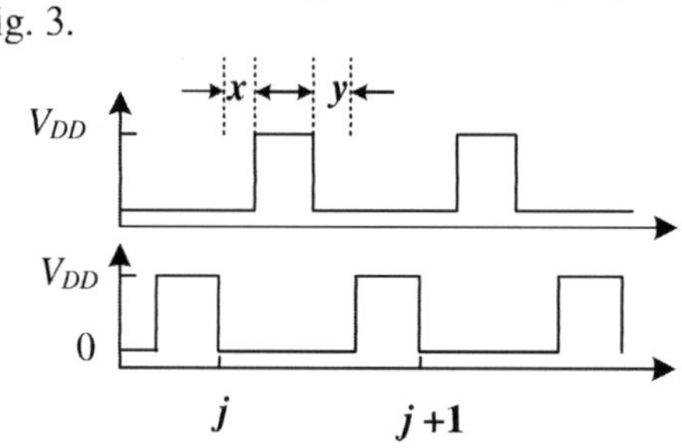

Figure 3 clock signal waveform

The circuit adds the working process that both X and Y clocks are zero, that is, clka = 0 and CLKB = 0. At this time, all capacitors are reset, and the capacitor is charged and discharged after fully discharged, so as to avoid the mismatch between the charge and discharge of the two capacitors.

3 Simulation Analysis

0.35 is adopted in this paper μ M CMOS process model, the designed charge pump circuit is simulated with HSPICE software. The simulation environment is set as follows: the simulation temperature is 25 °C, the power supply voltage is 5V, and the simulation time is 700 μ s.

Firstly, the Dickson charge pump shown in Fig. 1 is simulated to obtain the clock signal φ and φ/ Is a non overlapping clock with a duty cycle of 50%. The simulation results are shown in Figure 4.The figure (a) shows the initial waveform of the simulation, VA is the voltage of the first node of the charge pump, VB is the voltage of the second node of the charge pump, and VH is the output voltage of the charge pump. the figure (b) shows the complete waveform of the simulation. the figure (c) shows the amplified waveform of VH.

As can be seen from figure (a), in the first clock cycle (J = 1), the clock φ Is high level, VA (1) = 5.640v, VB (1) = 3.710v.

As can be seen from figure (b), when φ and φ/ At low level, the charge pump is turned off, and VH and VA are 5V; When φ and φ/ When it is a two-phase non overlapping clock, the charge pump is turned on, VA and VB oscillate and rise, and the output voltage VH also rises gradually.

As can be seen from figure (c), VH is at 65 μ When VH rises to 20V, the boost speed is very slow and almost stable. From the reading, VH is 142.680 μ When it reaches 20V at s, it takes 77.680 for VH to boost from 0V to 20V μ s.

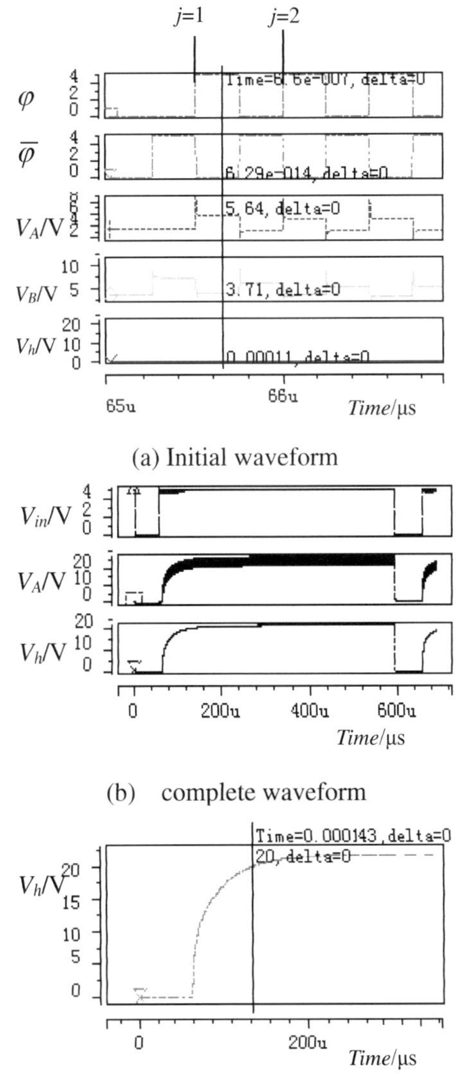

(a) Initial waveform

(b) complete waveform

(c)Vh amplification waveform

Figure 4 simulation waveform of Dickson charge pump

The optimized Dickson charge pump shown in Figure 2 is simulated, and the clock signal φ and φ/ The duty cycle is 30%. The simulation results are shown in Figure 5. The figure (a) shows the initial waveform of the simulation. VA is the voltage of the first node of the charge pump, VB is the voltage of the second node of the charge pump, and VH is the output voltage of the charge pump. the figure (b) shows the amplified waveform of VH.

As can be seen from figure (a), in the first clock cycle (J = 1), the clock φ Is high level, VA (1) = 6.74v, VB (1) = 4.58v. Compared with the traditional charge pump shown in Fig. 4 (a), the initial voltage is about 1V higher. It can be seen that the introduction of precharge tube on the basis of traditional circuit can improve the initial voltage.

Journal of Physics: Conference Series **2195** (2022) 012034 doi:10.1088/1742-6596/2195/1/012034

As can be seen from figure (b), VH is at 65 μ When VH rises to 20V, it almost reaches stability. From the reading, VH is 116.650 μ When it reaches 20V at s, it takes 51.650 for VH to boost from 0V to 20V μ s. From 5V to 20V, the improved charge pump boosts 26.03 faster than the traditional charge pump μ s. It can be seen that the charge pump circuit can speed up the boost speed by using the clock signal with a duty cycle of 30%.

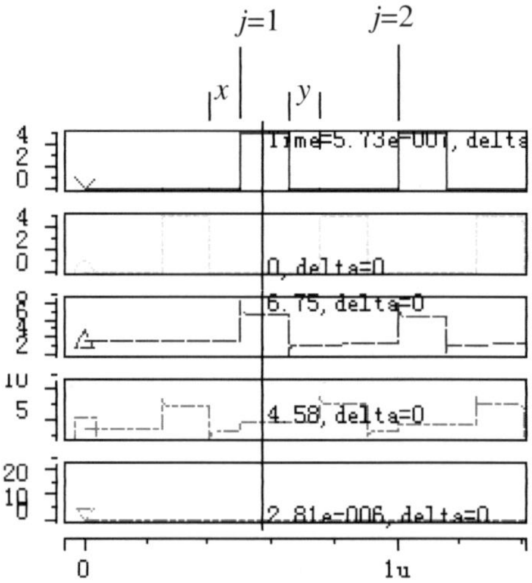

(a) Initial waveform

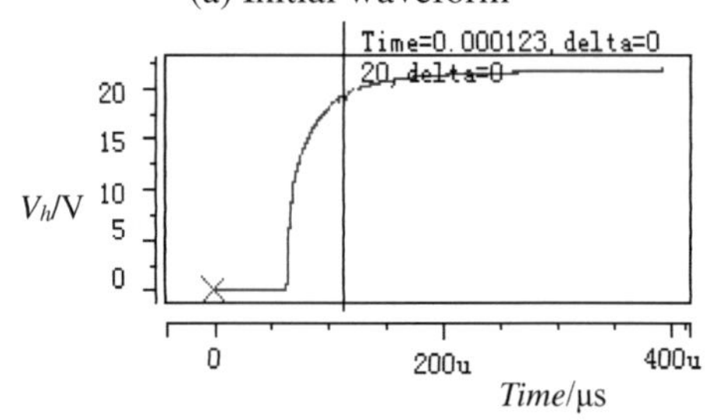

(b) Vh amplification waveform

FIG. 5 Simulation waveform of improved charge pump

4. conclusion

A stable charge pump boost circuit structure applied to EEPROM is designed. The circuit uses a stable reference voltage to drive the oscillation circuit to produce an accurate clock with a duty cycle less than 50%, which greatly improves the boost speed of the charge pump; A precharge tube structure is added to the first two stages of the charge pump core, which further improves the boost speed of the charge pump; The combination of charge pump and voltage regulating circuit stabilizes the output voltage and reduces it to the required value. Under the working voltage of 5V, it only takes 56.256 for the charge pump to boost to the high voltage of 15.962v μ s. And can achieve stable high-voltage output. The simulation results show that the charge pump has fast boost speed and high reliability, and can provide stable voltage for EEPROM reading, writing and erasing.

Reference

[1] Dickson J F. On-chip high-voltage generation in NMOS integrated circuits using an improved voltage multiplier technique [J]. IEEE J Sol Sta Circ, 1976;11(3) : 374-378.

[2] Jong-shin Shin, In Yong Chung , Young June Park, etal. A new charge pump without degradation in threshold voltage due to body effect [J] . IEEE J Solid State Circuits, 2000; 35 (8): 1 227-1 230.

[3] RICHARD J F , SAVARIA Y. High Voltage Charge Pump Using Standard CMOS Technology [J]. IEEE Regular Session G: High Voltage Techniques & Continuous Time Filtering, 2004: 317-320.
[4] Mantooth H A, Duliere J L. A Unified Diode Model for Circuit Simulation [J]. IEEE Trans. on Power Electronics, 1997, 12(5): 816-823.
[5] Shin Jongshin, CHUNG In-Young, Young June Park, Hong Shick Min. A New Charge Pump without Degradation in Threshold Voltage Due to Body Effect [J]. IEEE Journal of Solid-State Circuits, 2000, 35(8): 1227-1230.
[6] JOHAN S Witters, GUIDO Groeseneken. Analysis and Modeling of On-Chip High-Voltage Generator Circuits for Use in EEPROM Circuits [J]. IEEE Journal of Solid-State Circuits, 1989, 24(5): 1372-1380.
[7] MARTINE J Silva. A Switched Capacitor Double Voltage Generator [J]. IEEE Proc. Mid-West Symp. Circuits and Systems, 1994, 1: 177-180.
[8] FAVRAT P. High-Efficiency CMOS Voltage Doubler [J]. IEEE Journal of Solid-State Circuits, 1998, 33(3): 410-416.
[9] TORU Tanzawa, TOMOHARU Tanaka. A Dynamic Analysis of the Dickson Charge Pump Circuit [J]. IEEE Journal of Solid-State Circuits, 1997, 32(8): 1231-1240.
[10]ZHANG Ming, LLASER Nicolas. Dynamic Analysis of Dickson Charge Pump Circuits with a Resistive Load [J]. IEEE ICECS, 2003: 431-434.
[11]Zhu Y , Sang X, Kim YH. Analysis and Design of an Exponential Charge Pump.Advanced Materials Research. 2021,vol 753-755,1483-1488.
[12]Ngueya WS , Mellier Julien, Ricard Stephane. Ultra Low Power Charge Pump with Multi-Step Charging and Charge Sharing. IEEE International Memory Workshop.2016

Research on Commutation Switch to Control Three-phase Unbalance and Azimuth Configuration

Zhenwei Li[1], Jianjun Wu[1], Kefei Long[1], Jiajun Li[1], Jiao Nie[2], Yu Wang[2] and Yi Zhang[2*]

[1] Enshi Power Supply Company, State Grid Hubei Electric Power Company, (Enshi, Hubei Province, 445000, China)

[2] Wuhan Haomai Electric Power Automation Technology Limited Liability Company (Wuhan, Hubei Province, 430074, China)

*Corresponding author's email: sales@haomai.net

Abstract-The randomness of each phase load in the station combined with the distribution of new energy into the network has caused the three-phase unbalance to become more and more serious. The use of commutation switch to switch the three-phase unbalance load effectively solves the problem. Considering the number of commutation switches, the particle swarm algorithm is used to optimize the configuration orientation of the commutation switch and to improve the accuracy of the installation location of the commutation switch. Effectively reduces the three-phase unbalance in the platform area.

1. Introduction

Three-phase unbalance is one of the power quality problems that have long plagued the desk area. Serious unbalance will cause high line loss[1], uneven heat and output of transformer, motor vibration and heat, which endanger the safety of distribution network and electrical equipment[2]. In order to solve the three-phase unbalance problem, experts and scholars at home and abroad have done a lot of work. Literature[3] proposed a three-phase line current unbalance suppression algorithm based on distributed static series compensators (DSSCs). Literature[4] selected three-phase four-arm topology and real-time detection method of three-phase current based on instantaneous reactive power theory. Literature[5] the active unbalance control technology with power electronic properties is proposed to control the three-phase unbalance. Document[6] combines a commutation switch with an energy controller for treatment. The commutation switch is currently the most economical and effective way to eliminate the three-phase unbalance. It changes the phase sequence of the access load by switching the phase of the load mechanically, so as to switch the phase of the heavy load to the phase of the light load. In the actual installation process of a project, the number of installations in a desk area is usually preset. It is worth studying where to install it to improve the three-phase unbalance of the desk area with maximum efficiency. In this paper, the load data of the desk area is analyzed by particle swarm algorithm, the configuration orientation of the commutation switch is optimized, and the method to find the optimal location is analyzed. The three-phase unbalance of the platform can be better improved by using the commutation switch and reasonable installation position.

2. Principle of commutation switch

2.1. Overview of three-phase unbalance

In the distribution network of power system, the distribution network of 0.4kV voltage level station often uses the three-phase four-wire system to supply power to the users. After the secondary output of the three-phase transformer, the users will only use one phase, which is affected by wiring, time and season. Three-phase unbalance will occur when one phase user consumes more power while the others are smaller. The standard three-phase voltage is a three-phase symmetric sine wave as shown in figure1. unbalanced three-phase voltage can exhibit waveform distortion as shown in figure 2.

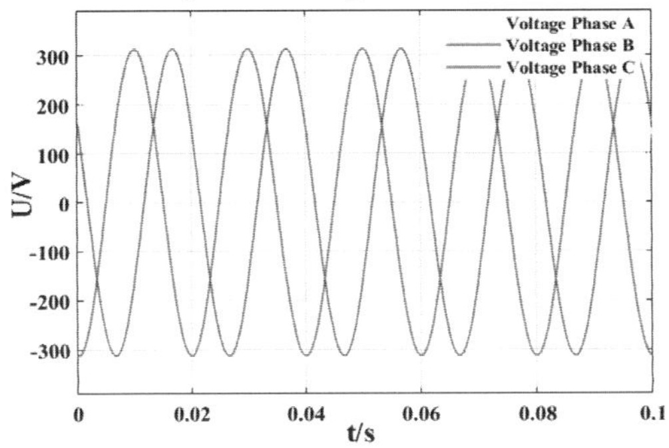

Figure 1. Symmetrical three-phase voltage

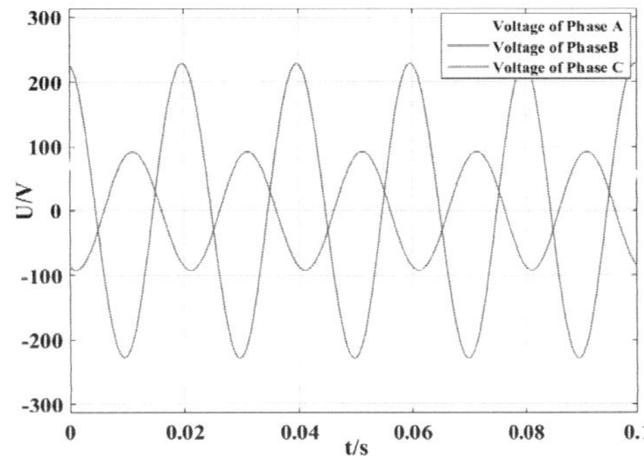

Fifure 2. Unbalanced three-phase voltage

The three-phase vector diagram can directly reflect the neutral point drift distortion of unbalanced voltage, as shown in figure 3, and the power quality will be affected.

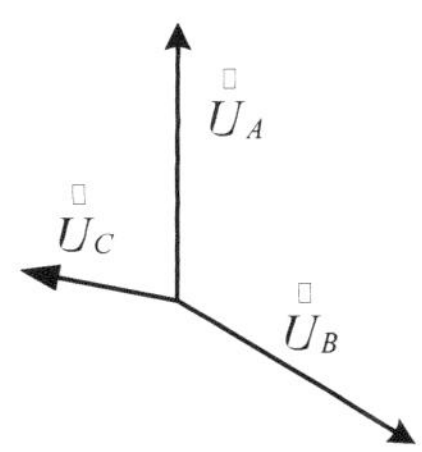

Figure 3. Three phase voltage distortion

The most intuitive impact of three-phase load imbalance is the imbalance of three-phase current and the increase of neutral line current. Describe the unbalance degree of three-phase current mathematically:

$$\beta = \frac{max\{I_A,I_B,I_C\} - min\{I_A,I_B,I_C\}}{max\{I_A,I_B,I_C\}} \times 100\% \quad (1)$$

In equation (1) β is the three-phase unbalance degree, $max\{I_A, I_B, I_C\}$ is the largest phase of three-phase current, and $min\{I_A, I_B, I_C\}$ is the smallest phase of three-phase current. The surge of current at the neutral point will lead to high line loss. The equivalent power supply diagram is shown in figure 4.

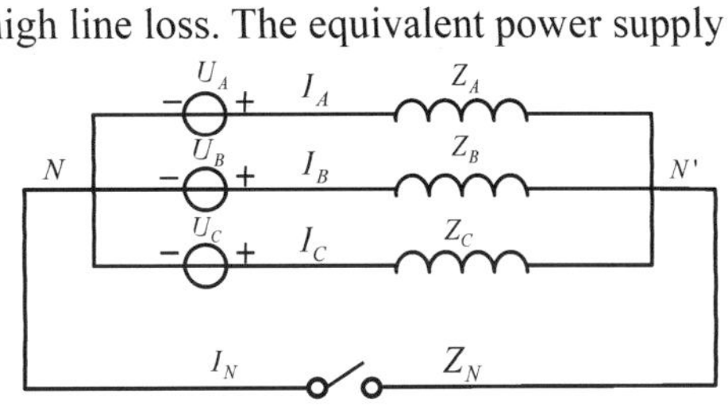

Figure 4. Equivalent circuit diagram of three-phase four wire power supply

The impedance values of phaseA, B and C lines are respectivelyZ_a,Z_b,Z_c. represents the voltage and current of each phase in U_a, I_a,U_b, I_b,U_c, I_c indicates that the current on the neutral line is expressed by I_N denotes impedance, and Z_n denotes impedance represents the line loss per unit length, which can be expressed by the following formula:

$$\Delta P = {I_a}^2 Z_a + {I_b}^2 Z_b + {I_c}^2 Z_c + {I_n}^2 Z_n \quad (2)$$

ΔP represents line loss, I_N in case of three-phase balance the value is zero, and the loss is the basic line loss. However, when the three-phase is unbalanced, the value of the neutral line is not zero, and the value is large. Because the radius of the neutral line is generally smaller than that of the phase line, the resistance will be much higher, resulting in more line losses.

2.2. Management of commutation switch

The commutation switch is internally composed of thyristor and relay, as shown in figure 5. A single thyristor and relay are connected in parallel to form a group of change-over switches. Three groups are connected in parallel to form the core components of the commutation switch. The switch is input by the three-phase incoming lines of A, B and C, and outputs any one of the three phases of A, B and C to the user.

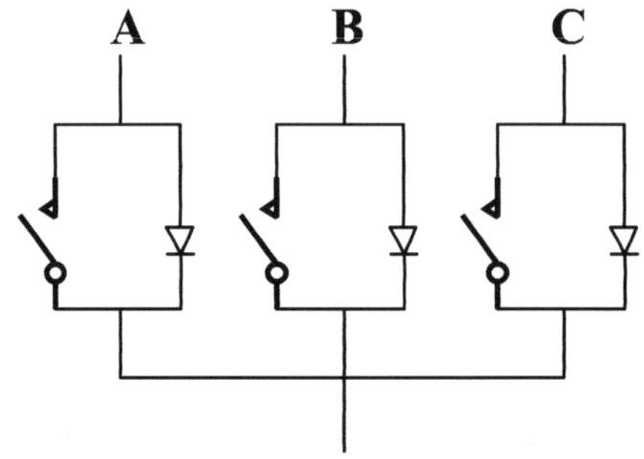

Figure 5. Equivalent circuit diagram of commutation switch

When phase A load is too heavy and phase B load is light, the three-phase unbalance in the bench area is serious. In the setting area, phase a flows through 80A, phase B flows through 30A, and phase C flows through 50A. According to formula (1), the unbalance degree of three-phase current is calculated as $\beta = \frac{max\{I_A,I_B,I_C\} - min\{I_A,I_B,I_C\}}{max\{I_A,I_B,I_C\}} \times 100\% = \frac{80-30}{80} \times 100\% = 62.5\%$ at this time, commutation measures shall be taken to switch part of the load of phase A to phase B. It is detected that the installation position of the commutation switch is a single-phase user. At present, phase A power supply mode is used, and the effective value of current flow is 25A. The user shall be switched from phase a power supply to phase B power supply. At present, the switch of phase A circuit is on, and then give a command to the thyristor of phase a to turn it on; In order to better disconnect phase A, disconnect phase A relay while maintaining the conduction of phase a thyristor; The magnetic holding relay has been disconnected, and all the current flows through the thyristor. At this time, the trigger signal of phase A is removed, and then the thyristor of phase A will be turned off when the current crosses zero. Turn on the phase B thyristor for the phase B trigger signal.The magnetic holding relay undertakes the normal working current to complete the positive sequence commutation, and completely turn off the phase A thyristor before giving the phase B trigger signal.

Commutation is to forcibly change the user's access phase sequence with a mechanical switch. If the commutation time is too long, it will lead to power failure or even phase to phase short circuit, which is absolutely unacceptable. According to relevant regulations of the State Grid, the commutation voltage loss time shall not exceed 20ms. The commutation process is closely coordinated by thyristor, relay and control signal. As shown in figure 6, the waveform capture of oscilloscope in the commutation process is shown, and the single commutation time is 6.52ms, meeting the requirements.

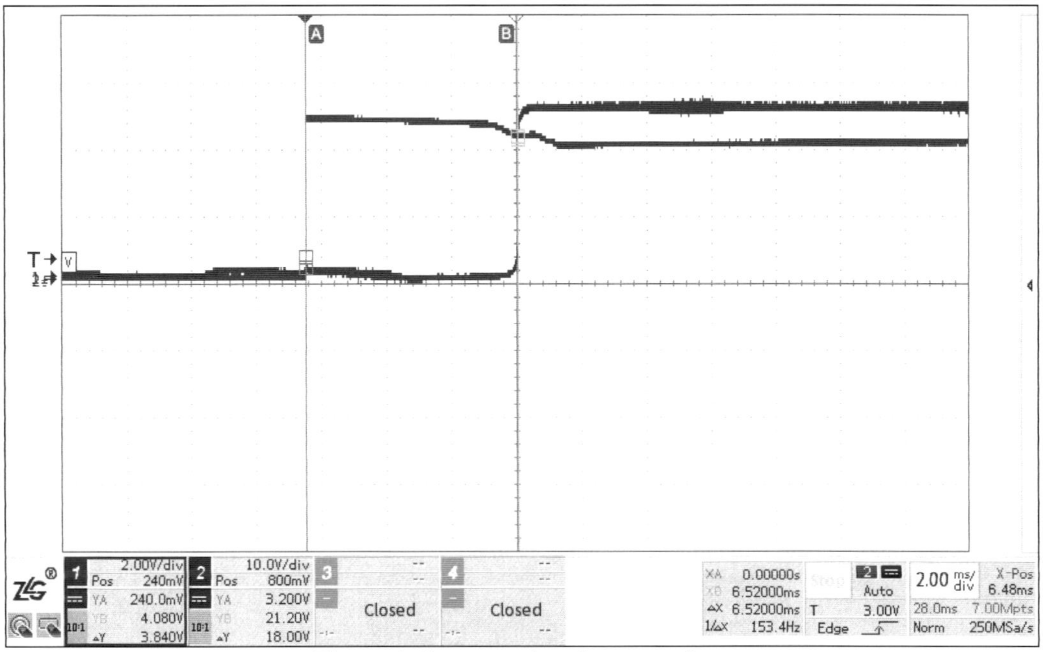

Figure 6. Commutation time waveform

After 25A of phase A is switched to phase B, the current of phase B is 55A, the current of phase A is 55A, and the current of phase C is 50A. The calculated three-phase unbalance after phase commutation is 9%, which meets the relevant provisions of national standards and the requirements of power quality.

3. Optimization of installation position of commutation switch

3.1. Historical data analysis of station area

When engineers decide to use commutation switch to manage a specific station area, considering economy and construction convenience, it is impossible to install a commutation switch for each user load. Therefore, it is particularly important to consider the installation position of commutation switch. According to the relevant provisions of the state grid operation inspection, the number of commutation switches shall be limited by the rated capacity of the transformer, and the number of 100, 200 and 400kVA shall not exceed 6, 9 and 12. The change law of power consumption in the station area will not change much in a year. Select the imbalance degree analysis in several days to give the time unbalance degree change curve, as shown in figure 7.

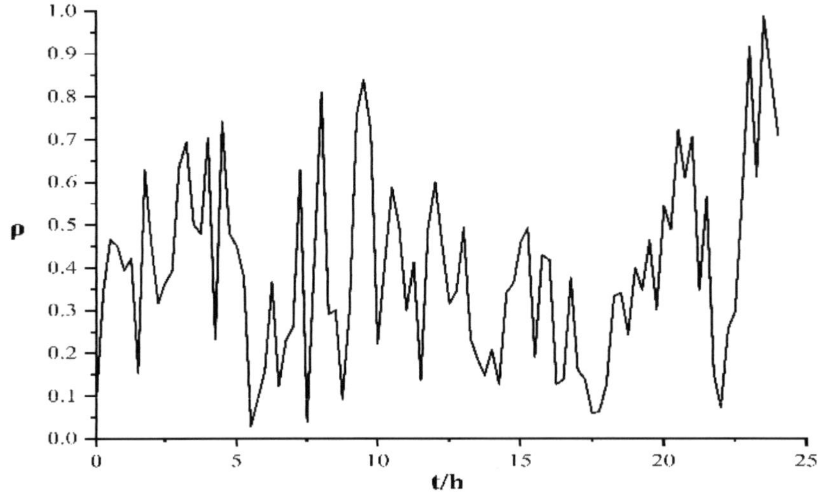

Figure 7. Variation curve of unbalance degree

Analyzing the historical data of the station area has important reference significance for the installation of commutation switch. Particle swarm optimization algorithm is used to analyze the historical three-phase unbalance in the station area, establish the position optimization model and iterative optimization [7]. The core formula of particle swarm optimization algorithm is as follows:

$$V_{id}(t+1) = \omega(t)V_{id}(t) + C_1r_1[P_{id}(t) - X_{id}(t)] + C_2r_2[P_{gd}(t) - X_{id}(t)] \tag{3}$$

$$X_{id}(t+1) = X_{id}(t) + V_{id}(t+1) \tag{4}$$

$$\omega(t) = \frac{\omega_{max} - t(\omega_{max} - V_{min})}{n} \tag{5}$$

$\omega(t)$ is the inertia factor, $V_{id}(t)$ represents the velocity in dimension d when the i'th particle iterates t times, $X_{id}(t)$ represents the position of the i'th particle on dimension d when iterating t times. Similarly, $P_{id}(t)$ represents the individual extreme value, $P_{gd}(t)$ is the global extremum, C_1C_2 is the learning factor, r_1r_2 is a random factor with a value range of (0,1).

3.2. Simulation analysis

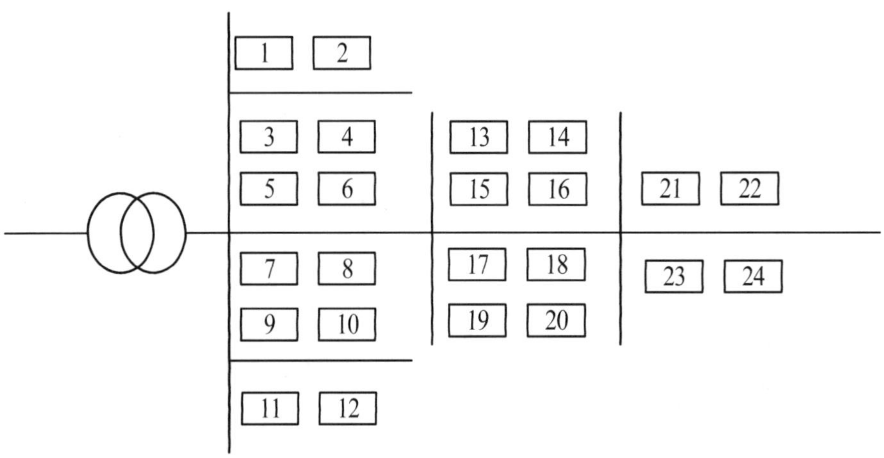

Figure 8. Load topology of station area

The station area model is established. There are 24 load points in the station area, which are distributed within the power supply range of the station area, as shown in figure 8. The load power consumption difference is analyzed to judge the three-phase imbalance. The installation position of the commutation switch is iterated continuously in 24 loads to generate $1 \times n$ based on the n-dimensional vector matrix, analyze the measures of the optimal commutation command, compare the unbalance degree of each load after commutation, the number of times involved in commutation and the number of switching actions, constantly compare the old position with the new position, and continuously iterate according to different data in different stations to obtain the best state, and finally find the appropriate installation position, and the optimization convergence is shown in figure 9.

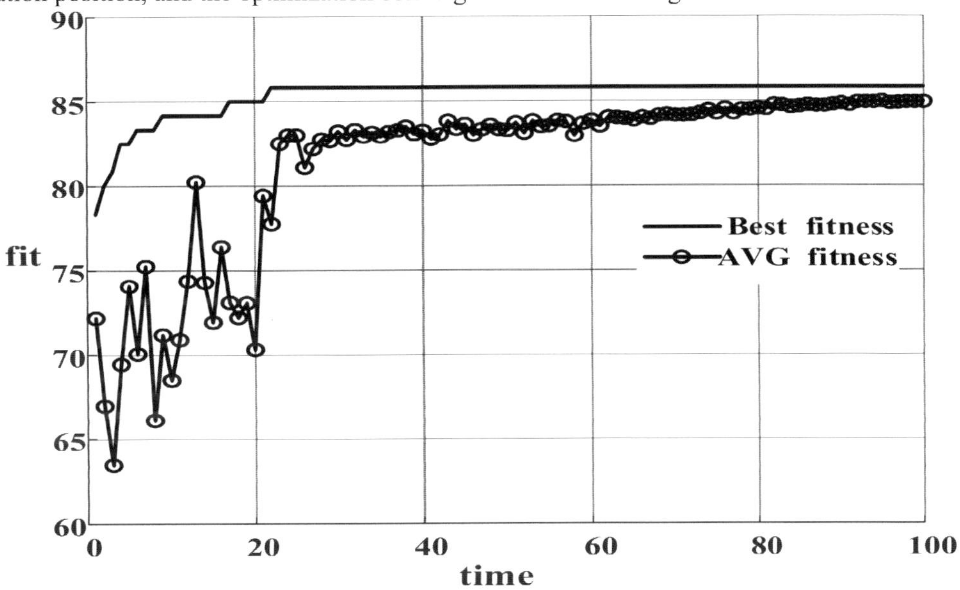

Figure 9. Optimization fitness curve

4. Conclusion

This paper introduces the basic concept of three-phase imbalance, simulates and analyzes the process of using commutation switch to control three-phase imbalance in the station area, and introduces the method of using particle swarm optimization algorithm to find the best installation position of commutation switch, which has reference significance for pursuing efficiency and economic value in engineering practice.

References

[1] Hao, S.,Cai, S., Zhang, Y., Liu, H., Chen, G., Chen,X., (2021) Quantitative Analysis Between Three-phase Unbalance and Line Losses. Power System Technology, 45(04):1547-1552.

[2] Yan, Z., Lei, X., He, J., Zhou, W., Ye, H., Zhao, L., (2018) Study on the influence of three-phase unbalance on the load capacity of distribution transformer. Electrical Measurement & Instrumentation, 55(08):51-57.

[3] Yoon, H., Cho, Y., (2019) Imbalance Reduction of Three-phase Line Current Using Reactive Power Injection of the Distributed Static Series Compensator. Journal of Electrical Engineering & Technology, 14(3).

[4] Zhang, C., Fu, Z., Ye, H., Niu, D., Study on three-phase unbalanced control of low voltage distribution network. In: *Proceedings of 2019 the 5th International Conference on Electrical Engineering, Control and Robotics (EECR 2019)*.Guang Zhou China. 2019, 77-83.

[5] Lu, Z., (2021) Research on the Three-phase Unbalanced Treatment Technology in Distribution Area. Power Electronics, 55(09):40-42.

[6] Yu, Z., Zhu, D., Meng, L., Teng, P., (2021) Research on Three-Phase Imbalance Treatment Based on Commutation Switch and Energy Controller. Applied Technology, (06):62-64+67.

[7] Yang, W., Li, Q., (2004) Survey on Particle Swarm Optimization Algorithm. Engineering Scinece, (05):87-94.

Journal of Physics: Conference Series **2195** (2022) 012036 doi:10.1088/1742-6596/2195/1/012036

Performance analysis of EES-based transcritical carbon dioxide two-stage compression/ejector refrigeration system for shipboard cold chambers

Yang Li[1,2,3,4#], Dazhang Yang[1,2,3,4#], Qing Zhang[1,2,3,4] and Jing Xie[1,2,3,4*]

[1]College of Food Science and Technology, Shanghai Ocean University, Shanghai, 201306, China

[2]Shanghai Professional Technology Service Platform on Cold Chain Equipment Performance and Energy Saving Evaluation, Shanghai, 201306, China

[3]Quality Supervision, Inspection and Testing Center for Cold Storage and Refrigeration Equipment, Ministry of Agriculture, Shanghai 201306, China

[4]National Experimental Teaching Demonstration Center for Food Science and Engineering (Shanghai Ocean University), Shanghai, 201306, China

Yang Li and Dazhang Yang contributed equally to this study

Corresponding author's Email: jxie@shou.edu.cn

Abstract. Carbon dioxide (CO_2) as a natural refrigerant has a broad application prospect in the refrigeration system of shipboard cold chambers. In this work, a novel transcritical CO_2 two-stage compression refrigeration system equipped with a two-phase ejector for shipboard cold chambers is proposed. At the same time, simulation of the two-phase ejector and the system was carried out by Engineering Equation Solver (EES). Additionally, The effects of high pressure stage exhaust pressure, intermediate pressure and evaporating temperature on system performance were also investigated. The results showed that the system COP increased with the evaporating temperature increased from -35°C to -25°C. The system COP decreased with the increase in exhaust pressure, when the evaporating temperature is maintained at -35°C. The optimal intermediate pressure is 4.75 MPa, when the evaporating temperature is fixed at -35°C.

1. Introduction

At present, ships are responsible for 90% of the world's cargo trade transportation, and the annual emission of greenhouse gases during the transportation of ships is about 1 billion tons, accounting for about 3% of the total global CO_2 equivalent emissions [1]. The shipping industry is growing rapidly, bringing prosperity to the world economy, but also exacerbating air pollution. In September 1987, the Montreal Protocol proposed ozone-depleting substances and a phase-out schedule. With the increasing global emphasis on environmental protection, the international community has proposed further restrictions on the greenhouse effect of refrigerants in ships [2]. In April 2018, the International Maritime Organization (IMO) determined the greenhouse gas emission reduction strategy for the world's first shipping industry. The main content is that by 2050, the annual total greenhouse gas emissions will be reduced by at least 50% compared with 2008. The Kigali Amendment to the Montreal Protocol, which entered into force on January 1, 2019, regulates and reduces hydrofluorocarbons (HFCs) and their refrigerant blends with a high global warming potential (GWP). In May 2019, the 74th session of the IMO Marine Environment Protection Committee (MEPC) developed a process for assessing the impact of greenhouse gas emission reductions from ships and a follow-up consideration program for emission reductions to further promote the low-carbon process [3-5].What's more, as a natural refrigerant, CO_2 has no environmental problems and has broad application prospects in ship refrigeration system [6-8]. CO_2 as a refrigerant has many advantages: ①Cheap and good for large refrigeration equipment; ②Non-combustible, non-toxic, adaptable to various lubricants and mechanical equipment materials; ③High pressure and high density, CO_2 system can be more compact and lighter under the same cooling and heating requirements; ④Good heat transfer performance.

Generally, the commonly used shipboard refrigeration system mainly consists of four main components: compressor, condenser, throttling valve and evaporator, which is an important auxiliary equipment of the ship and a guarantee for the normal operation of the equipment, which can improve people's living and working environment and ensure the storage conditions of food [9]. However, this shipboard refrigeration system is a trancritical CO_2 two-stage compression system equipped with a two-phase ejector. The purpose of introducing the two-phase ejector in the shipboard cold chamber refrigeration system is to reduce the throttling loss and enhance the performance of the system. The structure of the two-phase ejector is mainly composed of four parts: working nozzle, receiving chamber, mixing chamber and diffuser [10]. The working fluid enters the working nozzle and then decreases the pressure and increases the speed to a higher flow rate from the nozzle and enters the receiving chamber, and then the low-pressure fluid from the evaporator is induced, and the two enter the mixing chamber, where the working fluid and the induced fluid mix with each other and exchange momentum and energy to equalize the speed, while along with the increase in pressure, the mixed fluid enters the diffuser and the pressure continues to increase, finally forming a mixed fluid with a pressure between the working fluid pressure and the induced fluid pressure. The final formation of a mixed fluid with a pressure between the working fluid pressure and the intermediate pressure of the pilot fluid pressure [11, 12]. Throughout the process, the ejector is able to increase the pressure of the induced fluid without directly consuming mechanical energy, which is the main feature of the ejector [13].

In this work, a novel transcritical two-stage compression refrigeration system equipped with a two-phase ejector for shipboard cold chambers is proposed. At the same time, simulation of the ejector was also carried out through the EES. In addition, the effects of high pressure stage exhaust pressure, intermediate pressure and evaporative temperature on the system coefficient of performance (COP) were explored.

2. System description

The transcritical CO_2 two-stage compression/ejector refrigeration system in the shipboard cold chamber and its pressure-enthalpy diagram are illustrated in Figure 1. The system is composed of an evaporator (EVA), a low-pressure expansion valve (EXP II), a gas-liquid separator (SEP), a two-phase ejector (EJE), a low-pressure compressor (LCM), a high-pressure compressor (HCM), an intercooler (ICL), a high-pressure expansion valve (EXP I) and a gas cooler (GC).

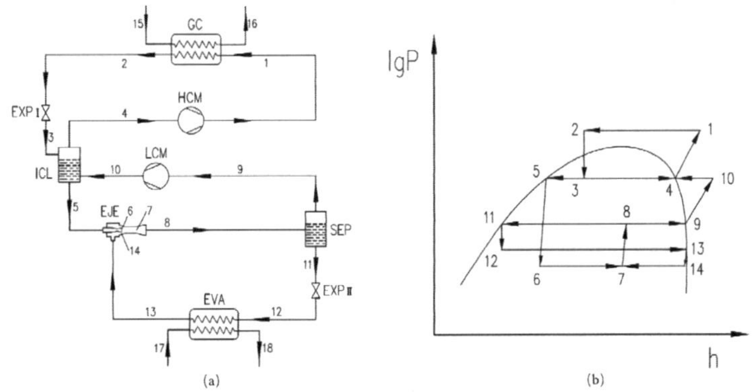

Figure 1. System schematic (a) and its pressure enthalpy diagram (b).

The working principle of the two-stage compression/ejector refrigeration system for shipboard cold chambers is as follows: the saturated steam (state point 9) enters the low-pressure compressor and is compressed into medium temperature superheated steam (state point 10). After that, it enters the intercooler and is cooled to saturated steam (state point 4) by heat conduction with two-phase CO_2 fluid from the high-pressure expansion valve. Subsequently, it enters the high-pressure compressor and is compressed into a supercritical fluid with high temperature and high pressure (state point 1), and is cooled under constant pressure through a gas cooler (state point 2) to release sensible heat to the surrounding environment. The medium-temperature supercritical fluid from the gas cooler enters the high-pressure expansion valve, and becomes two-phase CO_2 fluid (state point 3) through throttling and decompression. Then it enters the intercooler and superheated steam (state point 10) for heat exchange. The saturated steam (state point 4) enters the high-pressure compressor, and the saturated liquid (state point 5) enters the working nozzle of the two-phase ejector as the main fluid. The ejector leads the saturated steam from the evaporator outlet (state point 13) to the suction chamber of the ejector (state point 1). The two fluids are mixed at medium pressure in the mixing chamber (state point 7), and then enter the diffuser for pressure. Finally, the main fluid leaves the ejector in two-phase state (state point 8) and is separated into saturated steam (state point 9) and saturated liquid (state point 11) in the gas-liquid separator. The saturated steam enters the low-pressure compressor for recompression, while the saturated liquid is decompressed through the low-pressure expansion valve (state point 12), and then enters the evaporator to produce refrigeration effect. This completes a refrigeration cycle.

3. Mathematical modeling and simulation

3.1. Two-phase ejector model

To simplify the modeling process for the two-phase ejector, the assumptions are as follows:

- The mixing process in the mixing chamber of the two-phase ejector is carried out at constant pressure.
- The two-phase ejector is insulated from the surrounding environment.
- The two-phase ejector uses a one-dimensional steady-state model.
- Neglect the variation of kinetic energy at the inlet and outlet of the injector.
- The efficiencies of the working nozzle, suction chamber, mixing chamber and diffuser remain unchanged.

The entrainment ratio μ can reflect the performance of the ejector, given as:

$$\mu = \frac{\dot{m}_{13}}{\dot{m}_5} \tag{1}$$

where $\dot{m}_{13}$ and $\dot{m}_5$ stands for the mass flow rates of the secondary and primary flow of the two-phase ejector, respectively.

The quality of the mixed flow x exiting the ejector should satisfy the following relationship:

$$x = \frac{1}{1+\mu} \tag{2}$$

Additionally, Figure 2 illustrates the flow chart of ejector simulation calculation. A simulation model of the two-phase ejector was built in EES according to the equations in Figure 2, as shown in Figure 3.

3.2. Energetic model

The energy analysis model is based on the first law of thermodynamics. To simplify the model, assumptions are as follows:

- System maintains steady-state operation.
- Ignore the pressure drop and heat loss of pipes and heat exchangers such as gas cooler, evaporator.
- The compression process of the compressor is adiabatic and non-isentropic.
- The leaving fluid at the separator and evaporator are saturated.

The COP of this transcritical CO_2 two-stage compressor/ejector refrigeration system can be calculated as:

$$COP = \frac{Q_0}{W_{CM}} = \frac{Q_0}{W_{HCM} + W_{HCM}} = \mu \frac{h_{13} - h_{12}}{(h_{10} - h_9) + (h_1 - h_4)} \tag{3}$$

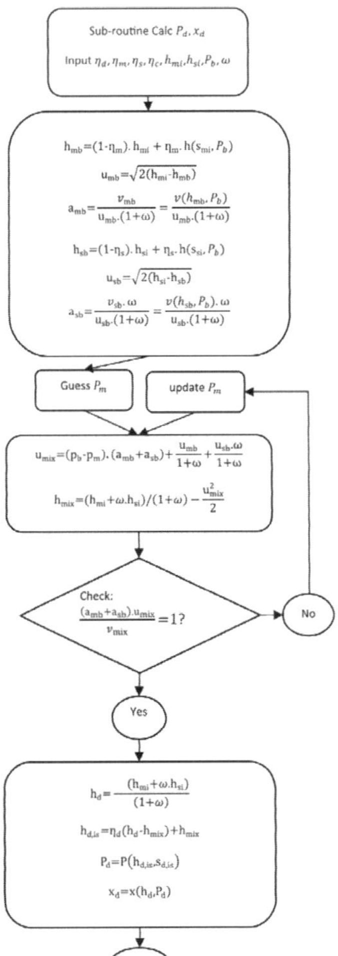

Figure 2. Flow chart of ejector simulation calculation [14].

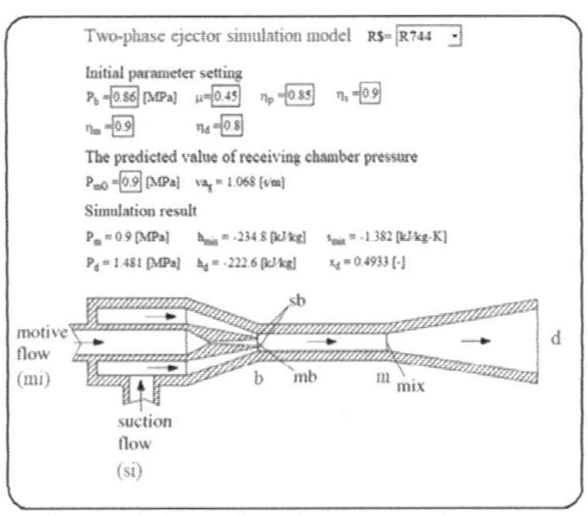

Figure 3. The EES-based injector simulation interface.

4. Results and discussion

This section discusses the effects of high pressure stage exhaust pressure, intermediate pressure and evaporating temperature on system performance.

Figure 4 illustrates the variation of the system COP with different exhaust pressure at different evaporation temperatures. It can be found that the system COP decreases significantly with the increase of the exhaust temperature. At an evaporation temperature of -35°C, the COP of the system decreases from 1.241 to 1.074 (with a reduction of 13.46%) as the exhaust pressure changeing from 8 MPa to 10 MPa. In addition, When the evaporation temperature increases, the system COP does not change significantly.

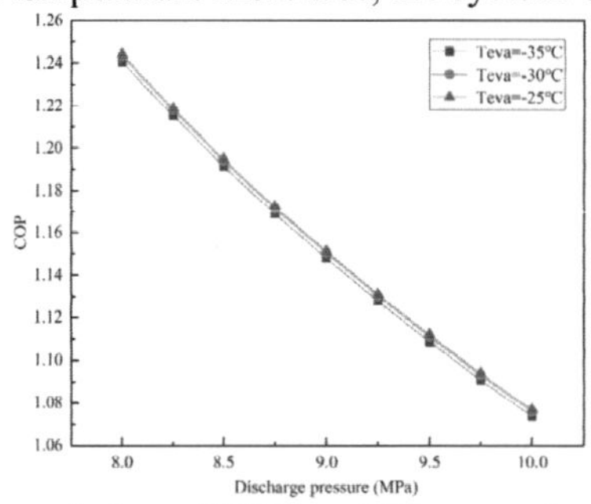

Figure 4. The variation of COP at different exhaust pressures.

Figure 5 illustrates the variation of the system COP with different intermediate pressure at different evaporating temperatures. It can be find that there are two peaks in the system COP when the intermediate pressure changes from 3.0 MPa to 5.0 MPa. At an evaporation temperature of -35°C, the two peaks are 1.126 and 1.130, corresponding to intermediate pressures of 3.75 MPa and 4.75 MPa, respectively. It follows that the optimal intermediate pressure that maximizes the system COP in the above range is 4.75 MPa. At the same time, the COP increases with the augment of evaporating temperature, but the change is small.

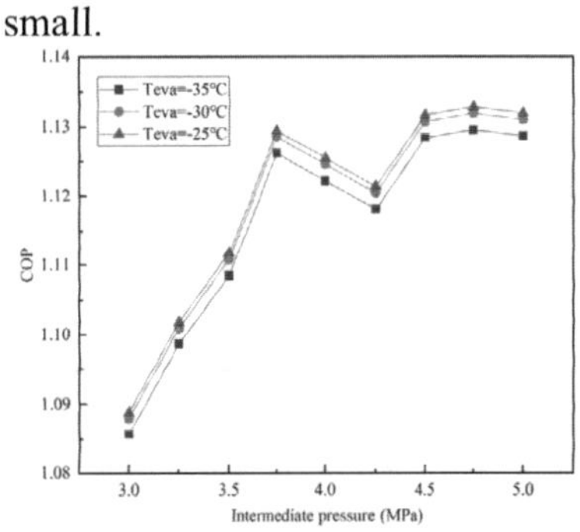

Figure 5. The variation of COP at different intermediate pressures.

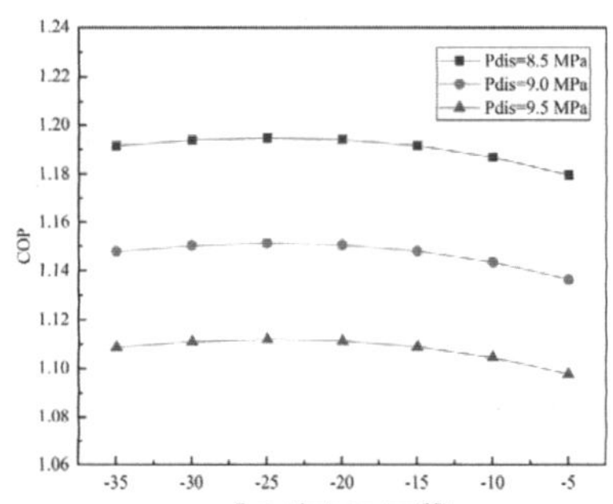

Figure 6. The variation of COP at different evaporating temperature.

Figure 6 illustrates the variation of the system COP with different intermediate pressure at different discharge pressures. It can be found that the system COP increases and then decreases with the increase of evaporation temperature. More importantly, as the exhaust pressure increases, the system COP decreases significantly. At an exhaust pressure of 8.5 MPa, the system COP raises from 1.191 to 1.195 when the evaporation temperature changes from -35°C to -25°C, and then decreases slowly as the evaporation temperature increases.

5. Conclusion

In this work, a novel transcritical CO_2 two-stage compression/ejector refrigeration system equipped with a two-phase ejector for shipboard cold chambers is proposed. At the same time, the simulation of a two-phase ejector was performed by EES. Finally, the effect of the system parameters: exhaust pressure, intermediate pressure and evaporating temperature on the COP of this system was investigated. The main findings are summarized as follows:

- With the evaporating temperature increases from -35°C to -25°C, the system COP shows an increasing trend; after that, it slowly decreases with the augment of evaporating temperature.
- With the evaporating temperature is maintained at -35°C, the system COP decreases with the increase in exhaust pressure.
- There exists an optimal intermediate pressure of 4.75 MPa, which makes the COP of the system reach a maximum of 1.130, when the evaporating temperature is fixed at -35°C.

Acknowledgement
This paper is supported by the Science and technology innovation action plan of Shanghai Science and Technology Commission (19DZ1207503) and Shanghai Professional Technology Service Platform on Cold Chain Equipment Performance and Energy Saving Testing Evaluation(17DZ2293400).

References

[1]. Wang S, Wang Y, Peng J, Lin R, Chen Y. (2021) Application Status and Development of Marine Refrigerant. Ship & Boat, 32: 55-62
[2]. Pigani L, Boscolo M, Pagan N. (2016) Marine refrigeration plants for passenger ships: Low-GWP refrigerants and strategies to reduce environmental impact. Int. J. Refrig., **64**: 80-92
[3]. Heredia-Aricapa Y, Belman-Flores JM, Mota-Babiloni A, Serrano-Arellano J, García-Pabón JJ. (2020) Overview of low GWP mixtures for the replacement of HFC refrigerants: R134a, R404A and R410A. Int. J. Refrig., **111**: 113-123
[4]. Harby K. (2017) Hydrocarbons and their mixtures as alternatives to environmental unfriendly halogenated refrigerants: An updated overview. Renewable Sustainable Energy Rev., **73**: 1247-1264
[5]. Granryd E. (2001) Hydrocarbons as refrigerants — an overview. Int. J. Refrig., **24**: 15-24
[6]. Wang A, Yin X, Fang J, and Cao F. (2021) Refrigerant distributions and dynamic migration characteristics of the transcritical CO2 air conditioning system. Int. J. Refrig., **130**: 233-241
[7]. Mitsopoulos G, Syngounas E, Tsimpoukis D, Bellos E, Tzivanidis C, Anagnostatos S. (2019) Annual performance of a supermarket refrigeration system using different configurations with CO2 refrigerant. Energy Convers. Manage., **1**: 100006
[8]. Tamura T, Yakumaru Y, Nishiwaki F. (2005) Experimental study on automotive cooling and heating air conditioning system using CO2 as a refrigerant. Int. J. Refrig., **28**: 1302-1307
[9]. Guo Z, Kan A, Meng C, Zhang C.(2016) Research Status and Prospect of Energy Saving Technology of Marine Reefer Container. Chinese J. Refrig. Tech., **36**: 53-59
[10].Bai T, Yan G, Yu J. (2015) Thermodynamic analyses on an ejector enhanced CO2 transcritical heat pump cycle with vapor-injection. Int. J. Refrig., **58**: 22-34
[11].Xing M, Yu J, Liu X. (2014) Thermodynamic analysis on a two-stage transcritical CO2 heat pump cycle with double ejectors. Energy Convers. Manage., **88**: 677-683
[12].Manjili FE, Yavari MA. (2012) Performance of a new two-stage multi-intercooling transcritical CO2 ejector refrigeration cycle. Appl. Therm. Eng., **40**: 202-209
[13].Yadav SK, Murari Pandey K, Gupta R. (2021) Recent advances on principles of working of ejectors: A review. Mater. Today: Proc., **45**: 6298-6305
[14].Eskandari Manjili F, Cheraghi M. (2019) Performance of a new two-stage transcritical CO2 refrigeration cycle with two ejectors. Appl. Therm. Eng., **156**: 402-409

Journal of Physics: Conference Series **2195** (2022) 012037 doi:10.1088/1742-6596/2195/1/012037

Performance analysis of multi-refrigerant multi-variable environment refrigeration system based on marine cold chamber

Jie Zhu[1,2,3,4#], Dazhang Yang[1,2,3,4#], Qing Zhang[1,2,3,4] and Jing Xie[1,2,3,4*]

[1]College of Food Science and Technology, Shanghai Ocean University, Shanghai, 201306, China

[2]Shanghai Professional Technology Service Platform on Cold Chain Equipment Performance and Energy Saving Evaluation, Shanghai, 201306, China

[3]Quality Supervision, Inspection and Testing Center for Cold Storage and Refrigeration Equipment, Ministry of Agriculture, Shanghai 201306, China

[4]National Experimental Teaching Demonstration Center for Food Science and Engineering (Shanghai Ocean University), Shanghai, 201306, China

Jie Zhu and Dazhang Yang contributed equally to this study

Corresponding author's Email: jxie@shou.edu.cn

Abstract. This paper is based on the engineering equation solver to calculate the performance of refrigeration system under multi-refrigerant multivariable environment, which can simulate the operating performance of the system under different working conditions and the view of each state point, and provide help for the design and optimization of trans-critical CO2 refrigeration cycle refrigeration units. Using thermodynamic analysis, the system performance is observed by using the equations of conservation of energy, conservation of momentum and conservation of mass as the basis for the change of evaporator pressure, gas cooler pressure and intermediate pressure in the system; the state of each position in the ejector part of the system is calculated; after the calculation is completed, the pressure-enthalpy diagram of the system can be plotted; according to the different data parameters selected, the COP of the system can be plotted with the change of certain data parameters for different parameters.The results showed that the calculation method used is suitable for different refrigerants, compared with the traditional calculation and selection, the advantage of the EES software is that it can use the laws of thermodynamics for calculation and iteration.

1. Introduction

Ship borne transportation has played an important role in the development of the world economy since its development. International shipping traffic, especially pelagic fishery, accounts for a large proportion of the total traffic[1, 2]. Due to people's increasing attention to the environment and economy, the concept of cold chain transportation has been put forward, and the consumption of marine cold chamber has increased sharply. With the prominence of environmental problems, the demand for environmental protection and energy-saving marine cold chamber is very urgent[3]. Generally speaking, from the perspective of energy operation, there are two types of energy during ship operation: mechanical energy and thermal energy. That is, the mechanical energy in the compressor and the thermal energy in the evaporator and condenser[4]. For marine refrigeration, the working environment and energy utilization are a very special research field. Compared with civil and household refrigeration, the marine cold cabin generally needs lower evaporation temperature and endless seawater as cooling water. At the same time, there are particularly strict requirements for safety, weight and size in industrial equipment. Marine cold tanks are mostly used when Freon refrigerants are prosperous. Since many international documents and the damage of high GWP refrigerants to the environment is becoming more and more obvious. The world organization restricts the use of high GWP refrigerants and stipulates the final service life of some refrigerants, which limits the subsequent operation of the refrigeration system of the refrigerant and has a great impact on the existing marine cold chain transportation[5], according to the provisions of international law, this aspect has been limited[6]. Therefore, a refrigerant that can analyse the performance of multiple refrigerants in new systems is important. When the evaporator pressure is too low, the refrigerant volume increases, and when the theoretical pressure volume remains unchanged, the mass flow of the system decreases relatively; When the evaporation temperature is too low, the inlet and outlet pressure difference of the compressor is large, the compressor efficiency is reduced, which is unfavourable to the operating life of the compressor and leads to the reduction of the overall performance of the system[7, 8]. In addition, when the system's condensing temperature rises, the condensing pressure will rise with the condensing temperature, the exhaust temperature of the refrigeration compressor also rises; the unit refrigeration capacity of the refrigerator is reduced, the suction ratio volume remains unchanged, the unit volume refrigeration capacity decreases; the theoretical work of the refrigerator increases; if you ignore the change in the volumetric efficiency of the compressor does not count, The refrigerant mass flow remains unchanged, which leads to an increase in the theoretical power of the compressor[9-12].

CO2 is selected as refrigerant in this paper. The environmental protection characteristics of CO2 - zero oxygen consumption potential and negligible GWP have attracted more and more attention[13, 14]. In addition, as the working medium in the refrigeration system, carbon dioxide also has excellent properties compared with Freon refrigerants in use[15]. However, the main reason why CO2 refrigeration system cannot be widely used is its inefficiency. The main reason is that the huge pressure difference between HP and LP stage results in serious throttle loss before and after throttle valve[16]. Therefore, in order to improve the performance of the trans-critical CO2 refrigeration cycle, it is necessary to retrofit the traditional trans-critical CO2 refrigeration cycle. For example, in order to reduce the pressure difference between the front and rear of the compressor, a two-stage compression system is adopted to reduce the working load of the compressor, or in order to recover the expansion work consumed by throttling loss, an expander or ejector is used in the system to recover the expansion work[12, 13, 17].

In this paper, the performance of a trans-critical CO2 refrigeration system is studied based on the characteristics of a refrigeration system with multiple working fluids and variable environment in an offshore environment. Because of the changeable marine environment, the ambient temperature of ships will change with the change of latitude and longitude, and the refrigerants used by different ships will be different. This makes it extremely inconvenient to use a common manual calculation method, which makes it impossible to update the calculation of the cooling capacity of the condenser and the power consumption of the compressor in operation in offshore systems from time to time. In addition, intermediate pressure is particularly important in refrigeration systems containing two-stage compressors, each with its own optimum intermediate pressure value. From the variation of evaporation pressure (P_e), gas cooler pressure (P_{gc}) and intermediate pressure (P_i), this paper analyses the variation rule of system performance due to the variation of main parameters in operation. This includes the COP of the system, cooling capacity and compressor power consumption.

2. Cycle description and modelling

As shown in Figure 1 and Figure 2, this paper deals with the calculation of a multi-refrigerant multivariable environment refrigeration system for marine use, which includes an evaporator (EVA), a low-pressure compressor (LPC), an ejector (EJE), a high-pressure compressor (HPC), a gas cooler

(GC), an expansion valve (EV), and an intercooler (ITC). Compared with the conventional same, the expansion valve used for throttling is replaced by an injector, and the refrigerant from the intercooler to the compressor is injected by the injector. The flow process of refrigerant in the system cycle is as follows: the hot refrigerant first passes from the outlet of the HPC through the GC (state point 2), then is cooled by the GC (state point 3), then draws the saturated vapor CO2 refrigerant directly through the EJE from the ITC (state point 1), and then enters the ITC in a mixed two-phase fluid state (state point 4) after working through the EJE. The refrigerant leaving the ITC is divided into two parts. For saturated vapor refrigerant (state point 1), one part passes through the HPC and the other part is directed to the EJE. The saturated liquid refrigerant coming out of the ITC (state point 5) passes through the EVA (state point 6) after being throttled by EV (state point 7) and then enters the LPC. Finally, the vapor refrigerant is compressed by the LPC (state point 8) and mixed with the two-phase fluid refrigerant in the ITC. As mentioned above, the fundamental reason for using an EJE in the cycle is to replace the EV in a conventional system and to draw some of the saturated vapor refrigerant from the ITC, unlike directly into the HPC. Another part of the refrigerant from the new ITC is in saturated liquid under medium pressure, rather than supercooled liquid under high pressure under the traditional ITC. In this new cycle, the refrigeration capacity of EVA always increases and the compression work of HPC decreases. As a result, the refrigerant mass of the LPC is reduced and the corresponding compression work of the LPC is reduced. Therefore, when two systems have the same refrigeration requirement, the mass flow rate of the LPC in the system is reduced and thus the LPC power is reduced, thus improving the cycle performance. The refrigerant entering the HPC can also be controlled through the EJE inlet, making the system more adjustable.

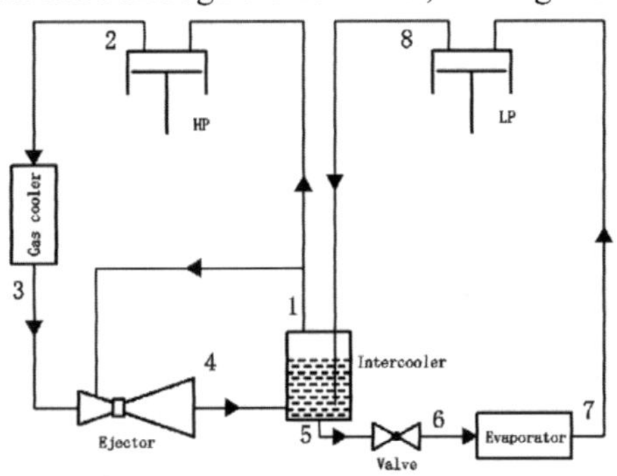

Figure 1. Schematic diagram of marine cold chamber system

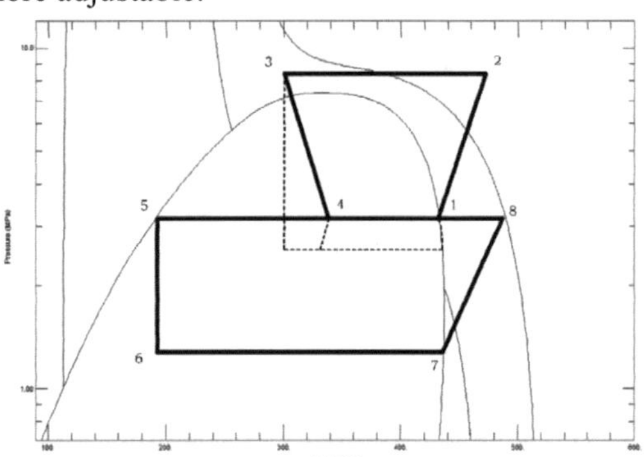

Figure 2. Pressure enthalpy diagram of marine cold chamber system

3. Theoretical modelling

In order to simplify the system, a thermodynamic model is established according to the laws of thermodynamics, and the following assumptions are made[18]:

(1) The system is in steady-state and steady-state flow process during operation.

(2) The pressure loss and kinetic energy loss in the connecting pipe are ignored.

(3) The refrigerant has no enthalpy difference before and after working in the throttle valve.

(4) The pressure drops and heat loss of refrigerant in pipes and equipment are ignored.

In the above assumptions, the thermodynamic modelling of each component and process of the system is carried out according to the laws of thermodynamics.

The power consumption of the compressor is calculated by the following formula:

$$W_{coml} = m_l(h_2 - h_1) = m_l(h_{1,is} - h_1) / \eta_{coml} \tag{1}$$

$$W_{comh} = m_h(h_8 - h_7) = m_h(h_{7,is} - h_1) / \eta_{comh} \tag{2}$$

where η_{coml} and η_{comh} are the isentropic efficiency of LPC and HPC. It is known from the article (Chen et al., 2017) [10]that this value is related to the relationship between the inlet and outlet pressure of the compressor:

$$\eta_{coml} = 0.815 + 0.022(P_2 / P_1) - 0.0041(P_2 / P_1)^2 + 0.0001(P_2 / P_1)^3 \quad (3)$$

$$\eta_{coml} = 0.815 + 0.022(P_8 / P_7) - 0.0041(P_8 / P_7)^2 + 0.0001(P_8 / P_7)^3 \quad (4)$$

For ejector, according to the conservation of energy:

$$h_4 \cdot (1+w) = h_3 + w \cdot h_1 \quad (5)$$

For the intercooler, the mass relationship can be calculated by the conservation law of mass and energy:

$$m_4 = m_h + m_1 = m_h + w \cdot m_h \quad (6)$$

$$m_4 \cdot h_4 + m_l \cdot h_8 = m_4 \cdot h_1 + m_l \cdot h_5 \quad (7)$$

For evaporator, the refrigeration capacity is calculated using the following equations:

$$Q_e = m_l \cdot (h_7 - h_6) \quad (8)$$

The coefficient of performance is:

$$COP = \frac{Q_e}{W_{coml} + W_{comh}} \quad (9)$$

4. Results and discussion

Calculate the refrigeration system according to the basic state shown in Table 1. On this basis, the main influencing parameters in the system are changed to observe the performance of the system. Variation range[19]: the pressure P_{gc} of the gas cooler is in the range of 8.3 - 10 MPa, the evaporation pressure of the LP evaporator (P_e) is from 1.2 – 1.5 MPa, the outlet temperature (T_{gc}) of the gas cooler is in the range of 35 to 44. It is assumed that the refrigerant mass flow at the inlet of the low-pressure compressor is 1.0 kg · s^{-1}. In addition, each section of the ejector part of the system is composed of main flow nozzles (the main body of the ejector enters through the nozzle, resulting in pressure drop and large kinetic energy vortex), the pilot jet chamber, the mixing chamber where two fluids are mixed, and the diffuser section where the mixed fluid passes through. The kinetic energy vortex generated after the main body passes through the nozzle can introduce the fluid from the low-pressure section. In the assumption of Table 1, the ejector efficiency at the nozzle, the mixing chamber and the diffuser section is on this basis, the thermodynamic analysis and calculation of trans-critical CO2 refrigeration system have been carried out.

Table 1. Parameters of the system in the assumed basic case.

Parameters	Values
η_d	0.8
η_m	0.9
η_n	0.9
P_e /MPa	1.2
P_{gc} /MPa	9.3
P_i /MPa	4.4

According to the parameters in the basic state, it is calculated in the EES software. The parameters of each state point after calculation are shown in Table 2. The state parameters include pressure, enthalpy, temperature and entropy. According to the data in the table, thermodynamic analysis of the trans-critical CO2 refrigeration system is carried out based on the thermodynamic model constructed to observe the changes of the system's performance. It should be noted that the parameter is invoked from the EES database.

Journal of Physics: Conference Series **2195** (2022) 012037 doi:10.1088/1742-6596/2195/1/012037

Table 2. The thermodynamic parameters of all state point

Point \ Parameters	Pressure/MPa	Enthalpy/ $kJ \cdot kg^{-1}$.	Temperature/°C	Entropy/ $kJ\cdot(kg\cdot K)^{-1}$
1	4.4	-82.9	9	-0.948
2	9.3	-48.6	69	-0.928
3	9.3	-211.6	35	-1.437
4	4.4	-166.0	9	0.586
5	4.4	-283.5	9	-1.659
6	1.2	-283.5	-35	-1.610
7	1.2	-70.6	-35	-0.716
8	4.4	1.8	67	-0.672

Figure. 3 can observe the variation curve of COP of NERC with P_{gc} at different P_e. As expected, each evaporator temperature has an optimal HP and a COP maximum value. As the P_{gc} increases, the specific work of the compressor increases. However, it can be seen from the pressure enthalpy diagram in Figure 2 that when the P_{gc} is low, the refrigeration effect (h1-h4) of the two-phase liquid in the ITC is low, which requires increasing the mass flow of the refrigerant to achieve the desired refrigeration effect. In addition, under any P_{gc}, the COP value increases with the increase of the P_e. From the results, the optimal pressure in the COP change curve of NERC system with the change of P_{gc} under different P_e shows that the curve trend before and after the optimal pressure decreases and is different This phenomenon exists under evaporation pressure. Therefore, it can be inferred that there is an optimal P_{gc} under any condition to make the system performance reach the best state, and the existence of this value should be emphatically measured in the experiment, so as to make the system in the best state during actual operation.

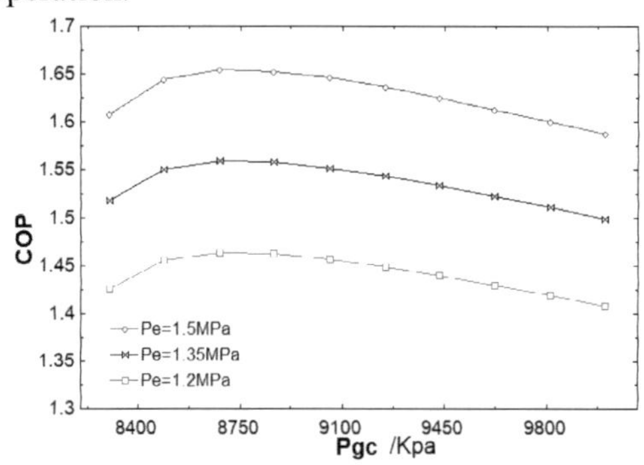

Figure 3. Effect of P_{gc} on COP of different P_e.

The influence of system COP with P_i under different P_{gc} is shown in Figure 4. From the figure that there is an optimal intermediate pressure P_i for any P_{gc}, and according to the figure, it can be observed that the optimal intermediate pressure gradually decreases with the increase of gas cooler pressure P_{gc}, and the COP of the system decreases with the increase of gas cooler pressure P_{gc}.

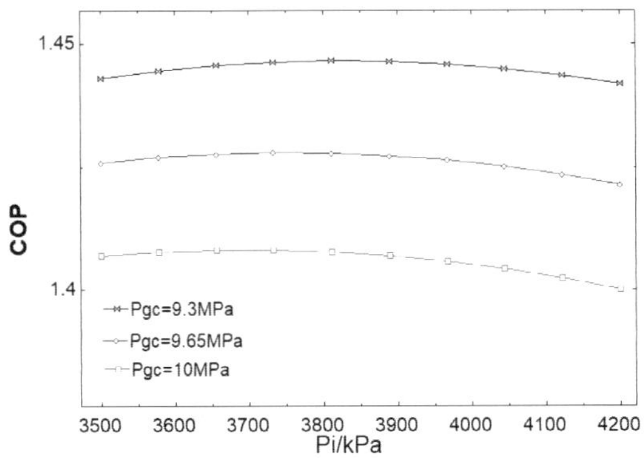

Figure 4. Effect of P_i on COP of the different Pgc.

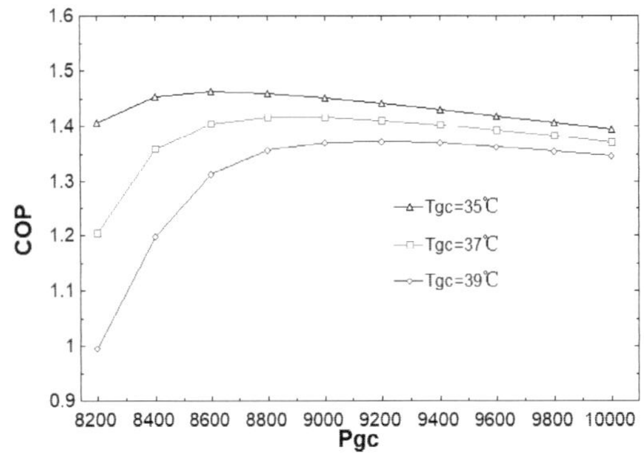

Figure 5. Effect of P_{gc} on COP of the different T_{gc}

The influence of system COP with P_i under different P_{gc} is shown in Figure 4. there is an optimal P_i for any P_{gc}, and according to the figure, the optimal P_i gradually decreases with the increase of P_{gc}, and the COP of the system decreases with the increase of P_{gc}. With the change of P_{gc}, the change of system COP under different T_{gc} is shown in Figure 5. Each T_{gc} has an optimal P_{gc}, and the system performance tends to decline before and after the optimal value. For any P_{gc}, the cop decreases with the increase of T_{gc} due to the high heat loss in the gas cooler. As shown in the figure, under different T_{gc}, the COP of the system changes obviously with the change of P_{gc}, and there is obviously an optimal P_{gc} to make the system reach the optimal state. This conclusion also verifies that the system has an optimal P_{gc} under any state, and the system has a downward trend before and after the optimal P_{gc}. Different from the above, under different T_{gc}, the system COP varies greatly at low pressure. When the temperature reaches a high temperature, the system COP drops to a very low level, which indicates that the system should pay attention to maintaining the system T_{gc} at a low level during actual operation.

5. Conclusions

In this paper, based on EES software, the thermodynamic modelling of the new trans-critical CO2 refrigeration system designed by the system is carried out to analyse the performance of the system. CO2 is used as circulating refrigerant. Through this study, it can also be applied to different refrigerants. The effects of the main parameters of the refrigeration cycle on the system performance are studied. The main results are as follows:

1. The calculation method used is suitable for different refrigerants. The system cycle can be analyzed by selecting different refrigerants in the software.
2. Compared with the traditional calculation and selection, the advantage of the EES software mentioned in this paper is that it can be directly imported into the software in the form of equations by using the laws of thermodynamics for calculation and iteration, and different trend change diagrams can be derived according to the changes of different parameters.
3. According to the method provided in this paper, the system loss can also be calculated, and the system loss can also be calculated and analyzed under different parameter changes.

Acknowledgments
This paper is supported by this research has been supported by Shanghai Professional Technology Service Platform on Cold Chain Equipment Performance and Energy Saving Testing Evaluation (20DZ2292200).

References

[1] Zhang Q, Luo Z, Zhao Y, Pavel S. (2021) Thermodynamic analysis and multi-objective optimization of a transcritical CO2 waste heat recovery system for cruise ship application. Energy Conversion and Management, 227: 113612.
[2] Artuso P, Rossetti A, Minetto S, Marinetti S, Moro L, Col D D. (2019) Dynamic modeling and thermal performance analysis of a refrigerated truck body during operation. International Journal of Refrigeration, 99: 288-299.
[3] Yan L, Chouw N. (2015) Effect of water, seawater and alkaline solution ageing on mechanical properties of flax fabric/epoxy composites used for civil engineering applications. Construction & Building Materials, 99: 118-127.
[4] Huang Y, Jiang P, Zhu Y. (2020) Experimental and modeling studies of thermally-driven subcritical and transcritical ejector refrigeration systems. Energy Conversion and Management, 224: 113361.
[5] Pigani L, Boscolo M, Pagan N. (2016) Marine refrigeration plants for passenger ships: Low-GWP refrigerants and strategies to reduce environmental impact. International Journal of Refrigeration, 80-92.
[6] Li D, Groll E A. (2005) Transcritical CO2 refrigeration cycle with ejector-expansion device. International Journal of Refrigeration, 28: 766-773.
[7] Liang Y, Sun Z, Dong M, Lu J, Yu Z. (2020) Investigation of a refrigeration system based on combined supercritical CO2 power and transcritical CO2 refrigeration cycles by waste heat recovery of engine. International Journal of Refrigeration, 118: 470-482.
[8] Gullo P, Hafner A, Banasiak K. (2018) Transcritical R744 refrigeration systems for supermarket applications: Current status and future perspectives. International Journal of Refrigeration, 93: 269-310.
[9] Xing H, Spence S, Chen H. (2020) A comprehensive review on countermeasures for CO2 emissions from ships. Renewable and Sustainable Energy Reviews, 134: 110222.
[10] Chen G, Volovyk O, Zhu D, Ierin V, Shestopalov K. (2017) Theoretical analysis and optimization of a hybrid CO2 transcritical mechanical compression – ejector cooling cycle. International Journal of Refrigeration, 74: 86-94.
[11] Kimura De Carvalho B Y, Melo C, Pereira R H. (2019) An experimental study on the use of variable capacity two-stage compressors in transcritical carbon dioxide light commercial refrigerating systems. International Journal of Refrigeration, 106: 604-615.
[12] Purohit N, Gupta D K, Dasgupta M S. (2016) Effect of Inter-stage Pressure on the Performance of a Two Stage Refrigeration Cycle Using Inter Cooler. Energy Procedia, 90: 171-178.
[13] Zhang Z, Wang H, Tian L, Huang C. (2016) Thermodynamic analysis of double-compression flash intercooling transcritical CO2 refrigeration cycle. The Journal of Supercritical Fluids, 109: 100-108.
[14] Yu B, Yang J, Wang D, Shi J, Chen J. (2019) An updated review of recent advances on modified technologies in transcritical CO2 refrigeration cycle. Energy, 189: 116147.
[15] Sun Z, Wang C, Liang Y, Sun H, Liu S, Dai B. (2020) Theoretical study on a novel CO2 Two-stage compression refrigeration system with parallel compression and solar absorption partial cascade refrigeration system. Energy Conversion and Management, 204: 112278.
[16] Nemati A, Mohseni R, Yari M. (2018) A comprehensive comparison between CO2 and Ethane as a refrigerant in a two-stage ejector-expansion transcritical refrigeration cycle integrated with an organic Rankine cycle (ORC). The Journal of Supercritical Fluids, 133: 494-502.

[17] Cecchinato L, Chiarello M, Corradi M, Fornasieri E, Minetto S, Stringari P, Zilio C. (2009) Thermodynamic analysis of different two-stage transcritical carbon dioxide cycles. International Journal of Refrigeration, 32: 1058-1067.

[18] Liu Y, Liu J, Yu J. (2020) Theoretical analysis on a novel two-stage compression transcritical CO2 dual-evaporator refrigeration cycle with an ejector. International Journal of Refrigeration, 119: 268-275.

[19] Eskandari Manjili F, Cheraghi M. (2019) Performance of a new two-stage transcritical CO2 refrigeration cycle with two ejectors. Applied Thermal Engineering, 156: 402-409.

Journal of Physics: Conference Series **2195** (2022) 012038 doi:10.1088/1742-6596/2195/1/012038

The research of Dual MOSFET Parallel technology in Servo drive power

Zengliang Ping [1,a], Shaonan Wu[1,b], Yuhua Ma[1,c*]

[1]CSSC Haiwei Zhengzhou HighTech Co., Ltd, Zhengzhou, Henan, China

[a] email: buaapzl@163.com, [b]email: wsn@cumt.edu.com,

[c*]email: 18838159601@163.com

Abstract—With the development of science and technology, the capacity of high-power power electronic conversion equipment is increasing. The growth rate of current carrying capacity of power electronic converter is lower than that of capacity of high-power power electronic converter. Therefore, the parallel technology of power electronic devices is gradually emerging. In this paper, a double MOSFET parallel technology is proposed, and the feasibility of the scheme is verified by simulation. The scheme can be used in the internal control power supply of power electronic conversion equipment such as driver to realize reliable operation without external control power supply.

1. Introduction

MOSFET, IGBT and other switching devices are more and more widely used in high-power power electronic conversion equipment. The power supply of the control core of power electronics is usually generated by the switching power supply [1]. As the main switch tube of the switching power supply, MOSFET is also more and more widely used. With the increasing demand of high-power power electronic conversion equipment for controlling power supply capacity, increasing power density has become a major problem [2]. On the one hand, it can be realized by parallel power supply of multiple power sources, and on the other hand, it can be realized by optimizing the topological structure of power sources [3,4]. MOSFET parallel scheme [5], half-bridge, full-bridge and other topologies [2,3] can meet the above requirements. In view of the positive temperature effect of MOSFET, the balanced current function can be realized [6], which promotes the development of MOSFET parallel technology and promotes the application of this topology in power supply [7].

In this paper, a technical scheme of flyback power supply with parallel MOS transistors is designed. The double MOS parallel scheme is simulated by MULTISIM software, and the state information such as DS voltage and power output voltage during the switching period of MOS transistors are measured to ensure the dual-mosfets-paralleled topology .

2. Double MOS parallel topology structure

2.1 Switchmode source Working principle

The principle diagram of switch-mode power supply is shown is in Fig.1 below. The AC source The AC power supply is changed to DC power by the rectifier filter circuit. The action of opening and closing of the transistor realizes the storage and transmission of energy through the transformer. The MCU acquisits output voltage of output feedback, compare it with the reference voltage in each period and determines the on-and-off action of the transistor. As a result, the high-level DC voltage is converted to a lower voltages, which can supply power to the MCU to realize power electron conversion

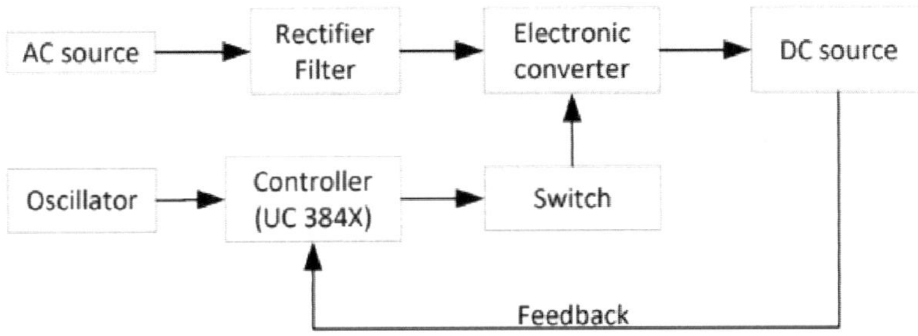

Fig.1 Working principle

2.2 Parallelled mosfet topology

Parallelled mosfet topology is shown is in Fig.2.The same control voltage controls the opening and closing of multiple MOS tubes simultaneously. The difference of MOSFET devices will lead to the difference of turn-on and turn-off speed. This difference is eliminated by connecting resistances in the gate of each MOSFET. At the same time, the resister between the gate and source can save the MOSFET from broken when it is powered on or off.

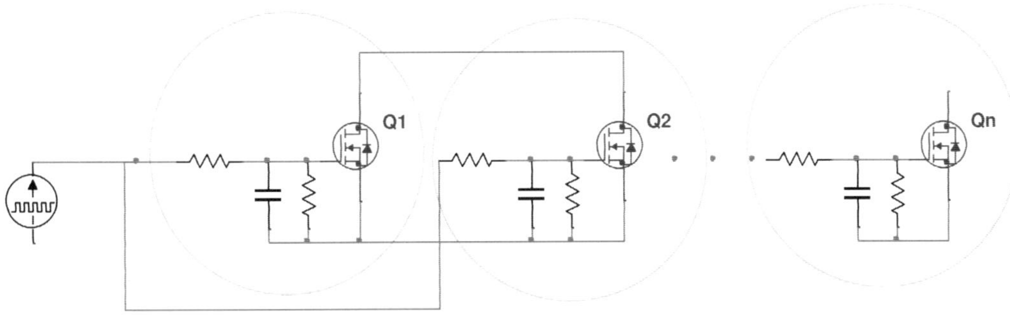

Fig.2 Parallelled mosfet topology

According to the relevant device data sheet[1], the static dirain-source on resistance of MOSFET is positively-correlated with the on-state current, operating junction temperature, etc. Under the conduction state, the voltage between the drain and the source of MOSFET is positively correlated with the on-state current Id and Rds.

If the voltage of the MOSFET's source electrode is set to the voltage reference point, the voltage of the drain electrode(vd) can be shown below.

$$v_d = ia \times Ra(T1) \quad (1)$$

$$v_d = ib \times Rb(T2) \quad (2)$$

$$\cdots\cdots$$

$$v_d = ix \times Rx(Tx) \quad (3)$$

$$i = ia + ib + \cdots + ix \quad (4)$$

Where, ix, Rx, Tx are the static drain current, drain-source on resistance, junction temperature of the n-st mosfet, respectively.

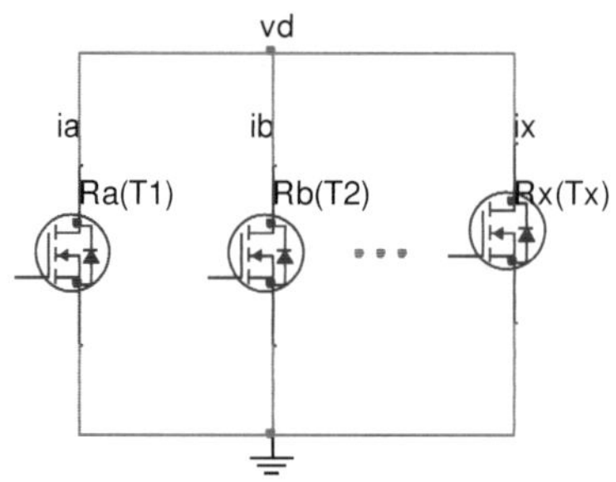

Fig.3 Equivalent circuit model of paralleled MOSFET topology

The relation of the drain-source on resistance and the junction temperature is shown in fig.4. The increase of junction temperature leads to the increase of on-state resistance. At a determined conduction voltage drop, with the conduction resistance of the tube increasing, the current flowing through the MOSFET whose junction temperature is higher should be decreased, leading to lower junction temperature at the next moment. At the next moment, with the junction temperature decreasing, drain-source on resistance of the tube is reduced, and the current flowing through the MOSFET should be increased. The dynamic balance of heat generation between the mosfets can realize the automatic balance of conduction current. Better current balancing can be achieved if all of the mosfet can be designed with the same radiator

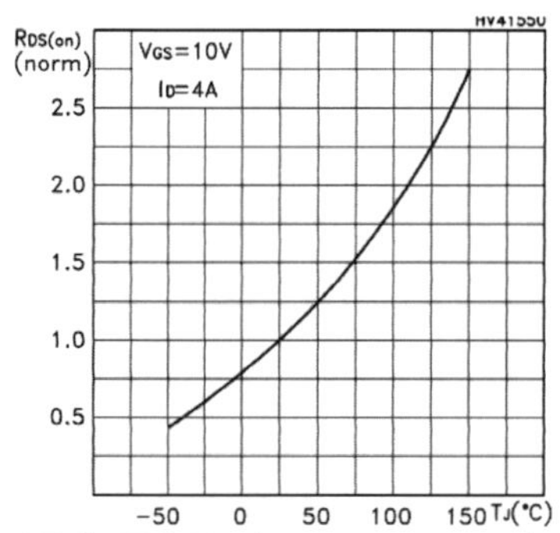

Fig.4 Rds *V.S.* Drain-source on resistance

2.3 Dual-MOSFET-paralledled scheme

The Dual-MOSFET-paralledled scheme is based on the above theory. In the same switch-mode source, a new MOSFET which is the same with the original one can be set in it to achieve the purpose of Dual-MOSFET-paralledled scheme. The double scheme is shown in Fig.5 below. The existence of reflected voltage and leakage inductance during the turn-off of the MOS tube will cause higher voltage spikes on the Drain electrode of the MOSFET, so RCD combination circuits are designed to suppress it to prevent the MOSFET from over-voltage breakdown.

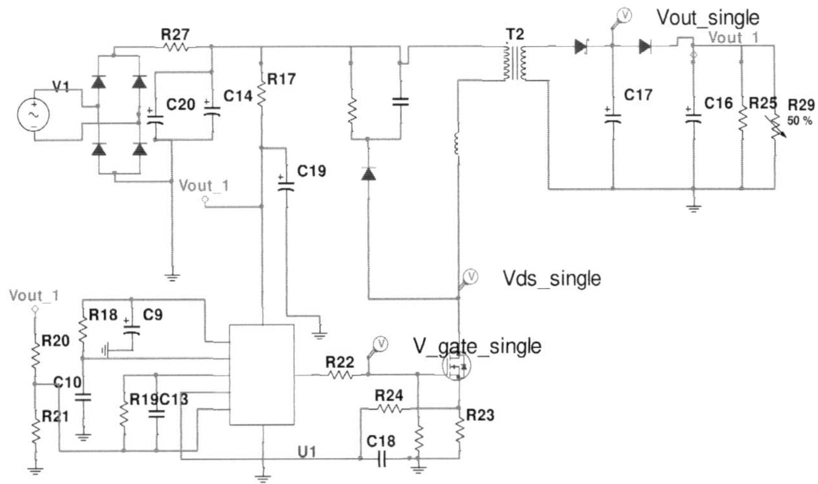

(a) The single MOSFET scheme

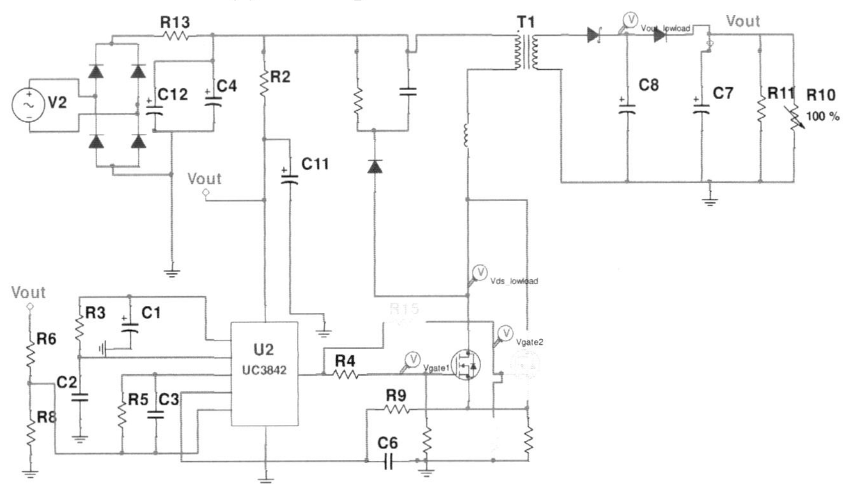

(b) The Dual-MOSFET-paralleled scheme

Fig.5 the difference between the two scheme

From the Fig.5, it can be seen that only a few elements are Added to the circuit diagram, which is labeled with green color. In the scheme, load resistance R11 is the virtual load to Ensure normal operation of switching power supply, and R10 is the variable resistance to simulate the simulated variable load.

3. Simulation Results and Discussions

A simulation based on the above model is carried out to verify the effectiveness of the dual-MOS-paralleled scheme.

(1)With the same load, the voltage between the drain and source electrode can be calculated. With the simulation, the two results is shown in Fig.6. The two voltage waveforms are approximate to each other.

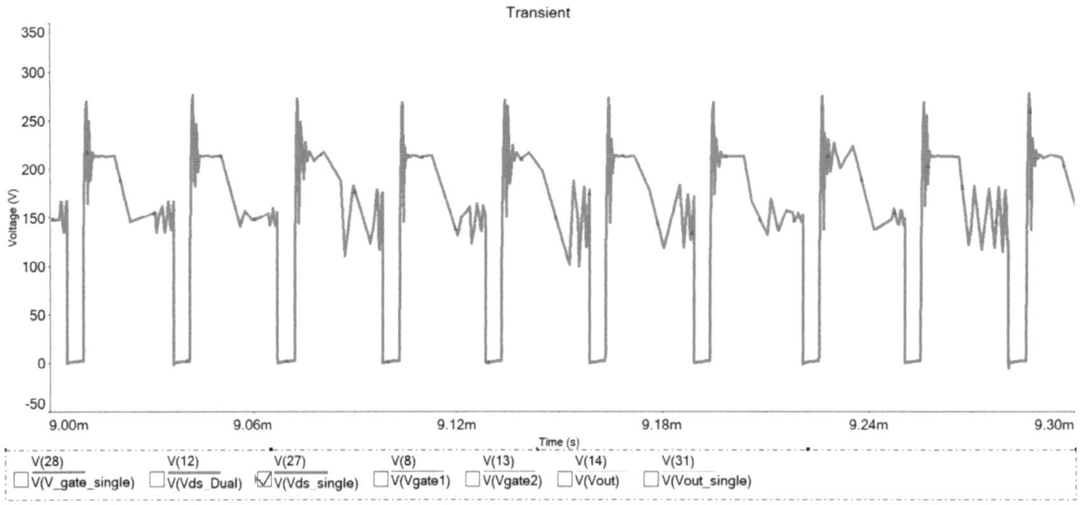

(a) Vds in the single MOSFET scheme

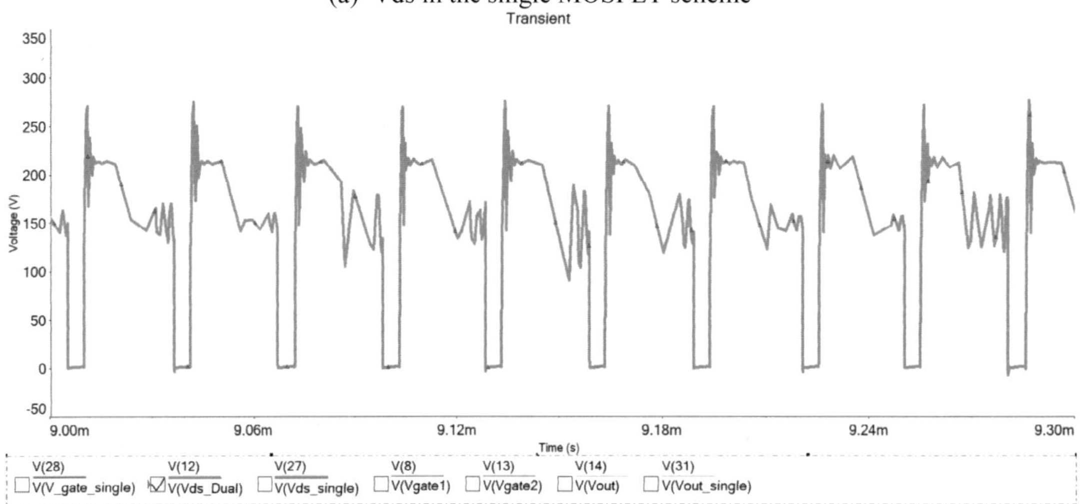

(b) Vds in the dual MOSFET scheme

Fig.6 Vds simulation results

(2) With the same load, the output voltage between the two schemes can be calculated. With the simulation, the two results is shown in Fig.7. Both one or two MOSFETS, the output voltages are consistent and similar to each other.

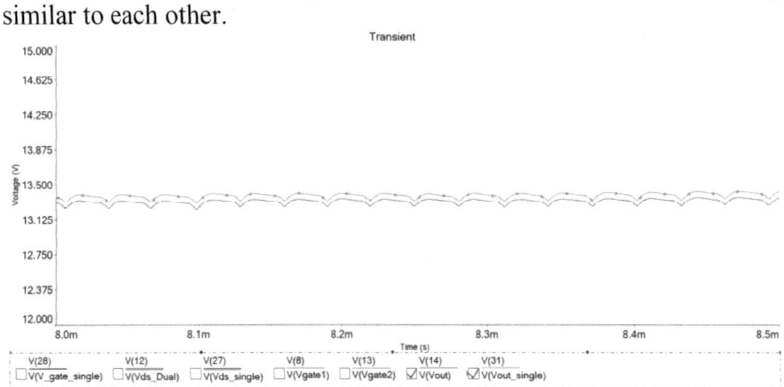

Fig.7 Vout simulation results

(3) In the dual-MOS-paralleled scheme, with the different loads, the voltage between drain and source, and the output is shown in Fig.8. Under different loads, the dual MOS-paralleled scheme can provide stable output voltage. And the voltage between drain and source is becoming higher with the increasing loads.

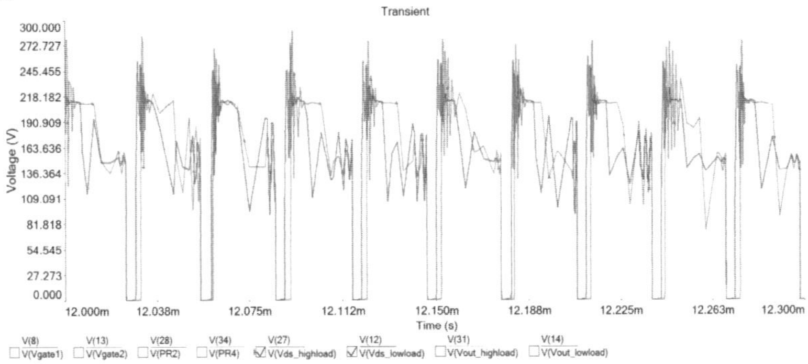

(a)Vds V.S.load

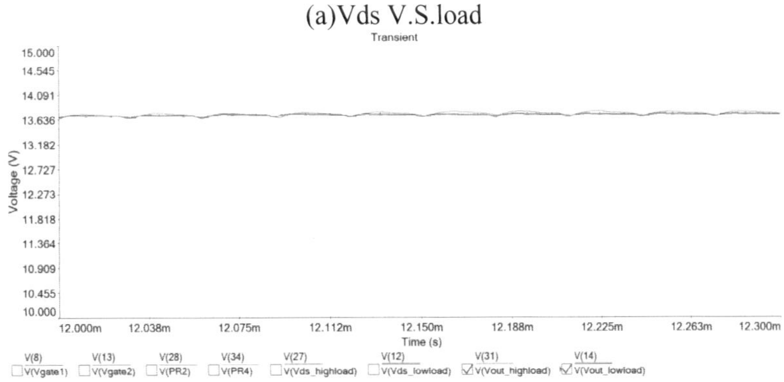

(b) Vout V.S. load

Fig.8 The influence of load on the Dual-Mos-Paralleled Scheme

4. Conclusion

In this paper, the simulation based on the dual MOS paralleled scheme is carried out. It can be concluded that the dual MOS paralleled scheme can supply stable output voltage, and is practicable and effective to increase the power capacity, especially in the case of insufficient MOS tube flow capacity.

References

[1] Rong, G. X.. (2010) Power Mosfet Paralleling Application. Application of Electronic Technique.

[2] Liu, T., Cui, Y. B. (2019) Design and Simulation of Active Current Sharing of Paralleled Power Mosfets. Navigation Positioning and Timing.

[3] Li W., Shu G. (2010) Simulation of Flyback Switch Power SupplyBased on Multisim [J]. Electrical Automation.

[4] Kimball J. W., Chapman P. L. (2004) Evaluating Conduction Loss of a Parallel IGBT-MOSFET Combination[C]. Industry Applications Conference. IEEE.

[5] Hui, C., et al (2018) Research on Current Sharing Method of SiC MOSFET Parallel Modules. IEEE International Conference on Electron Devices and Solid State Circuits (EDSSC) IEEE.

[6] Qiang X., Dong Z., Wu X., et al (2015) A 10kV/200A SiC MOSFET Module with Series-parallel Hhybrid Connection of 1200V/50A Dies[C]. International Symposium on Power Semiconductor Devices & IC's (ISPSD). IEEE.

[7] Lin, B. R., & Liu, C. H.. (2012) Implementation of a Parallel Zero-voltage Switching DC-DC Converter with Fewer Active Switches[J]. Iet Power Electronics, 5(9):1651-1659.

Dynamic Analysis and Model Predictive Control of Excavator's Working Device

Penglong Zheng[1,a]*, Jian Liu[2,b], WeiWei Wu[3,c]

[1,2,3]Institute of Intelligent Control Technology, XCMG Research Institute, Xuzhou, China

[a*] Corresponding author: zhengpenglong111@hotmail.com

[b]superlj168@126.com, [c]wu9519@sina.com

Abstract—Aiming at the development requirements of automation and intelligence of traditional hydraulic excavator, the robot modeling theory is applied to the modeling process of it, and the general forward and inverse kinematics models of excavator working device including the hydraulic drive structure were given. Based on Lagrange equation, the dynamics model of excavator is established, and a trajectory tracking control method is proposed for this model. The tracking effect of model predictive control method on preset trajectory under different working conditions is studied by computer simulation technology. The research results show that the model predictive control can give a more accurate feedforward control force under complex working conditions, realize accurate tracking of the trajectory, and effectively reduce the control delay caused by large inertia working conditions, and has good robustness.

1. INTRODUCTION

Hydraulic excavator is widely used in bridge construction, mining, tunnel construction and other aspects, is one of the important mechanical equipment of earthwork, plays a crucial role in national economic construction.

In recent years, China's construction machinery industry has made great progress, XCMG, Sany, Zoomlion and other products have gradually opened the domestic and foreign markets, and can compete with caterpillar, Japan komatsu and other traditional strong enterprises. However, the control system of existing hydraulic excavators mostly adopts open-loop control, or adopts simple PID control model [1,2], which requires personnel operation, low control accuracy and poor stability, and is inconsistent with the development trend of unmanned, automation and intelligence.

In this paper, the robot modeling theory [3] is applied to the excavator, the forward and inverse kinematics equations of the working device are given, and the Lagrange [4] method is used to derive the dynamics equations, and the corresponding model predictive controller is designed [5,6].

2. KINEMATICS AND DYNAMICS MODELING OF EXCAVATOR

2.1.Forward Kinematics Analysis

Without considering the walking structure of the excavator, the main working parts (moving parts) are rotary mechanism and mechanical arm, and the mechanical arm includes the maneuvering arm, bucket rod and bucket. Therefore, the working parts have four degrees of freedom in total, and the general structural form is shown in Fig. 1.

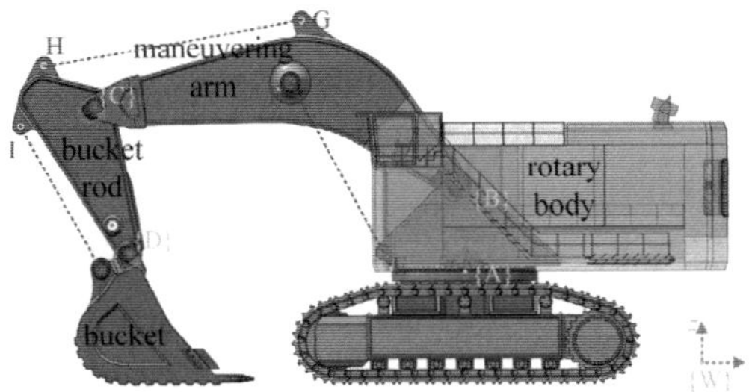

Figure 1. Establishment of frames of excavator working parts

As shown in the Fig. 1, establish the world frame {W}, select vertical upward as the positive direction of z axis, horizontal right as the positive direction of x axis, and y axis is determined by the right hand system. The walking mechanism of excavator realizes three degrees of freedom in plane motion through differential speed on both sides, do not discuss here. Therefore, it is assumed that the excavator walking mechanism is static, and is overlapped with the world frame.

It is only assumed that the initial direction of the coordinate axes in the above frames is the same as that in the world frame, and the above frames are all moving frames, then the movement process of the bucket in the world frame is described as follows:

(1) The translation distance along the z-axis is z_1, then the rotation about the z-axis is θ_1, and it becomes frame {A}.

(2) Moves z_2 along the z-axis, x_2 along the x-axis, rotates θ_2 around the y-axis, and becomes frame {B}.

(3) Moves z_3 along the z-axis, x_3 along the x-axis, rotates θ_3 around the y-axis, and becomes frame {C}.

(4) Moves z_4 along the z-axis, x_4 along the x-axis, rotates θ_4 around the y-axis, and becomes frame {D}.

The above process can be expressed in homogeneous coordinate form as:

$$T_1 = \begin{bmatrix} c\theta_1 & -s\theta_1 & 0 & 0 \\ s\theta_1 & c\theta_1 & 0 & 0 \\ 0 & 0 & 1 & z_1 \\ 0 & 0 & 0 & 1 \end{bmatrix} \quad T_2 = \begin{bmatrix} c\theta_2 & 0 & s\theta_2 & x_2 \\ 0 & 1 & 0 & 0 \\ -s\theta_2 & 0 & c\theta_2 & z_2 \\ 0 & 0 & 0 & 1 \end{bmatrix}$$
$$T_3 = \begin{bmatrix} c\theta_3 & 0 & s\theta_3 & x_3 \\ 0 & 1 & 0 & 0 \\ -s\theta_3 & 0 & c\theta_3 & z_3 \\ 0 & 0 & 0 & 1 \end{bmatrix} \quad T_4 = \begin{bmatrix} c\theta_4 & 0 & s\theta_4 & x_4 \\ 0 & 1 & 0 & 0 \\ -s\theta_4 & 0 & c\theta_4 & z_4 \\ 0 & 0 & 0 & 1 \end{bmatrix} \tag{1}$$

Therefore, the pose of the frame {D} of the last end bucket is:

$$T = T_1T_2T_3T_4$$
$$= \begin{bmatrix} c_1c_{234} & -s_1 & c_1s_{234} & c_1\left(x_2 + x_4c_{23} + z_4s_{23} + x_3c_2 + z_3s_2\right) \\ s_1c_{234} & c_1 & s_1s_{234} & s_1\left(x_2 + x_4c_{23} + z_4s_{23} + x_3c_2 + z_3s_2\right) \\ -s_{234} & 0 & c_{234} & z_1 + z_2 + z_4c_{23} - x_4s_{23} + z_3c_2 - x_3s_2 \\ 0 & 0 & 0 & 1 \end{bmatrix} \tag{2}$$

The driving part of the excavator is the hydraulic cylinder, and the linear movement of the hydraulic rod is transformed into the rotation of the joint by the restriction of the mechanical structure. As shown in fig. 1, EF, GH and IJ are hydraulic cylinders in $\triangle BEF$, $\triangle CGH$ and $\triangle DIJ$. Assume that the initial lengths of the hydraulic cylinders are l_{EF0}, l_{GH0} and l_{IJ0} respectively, the elongations are l_{EF}, l_{GH} and l_{IJ} respectively. Select $\triangle BEF$ as the research object:

$$\theta_2 = \text{acos}\left(\frac{l_{BF}^{\ 2} + l_{BE}^{\ 2} - (l_{EF0} + l_{EF})^2}{2l_{BF}l_{BE}}\right) - \text{acos}\left(\frac{l_{BF}^{\ 2} + l_{BE}^{\ 2} - l_{EF0}^{\ 2}}{2l_{BF}l_{BE}}\right) \quad (3)$$

$\triangle CGH$, $\triangle DIJ$ are processed in the same way.

2.2.Inverse Kinematic Analysis

Suppose the pose of the end bucket {D} frame is:

$$T = \begin{bmatrix} a_x & b_x & c_x & d_x \\ a_y & b_y & c_y & d_y \\ a_z & b_z & c_z & d_z \\ 0 & 0 & 0 & 1 \end{bmatrix} \quad (4)$$

By comparing formulas (2) and (4) it can be found that the position of the {D} frame of the end bucket is only related to θ_1 , θ_2 and θ_3 .

$$\frac{s_1}{c_1} = \frac{d_y}{d_x} \Rightarrow \theta_1 = \text{atan2}\left(d_y, d_x\right) \quad (5)$$

At the same time, only the {D} position of the end bucket is considered, which can be described as:

$$\begin{aligned} d_x &= c_1\left(l_{AB}c_0 + l_{BC}c_2 + l_{CD}c_{23}\right) \\ d_y &= s_1\left(l_{AB}c_0 + l_{BC}c_2 + l_{CD}c_{23}\right) \\ d_z &= z_1 + l_{AB}s_0 + l_{BC}s_2 + l_{CD}s_{23} \end{aligned} \quad (6)$$

where

$$l_{AB}s_0 = z_2 \quad l_{BC} = \sqrt{x_3^{\ 2} + z_3^2} \quad l_{AB} = \sqrt{x_2^{\ 2} + z_2^2} \quad l_{CD} = \sqrt{x_4^{\ 2} + z_4^2}$$

Where θ_0 is defined as the Angle between AB and the horizontal plane.

Solve the equation (6) and the result is:

$$\theta_3 = \text{acos}\left(\left(\left(\frac{d_x}{c_1} - l_{AB}c_0\right)^2 + (d_z - z_1 - l_{AB}s_0)^2 - \left(l_{BC}^{\ 2} + l_{CD}^{\ 2}\right)\right)/2l_{BC}l_{CD}\right) \quad (7)$$

θ_2 is determined by θ_3 .

2.3.Dynamics Analysis

Assume that the centroids of the rotary body, maneuvering arm, bucket rod and bucket are m1, m2, m3 and m4, the distances to frame {A}, {B}, {C} and {D} are l_{m1} , l_{m2} , l_{m3} and l_{m4} respectively, the coordinates of the center of mass are:

$$m1: \begin{pmatrix} l_{m1}c_{m1}c_1 \\ l_{m1}c_{m1}s_1 \\ z_1 + l_{m1}s_{m2} \end{pmatrix} \quad m2: \begin{pmatrix} c_1\left(l_{AB}c_0 + l_{m2}c_{m2+2}\right) \\ s_1\left(l_{AB}c_0 + l_{m2}c_{m2+2}\right) \\ z_1 + l_{AB}s_0 + l_{m2}s_{m2+2} \end{pmatrix}$$

$$m3: \begin{pmatrix} c_1\left(l_{AB}c_0 + l_{BC}c_2 + l_{m3}c_{m3+23}\right) \\ s_1\left(l_{AB}c_0 + l_{BC}c_2 + l_{m3}c_{m3+23}\right) \\ z_1 + l_{AB}s_0 + l_{BC}s_2 + l_{m3}s_{m3+23} \end{pmatrix} \quad (8)$$

$$m4: \begin{pmatrix} c_1\left(l_{AB}c_0 + l_{BC}c_2 + l_{CD}c_{23} + l_{m4}c_{m_4+234}\right) \\ s_1\left(l_{AB}c_0 + l_{BC}c_2 + l_{CD}c_{23} + l_{m4}c_{m_4+234}\right) \\ z_1 + l_{AB}s_0 + l_{BC}s_2 + l_{CD}s_{23} + l_{m4}s_{m_4+234} \end{pmatrix}$$

Where, c_{m2+2} represents the cosine of the Angle between the center of mass and the horizontal plane, θ_{m2}, rotated by θ_2 . The centroid velocity can be obtained by taking the time derivative of the joint variables in the above equation.

So the Lagrangian L of the system is:

$$L = T - V \quad (9)$$

Where T is the system kinetic energy and V is the system potential energy. They are calculated separately from the following equations:

$$T=\frac{1}{2}\sum_{i=1}^{4} m_i \left\|V_{mi}\right\|^2 \quad V=g\sum_{i=1}^{4} m_i z_{mi} \tag{10}$$

The equation of motion of the system is expressed by Lagrange equation:

$$\frac{d}{dt}\left(\frac{\partial L}{\partial \dot{q}_i}\right)-\frac{\partial L}{\partial q_i}=\tau \tag{11}$$

Where, q_i represents the generalized coordinates of $\theta_1 \sim \theta_4$ and $\dot{q}_i$ represents the generalized velocities of $\dot{\theta}_1 \sim \dot{\theta}_4$.

Combine formula (8)-(11) and sort it out:

$$M(q)\ddot{q}+C(q,\dot{q})\dot{q}+G(q)=\tau \tag{12}$$

Where $M(q)$ is the 4 × 4 mass matrix, $C(q,\dot{q})\dot{q}$ is a 4 × 1 vector of centrifugal and Coriolis terms, $G(q)$ is a 4 × 1 vector of gravity terms, τ is the joint generalized driving force.

3. Model predictive control

3.1. Establishment of Predictive Model

Assuming that during operation, the external force on the excavator is *f*, which is introduced into the dynamic equation of the excavator, it can be obtained:

$$\ddot{q}=\left(\tau+J^{\mathrm{T}}f-\left(C(q,\dot{q})\dot{q}+G(q)\right)\right)/M(q) \tag{13}$$

Where J^{T} is the transpose of the excavator Jacobian matrix.

Select state vector $Q=[q \quad \dot{q}]^{\mathrm{T}}$, and establish the system state equation:

$$\dot{Q}=AQ+B\tau+C \tag{14}$$

Where $A=\begin{bmatrix} 0 & 1 \\ -\frac{C}{M} & 0 \end{bmatrix}$, $B=\begin{bmatrix} 0 \\ \frac{1}{M} \end{bmatrix}$, $C=\begin{bmatrix} 0 \\ \frac{J^{\mathrm{T}}f-G}{M} \end{bmatrix}$.

3.2. Discretization of Prediction Models

Assuming the sampling period is T, using the forward Euler method to discretize the state equation:

$$\begin{aligned} &\dot{Q}:=\frac{Q(k+1)-Q(k)}{T}=AQ(k)+B\tau(k)+C \\ &\Rightarrow Q(k+1)=\bar{A}Q(k)+\bar{B}\tau(k)+\bar{C} \end{aligned} \tag{15}$$

Where $\bar{A}=\begin{bmatrix} 1 & T \\ -\frac{TC(q(k),\dot{q}(k))}{M(q(k))} & 1 \end{bmatrix}$, $\bar{B}=\begin{bmatrix} 0 \\ \frac{T}{M(q(k))} \end{bmatrix}$, $\bar{C}=\begin{bmatrix} 0 \\ \frac{T\left(J^T f-G(q(k))\right)}{M(q(k))} \end{bmatrix}$, k is the k-th sampling period.

3.3. Prediction

The system state predicted in the future p sampling periods is:

$$X_k=\left[Q(k+1|k)^T \quad Q(k+2|k)^T \quad \ldots \quad Q(k+p|k)^T\right] \tag{16}$$

Where p is the prediction time domain, $(k+1|k)$ indicates the prediction at k-th sampling period of the system state at (k+1)-th sampling period, and so on.

In addition, to predict the future state of a dynamic system, it is necessary to know the control quantity U_k in the prediction time domain:

$$U_k = \left[\tau(k|k)^T \quad \tau(k+1|k)^T \quad \dots \quad \tau(k+p-1|k)^T \right]^T \tag{17}$$

Through the analysis of the previous chapter, it can be found that the dynamic system is a nonlinear system, and the coefficient of the discrete system is related to the system state at the sampling time. Therefore, we must make necessary simplification, it is assumed that the coefficients $\overline{A}$, $\overline{B}$ and $\overline{C}$ vary linearly over p prediction time periods (therefore the coefficient p should be as small as possible), take the value of the intermediate prediction period, and maintain the value within the period $[k, k+p]$:

$$\begin{aligned} \overline{A} &= \begin{bmatrix} 1 & T \\ -\dfrac{TC(q(k+n|k),\dot{q}(k+n|k))}{M(q(k+n|k))} & 1 \end{bmatrix} \\ \overline{B} &= \begin{bmatrix} 0 \\ \dfrac{T}{M(q(k+n|k))} \end{bmatrix} \\ \overline{C} &= \begin{bmatrix} 0 \\ \dfrac{T(J^T f - G(q(k+n|k)))}{M(q(k+n|k))} \end{bmatrix} \end{aligned} \tag{18}$$

Where n is $p/2$ if p is even; When p is odd, n is $(p+1)/2$.

Now, the state of the system for p control cycles in the future can be predicted successively through the discretized state equation:

$$\begin{aligned} Q(k+1|k) &= \overline{A}Q(k) + \overline{B}\tau(k|k) + \overline{C} \\ Q(k+2|k) &= \overline{A}^2 Q(k) + \overline{A}\overline{B}\tau(k|k) \\ &\quad + \overline{B}\tau(k+1|k) + (\overline{A}+I)\overline{C} \\ Q(k+3|k) &= \overline{A}^3 Q(k) + \overline{A}^2\overline{B}\tau(k|k) + \overline{A}\overline{B}\tau(k+1|k) \\ &\quad + \overline{B}\tau(k+2|k) + (\overline{A}^2 + \overline{A} + I)\overline{C} \\ &\vdots \\ Q(k+p|k) &= \overline{A}^p Q(k) + \sum_{i=0}^{p-1} \overline{A}^{p-1-i}\overline{B}\tau(k+i|k) + \sum_{i=0}^{p-1} \overline{A}^{p-1-i}\overline{C} \end{aligned}$$

Integrate into matrix form:

$$X_k = \Psi Q(k) + \Theta U_k + \Xi \tag{19}$$

Where $\Psi = \left[\overline{A}^1 \quad \overline{A}^2 \quad \dots \quad \overline{A}^p \right]^T$, $\Theta = \begin{bmatrix} \overline{A}^{1-1}\overline{B} & \dots & \mathbf{0} & \mathbf{0} \\ \overline{A}^{2-1}\overline{B} & \overline{A}^{2-2}\overline{B} & \dots & \mathbf{0} \\ \vdots & \vdots & \ddots & \vdots \\ \overline{A}^{p-1}\overline{B} & \overline{A}^{p-2}\overline{B} & \dots & \overline{A}^{p-p}\overline{B} \end{bmatrix}$, $\Xi = \begin{bmatrix} \overline{A}^{1-1}\overline{C} & \dots & \mathbf{0} & \mathbf{0} \\ \overline{A}^{2-1}\overline{C} & \overline{A}^{2-2}\overline{C} & \dots & \mathbf{0} \\ \vdots & \vdots & \ddots & \vdots \\ \overline{A}^{p-1}\overline{C} & \overline{A}^{p-2}\overline{C} & \dots & \overline{A}^{p-p}\overline{C} \end{bmatrix}$.

The lower triangle form in the above formula can directly reflect the causal relationship of the system in time, that is, the future time has no influence on the current output, and the output of the system at the future time is completely determined by the current system state and system input.

3.4. Optimization

The goal of the control system is to make the system state of excavator track the desired trajectory, define the desired trajectory as the reference value, and select the reference value sequence in the prediction time domain p:

$$R_k = \left[r(k+1)^T \quad r(k+2)^T \quad \dots \quad r(k+p)^T \right]^T \tag{20}$$

We hope to find the best control quantity U_k, so that the state output in the time domain of control prediction is as close as possible to the reference value. This is an open-loop optimal control problem, which can provide more accurate feedforward input value of the system after optimization. Therefore, we define the cumulative error between the predicted state vector and the reference value as the optimization objective function of the system:

$$J(U_k) = (X_k - R_k)^{\mathrm{T}} V (X_k - R_k) \tag{21}$$

In general, we don't expect the control instruction to be too large, so we can add constraints on the control quantity in the optimization objective function:

$$J(U_k) = (X_k - R_k)^{\mathrm{T}} V (X_k - R_k) + U_k^{\mathrm{T}} W U_k \tag{22}$$

Therefore, the optimization problem can be described as:

$$\begin{aligned} &\min_{U_k} J(U_k) \\ &s.t.\ \left|\tau(k+i\,|\,k)\right| \le \tau_{max}, i = 0,1,2,\ldots,p-1 \end{aligned} \tag{23}$$

Expand formula (22) and unite like terms:

$$\begin{aligned} J(U_k) &= (X_k - R_k)^{\mathrm{T}} V (X_k - R_k) + U_k^{\mathrm{T}} W U_k \\ &= \Big(\underbrace{\Psi Q(k) + \Xi - R_k}_{E} + \Theta U_k\Big)^{\mathrm{T}} V \Big(\underbrace{\Psi Q(k) + \Xi - R_k}_{E} + \Theta U_k\Big) \\ &\quad + U_k^{\mathrm{T}} W U_k \\ &= (E + \Theta U_k)^{\mathrm{T}} V (E + \Theta U_k) + U_k^{\mathrm{T}} W U_k \\ &= E^{\mathrm{T}} V E + (\Theta U_k)^{\mathrm{T}} V (\Theta U_k) + 2E^{\mathrm{T}} V (\Theta U_k) + U_k^{\mathrm{T}} W U_k \\ &= U_k^{\mathrm{T}} \left(\Theta^{\mathrm{T}} V \Theta + W\right) U_k + \left(2E^{\mathrm{T}} V \Theta\right) U_k + E^{\mathrm{T}} V E \\ &= \frac{1}{2} U_k^{\mathrm{T}} H U_k + f^{\mathrm{T}} U_k + E^{\mathrm{T}} V E \end{aligned} \tag{24}$$

In the above formula, $E^{\mathrm{T}}VE$ is constant and has no effect on the minimization, $H = 2\left(\Theta^{\mathrm{T}} V \Theta + W\right)$, $f^{\mathrm{T}} = 2E^{\mathrm{T}} V \Theta$, the above equation is a standard quadratic optimization objective function. The optimization sequence U_k can be solved and the first element of U_k can be applied to the control at time k.

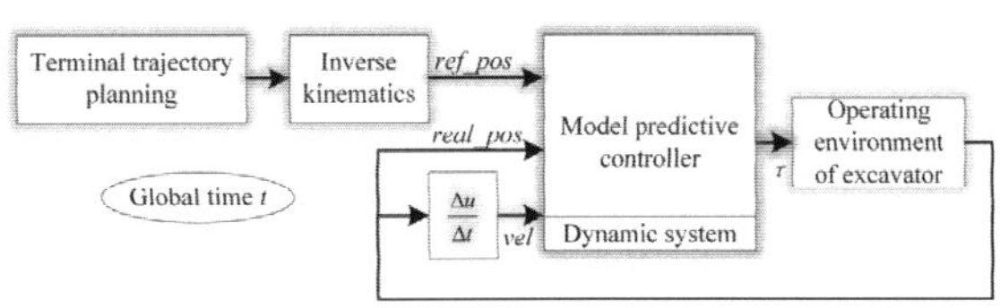

Figure 2. Model predictive control block diagram of excavator.

4. Simulation Establishment and Result Analysis

In order to quickly verify the effectiveness of the control algorithm, a control program is built in Matlab Simulink.

Mining loading action is adopted in simulation. Mining loads a cyclic simulation action model referring to industry standards, and the action process is shown in Fig. 4. In order to better simulate the effectiveness of the algorithm, no-load and load excavation experiments are carried out, in which the load excavation experiment assumes that the earthwork weight is known in the excavation process.

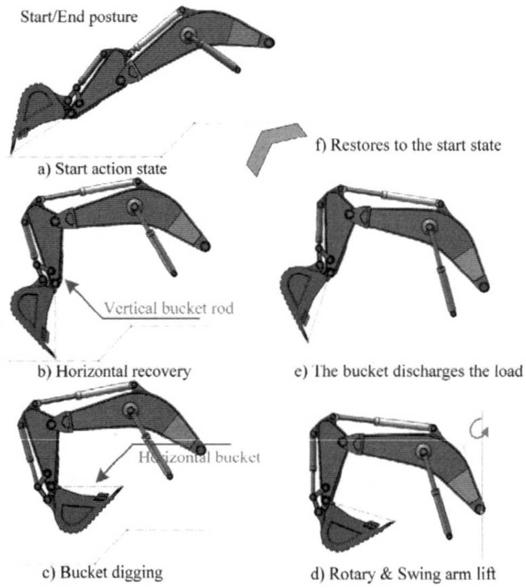

Figure 3. Mining action of excavator

4.1.Mining simulation experiment

The difference between the experimental process in this section and the mining process in section 3.1:

In stage b), increasing horizontal resistance and vertical force are added to the bucket to simulate the accumulation process of earthwork excavation;

In stage c), the decreasing horizontal resistance is increased to 0 and the increasing vertical force is added to the bucket;

Maintain maximum vertical force in stage d);

The vertical force is reduced to 0 in the stage e).

Assume that the above processes are uniformly applied, as shown in the figure:

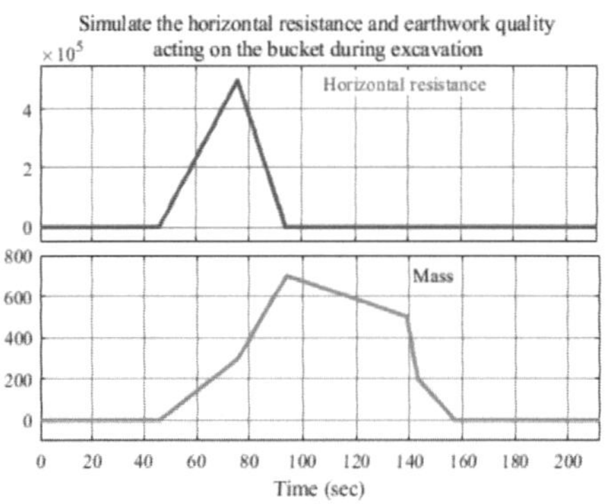

Figure 4. Mining simulation experiment load time sequence diagram of external force

4.2.Simulation results and analysis

In the working condition of section 3.1, the track tracking effect of each joint of excavator is good. The comparison of two experiments shows that the trajectory control algorithm has universality and can adapt to different working conditions. The change of force curve is also stable, and the difference of control force between the two simulations is large, indicating that the change of terminal working condition will be transmitted to the joint successively through the series arm, which is consistent with the force derivation of Newton-Euler recursive method in the mechanical arm, and the control effect of this control algorithm is good.

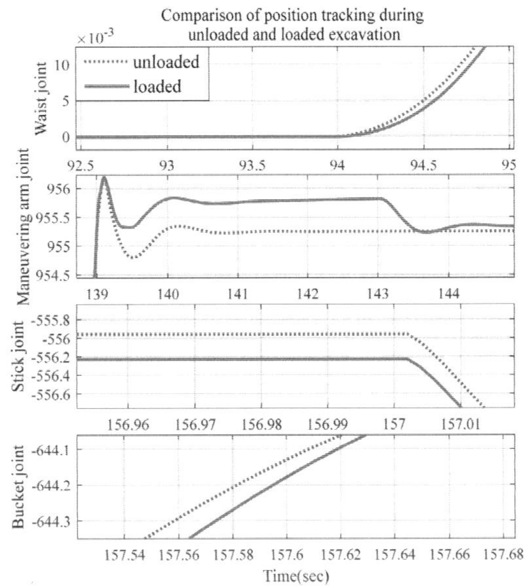

Figure 5. Comparison of no-load and loaded mining process position tracking

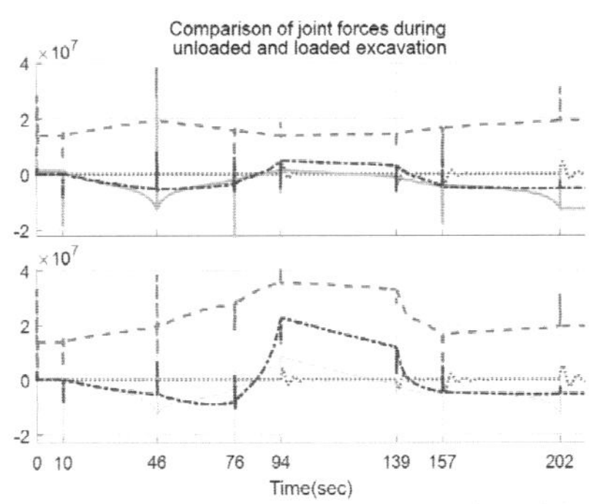

Figure 6. Comparison of joint forces in no-load and loaded excavation process

5. Conclusions

The following conclusions are obtained: (1) Ignoring the influence of hydraulic rod, the dynamic modeling of excavator can be adopted in the same way as that of mechanical arm; (2) The control effect of model predictive control is good, but it needs a specific dynamic model and requires a large amount of calculation;(3) Model predictive control can effectively eliminate the jitter caused by the traditional control method, and the control is more rapid and accurate; (4) The algorithm can adapt to different working conditions, but it needs to detect the working conditions, and can be obtained by parameter identification.

Acknowledgment

This work was supported by the National key research and development plan "Construction machinery energy consumption analysis and optimization control software" No. 2020YFB1709900.

References

[1] Hao Feng, Chen-Bo Yin, Wen-wen Weng, Wei Ma, Jun-jing Zhou, Wen-hua Jia, Zi-li Zhang, Robotic excavator trajectory control using an improved GA based PID controller, Mechanical Systems and Signal Processing,Volume 105,2018,Pages 153-168.
[2] Shuzhong Zhang, Tatiana Minav, Matti Pietola, Heikki Kauranne, Jyrki Kajaste, The effects of control methods on energy efficiency and position tracking of an electro-hydraulic excavator

equipped with zonal hydraulics, Automation in Construction, Volume 100, 2019, Pages 129-144.

[3] Dong Guan, Nan Yang, Jerry Lai, Ming-Fung Francis Siu, Xingjian Jing, Chi-Keung Lau, Kinematic modeling and constraint analysis for robotic excavator operations in piling construction, Automation in Construction, Volume 126, 2021.

[4] Mitrev, Rosen AND Janošević, Dragoslav AND Marinković, Dragan, Dynamical modelling of hydraulic excavator considered as a multibody system, Technical Gazette, volume 24, 2017.

[5] Z. A. Leman, M. Hatta Mohammad Ariff, H. Zamzuri, M. A. Abdul Rahman and S. Amri Mazlan, "Model Predictive Controller for Path Tracking and Obstacle Avoidance Manoeuvre on Autonomous Vehicle," 2019 12th Asian Control Conference (ASCC), 2019, pp. 1271-1276.

[6] F. A. Bender, S. Göltz, T. Bräunl and O. Sawodny, "Modeling and Offset-Free Model Predictive Control of a Hydraulic Mini Excavator," in IEEE Transactions on Automation Science and Engineering, vol. 14, no. 4, pp. 1682-1694, Oct. 2017.

Journal of Physics: Conference Series 2195 (2022) 012040 doi:10.1088/1742-6596/2195/1/012040

Study on insulation performance optimization of EMU high-voltage equipment box

Chao Li[1a*], Lin Lin[1b*], Weidong Qu[1c]

[1]Skills Training Center of State Grid North Hebei Electric Power Co., Ltd, Baoding 071051, China

[a*]731817442@qq.com, [b*]15933930851@163.com, [c]39400920@qq.com

Abstract—The EMU high-voltage electrical equipment on the roof not only bears the erosion of various harsh and extreme environments, but also bears the impact of various over-voltage, and the insulation performance of the electrical equipment on the roof is seriously threatened. This paper studies the insulation optimization design method of EMU high-voltage electrical equipment, puts forward the method of adding a certain length of insulating sheath on the electrical equipment to improve the insulation performance of high-voltage equipment box, and tests the insulation optimization measures on high-voltage circuit breaker and EMU high-voltage cable. The result shows that the installation of insulating sheath is feasible to improve the insulation performance of EMU high-voltage equipment box .

1. Introduction

With the development of high-speed railway, the running speed of high-speed EMU has reached 385km / h. In addition, the temperature difference of railway lines is large and the operating environment is poor, the insulation problem of EMU high-voltage electrical equipment is becoming increasingly prominent. In addition to environmental factors, insulation problems are also closely related to insulation design. Therefore, it is necessary to strengthen the research on insulation design and improve the insulation performance of EMU high-voltage electrical equipment[1] .

At present, a lot of investment has been made at home and abroad to improve the insulation performance of EMU high-voltage equipment. Part of the research is devoted to the selection of insulating materials. In the emerging research on electrical equipment insulation, outdoor insulating sheath is widely used, and its application in other devices needs to be further explored. After the test operation of "anti flashover insulating sheath" , there was no flashover accident on the connected line [2]. Literature [3] introduces the use of insulating metal sheath. The advantages of insulating sheath are introduced in literature [4]. Literature [5] proposed a method to install a certain length of insulating sheath on the transmission line at the end of high-voltage composite insulator to improve the electric field strength at the high-voltage end. Literature [6] concluded that the pollution flashover characteristics of composite insulators can be improved by optimizing sheath parameters. Through the power frequency withstand voltage test in literature [7], it is proved that the installation of insulating sheath can reduce the voltage borne by insulators. Based on the above research, this paper explores the application of insulating sheath in EMU high-voltage equipment.The paper adds a certain length of insulating sheath on EMU high-voltage equipment, and tests the lightning impulse withstand voltage of the equipment. The results prove the feasibility of adding insulating sheath to improve the insulation performance of EMU high-voltage equipment box.

2. Insulation optimization method

The insulation level of electrical equipment refers to the test voltage (withstand voltage) that the equipment insulation can withstand. Under the action of this voltage, flashover and breakdown or other damage will not occur.

(1)Selection of insulating materials

According to different insulating materials, high-voltage cables can be divided into high-voltage oil paper cables, high-voltage plastic cables and rubber insulated cables. Insulators can be divided into porcelain insulators, glass insulators[8], silicone rubber composite insulators[9] and epoxy resin composite insulators.

(2)Shielding

By improving the electrode shape, the potential distribution along the surface of solid medium is homogenized, the maximum potential gradient is reduced, and the surface flashover voltage can also be increased. This method is called shielding. There are many kinds of shielding, including shielding cover, shielding ring, shielding layer and insulating sheath. The function of shielding ring and shield cover is to control corona on electrical equipment. The insulating sheath is made of silicone rubber material by high-temperature molding, which has excellent electrical performance, aging resistance, high and low temperature resistance. Installing the insulating sheath on the high-voltage electrical equipment can improve the electric field strength at the high-voltage end, increase the effective leakage distance of the equipment, improve the corona voltage and improve the pollution flashover characteristics of the equipment.

(3)Barrier

The barrier is a solid insulating medium placed in the electric field and has prominent edges in the equipotential plane direction of the electric field. The barrier can significantly improve the electric field distribution and increase the surface flashover voltage.

3. Verification of impact withstand voltage test

The insulation level of high-voltage electrical equipment is determined by the most serious one of lightning over-voltage, operating over-voltage and maximum working voltage acting on the insulation. Since the system voltage is lower than 220kV, it is uneconomical to limit the lightning over-voltage to lower than the operating over-voltage in the EMU high-voltage electrical system . Therefore, the insulation level of the EMU high-voltage equipment box is determined by the lightning over-voltage. The EMU high-voltage electrical equipments include circuit breaker, high-voltage cable, insulator, high-voltage transformer, lightning protector,isolating switch and traction transformer[10]. Taking CRH380 EMU as an example, the lightning impulse withstand voltage test of high-voltage circuit breaker and high-voltage cable of EMU is carried out to obtain the insulation performance of the equipment in this paper. Through the impact withstand voltage test of EMU high-voltage circuit breaker and high-voltage cable before and after the installation of insulating sheath, the influence of insulating sheath on the insulation performance of EMU high-voltage equipment under over-voltage is explored.

3.1 Test operation instructions

This paper tests the impulse withstand voltage of the high-voltage circuit breaker and high-voltage cable respectively. The test object sequence is circuit breaker, circuit breaker with insulating sheath installed on the head, cable and cable with insulating sheath installed on the terminal. The test operation process is as follows:

(1) Conduct wiring according to the experimental schematic diagram, select appropriate wave head and wave tail resistance, and make the impulse voltage generator output lightning impulse voltage;

(2) After confirming that the wiring is correct, all personnel shall leave the experimental area;

(3) Remove the grounding rod from the impact body capacitance, determine that there is no one in the experimental area, and close the experimental area;

(4) Turn on the power supply of the impulse voltage generator experimental console, and set the

charging voltage of each stage as the design value on the control operation interface;

(5) Press the charging button;

(6) Record the waveform after synchronous discharge of the impact machine;

(7) Click the grounding button on the console, turn off the power supply, and hang the grounding rod on the non grounding end of the first stage capacitor of the impact generator;

(8) Gradually increase the voltage applied to the test object, repeat steps 5-7, and observe the change of output voltage waveform until the test object is broken down;

(9) Replace the test object and retest;

(10) Disconnect the main power supply of the experiment and restore the experimental site clean and tidy after the experiment, .

The field physical drawing of the equipment before and after the installation of insulating sheath is as follows. The left side of Fig.1 is the head of the circuit breaker in the initial state and the right side is the head of the circuit breaker after the installation of insulating sheath. The left side of Fig.2 is the cable terminal in the initial state and the right side is the cable terminal after the installation of insulating sheath. Fig.3 is the holistic model of CRH380 EMU high-voltage equipment.

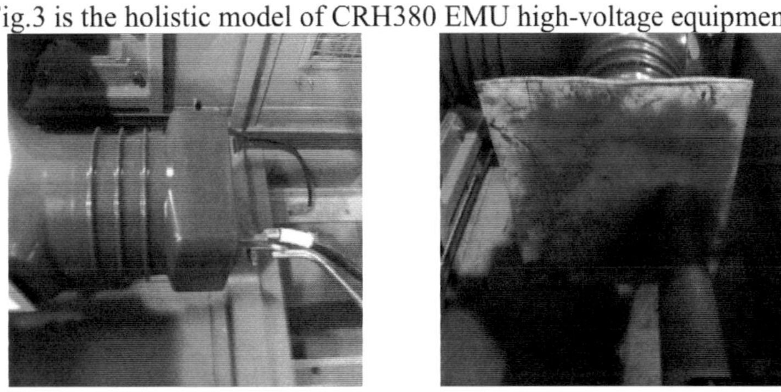

Fig.1 Circuit breaker head before and after installing insulating sheath

Fig.2 Cable terminals before and after installing insulating sheath

Fig.3 The holistic model of CRH380 EMU high-voltage equipment

In this paper, the measured wave voltage waveform is selected as the injury indication method in lightning impulse voltage test. Whether the tested object is broken down can be determined by the dynamics of the waveforms.The voltage divider with oscilloscope, peak voltmeter and digital recorder are used to measure the voltage peak and waveform. The tested voltage waveform is output after being reduced in a certain proportion.

3.2 Test results

3.2.1 High-voltage circuit breaker

First, the impulse withstand voltage test is made on the circuit breaker before installing the insulating sheath, and then on the circuit breaker after installing the insulating sheath. The impulse voltage applied to the circuit breaker before and after installing the insulating sheath is recorded in Tab.1. Fig.4 shows the voltage waveform before installing insulating sheath. Fig.5 shows the voltage waveform after installing insulating sheath.

Tab.1 Impulse voltage applied to circuit breaker before and after installing insulating sheath

	Test voltage 1	Test voltage 2	Test voltage 3	Test voltage 4	Test voltage 5	Test voltage 6
Initial state	139.2kV	145.6kV	150.4kV	163.2kV	179.2kV	--
After installing the insulating sheath	182.4kV	185.6kV	192kV	212.8kV	220.8kV	228.8kV

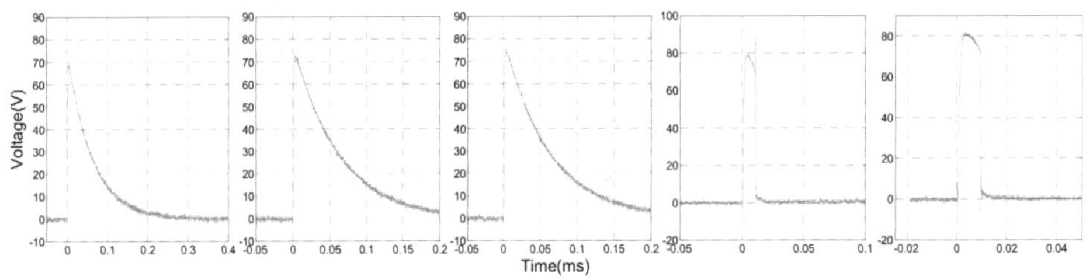

Fig.4 Voltage waveform before installing insulating sheath

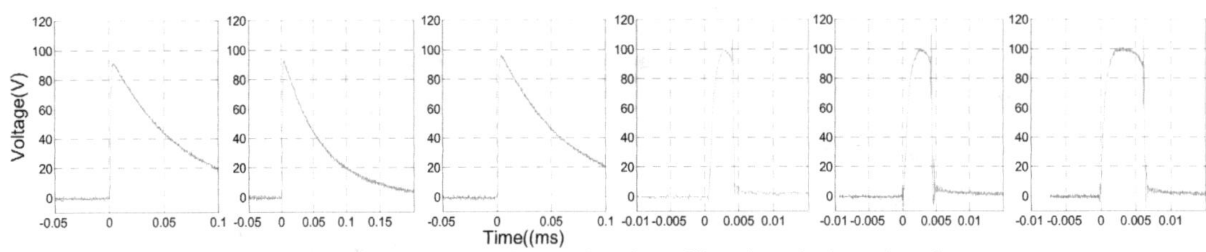

Fig.5 Voltage waveform after installing insulating sheath

3.2.2 High-voltage cable

Firstly, the impulse withstand voltage test of high-voltage cable before installing the insulating sheath is conducted. The waveform recorded is shown in Figure 6. Next is the cable installed with insulating sheath. The waveform is shown in Figure 7. The impulse voltage applied to the high-voltage cable before and after installing the insulating sheath is recorded in Table 2.

Tab.2 impulse voltage applied to the cable before and after installing insulating sheath

	Test voltage 1	Test voltage 2	Test voltage 3	Test voltage 4	Test voltage 5	Test voltage 6
Initial state	122.4kV	139.2kV	177.6kV	--	--	--
After installing the insulating sheath	168kV	176kV	187.2kV	195.2kV	201.6kV	243.2kV

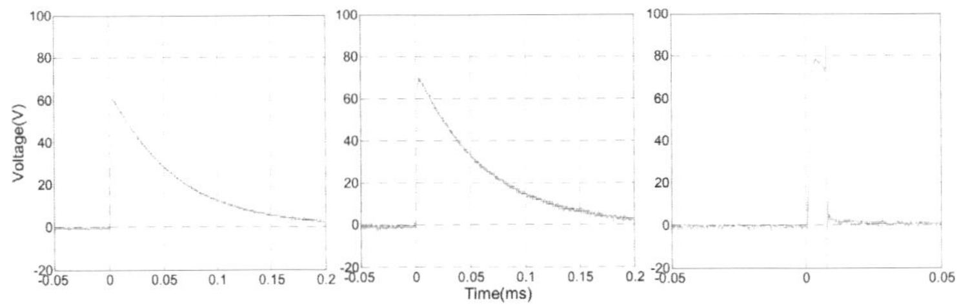

Fig.6 Voltage waveform before installing insulating sheath

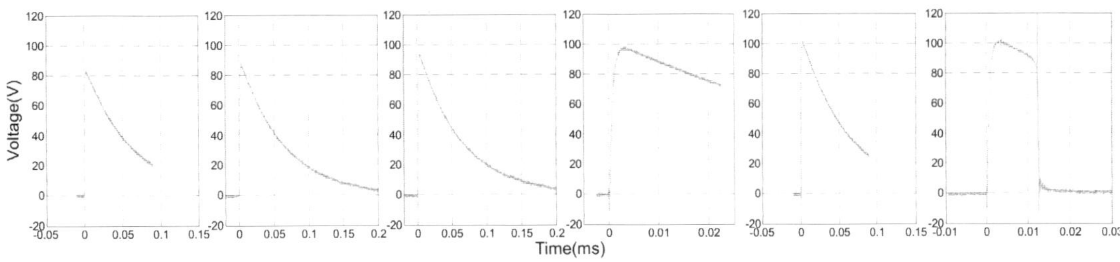

Fig.7 Voltage waveform after installing insulating sheath

3.3 Analysis of test results
It can be seen from the test results that the installation of insulating sheath for EMU high-voltage equipment does improve the insulation performance of the equipment. For the high-voltage circuit breaker, the installation of insulating sheath increases its impulse withstand voltage value from 163kV to 213kV. For the high-voltage cables, the installation of insulating sheath increases the impulse withstand voltage value from 178kV to 243kV.

4. Conclusion
Due to the existence of insulating sheath, the withstand voltage on the circuit breaker and cable is lower than that added by the impulse voltage generator to the tested object, and the reduced part of the voltage is borne by the insulating sheath. The insulating sheath plays a role in sharing the impact high voltage during the test. By changing the equivalent capacitance of the insulating sheath, the improvement degree of the insulation performance of the equipment installed with the insulating sheath can be changed. But first of all, the voltage shared on the insulating sheath cannot exceed its puncture voltage. And the different of installation positions has different effects on the optimization of insulating performance. Therefore, it is necessary to reasonably configure the parameters of insulating sheath to achieve the best effect.

References
[1] Guo Chenxi. Discussion on high voltage insulation design of electric locomotive / EMU [J]. China Academy of Railway Sciences. 2010.
[2] Wang Jianhong. "Anti flashover insulating sheath" fills the gap in China [n]. Ningxia daily, December 17, 2008 (006).
[3] Popovic，Ljubivoje，M. Comparative analysis of grounding systems formed by feeders in one case with uninsulated and in the other case with insulated metallic sheath [J]. 2005，1-5.
[4] Lu Zhigang. Application of insulating sheath in oilfield power system [J]. Instrument electrical, 2012, 31 (4): 51-51.
[5] Mei Hongwei, Chen Jinjun. Study on installing insulating sheath on high voltage transmission line of composite insulator [J]. Chinese Journal of electrical engineering, 2011, 31 (1): 109-116.
[6] Mei Hongwei, Peng GONGMAO. Pollution flashover characteristics of composite insulator suspension conductor after installing insulating sheath [J]. Chinese Journal of electrical engineering, 2011, 31 (31): 189-195.

[7] Ma Kui, Wang Hailong. Analysis of flashover protection effect of anti flashover insulating sheath on bird droppings [J]. Ningxia electric power, 2012, (2): 20-24.
[8] Ma Limin, Cui Wei, Yu Demai. Analysis and research progress of three insulator materials [J]. New chemical materials, 2011, 39 (12): 28-30.
[9] Wang Shuqin. Analysis of insulators for CRH2 EMU Roof [J]. Shanghai Railway Science and technology, 2009, (4): 43-44.
[10] Zhang Zhaohui, Wu Mingli. Common fault analysis of high voltage electrical system on the roof of EMU [C]. International Academic Conference on rail transit electrical and information technology, 2013.

Analysis of wellbore stability of gas underbalanced drilling technology

Lei Zhao[1,a*], Aihua Bao[2,b], Jingxin Cao[3,c]

[1]No.4 Drilling Engineering Company, BHDC, Cangzhou 062552, China

[2]No.3 Drilling Engineering Company, BHDC, Cangzhou 062552, China

[3]No.4 Drilling Engineering Company, BHDC, Cangzhou 062552, China

[a*]Corresponding author: zhaolei9@cnpc.com.cn

[b]pxb_cbm@petrochina.com.cn, [c]caojingxin-bh@cnpc.com.cn,

Abstract—Gas underbalanced drilling technology has the advantages of high ROP and little damage to oil and gas reservoirs. In recent years, it has been widely used in the exploration and development of oil and gas fields in China. Under normal air drilling conditions, the pore pressure is basically zero, so it is necessary to establish the wellbore stability model under the condition of zero pore pressure to provide theoretical support for the wellbore stability analysis of air drilling. However, gas underbalanced drilling also has wellbore instability, more sand deposition at the bottom of the well, and even the malignant accident of sticking. Firstly, through the analysis of the traditional wellbore stability model, the influence of pore pressure is removed, and the wellbore stability model suitable for air drilling is obtained. At the same time, it is proposed that the sand deposition is mainly caused by the collapse material when the collapse ellipse is formed on the shaft wall, and the formation of the collapse ellipse is conducive to maintaining the stability of the shaft wall. For large and medium-sized gravel layers, the influence of cementation strength on wellbore stability should be considered.

1.PREFACE

Air drilling has great advantages in tight lithologic strata and difficult to drill strata. Gas drilling can greatly improve drilling speed and increase the footage of a single bit, which is of great significance for improving drilling speed in tight strata and difficult to drill strata. Since the 1960s, the world energy demand has entered an era dominated by oil and natural gas. With the development of economy and the growth of population, the demand for oil and natural gas is increasing rapidly [i]. The world oil industry is facing a very severe situation, under the new situation, the exploration and development of oil and gas fields have the following characteristics: The difficulty of exploration and development is getting bigger and bigger, the investment cost is getting higher and higher, and the risk is getting higher and higher. In order to meet the requirements of these new exploration and development situations, it is necessary to improve the exploration and development ideas and technologies, so as to reduce the exploration and development costs and risks and meet the needs of the current international energy demand [ii]. In the process of oil and gas field development, drilling investment often accounts for about two-thirds of the total cost. Therefore, improving drilling technology and reducing drilling costs have become an important means to reduce the whole exploration and development costs, and underbalanced drilling technology came into being under this background [iii].

Wellbore stability is an important problem in oil drilling engineering. Every year, the major oil companies in the world consume a lot of manpower, material resources and financial resources to solve various problems caused by wellbore instability. The safety and effectiveness of drilling operation can not be effectively guaranteed because of wellbore instability, resulting in huge economic losses[iv]. There are two main problems of wellbore stability, one is tensile fracture of wellbore surrounding rock due to the high liquid column pressure in wellbore which is greater than the fracture pressure of wellbore rock; the other is the shear failure caused by the low drilling fluid pressure which can not provide effective support to the wall rock. The former leads to lost circulation while the latter leads to the collapse of the wall and the enlargement of the well diameter[v]. The factors affecting wellbore stability can be summarized as man-made and natural. Natural factors include: stratum rock properties, geological structure types, in-situ stress, clay-bearing mineral types, soil mineral content, fracture development course, rock cementation, rock porosity, permeability and so on; human factors include the composition of drilling fluid, drilling fluid properties (fluid loss, viscosity, rheological property, density), bare time of wellbore, erosion of wellbore caused by annulus return velocity of drilling fluid, impact of drilling string vibration on wellbore, etc[vi].

Gas underbalanced drilling technology has the advantages of high penetration rate and little damage to oil and gas reservoirs, and has been widely used in oil and gas field exploration and development in China in recent years. In the drilling process, there is no liquid column pressure in the well to support the wellbore and balance the formation pressure, and the formation fluid flows into the well under the action of negative pressure difference, thus changing the stress distribution around the well, so the wellbore stability problem is more prominent and becomes the biggest factor restricting the application of gas underbalanced drilling technology.

2.Materials and Methods

(1) The stress model of borehole lining under the condition of air drill

In the traditional wellbore stability model, when the cylindrical coordinate system is adopted for vertical wells, the effective stress around the wellbore can be expressed as:

$$\sigma_{rr} = \frac{1}{2}(S_{Hmx} + S_{Hmin} - 2p_0)\left(1 - \frac{R^2}{r^2}\right) + \frac{1}{2}(S_{Hmax} + S_{Hmin}) \times \left(1 - \frac{4R^2}{r^2} + \frac{3R^4}{r^4}\right)\cos 2\theta + \frac{p_0 R^2}{r^2} \quad (1)$$

$$\sigma_{\theta\theta} = \frac{1}{2}(S_{Hmax} + S_{Hmin} - 2p_0)\left(1 + \frac{R^2}{r^2}\right) - \frac{1}{2}(S_{Hmax} - S_{Hmin}) \times \left(1 + \frac{3R^4}{r^4}\right)\cos 2\theta - \frac{p_0 R^2}{r^2} - \sigma^{\Delta T} \quad (2)$$

$$\tau_{r\theta} = \frac{1}{2}(S_{Hmax} - S_{Hmin})\left(1 + \frac{2R^2}{r^2} - \frac{3R^4}{r^4}\right)\sin 2\theta \quad (3)$$

The new wellbore stability model under the condition of air drilling is as follows, because the formation water must be avoided in the stage of horizon optimization under gravel drilling condition, and the new wellbore stability model under air drilling condition is as follows:

$$\sigma_{rr} = \frac{1}{2}(S_{Hmx} + S_{Hmin})\left(1 - \frac{R^2}{r^2}\right) + \frac{1}{2}(S_{Hmax} + S_{Hmin}) \times \left(1 - \frac{4R^2}{r^2} + \frac{3R^4}{r^4}\right)\cos 2\theta + \frac{R^2}{r^2} \quad (4)$$

$$\sigma_{\theta\theta} = \frac{1}{2}(S_{Hmax} + S_{Hmin})\left(1 + \frac{R^2}{r^2}\right) - \frac{1}{2}(S_{Hmax} - S_{Hmin}) \times \left(1 + \frac{3R^4}{r^4}\right)\cos 2\theta - \frac{R^2}{r^2} - \sigma^{\Delta T} \quad (5)$$

$$\tau_{r\theta} = \frac{1}{2}(S_{Hmax} - S_{Hmin})\left(1 + \frac{2R^2}{r^2} - \frac{3R^4}{r^4}\right)\sin 2\theta \quad (6)$$

Where: σ_{rr}-radial stress around the wellbore, MPa; $\sigma_{\theta\theta}$-circumferential stress around the wellbore, MPa; $\tau_{r\theta}$-tangential stress around the wellbore, MPa; S_{Hmax}-the magnitude of the horizontal maximum principal stress, MPa; S_{Hmin}-the magnitude of the horizontal minimum principal stress, MPa; θ-the angle between the azimuth of the horizontal maximum principal stress and the specified direction, °; p_0-formation pore pressure, MPa; $\sigma^{\Delta T}$-stress due to temperature, MPa; ΔT-the difference between bottom hole temperature and formation temperature, °C; R-wellbore radius, m; R-the distance between a point in the formation and the center of the wellbore, m.

(2) The establishment of model and the process of wellbore instability

Because of the heterogeneity distribution of gravel formation, the whole failure process can not be effectively characterized by traditional software simulation. In order to solve this problem, the discrete element method is used to simulate the instability process of gravel wellbore lining. Based on the geology and engineering parameters of gravel stratum, the discrete element numerical model is established. The stratum depth is 2000m; gravel characteristics: medium and fine conglomerate; maximum radius of gravel: 50mm; minimum radius of gravel: 6mm; radius of filling matrix: 1 ~ 2mm; cementation degree: poor; formation porosity is 18%, and there is no fluid in the formation; maximum horizontal principal stress: 45MPa; minimum horizontal principal stress: 35MPa; borehole size: 431.8 mm.

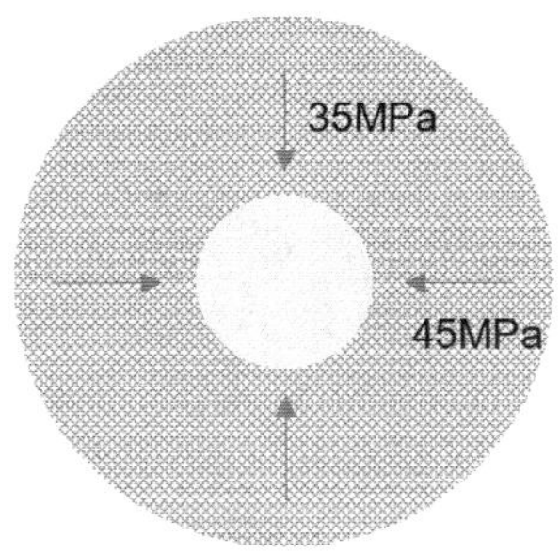

Fig. 1 Stress loading and crack initiation

Fig. 1 shows the crack initiation process. In the left picture, the red line represents the stress loading, with the horizontal direction corresponding to the maximum horizontal principal stress and the vertical direction corresponding to the minimum horizontal principal stress. In the process of fracture propagation, the green dots near the borehole indicate tensile fractures, and the red dots indicate shear fractures. Fractures are first formed near the borehole wall, and the number of fractures in the figure is relatively small and not obvious. The number of fractures counted by program shows that tensile fractures are more than shear fractures. When there are large gravel particles around the borehole wall, tensile fractures are formed around the large gravel particles, and then a large number of shear fractures are gathered around the gravel particles, and the fractures are connected with each other, resulting in the shedding of gravel particles. In order to observe the fracture propagation form around the borehole wall, the integrity of the formation failure area is guaranteed by setting program commands, but in fact, the area where the fracture propagates has collapsed.

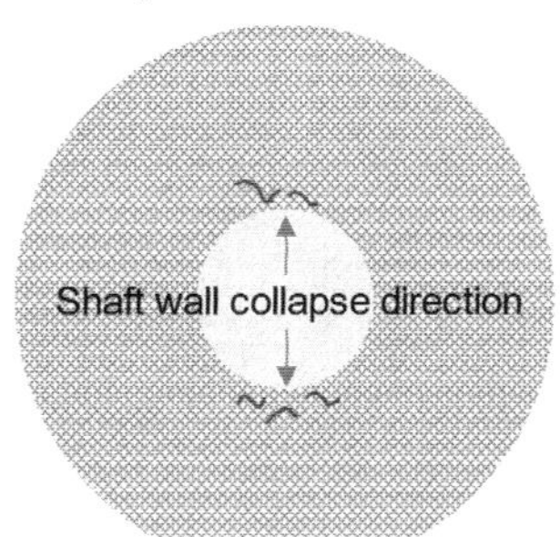

Fig. 2 Crack propagation process

After fractures are formed around large gravel particles around the well, they quickly penetrate with nearby fractures. Under the action of in-situ stress, a large number of accumulated fractures will propagate along the direction of minimum principal stress (see Figure 2). The minimum in-situ stress direction in the model of Figure 2 is along the longitudinal direction, so an elliptical wellbore with long axis in the longitudinal direction will be formed (red ellipse in Figure 3). Fig. 3 shows the failure mode of shaft lining, in which the thickness of black line represents the contact force between particles, and the part without contact force indicates that falling and collapse have occurred. It is worth noting that a large number of falling blocks (blue circles) are also caused around gravel particles on the right side of the wellbore, while the left side of the wellbore is not damaged, because there are many large gravel particles distributed on the right side of the wellbore, which are poorly cemented with cements and have micro-fractures. Under the action of shear stress, the fractures expand and penetrate, reducing the rock strength of the wellbore, resulting in falling blocks on the wellbore.

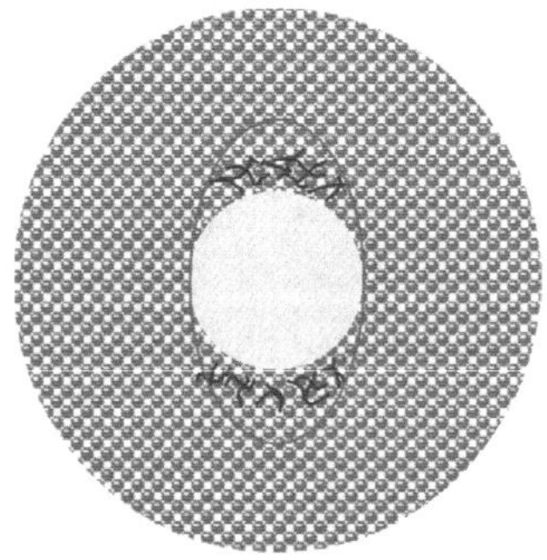

Fig. 3 Wellbore failure mode

3.RESULTS

(1) Effect of particle size characteristics on wellbore instability

Based on the statistics of particle size distribution, the radii of gravel particles are 0.5 ~ 1.5 mm, 1.5 ~ 3mm, 3 ~ 4mm and 4 ~ 5mm. The calculation time of the model is 2 × 10 s.

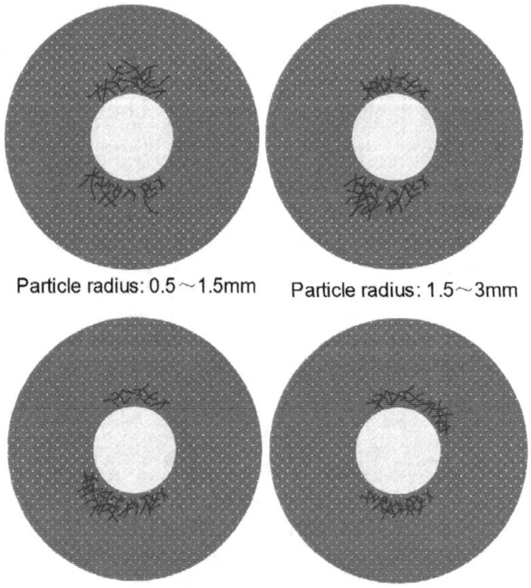

Fig. 4 Wellbore instability characteristics of different gravel particle size distributions

Fig. 4 shows the failure characteristics of four different gravel particle sizes in phase wellbore model, the red dots represent cracks, the red arrow represents the crack propagation direction, the aggregation of red dots represents the formation of fracture zone and the spalling of surrounding rock blocks. The large gravel particles around the well reduce the compressive strength of surrounding rock. Under the action of borehole shear stress, the confining pressure of gravel layer produces blocks, forming an irregular caving ellipse with the direction of minimum horizontal principal stress as its long axis. At the same time, the fracture propagation direction of wellbore models with different gravel particle size distribution is slightly different. The larger the gravel particle size, the greater the influence of gravel location distribution on fracture propagation, and the failure area around the well extends to the direction of large gravel particle size. When the radius of gravel particles is 0.5 ~ 1.5 mm, the collapse area of borehole wall is near the borehole, the shedding of gravel particles and the fracture of cement almost occur at the same time, the distribution of gravel around the borehole has little influence on the fracture propagation direction, and the fracture propagates almost along the minimum principal stress direction, that is, the Y-axis direction, but the distance of the failure area in the Y-axis direction is short; when the radius of gravel particles is 1.5 ~ 3mm, the failure area around the well covers the area where large gravel particles are located in the X-axis direction, and the cementation surface around the large gravel particles has low strength, which leads to preferential failure and forms a fracture aggregation area. At the same time, multiple fracture aggregation areas penetrate each other, which accelerates the fracture propagation process, and the failure area has a long distance in the Y-axis direction; when the radius of gravel particles is 3 ~ 4mm, the micro-cracks propagate along the direction of maximum principal stress, and gradually connect with the fracture aggregation area of large-size gravel particles nearby, and the propagation process is faster, and the failure area is farther away in the Y-axis direction; when the radius of gravel particles is 4 ~ 5mm, the gravel particles peel off from the cement and form irregular wellbore, and the wellbore instability is more prominent in the area where large gravel particles gather around the wellbore. For the gravel layer with the same borehole size, under the action of non-uniform tectonic stress, the particle size distribution of gravel particles affects the fracture propagation process around the well. The larger the particle size (larger than 3mm), the multiple fracture aggregation areas penetrate each other, and the fracture propagation is farther, and the wellbore failure area extends to the depth around the well.

The results of discrete element simulation show that the strength between cement and gravel and the size of gravel particle play a decisive role in wellbore stability. When the gravel layer is unstable and destroyed, only cement is destroyed, and then gravel particles peel off from the borehole wall. When air drilling is used, it is necessary to ensure that gravel particles can be carried out of the wellbore; when gravel particles cannot be taken out, it is easy to cause sand settling at the bottom of the hole and even sticking.

(2) Field application

1) Analysis of strength and particle size of cement in gravel layer

When the gravel content is the same, the larger the gravel particle size, the lower the rock strength; the greater the strength of cement, the greater the strength of rock. Through statistical analysis, it is found that the larger the gravel particle size is, the easier it is to form collapse and block falling. When the gravel particle size is larger (generally larger than 5mm), it is difficult for cuttings to be taken away from the bottom of the well, resulting in excessive sand settling, resistance in tripping and sticking.

2) Law of sand settling

The law of sand settling in a 333. 375 mm borehole is 8m/100m (i.e. 7m sand settling per 100m formation), and that in a 333. 375 mm borehole is 17.5 m/100m. The main factors affecting the thickness of grit settling are cement strength and gravel particle size (Figure 5).

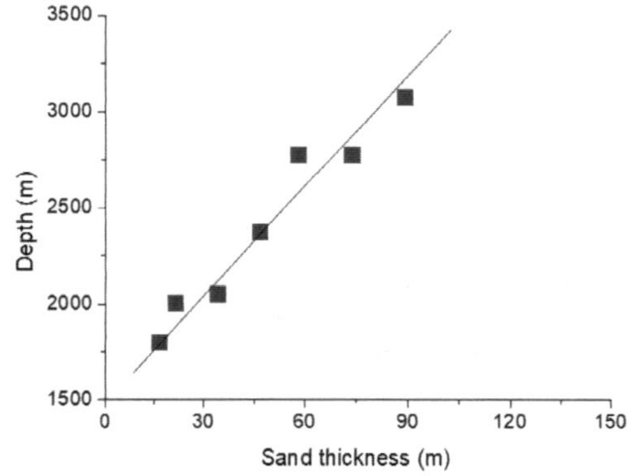

Fig. 5 Characteristics of sand settling law

3) Cause analysis of sand settling

According to the simulation of air drilling, when the wellbore diameter is 333.375 mm and the well depth is 4500m, the simulation results are shown in Figure 6. The 400m/min gas injection rate can only carry 5mm grain size sand when the wellbore is enlarged by 20%, and only 3mm grain size sand when the wellbore is enlarged by 40%.

After the drill bit opens the formation, a collapse ellipse is formed, and a large number of grit is formed, which is gravel. Therefore, the rock carrying capacity of the air injected into the wellbore will have a great influence on the grit speed and depth.

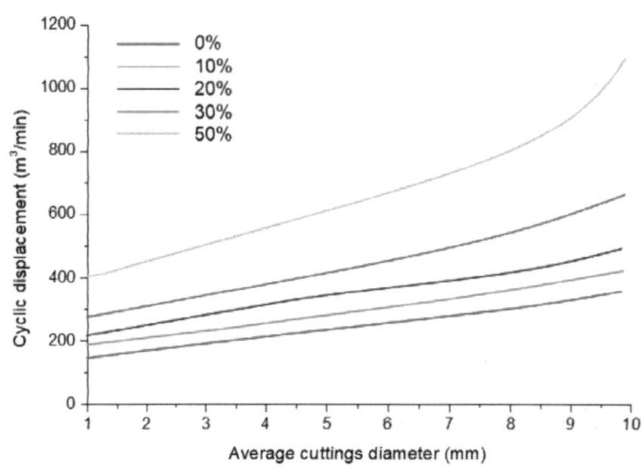

Fig. 6 Relationship between different cuttings diameters and displacement

4.CONCLUSION

1) When drilling in gravel layer, the collapse ellipse is beneficial to reduce the stress concentration around the borehole wall, and the main source of bottom hole sand settling is the rock forming the collapse ellipse.

2) The content and strength of cement and the size of gravel have great influence on wellbore stability. Air drilling is suitable for small conglomerate formation with uniform gravel distribution (the size of gravel is less than 5mm). For large and medium gravel layers, the influence of cementation strength on wellbore stability should be considered.

3) At present, the influence of irregular well diameter is not considered in the calculation of rock carrying capacity under air drilling conditions, so further research is needed. Strengthen and improve the data collection of upper gravel electrical measurement, further strengthen and refine the research on gravel content, particle size and distribution law, and provide basic data for air drilling well location optimization.

REFERENCES

[1] Gomar. M, Goodarznia. I, Shadizadeh. S.R, A transient fully coupled thermo-poroelastic finite element analysis of wellbore stability, 8rd ed., vol. 6. Arabian Journal of Geoences, 2015, pp. 3855-3865.

[2] S. He, W. Wang, M, Tang, B. Hu, W. Xue, Effects of fluid seepage on wellbore stability of horizontal wells drilled underbalanced, vol. 21. Journal of Natural Gas Science & Engineering, 2014, pp. 338-347.

[3] Baikov. V.P, Trukhin. B.V, Mironenko. A.A, 1989 Analysis of the stability and conditions for bulging of gas main line sections laid with an initial curvature, 21rd ed., vol. 9. Strength of Materials, 1989, pp. 1269-1273.

[4] Zhen. N, Xia. B, Zhou. L, Deng. J, Modeling of wellbore stability for gas drilling, 31rd ed., vol. 6. Natural Gas Industry, 2011, pp. 71-76.

[5] Michaelis. M.M, Notcutt. M, Cunningham. P.F, Drilling by gas lens focused laser, 59rd ed., vol. 5-6. Optics Communications, 1986, pp. 369-374.

[6] Liu. H, Meng. Y, Wan. S, Gao. L, Jiang. Z, Bin. L, Analysis of wellbore stability during gas drilling through high-pressure gas formations, 30rd ed., vol. 11. Natural Gas Industry, 2010, pp. 59-62.

User-transformer relation identification based on power balance model and adaptive AFSA

Jiawei Song[1*], Yan Jiang[1], Xueying Song[1], Zhiqiang Sheng[1],and Zibing Meng[1]

[1] Fibrlink Communications Co., Ltd., Beijing 100070, China

[*]2246792482@qq.com

Abstract. User-transformer relation identification plays an important role in the correct management of low-voltage area archives and the improvement of line loss. In order to obtain an accurate user-transformer relation identification, this paper proposes a user-transformer relation identification method in the low-voltage area based on power balance model and adaptive artificial fish swarm algorithm(AFSA). This method uses the summation relationship between the total meter of the transformer and the user's meter/meter box to fit the coefficients of the power balance equation through the AFSA, then we use the coefficients and related statistical values to judge the user-transformer relation. The main innovations are: this paper proposes a power balance model to solve the problem of user-transformer relation identification, which is simpler than previous methods and has strong operability; AFSA is used to fit the regression coefficients of the power balance equation, which has advantages in calculation accuracy and efficiency compared with the traditional least squares method; an improvement strategy of adaptive step length is proposed to make the ability of AFSA to find superior stronger. By selecting real station data for verification, the result shows that the method in this paper can quickly and accurately identify user's meters/meter boxes with abnormal user-transformer relationship, the method in this paper has high computational efficiency and recognition accuracy without additional labor and hardware costs.

1. Introduction

The line loss rate of the transformer is an important indicator to measure the operation and management level of the power company. Improving the calculable of line loss rate is the fundamental way to reduce loss and increase efficiency[1]. The correct relationship of user-transformer in low-voltage stations is conducive to the calculation and analysis of line losses. However, in reality, there are often deviations in the archives of transformers, which is caused by reasons such as untimely updating or loss of recorded information. This also resulted in the inability to identify the exact ownership of a certain electric meter during the inspection process in the transformers, which easily caused customer disputes, affected the social image of the power company, and reduced the efficiency of the power grid operation[2].

At present, there are two methods for user-transformer relation identification in low-voltage stations. The first method is to install a transformer identification device. This device mainly uses power carrier technology, but the carrier signal cannot be effectively isolated by the transformer, it often causes crossover communication problems in the transformer, which has reduced the accuracy of user-transformer relation identification. The literature [3] improves the accuracy of the user-transformer relation identification by applying deep learning to the detection of the transformer recognition instrument, but when using it, it is necessary to install some hardware devices such as current transformers at the outlet end of the transformer, which often faces some safety problems. The other method is to extract the electrical data of the transformer such as voltage, current, etc. in the power consumption information collection system, and use the relevant algorithm model to the user-transformer relation identification. The voltage zero-crossing drift data is clustered in the literature [4], and compared with the voltage zero-crossing drift data of the transformer meter to finally determine the user-transformer relationship, but the zero-crossing drift data needs to be measured in advance by professional equipment, and the algorithm model proposed in the literature is relatively complex and not practical; the literature [2] analyzes the correlation of voltage between the user meter and the transformer meter from the two dimensions of space and time, and then determines the user-transformer relationship. This method needs to manually obtain the attribution relationship between the transformer and some subordinate meters as sample for training, which brings additional labor costs. Therefore, it is important to improve a method for user-transformer relation identification ensures recognition accuracy and computational efficiency, which can also reduce labor and hardware costs.

This paper proposes a user-transformer relation identification method based on the power balance relationship. The user-transformer relationship is determined by the electricity additive relationship between the transformer meter and the user's electricity meter.

2. Theoretical Introduction

2.1. Power balance model

Without calculating the line loss, there is a summation relationship between the power consumption of the transformer meter and the daily power consumption of each user meter, as shown in the following formula:

$$y_t = \sum_{i=1}^{n} a_i \times x_i(t) \tag{1}$$

In formula (1), y_t represents the collected power on the Day t of the transformer meter to be analyzed($t = 1,2,3,\ldots,D$), and x_t represents the collected power on the Day t of the user's electricity meter i to be analyzed($i = 1,2,3,\ldots,D$). The analysis days is D , and there are n meters to be analyzed in the transformer. a_i is the coefficient of the user meter to be solved. Ideally, $a_i = 1$ indicates that the user's electricity meter i belongs to the transformer to be analyzed and the user-transformer relation identification is correct; $a_i = 0$ indicates that the user's electricity meter i does not belong to the transformer to be analyzed, and the user-transformer relation identification is wrong.

In normal power grid operation, line loss must exist. According to electrical engineering theory, in the same transformer, except for other factors, the normal line loss of each line is determined by the current of the line. The power consumption of user meters is closely related to the current, so there is a positive correlation between line loss and power consumption of user's meters[5]. Due to the different electrical distances of user's meters, the contribution ratio of the power consumption of each user's meter to the line loss is also different, so the additive relationship of the power consumption in the transformer is:

$$y_t = \sum_{i=1}^{n} w_i \times x_i(t)$$

$$w_i = a_i + \theta_i \tag{2}$$

In the formula (2), θ_i represents the contribution ratio of the user's electricity meter to the line loss of the transformer(the power consumption of users in the low-voltage station area has little fluctuation and θ_i can be approximated as a fixed value). w_i represents the actual coefficient of power consumption in the user's electricity meter considering the line loss. Ideally, when $w_i \geq 1$ and is close to 1, it means that the user's meter belongs to the transformer to be analyzed, and the archives is correct; $w_i = 0$ means that the user's meter does not belong to the transformer.

Normally, the user's meters are installed in the meter box. In view of the low possibility of abnormal affiliation between the electric meter box and the user's meters, this article will summarize the power consumption in the unit of the meter box. The power consumption of the user's meters is summarized as the power consumption of the meter box. By judging the relationship between the meter box and the transformer, the affiliation relationship between each user's meter and the transformer is determined. This article will focus on analyzing the affiliation relationship between the transformer and the subordinate meter box.

2.2. *Artificial fish swarm algorithm*

It can be seen from 2.1 that the coefficient fitting of the power consumption data requires the multiple linear regression method, so as to obtain the regression coefficients of each user's meter, and then determine the attribution relationship between the user's meter and the transformer. At present, the least square method is the commonly method to fit regression coefficients, but this method has certain limitations for data sets with complex structures and large scales[6]. There are many electrical data fields in the transformer area, and the amount of data is huge. Therefore, the traditional least squares method cannot achieve ideal results in the user-transformer relation identification. In recent years, intelligent algorithms such as bee colony algorithm[7] and particle swarm algorithm[8], which have been widely used to solve the parameter estimation problem of multiple linear regression, they have advantages of higher accuracy and faster convergence speed. Among these intelligent optimization algorithms, the artificial fish swarm algorithm is more robust, and has better optimization capabilities[9]. Therefore, this article chooses artificial fish swarm algorithm(AFSA) to solve the parameter estimation problem of multiple regression equation.

The AFSA is an intelligent algorithm proposed by studying the foraging methods of fish schools. Its principle is to continuously update the status of individuals through activities such as foraging, clustering and rear-end collision, using fitness function for evaluation, and iteratively find the optimal value[10].

The basic steps of the AFSA are as follows:

(1) Initialize settings

Initial setting of algorithm parameters: including the number of fish swarm N, the initial position of the individual X, the field of view $visual$, the step length $step$, the crowding factor δ, and the number of repetitions $Trynumber$.

(2) Calculate fitness value

The bulletin board is responsible for recording the optimal individual in the algorithm. After the initial settings, the initial fish swarm uses the evaluation function to evaluate the location of the individual, and transmits the information of the individual with the highest food concentration to the bulletin board.

(3) Execute behavior function

Executing behavioral functions including foraging, clustering, rear-end collision, and randomization for all individuals. The execution rules and the operation process of each behavior function are as follows[11]:

① Foraging behavior: this behavior is simulating individuals swimming to the direction of more food. The individual W_i randomly swims to a certain position W_j in the field of view.

$$W_j = W_i + rand() \times visual \tag{3}$$

$rand()$ generates a random number between 0-1. If φ_j is greater than φ_i (φ_j and φ_i are the fitness values of W_j and W_i respectively), then W_i moves a certain distance toward W_j.

$$W_i^{t+1} = W_i^t + \frac{W_j - W_i^t}{\|W_j - W_i^t\|} \times step() \times rand() \tag{4}$$

If the forward condition is not met, W_i continues to select the state W_j in its field of view, and executes random behavior when the number of attempt is reached.

② Grouping behavior: This behavior mainly learns that fish swarm often gathers to find food or avoid harm. Individual W_i finds the number n_f of other individuals in the field of view ($d_{ij} < visual$) and calculates the center position W_c, if $\varphi_c / n_f > \delta\varphi_i$, then W_i moves a certain distance toward the center position, otherwise perform foraging behavior.

$$W_i^{t+1} = W_i^t + \frac{W_c - W_i^t}{\|W_c - W_i^t\|} \times step() \times rand() \tag{5}$$

③ Rear-end behavior: This behavior simulates that when an individual tracks food, other individuals will also swim over. Individual W_i looks for the position W_j with the highest food concentration ($d_{ij} < visual$) in the field of view. If $\varphi_c / n_f > \delta\varphi_i$ indicates that the surrounding of the optimal individual is loose, W_i move a certain distance to the position of the optimal individual, otherwise perform foraging behavior.

$$W_i^{t+1} = W_i^t + \frac{W_j - W_i^t}{\|W_j - W_i^t\|} \times step() \times rand() \tag{6}$$

④ Random behavior: this behavior indicates that the individual walks randomly, and the range of swimming is limited to the size of the field of view. The individual W_i moves a distance randomly and arrives at a new location.

$$W_j = W_i + rand() \times visual \tag{7}$$

(4) Update fish swarm

The fish swarm in the previous round executes the various behavior functions of step (3) to generate a new fish school. Using the evaluation function to calculate the fitness value of all individuals in the new fish school. Compared with the bulletin board individuals, if there is a better fitness value, please update the bulletin board.

(5) Algorithm termination conditions

When the optimal solution reaches the set threshold or reaches the number of iterations, the algorithm stops running, otherwise, go to step (3).

Summarizing the above process, the overall flow of the AFSA is shown in Figure 1.

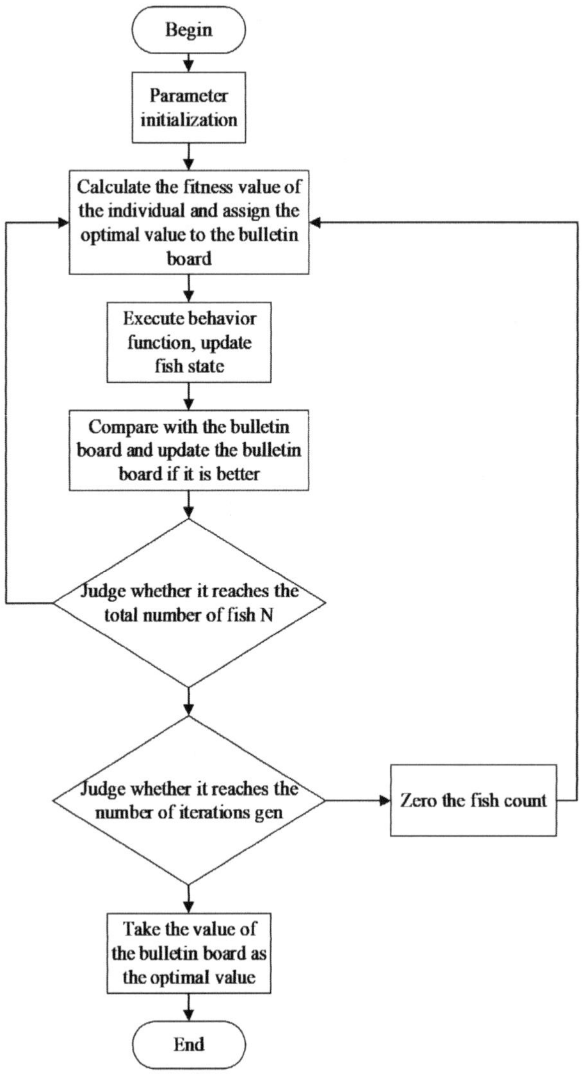

Fig.1 The flow chart of AFSA.

3. User-transformer relation identification

3.1. Adaptive artificial fish swarm algorithm

In the traditional AFSA, the search range of the individual is determined by the field of view, the convergence accuracy and speed are determined by the step size. These two parameters are fixed values in the original algorithm. So individuals cannot quickly converge to the optimal value in the later stage of the algorithm[12]. In order to quickly iteratively calculate the optimal solution, this article adopts a variable step size improvement idea to improve the traditional AFSA. The basic idea is:

(1) After each round of iteration is completed, all individuals will transmit their position and fitness value to the bulletin board.

(2) The field of view ($visual$) needs to be determined by the individual according to the current state of the fish swarm, which is a non-fixed value.

(3) Introduce a new parameter-visual step coefficient a ($0 < a \le 1$), the step length formula of individual is set to $step = a \times visual$.

(4) When individual executes rear-end and clustering behaviors, if the fitness value of individual in the optimal position or the center position in the field of view is greater than the current position, then move a certain distance to the position with the largest fitness value. At the same time, in the adaptive AFSA, the crowded state is not considered.

The step size setting rules of the adaptive AFSA: suppose Z_0 is the current artificial fish and the position is C ; Z_1 is the current optimal artificial fish; Z_2 is the nearest artificial fish to Z_0 . Z_0 uses two fields of view $visual1$ (distance between Z_1 and Z_0) and $visual2$ (distance between Z_2 and Z_0) to find the optimize position in the surrounding, and determines two positions A and B randomly, calculates and compares the fitness values of A、 B and C , if the fitness values of A and B are both smaller than C , at the same time the number of attempts is not reached, it will continue to search. If the fitness value of A or B is better than C or both better than C , the individual will swim to the point with the highest fitness value(A or B). The step length is set to $visual1 \times a \times rand()$ when moving forward to A, the step length is set to $visual2 \times rand()$ when moving forward to B, as shown in Figure 2. The improved algorithm can adaptively conduct selective foraging in different fields of view, which not only speeds up the convergence speed, but also improves the optimization effect.

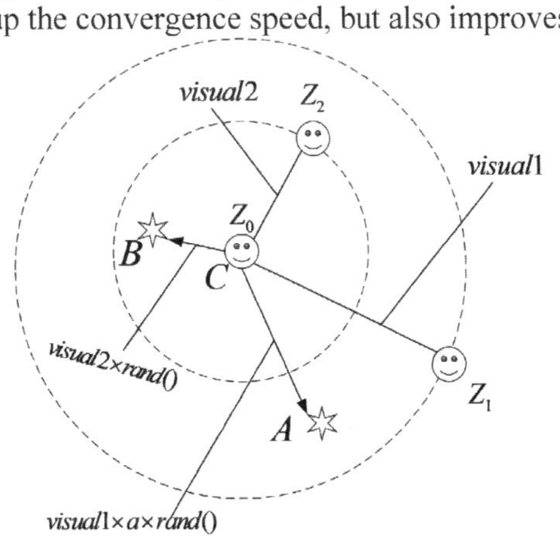

Fig.2 Description of foraging behavior with adaptive AFSA.

3.2. *User-transformer relation identification based on power balance model and adaptive AFSA*

The main steps of user-transformer relation identification method based on the power balance model and adaptive AFSA are as follows:

(1) Data collection

The power data used in this article are all obtained from the power consumption information collection system, and the main data information required is shown in Table 1.

Table 1. Data required for user-transformer relation identification.

Serial number	Data name
1	Date
2	Transformer number
3	Meter box number
4	Meter number
5	Daily power consumption

(2) Adjacent transformers division

Considering that user-transformer relationship errors often occur in adjacent transformers geographically, the scientific and efficient division of adjacent transfromers can reduce the scope of user-transformer relation identification and quickly improve the accuracy of recognition. In view of the standardization of transformer naming, this paper uses the similarity of transformer names as the basis for division, transformer with high similarity in names are divided into adjacent transformers, such as " "PMS_10KV XX Bay 1# Transformer T1" and "PMS_10KV XX Bay 1# Transformer T2" are a group of adjacent transformers, and each group of adjacent transformers usually contains 2-4 transformers.

(3) Data preprocessing

Due to the delayed collection or missing collection of the electricity collection equipment, the data collected from the system will meet the problems of missing some power consumption data and duplication of some data fields. Data preprocessing is required before the user-transformer relation identification is carried out. It is important to delete duplicate records and the meters with missing power consumption data.

(4) Multiple linear regression equation construction

The independent variable X of the regression equation is a matrix of $D \times n$, n represents the number of meter boxes under adjacent transformers, and D represents the number of days that can be actually calculated after data preprocessing. As shown in (8), each column of X represents the power consumption of the meter box in each day of D days.

$$X = \begin{bmatrix} x_{11} & x_{12} & \cdots & x_{1n} \\ x_{21} & x_{22} & \cdots & x_{2n} \\ \vdots & \vdots & \ddots & \vdots \\ x_{D1} & x_{D2} & \cdots & x_{Dn} \end{bmatrix} \tag{8}$$

The dependent variable Y is a matrix of $D \times s$, s represents the number of the transformers in adjacent transformers, and D has the same meaning as above. Each column of Y represents the power consumption of transformer meter in each day of D days.

$$Y = \begin{bmatrix} y_{11} & y_{12} & \cdots & y_{1s} \\ y_{21} & y_{22} & \cdots & y_{2s} \\ \vdots & \vdots & \ddots & \vdots \\ y_{D1} & y_{D2} & \cdots & y_{Ds} \end{bmatrix} \tag{9}$$

According to the constraint conditions of multiple linear regression analysis, in order to solve the equations, the number of independent equations of the power balance equations should be more than the number of coefficients to be solved. Due to the uncertainty of user load, there is basically no two sets of equations that are fully proportional to power consumption information(the two sets of equations are not linearly related), which only need to ensure that the number of days D available for electricity collection is more than the number of meter boxes N in the transformer to be analyzed. At the same time, each group of adjacent transformers needs to execute s times, and a column in Y and X need to be selected for regression calculation in each time.

(5) Use adaptive AFSA to fit multiple linear regression parameters

The goal of the problem to be solved is to find the optimal regression coefficients so that the fitting result of the power balance equation is close to the actual value. Therefore, the fitness function of the adaptive AFSA is set to calculate the deviation between the actual power consumption and the power consumption through the equation calculation of the transformer meter.

$$\varphi(\theta) = \sum_{i=1}^{n} \left[y_i - \left(w_1 x_1 + w_2 x_2 + \cdots + w_n x_n \right) \right]^2 \tag{10}$$

(6) Calculate the t-test statistic value of the multiple linear regression coefficient

The t statistic is used to characterize the sub-fitness of the regression equation. The larger the statistical value, the better the fitting degree of the regression equation and the more reliable the regression coefficient. The calculation formula for the t statistic as follows:

$$t = w_j / \left(\hat{\sigma} \times c_{jj} \right) \tag{11}$$

Where w_j is the regression coefficient, j is a natural number from 1 to p, $\hat{\sigma}$ is the standard deviation of w_j, c_{jj} is the element in the j row and j column of $\left(X^{'}X\right)^{-1}$, and X is the independent variable matrix of the regression equation.

(7) Judgment of abnormal meter box

Combined with the theory of power balance model, it can be seen that when the regression coefficient of the meter box approaches 0, and the regression coefficient with other transformer in the adjacent transformers approaches 1, it is considered that the meter box may be abnormal. In the actual calculation process, due to the existence of meter boxes with low power consumption or even zero power consumption, and the small number of samples that can be calculated, it will cause deviation when we use the coefficient discrimination method simplely.

In response to this problem, this article introduces the t statistic value as an additional basis for judgment based on the relevant theories of the significance test of multiple linear regression coefficients. When there are two transformers under a group of adjacent transformers. Suppose the regression coefficient between the meter box and the original transformer is w_1, and the t test statistic value is t_1; the regression coefficient with the other transformer is w_2, and the test statistic value is t_2. Then the abnormal judgment rule can be described in the following mathematical way:

$$\begin{cases} |w_1| \to 0 \\ |w_2| \to 1 \\ |t_2| \geq n|t_1| \end{cases} \tag{12}$$

In formula (12), the degree of w_1 approaches 0 needs to be determined according to the data quality after preprocessing (including the actual line loss rate, the proportion of low power consumption meter boxes, and the ratio of the number of samples that can be calculated to the number of meter boxes). The same is true for w_2; the value of n is an integer greater than or equal to 2, and the specific value also needs to be determined according to the data quality. When there are more than two transformers under a group of adjacent transformers, we need to calculate the coefficient and statistic of the same number.

4. Experimental results

All simulation experiments are based on Windows 10 Professional x64 system, and Anaconda3 development tools. First, select three typical Benchmark functions[13] to verify the effectiveness of the adaptive AFSA used in this paper. The related information of the test functions is shown in Table 2.

Table 2. Benchmark functions.

Function	Dimension	Expression	Feasible region	Optimal value	Accuracy threshold
Sphere	30	$f(x)=\sum_{i=1}^{n} x_i^2$	[-100,100]	0	10^{-4}
Schwefel	30	$f(x)=\sum_{i=1}^{n}\left(\sum_{j=1}^{i} x_j\right)^2$	[-5,5]	0	10^{-2}
Rosenbrock	30	$f(x)=\sum_{i=1}^{n-1}\left[100\left(x_i^2-x_{i+1}\right)^2+\left(x_i-1\right)^2\right]$	[-100,100]	0	10^{-1}

The test results of the ASFA and adaptive AFSA(A-AFSA) are shown in Table 3 The iteration stop condition is to reach the accuracy. Adaptive AFSA parameter settings: population size $N=60$,visual step coefficient $a=0.3$, number of attempts $Trynumber=15$. The test results of the two algorithms are shown in Table 3.

Table 3. Test results of AFSA and A-AFSA.

Algorithm	Function	Optimal value	Number of iterations
AFSA	Sphere	6.17×10^{-6}	216
	Schwefel	4.53×10^{-3}	371
	Rosenbrock	3.24×10^{-2}	297
A-AFSA	Sphere	3.59×10^{-7}	154
	Schwefel	1.72×10^{-3}	192
	Rosenbrock	6.58×10^{-3}	208

It can be seen from Table 2 that adaptive AFSA has faster convergence speed and optimization accuracy, which proves the effectiveness of the improvement measures proposed in this paper.

We select some test sets from Kaggle to verify the effect of the adaptive AFSA in multiple linear regression analysis, the relevant information of the test sets is shown in Table 4.

Table 4. The test sets of multiple linear regression.

Test sets	Number of samples	Number of independent variables
Boston House Price	506	13
Medical Cost Personal Datasets	1338	6
World Happiness	155	12

The adaptive AFSA and the least squares method LSM are used to perform multiple linear regression tests on the data set in Table 4. The parameters of adaptive AFSA: population size $N = 80$, visual step coefficient $a = 0.45$, and number of attempts $Trynumber = 20$. The experimental results are shown in Table 5 (the name of the test sets adopts the abbreviated form of Table 3).

Table 5. The experimental results of test sets.

Algorithm	Test sets	Root mean square error	Operation hours
LSM	BHP	4.128	9.5s
	MCPD	3.372	16.2s
	WH	5.627	7.3s
A-AFSA	BHP	3.485	6.9s
	MCPD	2.256	12.4s
	WH	5.018	5.2s

It can be seen that adaptive AFSA has advantages in calculation accuracy and efficiency in the fitting of multiple linear regression coefficients compared with the LSM. Then we will use the adaptive AFSA to identify the user-transformer relationship.

The power consumption data in this chapter comes from the electricity consumption information collection system of a prefecture-level city in Jiangsu Province from June 2020 to November 2020. It includes 20 transformers in two power supply stations, as shown in Table 6.

Table 6. The data size of user-transformer relation identification.

Power supply stations	Number of transformers	Number of adjacent transformers	Number of meter boxes
A	14	7	246
B	6	2	128

In view of the small proportion of missing and repeated values in Jiangsu Province, the ratio of the number of samples that can be calculated to the number of meter boxes in each group of adjacent transformers is generally about 2-3, and the number of samples that can be calculated is large. Moreover, the quality of archives management in Jiangsu Province is relatively good. Therefore, the meter box with w_1 in -0.1-0.1, w_2 in 0.9-1.1, and $|t_2| \geq 10|t_1|$ is judged to have an abnormal user-transformer relationship. After calculation, 4 groups of adjacent transformers are found to have a suspected abnormal meter box. The calculation results are shown in Table 7.

Table 7. The meter boxs with abnormal user-transformer relation.

Meter boxs number	w_1	w_2	t_1	t_2
No.1	-0.06631	1.01261	-0.34161	57.08045
No.2	-0.03994	1.03639	-1.63057	39.32766
No.3	0.02860	1.04189	1.04435	33.15632
No.4	-0.02121	1.06532	-0.56239	10.64927

Through verification by professionals, 3 meter boxes (No.1-No.3) are confirmed that there is an abnormal user-transformer relationship. The original records are incorrect.

5. Conclusions

In order to solve the problem of user-transformer relation identification in low-voltage area, this paper proposes a user-transformer relation recognition method based on power balance model and adaptive AFSA. It can quickly determine the meter box with an abnormal user-transformer relationship without using additional measurement devices, thereby solving the problem of archiving the original low-voltage area; effectively improving the level of management in the transformers. The disadvantage is that the method proposed in this paper is highly dependent on the quality of the electricity data. When the electricity data is severely missing or the overall power consumption of the meter box is small, the recognition effect of the method will be reduced, We will focus on improving the robustness of the model.

References

[1] Li Y L, Deng F, Gong J, et al. (2018)Identification And Adjustment Method Of User-transformer Relationship With Spatial Contraints. Computer Knowledge and Technology, 14: 191-194.

[2] Ye A L S, Gu J, Zhang X Q. (2018)Intelligent transformer room recognition technology based on temporal and spatial data correlation. Modern Electronics Technique, 41: 30-33.

[3] Li Y, Fan R L, Jiang W, et al. (2017)Research on the intelligent transformer area recognition method based on BP neural network. Electrical Measurement & Instrumentation, 54: 25-30.

[4] Yao L, Zhang J M, Ni L N. (2020)User-transformer relation identification based on quantum genetic and kernel fuzzy clustering. Electrical Measurement & Instrumentation, 57: 106-113.

[5] Lian Z K. (2020)Data-driven topology identification of low-voltage distribution network. ZheJiang University.

[6] Li Y. (2011)Solution to problems concerning AFSA-based multiple linear regression analysis. Journal of Bohai University(Natural Science Edition), 32: 168-171.

[7] Zhang J L. (2011)Application of Artificial Bee Colony Algorithm to Multiple Linear Regression. Journal of Guangdong Polytechnic Normal University, 32: 31-33.

[8] Wu X J, Li Y, Zhang D,et al. (2007)Application of Particle Swarm Optimization Algorithms in Multi -Linear Regression Analysis Problem. Journal of Liaoning University(Natural Sciences Edition), 34: 18-20.

[9] Zhu X H, Ni Z W, Cheng M Y. (2015)Self-adaptive Improved Artificial Fish Swarm Algorithm with Changing Step. Computer Science, 42: 210-216, 246.

[10] Yu Z X, Jin Y J. (2018)Parameters Estimation of Regression Model Based on Improved AFSA. Statistics & Decision, 34: 75-77.

[11] Chen J J, Liu S. (2021)An Adaptive Artificial Fish Swarm Optimization Algorithm Based on Taboo Search. Computer Technology and Development, 31: 8-13.

[12] Liu Y J, Jiang M Y. (2009)Improved artificial fish swarm algorithm based on adaptive visual and step length. Computer Engineering and Applications.Computer Engineering and Applications, 45: 35-37.

[13] He Q, Wei K Y, Xu Q S. (2019)Mixed strategy based improved whale optimization algorithm. Application Research of Computers, 36: 3647-3651, 3665.

Journal of Physics: Conference Series **2195** (2022) 012043 doi:10.1088/1742-6596/2195/1/012043

Probability Density Analysis of Wind Storage Output Based on Monte Carlo Simulation

Haiming HU [1,a*], Yan Yan[1,b]

[1]Shanghai Dianji University, Lingang, Pudong New Area, Shanghai, China

[a*]huhaiminga@21cn.com.com

Abstract: Due to the relatively large fluctuation amplitude of the output power of wind farms, when using energy storage to stabilize the output characteristics of wind power, the impact of the addition of energy storage on the probability density of wind power output needs to be further analyzed. This paper uses Monte Carlo simulation algorithm to analyze the probability density of active power output in consideration of the operating state of wind power and compares the change of the output probability density curve after adding energy storage. According to the output probability density curve obtained by the algorithm, the abscissa of the curve is shifted to the right by one energy storage capacity unit after the addition of energy storage, so that the output probability density curve of the wind storage system can reach the required value by adding energy storage.

1. Introduction

Nowadays, wind power has penetrated into many application fields, but the operation of wind power is very susceptible to the influence of environmental wind speed, so the power output has greater random fluctuations than other types of power generation devices. In order to improve the stable output characteristics of wind farms, in actual projects, wind farms are usually equipped with energy storage systems to stabilize the output characteristics of wind power. The addition of energy storage makes the influence of the probability density of wind power output have a close relationship with the output reliability of wind storage. Based on Monte Carlo sampling, this paper analyzes the output characteristics of energy storage boosting wind power.

Monte Carlo simulation is a commonly used research method in statistics. Its basic principle is to estimate the probability density of the sample by a given statistical sample set, and subject to a known specific probability distribution, from a randomly sampled numerical sequence. Literature[1] does not consider the time sequence of the system in wind farm output evaluation, so the non-sequential Monte Carlo simulation method can be used to establish a wind farm output probability model. Literature[2] considers the state of operation, shutdown and derating, establishes a multi-state fault model for wind turbines; combines the randomness of wind speed, wind farm wake effects and the failure of the wind turbine itself to establish a wind farm reliability model. Relevant programs are written in MATLAB by the sequential Monte Carlo method, and the reliability of the active output of the wind farm is evaluated. Literature[3] established a Monte Carlo simulation-based simulation program for the probability evaluation of the output of wind turbines in MATLAB, and evaluated the probability of the output of a wind farm, and gave the probability density of this type of wind turbine under different output states. Literature[4] uses the strategy of combining Monte Carlo sampling and the equivalent power function method to calculate the power supply reliability of the power system with a high proportion of wind power grid-connected under different network environments.

The active power output of a wind farm is affected by the scale of the wind farm, wind speed and the operating status of the unit. The addition of energy storage will change the probability density curve of wind farms, which is the focus of this paper, guiding us to improve the output characteristics of wind farms.

2. Energy storage system supports wind power output

The energy storage system of a wind farm is mainly composed of energy storage units, power conversion systems and control equipment. Energy storage unit is the main carrier of energy storage system to realize electric energy storage and release, and its capacity is directly related to the energy storage capacity of energy storage system. Restricted by factors such as low battery cell terminal voltage, limited energy density and power density, in order to expand the capacity, the battery system is generally composed of thousands of battery cells in series and parallel. The power conversion system is an important part of the energy storage system. The two-way energy transfer between the energy storage system and the external grid can be realized through the control equipment and the power conversion system.

The energy storage system provides additional power to the system when the wind farm's output is insufficient. The greater the power capacity of energy storage can compensate for the insufficient output of the wind farm, the greater the capacity. When the probability of continuous output power shortage of the wind farm is higher, the configuration of a certain capacity of energy storage can reduce the insufficient output of the wind farm. As shown in Figure 1, the 24-hour output power density curve of wind farms is relatively sparse.

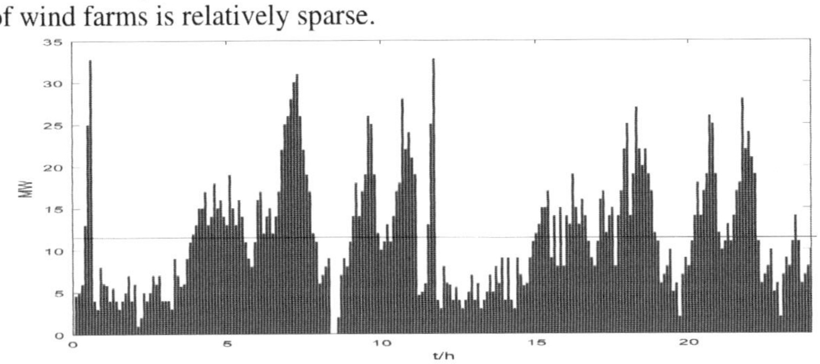

Fig. 1 24-hour output active power characteristics of wind farms

The distribution of the power curve that the wind farm can output to the outside world after the energy storage is configured is shown in Figure 2:

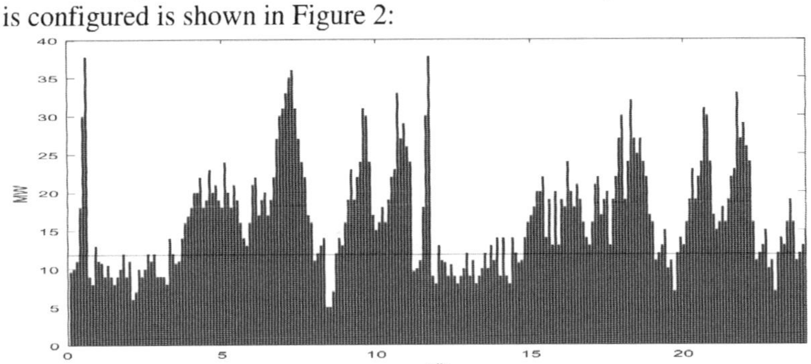

Fig. 2 24-hour active power output characteristics of the wind farm after the configuration of energy storage

The addition of energy storage increases the minimum output power of wind power by one unit of the maximum output capacity of energy storage.

3. Wind farm operation status and output

The change between the available state and the unavailable state of the wind power equipment in a wind farm is considered to be an independent event. Considering the natural aging of the equipment and the intervention of maintenance personnel, the random changes in these two states are considered. Analyzing the state of the system components and the transitions between the states of the system components need to be expressed by the conversion rate. Commonly used conversion rates include failure rate λ(t) and repair rate μ(t). The expression of failure rate is as follows:

$$\lambda(t) = \lim_{\Delta t \to 0} \frac{1}{\Delta t} P[t < T \le t + \Delta t | T > t] \tag{1}$$

In the formula: T is the failure time of the component. The expression of the repair rate is as follows:

$$\mu(t) = \lim_{\Delta t \to 0} \frac{1}{\Delta t} P[t < T_D \le t + \Delta t | T_D > t] \tag{2}$$

In the formula: T_D is the repair time of the element. The state model of the system components needs to choose a two-state model or a three-state model for analysis according to actual research needs.

The state probability of each element in the wind farm can be simulated with random numbers in the interval [0,1]. The research object has two independent states of failure and work. Let S_i denote the state of the system element i, Q_i denote the failure rate of the element, and generate a random number R_i in the interval [0,1] for the element i to obtain:

$$S_i = \begin{cases} 0, & R_i > Q_i \ (\text{Working status}) \\ 1, & 0 \le R_i \le Q_i \ (\text{Failure state}) \end{cases} \tag{3}$$

The state of the system with N elements is represented by the vector S, as shown in the following formula:

$$S = (s_1, \cdots s_i \cdots s_N) \tag{4}$$

When the state of the component is selected, the state is judged whether it is in the invalid state, and if it is, the value of the state is accumulated.

Since wind power does not always run at rated power, this paper considers the output state of wind power to be a three-state model, and outages and derating are considered random events. From the Markov state transition probability, the following formula can be obtained:

$$P_{fo} = \frac{\mu_1 \lambda_2}{\lambda_1 \mu_2 + \lambda_2 \mu_1 + \mu_1 \mu_2} \tag{5}$$

$$P_{do} = \frac{\mu_2 \lambda_1}{\lambda_1 \mu_2 + \lambda_2 \mu_1 + \mu_1 \mu_2} \tag{6}$$

In the formula: λ_1 and λ_2 are the fault migration rates of derating and outage respectively; μ_2 and μ_1 are the repair migration rates of outage and derating respectively. P_{fo} is the probability of outage state; P_{do} is the probability of derating state; refer to equation (3) A three-state probability model can be obtained, as follows:

$$S_i = \begin{cases} 0(\text{run}) & P_{do} + P_{fo} < U \le 1 \\ 1\ (\text{Out of service}) & P_{do} < U \le P_{do} + P_{fo} \\ 2(\text{Derating}) & 0 \le U \le P_{do} \end{cases} \tag{7}$$

In the formula: U is a random number randomly selected in the interval [0,1]; the output state of the wind farm can be obtained by formula (7) combined with the classic wind power output model[5]:

$$P_{Gi} = \begin{cases} P(U_w) & P_{do} + P_{fo} < U \le 1 \\ 0 & P_{do} < U \le P_{do} + P_{fo} \\ aP(U_w) & 0 \le U \le P_{do} \end{cases} \tag{8}$$

In the formula: P_{Gi} is the output of the i-th typhoon; a is the derating factor of the wind power。

4. Wind power output evaluation based on Monte Carlo simulation

According to the basic knowledge of probability statistics, when the required quantity X is the mathematical expectation $E(\zeta)$ of the random variable ζ, then the method of approximately determining X is to repeat the sampling of ζ for N times to generate a sequence of mutually independent ζ values $\zeta_1\zeta_2...\zeta_N$, and calculate its arithmetic mean:

$$\zeta_N = \frac{1}{N}\sum_{n=1}^{N}\zeta_n \tag{9}$$

From the theory of large numbers:

$$P(\lim_{N\to\infty}\bar{\zeta}_N = x) = 1 \tag{10}$$

Thus when N is sufficiently large, the following formula can be obtained:

$$\bar{\zeta}_N \approx E(\zeta) = \infty \tag{11}$$

The probability that this formula holds is 1. According to the above theory, it is assumed that the probability of a random event A is p, the value of the random variable ζ is 1, and the value of ζ is 0 when A does not appear. In N experiments, assuming that A appears V times, the frequency of its appearance is also considered as a random variable. Its mathematical expectation is $E(v)=N_p$, variance $\sigma^2(v)=N_{pq}$, where $q=1-p$. Let $\bar{p}=v/N$, which means the frequency of occurrence of the event. When N is sufficiently large, the following formula is obtained:

$$\bar{p} = \frac{v}{N} \approx E(\zeta) = p \tag{12}$$

The probability that the above formula holds is 1. Thus $\bar{p}=v/N$ can be obtained approximately to obtain the required quantity, and the sample variance is:

$$\sigma^2(\bar{p}) = \frac{\bar{p}(1-\bar{p})}{N-1} \tag{13}$$

The Monte Carlo algorithm continuously samples from the system state parameter S and counts the number of different states. When the number of states obtained by sampling reaches a certain value, the sampling frequency of the system state S is used as an unbiased estimation of the sampling, which can be obtained:

$$\hat{P}(S) = \frac{n(s)}{N_S} \tag{14}$$

In the formula: Ns is the sampling times of the system; $n(s)$ is the number of appearances of the state s; from the unbiased estimation of the system, the mean value of the reliability index of the system can be calculated, and the expression is:

$$\hat{E}(J_R) = \sum_{s\in fail} F(S)\times\hat{P}(S) \tag{15}$$

In the formula: J_r is the value of the reliability index; F(s) is the index function of the state S; P(s) is the probability of the appearance of the state S; The uncertainty is measured by the mean sample variance, as shown in the following formula:

$$V\left(\hat{E}(J_R)\right) = \frac{1}{N_s(N_s-1)}\sum_{K=1}^{N_S}\left(F_K(S)-\hat{E}(J_R)\right)^2 \tag{16}$$

The accuracy achieved by Monte Carlo simulation is measured by the coefficient of variance η, as shown in the following formula:

$$\eta = \frac{\sqrt{V\left(\hat{E}(J_R)\right)}}{\hat{E}(J_R)} \tag{17}$$

The specific steps of the Monte Carlo simulation wind power output evaluation algorithm process are as follows:

①Enter the relevant data of the wind farm: time series wind speed, number of wind power, wind power outage rate, derating rate, maximum sampling times and minimum variance.

②Randomly drawn values in the interval [0,1] are used as the values of U_w and U, where U_w represents the working state of the wind power, and U represents the sampled wind speed.

③Calculate the output of the K-th sampled wind farm, and determine the attribution range of the divided output state.

④Calculate the coefficient of variance and judge whether its accuracy meets the requirements, then the sampling ends; otherwise, K=K+1, go to step ② until the maximum sampling value.

⑤ Judge whether the evaluation of all output states is completed, calculate and output the probability of all output states.

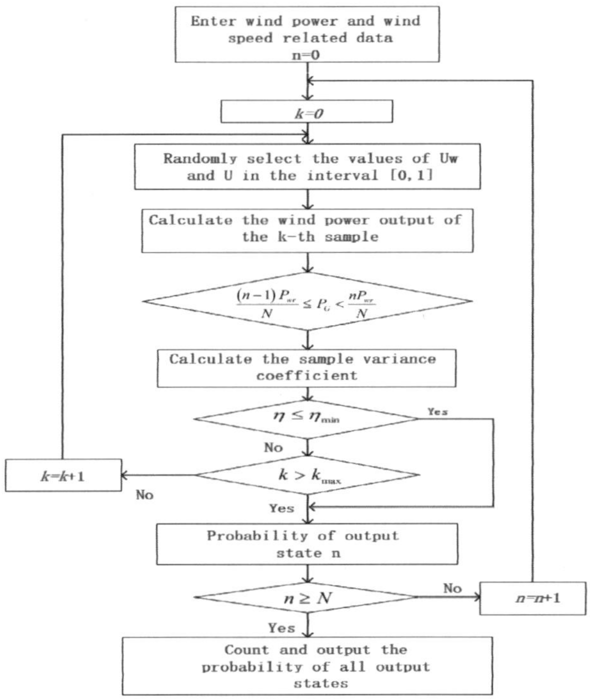

Fig. 3 Monte Carlo simulation algorithm flow

5. Simulation analysis

Take a wind farm in Inner Mongolia as an example, the wind farm scale is 30 3MW double-fed wind turbines, the cut-in wind speed of the wind power is 3m/s, the cut-out wind speed is 25m/s, the rated wind speed is 11.4m/s, and the wind power outage rate is 0.052 The derating rate is 0.25 and the derating factor is 0.6. It is considered that the arrangement of all wind power plants is a regular arrangement without obstruction, and the wind speed of the wind farm is fitted using the Wilber distribution [6] , the scale parameter c=5.218, the shape parameter k=1.637. The annual wind speed frequency distribution of this wind farm is shown in the figure below:

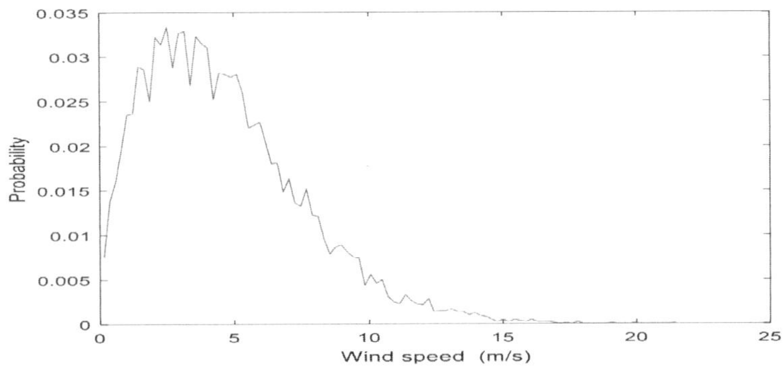

Fig. 4 Wind speed frequency distribution

According to MATLAB statistics of wind power output, the probability density of wind farm output can be obtained as shown in the figure below:

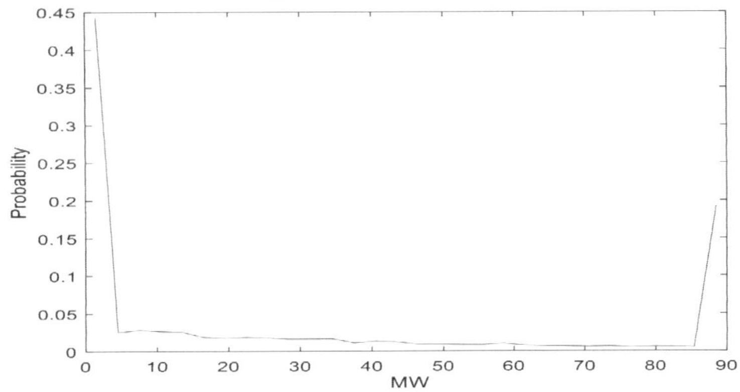

Fig. 5 Annual output power probability density distribution of wind farms

It can be seen from the above figure that the probability density of wind power output is lower in the power range of 5MW~85MW, and the probability density is higher in the power range of 0MW~5MW. After adding energy storage to the wind farm, the output probability density curve of the wind farm can be obtained as shown in the figure below:

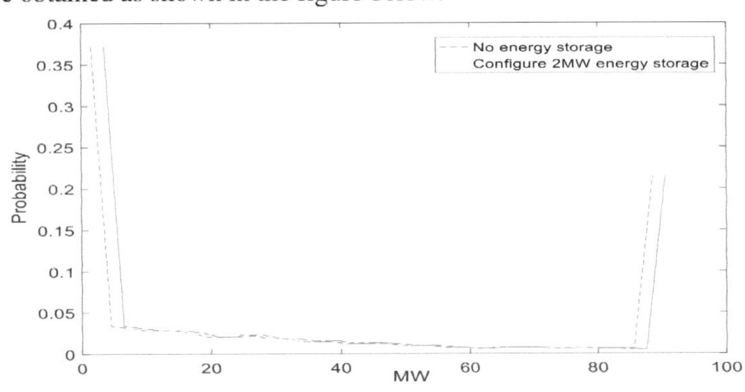

Fig. 6 Comparison of the output probability density of a wind farm with 2MW energy storage and that of a wind farm without energy storage

After the wind farm is equipped with 2MW energy storage, its probability density curve shifts to the right, and the trend of its probability density does not change. When the capacity or wind speed environment of the wind farm is fixed, the probability density curve of wind farm output can be moved to the required power range by increasing the energy storage capacity.

6. Conclusion

According to the above analysis, as the power capacity of energy storage increases, the output probability density curve of wind storage shifts to the right. The increase in energy storage capacity increases the basic output power of the wind farm at low wind speeds, and does not affect the output of the wind farm. The trend of the probability density curve. In order to make the output probability density curve of the wind storage power station reach the required power capacity range, increasing energy storage is a more direct method. Due to the high selling price and maintenance costs of energy storage units, investors in actual projects hope that the energy storage capacity is as economical as possible. By gradually increasing the energy storage capacity, the output curve reaches the power output range we need.

References

[1]Deng Shanshan. Probabilistic evaluation of wind farm output [D]; North China Electric Power University, 2013.

[2]Wang Xinwei, Zhang Jianhua, Jiang Cheng, et al. Wind farm active output reliability evaluation based on sequential Monte Carlo method [J]. Acta Solar Energy, 2014, 35(01): 106-12.

[3] Jiang Cheng, Liu Wenxia, Zhang Jianhua, et al. Risk assessment of power generation system for wind farms containing battery energy storage [J]. Acta Energia Sinica, 2014, 35(02): 207-13.

[4] Yu Haifeng, Pan Liqiang, Wu Yaru, et al. Energy storage enhances the reliability of high-proportion wind power power systems [J]. Power Grid and Clean Energy, 2020, 36(06): 92-8.

[5] Huang Tao, Xu Zhipeng, Ge Le. Optimal dispatch of hydrogen storage integrated energy system considering the random characteristics of wind power output [J]. Electrical Appliances and Energy Efficiency Management Technology, 2020, 05): 72-7.

[6] BOWDEN G J, BARKER P R, SHESTOPAL V O, et al. The Weibull distribution function and wind power statistics [J]. 1983, 7(2): 85-98.

Journal of Physics: Conference Series **2195** (2022) 012044 doi:10.1088/1742-6596/2195/1/012044

Research about the on-line system for monitoring the geological subsidence of transmission tower based on high-precision positioning of Beidou

Zhimei Cui[1,a*], Zhidu Huang[2,b], Siyuan Tan[3,c]

[1,2,3]Electric Power Research Institute of Guangxi Power Grid Co., Ltd. Nanning, China

[a]1651565356@qq.com

[b]huangzhidu@126.com

[c]453406524@qq.com

Abstract—This article proposes an innovative method to achieve Beidou high-precision positioning. Then, the online monitoring physical information system of transmission pole tower is constructed with Beidou high precision positioning and inclination measurement information as the data source. The perception layer is Beidou high-precision positioning sensor and inclination measurement sensor, the data transmission layer is the way of combining wireless private network with wired private network, and the application control layer is the data processing center, which presents the results of analyzing sensor data to the user in a visual interface. The system can send an early warning message to the user at the very early stage of the deformation of the transmission pole tower, So that the user can take measures in time.

1. INTRODUCTION

Overhead line is the main carrier of power transmission, and transmission tower is the structural foundation supporting overhead line. The factors affecting the structural stability of transmission tower include: geological disasters, hail, ice disaster and other meteorological factors, uneven operation, construction near the foundation and other human factors, bar tower design is unreasonable, component corrosion, bolt loosening and other pole tower itself, as well as external force damage and other unexpected circumstances, etc. During the life cycle of the transmission pole tower, the power pole tower will fluctuate and adjust its attitude under the action of external factors. When a certain limit is exceeded, there may be deformation, tilting or even dumping accidents, seriously affecting the stability and safety of power supply[1~3].

The monitoring of millimeter-level variables can be realized by using the ground base station to correct the positioning data of Beidou[4], and the attitude adjustment of the transmission pole tower can be accurately monitored by the use of Beidou positioning technology. Literature[5] puts forward a concept of using Beidou system positioning to realize on-line monitoring of transmission ower, but the basic positioning service used by Beidou can only reach 10m positioning accuracy. The monitoring of the millimeter-shaped variable of the tower is realized by using a single reference station, which requires the installation of a base station near the tower, which improves the cost of the system application while limiting the application conditions[6]. There has also been a lot of research on the installation of inclination sensors on transmission pole towers to monitor the attitude adjustment of pole towers[7-9]. The slide and settling of the tower can be obtained by locating the data in Beidou and the tilt and offset of the tower can be obtained by the inclination measurement data. The comprehensive positioning and inclination measurement information of Beidou can make a more comprehensive and comprehensive assessment of the attitude adjustment of the tower[4, 7~9].

In this paper, a method for high-precision positioning is proposed by using Beidou satellite navigation data, then the design and implementation method of the information physics system of the transmission tower online monitoring is explained, and the effectiveness of monitoring is analyzed by the application example of the system.

2. HIGH PRECISION POSITIONING TECHNOLOGY OF BEIDOU

Beidou Satellite Navigation System obtains the location information of the anchor point by calculating the signal transmission time between the anchor point and the navigation satellite to find the distance between the anchor point and the navigation satellite. However, the sub-meter positioning service provided by Beidou positioning system is difficult to meet the needs of monitoring the change of the attitude of the pole tower. This paper corrects the error of satellite positioning by adding a base station on the ground[10].

The s is the point of tower to be located, its spatial coordinates are assumed to be (x, y, z). The i is the number of navigation satellites(i=1,2,3,4). The locations of navigation satellites are assumed that their spatial coordinates are (x_i, y_i, z_i). The following relationships exist:

$$\begin{cases}(x-x_1)^2+(y-y_1)^2+(z-z_1)^2=(t_{c1}-(t_s-t_1)v_c)^2\\(x-x_2)^2+(y-y_2)^2+(z-z_2)^2=(t_{c2}-(t_s-t_2)v_c)^2\\(x-x_3)^2+(y-y_3)^2+(z-z_3)^2=(t_{c3}-(t_s-t_3)v_c)^2\\(x-x_4)^2+(y-y_4)^2+(z-z_4)^2=(t_{c4}-(t_s-t_4)v_c)^2\end{cases} \quad (1)$$

where, t_{ci} is the time of electromagnetic wave propagation between the locations of navigation satellite i and the positioning point of tower to be measured s; t_i is the time difference between the navigation satellite and the headquarters clock; t_s is the time difference between the positioning point to be measured and the clock of the headquarters; v_c is the speed of light. t_{ci} is computed from the relationship between i and s. (x_i, y_i, z_i), t_{ci}, t_i, v_c are known quantities. (x, y, z) and t_{ci} are unknown. The equations can be used to find the spatial coordinates (x,y,z) of s.

Because of the influence of ionosphere, troposphere on electromagnetic wave reflection and electromagnetic wave transmission channel, there are errors in t_{ci}.

The data in the ground base station data equation (1) is pushed back. For the ground base station, t_s are known. Time of electromagnetic wave propagation between navigation satellite i and base station t_{ci} is unknown. Four known quantities in the original equation group are converted to unknown quantities and four unknown quantities are converted to known quantities. The joint cube system finds the time of electromagnetic wave propagation ${t_{ci}}^1$ between the navigation satellite i and the base station. The time of transmission of electromagnetic waves between the navigation satellite and the base station is obtained by communication between the base station and the navigation satellite.

The error formula is as follows

$$\Delta t_{ci} = {t_{ci}}^1 - t_{ci} \quad (2)$$

The error t_{ci} of each anchor point is provided by four navigation satellites. Each anchor point is corrected by three nearby base stations using the same positioning satellite data。

Based on the geographic location information obtained by the location point to communicate directly with the navigation satellite, the system background matches the nearest three base stations using the same navigation satellite based on geographical location (Figure 1).

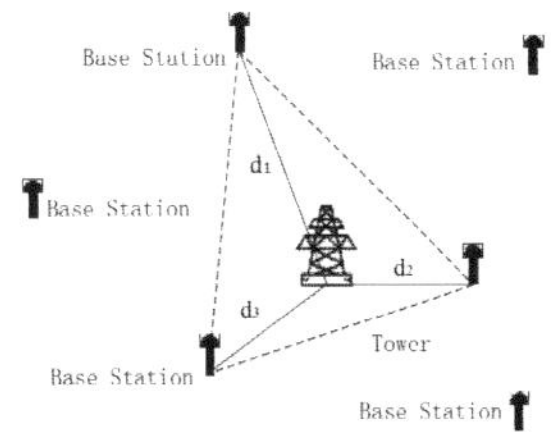

Fig 1 Schematic diagram of matching base station

According to the location information, the distance between the location point and the three base stations is d_1, d_2, d_3。(3) is the formula of the error correction value $\Delta t_{ci}{}^{0}$ which is the point to be located:

$$\Delta t_{ci}{}^{0} = \frac{d_2 d_3 \Delta t_{ci}{}^{1} + d_1 d_3 \Delta t_{ci}{}^{2} + d_1 d_2 \Delta t_{ci}{}^{3}}{d_1 d_2 + d_1 d_3 + d_2 d_3} \quad (3)$$

The time error of electromagnetic wave propagation with navigation satellite i obtained by 3 base stations are $\Delta t_{ci}{}^{1}$, $\Delta t_{ci}{}^{2}$, $\Delta t_{ci}{}^{3}$。

Furthermore, the corrected positioning point and the propagation time of electromagnetic wave of the navigation satellite can be shown in (4):

$$t_{ci}{}^{2} = t_{ci} + \Delta t_{ci} \quad (4)$$

By solving the position of the point to be measured again, the Beidou high precision positioning information of the point to be measured can be obtained. The displacement data of the point to be located can be obtained by comparing the difference of the positioning information before and after the point to be located.

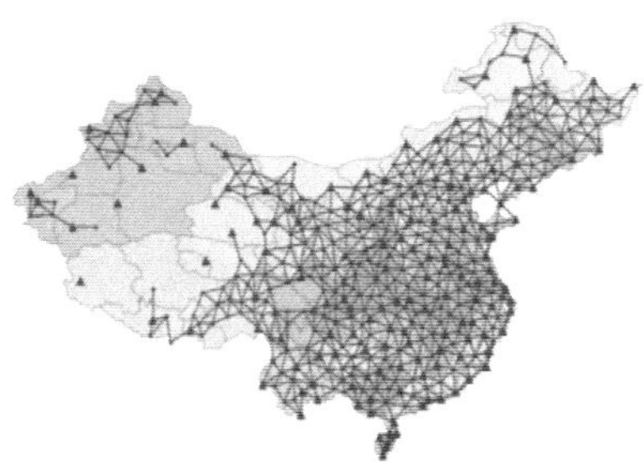

Figure 2 location distribution of base stations

The Beidou Auxiliary Positioning Base Station System relied on in this paper was officially put into operation on May 18, 2016. So far, there are more than 2700 working base stations in China. Up to now, cm-level positioning service can cover 20 provinces and cities in China. Specific to the areas of China Southern Power Grid, has achieved the millimeter service coverage.

Most of the transmission towers are located in remote areas with inconvenient transportation. It consumes a lot of labor costs while affecting the real-time monitoring. It is particularly necessary to build an information physical system covering on-line monitoring of transmission towers.[11]

The sensing layer consists of sensors installed on the transmission tower. The specific applications include the transmission tilt sensor of Beidou short message and the high-precision positioning sensor of Beidou.

The data transmission layer is a combination of wireless private network and wired private network. The transmission tower is connected to GPRS network by wireless communication module and the wireless network is connected by VPN agent. In the application, the mobile 4G private network is adopted to upload the navigation satellite communication data once in 10 seconds, and the required bandwidth is 5Mbps.The base station is accessed by wired private network, using carrier private line, and the required bandwidth is 10Mbps.

The application control layer is the data processing center. The functions of the data processing center include: obtaining Beidou high-precision positioning information according to Beidou sensor location information and base station information, comprehensive analysis of the information provided by the sensor, providing early warning information to users according to changing trends, change amount, mutation amount, etc., and presenting transmission tower history/real-time monitoring information and alarm information to the user in a visual interface on the client side.

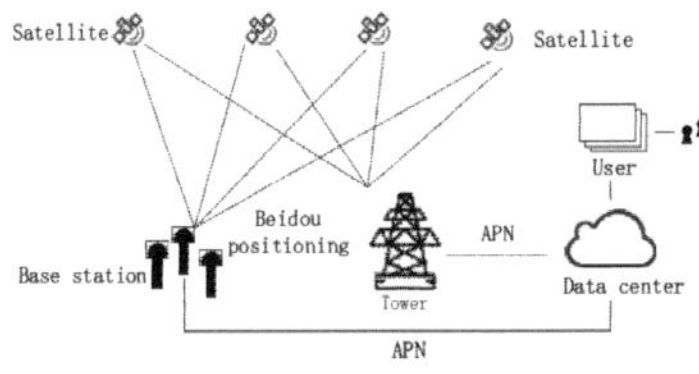

Figure 3 system structure diagram

3. SYSTEM APPLICATION EXAMPLES

A 110kV line tower in Guangdong is located on the hillside of hilly area. In late April 2019, because of the planning of the municipal department, there is road construction at 50m northeast of the tower, which requires tunnel excavation and blasting.

Figure 4 geographical location of tower

The transmission tower is located in the hilly area and the precipitation is large. The construction operation may cause the tower to slip and tilt. The transmission line operation and maintenance unit takes the tower as the key control object and provides it with an online monitoring system. Each tower body and slope has a Beidou high-precision positioning sensor for monitoring the displacement and settlement trend of the tower body and slope. A tilt sensor for the transmission of two-axis Beidou short messages is installed in the pole tower to monitor the attitude change of the pole tower.

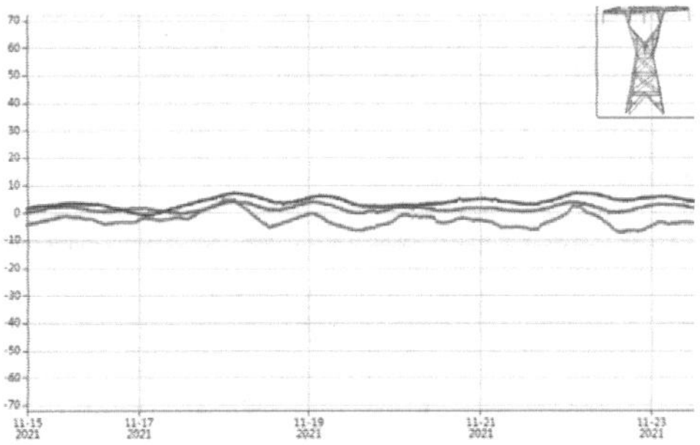

Figure 5 Tilt sensor data transmitted by Beidou short message(mm)

The variation of positioning data of transmission tower body and slope is less than 5mm, and the tower body and infrastructure are stable. According to the data obtained by the tilt sensor, the tower body is relatively stable along the line, the inclination angle of the tower body changes periodically perpendicular to the line, the inclination rate is less than 0.006, and the change of tower attitude is within the allowable range. The data obtained by the system is basically consistent with the data measured by line operation and maintenance.

4. CONCLUSIONS

This paper provides a construction idea of transmission tower on-line monitoring information physical system integrating Beidou high-precision positioning. The on-line monitoring function provides a guarantee for the operation of transmission equipment in controllable and controlled state.

Next step, we will use the established information physics system platform as the basis, according to the actual needs of the rod tower to add micrometeorine, video monitoring, wire arc prolapse measurement, wire pull measurement and other sensor modules, further improve the system perception function, to improve the operation of transmission system and the lean level of maintenance.

Reference

[1] Gan Yan, Zhou Wenfeng, Du Zhiye, et al. Research on tower strain real-time monitoring and failure early warning technology under ground settlement condition [J]. Electrical measurement and instrumentation, 2019 (9): 1-8

[2] Xu Bo, Yang Jingsheng, Bai Qiang, et al. Study on mechanical properties and design scheme of iron tower under foundation displacement [J]. Electric Power Survey and design, 2018 (1): 54-58

[3] Li Qingfeng, Fan Zheng, Wu Qiong, et al. National transmission line icing investigation and accident analysis [J]. Power grid technology, 2008,32 (9): 33-36

[4] Xiao Yugang, Jiang Weiping, Chen Hua, et al. Millimeter accuracy deformation monitoring algorithm and implementation of Beidou satellite navigation system [J]. Journal of Surveying and mapping, 2016, 45 (1): 16-21

[5] Zhu Bing, ye Shuiyong, Shao Mingsheng, et al. Research on tower online monitoring system based on Beidou system [J]. Shaanxi electric power, 2016,44 (4): 51-53

[6] Feng Zhiqiang, Hu Danhui, Yao Yao, et al. Monitoring transmission line tower foundation displacement by Beidou precise positioning technology [J]. Hubei electric power, 2017, 41 (11): 15-19

[7] Wang Jun, Yan Shiping, Gao Wei, et al. Application of tower inclination on-line monitoring system [J]. Shaanxi electric power, 2010, 38 (11): 91-93

[8] Li Huwei, Li shenglai, Li Hui. Design and research of power tower tilt early warning device [J]. Guangdong electric power, 2011,24 (2): 56-59

[9] Wu yuebin. Research and application of transmission tower inclination monitoring system based on high-precision sensing technology [J]. Automation and instrumentation, 2018 (9): 46-48

[10] Yang Yuanxi, Li Jinlong, Wang Aibing, et al. Preliminary evaluation of basic navigation and positioning performance of Beidou regional satellite navigation system [J]. Chinese Science: geoscience, 2014 (1): 72-81

Optical FBG-T Based Fault Detection Technique for EV Induction Machines

Wenping Cao[1], Belema P. Alalibo[2], Bing Ji[3], Xiangping Chen[4] and Cungang Hu[1]

[1]School of Electrical Engineering and Automation, Anhui University, Hefei, China, 230601

[2] College of Engineering & Physical Sciences, Aston University, Birmingham, UK, B4 7ET

[3]Department of Engineering, University of Leicester, Leicester, UK, LE1 7RH

[4]College of Engineering, Guizhou University, Guiyang, China, 550025

Abstract—Electric vehicles (EV) represent a key technology to achieve a low-carbon transportation objective, whist induction motors are one of the promising topologies. The reliability of these machines is crucial to minimize the downtime, cost and unwanted human lives. Although several techniques are utilized in the condition monitoring and fault detection of electrical machines, there is still no single technique that provides an all-round solution to fault detection in these machines and thus hybrid techniques are used widely. This paper presents a novel non-invasive optical fiber technique in condition monitoring of induction machines and in the process detecting inter-turn short circuit faults. Owing to optical fiber's immunity to magnetic flux, a composite FBG-T sensor formed by bonding a giant magnetostrictive transducer, Terfenol-D, onto a fiber Bragg grating is utilized to sense machines' stray flux as a signature to determine the internal winding condition of the machines. A tri-axial auto datalogging flux meter was used to obtain the stray magnetic flux and test results obtained via LabView were analyzed in MatLab. Experimental and numerical results agree with each other and how that the FBG-T sensor accurately and reliably detected the short-circuit faults. Bragg shifts observed under short-circuit faults were in 100s of picometre range under various operating frequencies compared to the mid-10s of picometre obtained under healthy machine condition. These provide much promise for future EVs.

1. Introduction

MANKIND must adapt to a more sustainable way of living to avoid the catastrophic impacts of climate change; the electric vehicles (EVs) provide a solution to reduce carbon emissions related to road transportation. Of all the industrial motors in use globally, induction motors are the most common type [1]. Transitioning to full mode electric transport – be it on land, air or sea; more renewable power generation and even more sustainable industrial practices would mean that electric machines are expected to undoubtedly play an important role. Thus, the reliability and continuous availability of these electric machines is pivotal to industry. The reality however is that electric machines are susceptible to unexpected breakdown due to the often harsh and highly stressed environments [2] under which they are operated in order to provide the goods and services humans need. Many electric machines run 24/7 all year round across various industrial plants with minimal downtime for scheduled or routine maintenance. Unexpected machine failures and consequent downtimes do have serious operational and financial adverse impacts including repair or replacement costs, stopping production processes with consequent economic losses, likely impact on the company's reputation, and customer dissatisfaction. Real time online condition monitoring of these machines is crucial now more than ever, as reliability of these machines is a sine qua non to boosting end user confidence in transiting to the dream world of net zero carbon emissions. Various condition monitoring and fault detection techniques are currently available each having its own limitations [3]; however, there exists no single signature or technique that is capable of diagnosing all possible machine failures [4]. Best industrial practice optimally utilizes a multi-monitoring approach where two or more of available techniques are simultaneously deployed for comprehensive real time condition monitoring and fault detection in electric machines. Given the crucial role of these machines both in the present and in future, and the vast and rapidly increasing number of usage of these machines globally, there will never be a superfluous number of condition monitoring and fault detection techniques. Therefore, increasing the pool of available machine condition monitoring options is rather an urgent and welcome development.

Optical fiber sensors continue to find applications in various fields owing to their high reliability, low cost and zero electromagnetic interference in addition to other numerous advantages. Fiber Bragg gratings (FBG) appear to be the most commonly used optical fiber sensor in recent times with most recent applications in various fields. The use of FBG sensors for electric machine condition monitoring have been invasive hitherto; with FBG sensors usually mostly installed within stator slots or circumferentially around stator windings. Optical fibers are generally very brittle and highly susceptible to breakage. With invasive techniques, sensor replacement would require downtimes and temporarily disconnecting the machines which would have economic and social consequences depending on the service being disrupted. Hitherto, there is no known FBG sensor that measures magnetic field in a non-intrusive manner utilizing external stray flux to determine an electric machine condition as well as detect faults. Although the external stray flux is typically attenuated by the machine cast frame, its small magnitude can be adequately detected by optics given that the order of magnitude of the broadband laser used for such sensors, are typically in the nanometer (nm) range. FBGs on its own are immune to electromagnetic field hence a transducer is needed to convert the stray magnetic field into strain before it can be optically detected. Terfenol-D, a giant magnetostrictive alloy is used to achieve this purpose where the FBG is bonded onto the alloy using cyanoacrylate adhesive which transfers the strain movement from the alloy to the FBG and then subsequently observed as a wavelength shift known as Bragg shift on an optical spectrum analyzer (OSA).

2. Principle of FBG Sensing

Light refraction occurs because air has a different refractive index from the medium through which the light now travels. In other words, due to the difference in atomic particulate composition of different materials, the speed at which light travels through different mediums is not the same resulting in the bending phenomenon. Although the frequency of the light wave does not change, its wavelength changes [5] and the relationships between its speed in air (c), its speed in a different medium (v), the refractive index of the medium (n), its wavelength (λ) and its frequency (f) are given by:

$$c = f\lambda \tag{1}$$

$$n = \frac{c}{v} \tag{2}$$

According to Snell's law of refraction, the ratio of the refractive indices of the materials (n_1 and n_2) at the boundary surface equals the ratio of the sine of the angles of incidence (θ_1) and transmission (θ_2). He also relates the refractive indices to the wavelengths in each medium λ_1 and λ_2 as follows:

$$\frac{n_1}{n_2} = \frac{\lambda_2}{\lambda_1} = \frac{sin\theta_2}{sin\theta_1} \tag{3}$$

Light reflection follows that the angle of incidence must equal the angle of refraction i.e. $\theta_1 = \theta_2$. The reflection of light at the interface (boundary) between two different medium is also known as Fresnel's reflection effect. Investigations in physics do show that the refracted angle tends to get smaller than the incident angle when light is travelling from a fast medium to a slow one (for example, air to water). Conversely, for light travel from water to air, for example, the incident angle tends to get larger but it cannot exceed 90° while remaining in the new medium. Hence this incident angle constraint suggests that the largest incident angle for refraction occurs when $n_2 < n_1$. However, beyond a certain incident angle known as critical angle, θ_c, there will be no refraction at all and the light wave will be totally reflected away from the new medium through which the light was intended to travel provided $n_2 < n_1$ still exists. This is called total internal reflection which is the phenomenon used in optical fibers. In a typical optical fiber silica cable, its core with a higher refractive index, n_1 is surrounded by a cladding whose refractive index, n_2 is only about 1% lower than that of the core in order to satisfy the condition for total internal reflection. For example, the core could have $n_1 \approx 1.4475$ and the cladding $n_2 = 1.444$ and provided the fiber cable is not excessively bent; the light will just bounce around in the cable with minimal loss [5].

Uniform FBG is the standard FBG in which the gratings have equal and evenly spaced distance called the period between them. They are fabricated usually using a phase mask technique [6] to expose a short length of photosensitive fiber to a periodic distribution of light such that the fiber's refractive index is permanently varied in accordance with the exposure of the light intensity as shown in Fig. 1.

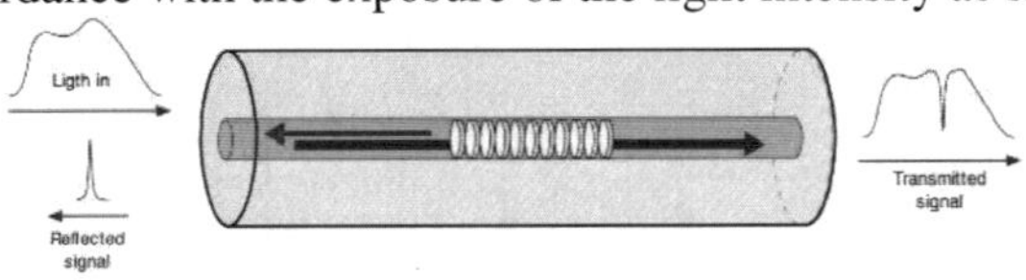

Fig. 1: FBG - Fibre cladding with evenly spaced gratings within its core [7]

When the light is guided along the fiber core comes in contact with the inscription of the gratings, it is weakly reflected by each grating plane by Fresnel effect [8]; and the reflected light from each grating plane then recombines with other reflections in the backward direction in either a constructive or destructive interference pattern depending on if the wavelength of the incoming light satisfies Bragg's law given by:

$$2dsin\theta = n\lambda \tag{4}$$

With the light's incident angle $\theta = 90^o$ and d, the distance between peaks of the interference pattern, then for vacuum with refractive index, n=1; λ=2d is the approximate wavelength of the reflection peak which means the fiber reflects part of the incoming spectrum [8]. For silica material used to make optical fiber, the distance travelled by light, d is affected by the material's refractive index, n, thus equation (4) can be adapted for silica as:

$$2n_{eff}\Lambda = \lambda_B \tag{5}$$

where λ_B is the Bragg wavelength, n_{eff} is the effective refractive index of the fiber and Λ is the periodicity of the grating. It is possible to change the Bragg wavelength also known as 'Bragg Shift' by either varying the fiber's effective refractive index or the grating periodicity or both. This is the fundamental principle of the FBG sensing. By designing the proper interface, desired measurement can be made to impose some external force or disturbance on the grating (such as temperature, vibration, pressure, and displacement) which will result in a Bragg shift proportional to the external force applied, as shown in Fig. 2.

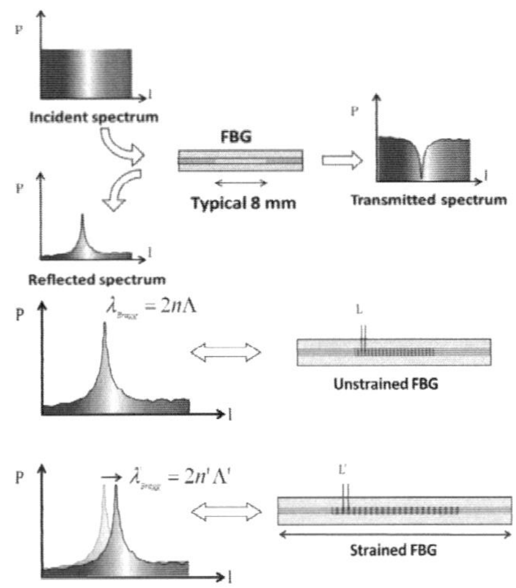

Fig. 2: a) Light spectrum – incident, transmitted and reflected; b): Bragg shift concept [9]

3. Experimental Method

Since optical fiber is immune to magnetic flux, if FBG were to be used for sensing magnetic field, it will require a transducer to convert the magnetic flux into one of the strains to be able to cause Bragg shifts. Terfenol-D is used in this research to achieve the transduction of magnetic flux into strain which is then impressed on the FBG via an adhesive acting as a transfer medium. Prior to carrying out experiments, the effects of the number of bonding points on metal-surface mount FBG sensors was investigated and reported in [10] which takes account of the impact of the adhesive used.

The complete set up used for this work is as shown in Fig. 3. The FBG-T test rig set up for both healthy and short circuit inter-turn fault conditions. Each of the test induction motors is used to drive a DC motor with the FBG-T sensor transversely positioned relative to the rotor shaft at the drive end (DE) of the induction motor. Stray magnetic flux data was acquired using an auto datalogger gaussmeter - HHG1394 - manufactured by Omega Engineering UK. The HHG1394 comes with proprietary software and a tri-axial sensing functionality which makes if suitable for low and extremely low frequency magnetic fields independent of measurement angle. The FBG-T sensor is then connected to a dual-functional broadband light source and optical spectrum analyzer (OSA) which is serially connected to a PC via National Instrument GPIB cable. This is to allow the use of LabView software to collect and store numerical spectral data that can be retrieved and analyzed in MatLab.

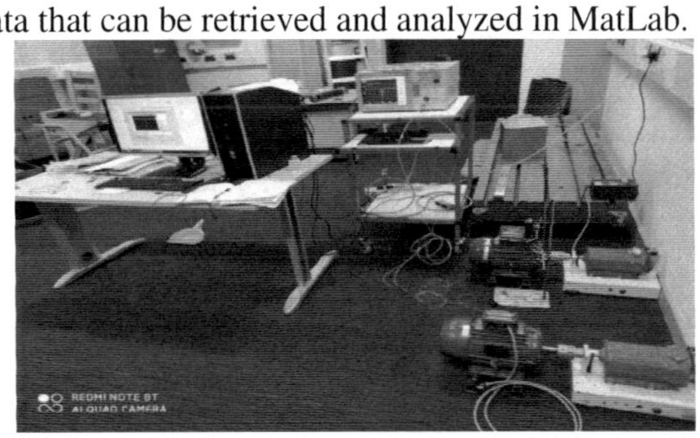

Fig. 3: Complete experimental test rig for FBG-T stray flux sensing

The motor was driven by a variable frequency drive (VFD). Each machine was run at similar frequency range between 5 – 30Hz with 5Hz interval initially for two hours without the auto data logger and then repeated for four hours with the auto data logger, after which the motor is then turned off each time. Prior to running the healthy motor, it was disassembled and observed to confirm its healthy state before reassembly. For short circuit inter-turn conditions, strands of copper windings were used to bridge turns within the motor as shown in Fig. 4.

Fig.4: Healthy and Inter-turn fault conditions prior to assembly

4. Results and Discussion

In order to carry out any experiments, it was imperative to determine the optimal location where the sensor should be positioned as well as to calibrate the sensor to determine its sensitivity.

4.1 Optimal FBG-T Location

Two major factors were taken into consideration in positioning the FBG-T: proximity to the main air gap flux and sensor surface area exposed to the external flux. The closest part of the machine to the main air gap flux is the flange coupling at the drive end (DE) of the machine. Hence, this was one of the positions considered.

Three possible axes to position the sensor at the DE are: transverse (perpendicular to rotor shaft), inclined (intermediate between parallel and perpendicular to rotor shaft) or axial or longitudinal (parallel to rotor shaft). The second option was very quickly dis-carded because magnetic fields are well known to be either at their maximum field strength or zero depending on whether the flux lines are cutting across a surface either perpendicularly or in parallel to the surface respectively. Transverse and longitudinal positions were later empirically tested to identify the optimal sensor position.

The variation of observable stray flux with frequency for both transverse and axial sensor positions showed that more flux will cut across the sensor when placed in the transverse position at the drive end of the machine compared to the axial position (Fig. 5). Although this has been attributed to reduced surface area of the flux-attenuating and corrugated yoke, one can argue that the transverse position is closer to the source of the magnetic flux which is in the air gap within the machine. This is because the air gap is between the rotor and the stator whose windings are wrapped internally close to the flanges of the machine. Based on the optical spectral response only, there was no observable difference in terms of sensor position; instead there were evidence that the FBG-T sensor was indeed sensing the stray magnetic flux present. From both the behavioral characteristics and observable magnetic flux density viewpoints, the transverse position became the experimental choice and thus was utilized throughout the rest of the investigation.

4.2 FBG-T Flux Calibration

For the magnetic field source, an openly-wound 4V DC motor whose windings were exposed was used with a DC power supply unit to provide the need current for different magnetic field strengths. A tri-axial auto data logger magnetic flux meter was used to measure and record the stray magnetic flux density for each applied current. Different amount of currents was applied generate magnetic field of different strengths measured by the auto data logging flux meter.

The optical spectra showed an increase in the Bragg shift with current given that the magnetic flux as measured by the auto data logger gauss meter also increased with current as expected (Fig. 6). Time varying plots of Bragg shifts showed a steady increase in the Bragg shifts with time until the maximum Bragg shift was attained. The rate at which the Bragg shifts occurred for each current value appeared to be faster than the rate observed for earlier tests performed with alternating stray flux. This could be attributed to the impact the sinusoidal nature of the ac flux has on the magnetic dipole of the Terfenol-D particles causing them to move faster to a new equilibrium state when magnetic flux is present. Fig. 6 confirms the sensitivity of the FBG-T sensor which was 19.3810pm/μT with room temperature reasonably constant with a recorded maximum change of about 0.13°C.

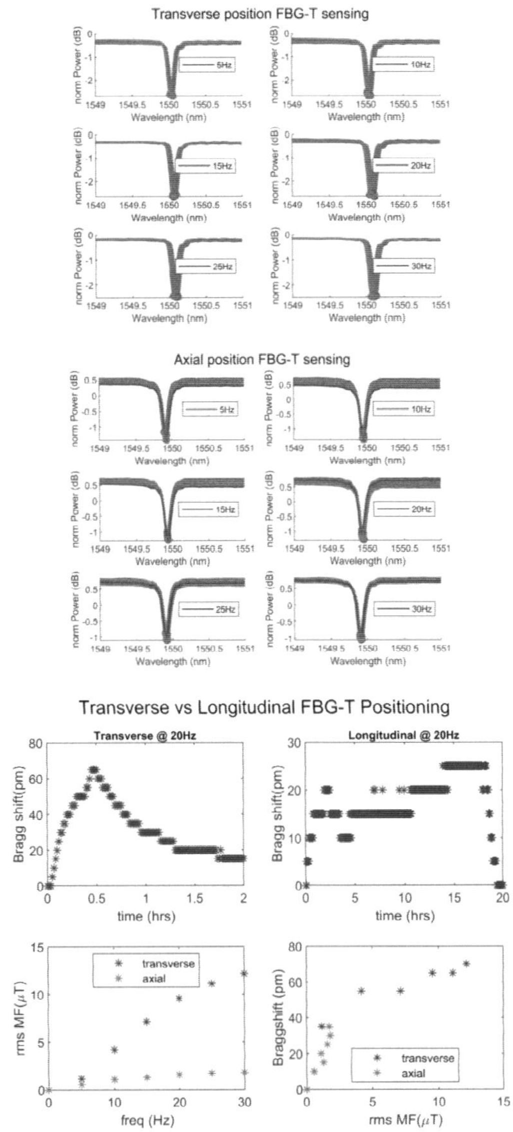

Fig. 5: Comparison between FBG-T sensed parameters when in transverse and axial sensor positions

4.3 FBG-T Short Circuit Inter-Turn Fault Detection

Observations confirmed significant temperature rise in the case of the short circuit inter-turn fault and a dramatic increase in Bragg shift accompanied by a set of rather near-normal magnetic flux readings for frequency between 5Hz and 15Hz. By 20Hz it was observed that both the temperature and magnetic flux had reduced unexpectedly developing some curiosity as to why. The unexpected reduction in temperature and magnetic field strength necessitated the need to confirm of the copper conductor strands were still in place, and the inter-turn fault condition had not been compromised. After the motor was disassembled it was observed that part of the stator winding had burnt out which was the reason for the reduced temperature and magnetic flux as a smaller number of conductors were carrying current.

Therefore, the investigation was halted at this stage as the FBG-T sensor has proved to be very reliable in detecting such serious fault condition non-intrusively using stray flux. A repeat test was later performed with a smaller number of strands to attempt to avoid burning the stator windings on an identical machine, and also substantiate this crucial observation.

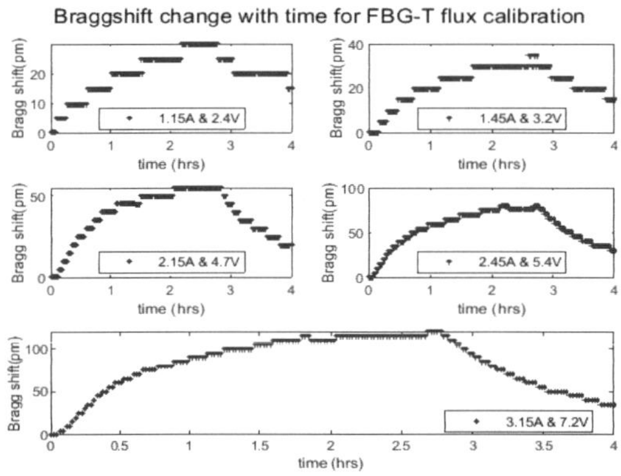

b) Braggshift variation with time at various currents

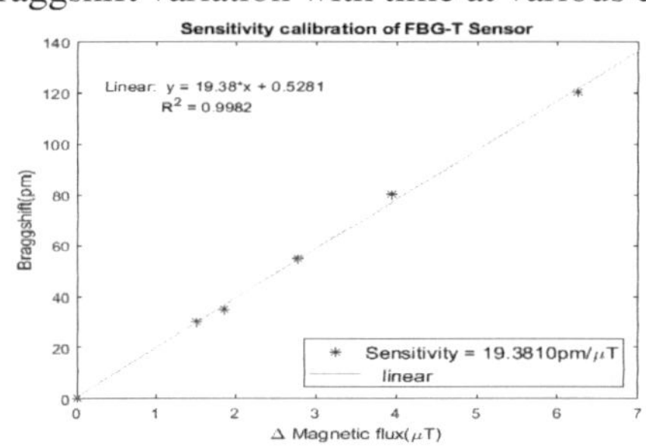

d) Sensitivity of the FBG-T sensor

Fig. 6: Magnetic flux FBG-T sensor calibration results

In order to detect the aberration from the healthy motor condition, data from both conditions were analyzed in Table 1. Bragg shifts observed with increase in frequency and in turn magnetic flux clearly showed a stark path between both motor conditions due to the severity of the inter-turn fault.

Whilst Bragg shifts were in the tens of picometer (pm) for the healthy condition, they were in the hundreds of picometer for the short circuit condition (Fig. 7) which is a succinct feature fundamental to the use of FBG-T sensor in fault detection.

TABLE 1: Comparison Of Braggshifts Changes With Temperature At Different Frequencies

Frequency (Hz)	**Healthy**	**Inter-turn**	**Healthy**	**Inter-turn**
	Braggshifts (pm)		**ΔTemp. (°C)**	
5	30	290	0.7615	3.3538
10	50	530	0.5600	4.2700
15	45	645	1.4692	5.5692

The disparity between the intensity of the stray magnetic flux was not large as only a few turns of extra copper conductor strands were added to emulate the fault condition. Fig. 8 shows a slight increase in the magnetic flux for the faulty condition compared to the healthy state due to a slightly higher current producing the magnetic flux. However, the temperature difference was very significant (Fig. 8)

Journal of Physics: Conference Series **2195** (2022) 012045 doi:10.1088/1742-6596/2195/1/012045

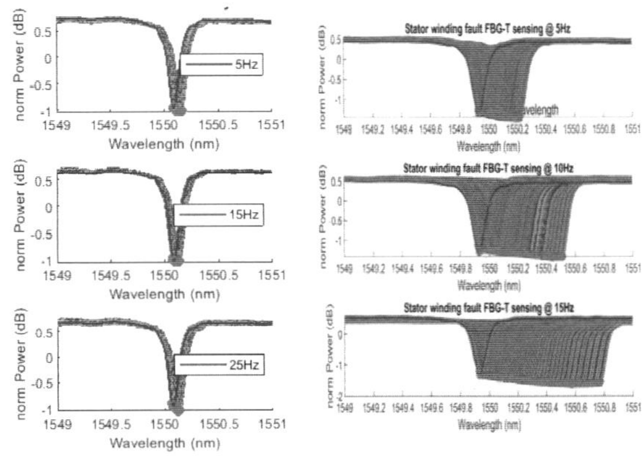

Fig. 7: Optical spectral response under healthy and inter-turn fault conditions

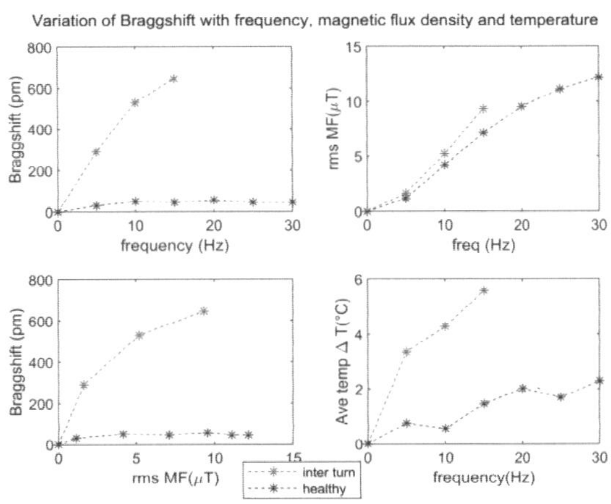

Fig. 8: Variation of frequency with braggshifts, external magnetic flux density and temperature under more short circuit faults

The noticeable large wavelength shifts are largely due to the over-temperature caused by the increased I^2R losses as a result of the inter-turn short circuit fault. These large Bragg shifts required several hours to return to their initial wavelength as shown in Fig. 9, clearly justifying the presence of a magnetic field and its effect on the Bragg shifts. It thus confirms that the Bragg shifts are due to both the magnetic flux and the over-temperature caused by the inter-turn short circuit condition as the result was consistent at all three frequencies. Another key observation is that regardless of machine condition, the FBG-T sensor exhibited reversible magnetization as shown in Fig. 9. This is because of the small amplitude of the stray flux which is insufficient is causing a permanent change of state of the magnetic dipoles of the Terfenol-D alloy. Thus, whenever the magnetic flux is removed, the magnetic dipoles lose their magnetization and attempt to return to their initial state of equilibrium.

5. REPRODUCIBILITY TEST: FBG-T INTER-TURN FAULT DETECTION WITH AUTO DATA LOGGER

Earlier a similar test was performed but with more copper conductor strands which resulted in burnt stator windings following the over-riding of the built-in overcurrent protection in the VFD. In order to avoid a repeat of such damage, less number of conductor strands was used to allow for testing over the entire test frequency range which was not achievable previously due to the damage to the motor. Motor was run from 5-30Hz with 5Hz interval as before but this time for the duration of four (4) hours and turned off for 16 hours per frequency to allow for sensor reset. Test results show the new induction motor with inter-turn fault emulated using less number of copper strands.

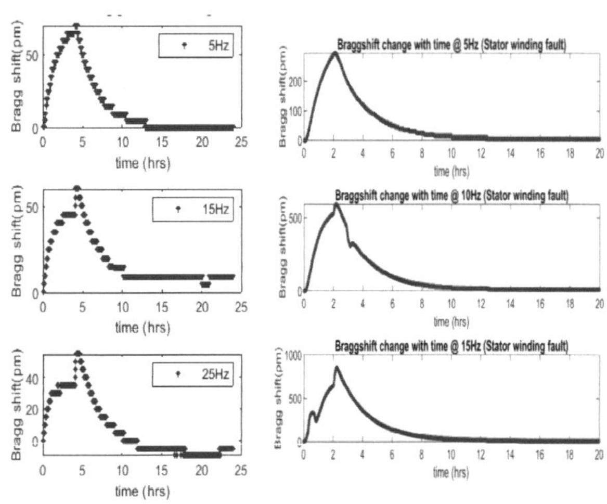

Fig. 9: Time varying braggshift response of FBG-T sensor under healthy and inter-turn fault conditions

With a longer duration of operation, there was consistency in the behavior of the FBG-T sensor with respect to Bragg shifts. The shift in wavelengths appear to rise and plateau within the four hours, and once the machine was turned off, a clear descent path back to the initial wavelength due to reversible magnetization as earlier explained. The FBG-T behavior described was observed when the motor was run at the different frequencies. Another key observation was that throughout the short circuit condition test, the FBG-T was observed to show less hysteresis when returning to its initial wavelength. This may be due to the spectral content of the stray magnetic flux as the other external factor (ambient temperature) which could affect the sensor was reasonably constant with minimal variation. It is worthy of mention that while this reduced hysteresis may be desired, it does not affect the use of the FBG-T sensor for fault detection because the underpinning principle of FBG sensing in general depends of the magnitude of Bragg shifts.

Table 2 shows that the order of magnitude of Bragg shifts observed during inter-turn fault condition are in hundreds of picometre which is clearly very high. These large Bragg shifts were succinct in the optical spectra at different frequencies where the wavelength shifts increased with frequency. In theory it would be easy to associate the large Bragg shifts to the increase in magnetic flux density or temperature or a combination of both factors. From Table 2 there is a direct but non-proportional increase in Bragg shifts with magnetic flux density as well as with maximum temperature change as frequency increased which agrees with theory. However, for such a large magnitude of Bragg shift when compared to corresponding values of magnetic flux density and temperature changes in healthy machine condition and during sensor calibration, it is difficult to convincingly suggest that only the amplitudes of these two measurands (flux and temperature) are responsible. With the aid of the auto datalogger, the magnetic flux obtained will be subjected to spectral analysis to further analyze its contents.

5.1 Healthy versus Inter-turn Conditions

Compared to the healthy condition where Bragg shifts are barely just over 50pm irrespective of the frequency at which the motor is run; under inter-turn fault condition the Bragg shifts are well over 100pm even up to over 400pm. In fact, the Bragg shifts directly responded to the change in the motor frequency. This was identical to the earlier observed when there were more number of shorted turns that burnt the stator winding initially. Though the experiment was halted after the motor damage incident, the trajectory for the both the healthy and the inter-turn conditions in the two tests were identical and clear in the accuracy of the FBG-T sensor to detect the short circuit faulty condition. The magnetic flux densities measured in both tests were of similar range as well and when compared with the frequency range showed an increase in amplitude as motor frequency increased.

Considering the magnitude of the stray flux is in micro-tesla range, the difference in flux densities between successive frequencies was less than 5μT which is really small to cause such large Bragg shifts. When compared to the healthy condition, the magnitude of the flux density of both healthy and inter-turn fault conditions where in the same range with increasing difference between corresponding values as frequency increased. For example, at 10Hz the difference in magnetic flux density between healthy and inter-turn conditions was about 0.145μT whereas at 20Hz and 30Hz the same difference was about 3.13μT and 4.16μT respectively. Same observation was made previously with a greater number of shorted turns and less operating time. With respect to the maximum temperature change observed in both reduced and higher number of turns the results were identical. As expected, due to the short circuit the maximum observed temperature change increased as frequency increased as the motor is made to do more work. Due to the difference in the number of turn in both tests, the temperature changes were slightly different. For instance, at 15Hz a maximum temperature change of 3.953°C was observed for this test with a smaller number of turns compared to 5.5692°C for the previous test with a greater number of turns. This is expected due to more I^2R losses in the previous test. When compared to the healthy machine state, identical results in both tests showed that the maximum temperature change was always less than 1.7°C across all the individual motor frequency of operation which is still lower than the lowest temperature change observed under short circuit condition for both tests. Another observation was that unlike in healthy state where there was no evidence of a direct relationship between temperature variation across operating frequencies; there was a clear and direct proportionality for the short circuit condition. As the motor frequency increased, the temperature also increased due to the additional short circuit I^2R losses which increase with loading on a motor.

6. Conclusions

A novel condition monitoring technique which utilizes FBG-T sensor to detect inter-turn faults in EV induction machines has been investigated. FBG-T sensing reliability and accuracy has been tested under healthy and short circuit machine conditions. From analytical and experimental results, it is evident that: 1) the FBG-T sensor can reliably and accurately monitor the condition of the machines; 2) the proposed method can unambiguously detect the short circuit faults with succinctly large Bragg shifts observable via the pre-analysed sensor's optical spectra; 3) Bragg shifts vary with time, frequency and temperature from the post-analysed plots and are thus used for detection and reasoning. However, the magnitude of the flux densities obtained is not enormously large to cause such significant Bragg shifts especially for the short circuit conditions. In the future work, a fast Fourier transform (FFT) method will be used in the data analytics for those acquired by the auto datalogger, and a study on harmonic contents of the stray flux from the FBG-T sensors.

References

[1] Schneider Electric, "Monitoring Induction Motors for Energy Savings (Part 2)," PowerLogic Solut., vol. 8, no. 2, 2006.

[2] P J Zhang, Y Du, T G Habetler, and B Lu, "A Survey of Condition Monitoring and Protection Methods for Medium-Voltage Induction Motors," Ieee Trans. Ind. Appl., vol. 47, no. 1, pp. 34–46, 2011.

[3] Israel Zamudio-Ramirez, Roque A.Alfredo Osornio-Rios, Jose Alfonso Antonino-Daviu, Hubert Razik, and Rene de Jesus Romero-Troncoso, "Magnetic Flux Analysis for the Condition Monitoring of Electric Machines: A Review," IEEE Trans. Ind. Informatics, 2021.

[4] Maria J. Picazo-Rodenas, Jose Antonino-Daviu, Vicente Climente-Alarcon, Rafael Royo-Pastor, and Ariel Mota-Villar, "Combination of Noninvasive Approaches for General Assessment of Induction Motors," IEEE Trans. Ind. Appl., vol. 51, no. 3, pp. 2172–2180, May 2015.

[5] Matthew Schwartz, "Refraction and Reflection," Lecture note. [Online]. Available: https://scholar.harvard.edu/files/schwartz/files/lecture15-refraction.pdf. [Accessed: 10-Oct-2021].

[6] National Instruments, "Fundamentals of Fiber Bragg Grating (FBG) Optical Sensing - National Instruments," 2016. [Online]. Available: http://www.ni.com/white-paper/11821/en/.

[Accessed: 27-Nov-2017].
[7] Carlos A. J. Gouveia, Jose M., and Pedro A.S., "Refractometric Optical Fiber Platforms for Label Free Sensing," in Current Developments in Optical Fiber Technology, InTech, 2013.
[8] Marcelo M., Regina C. S. B. Allil, Bessie A., and Fbio V. B. de Nazar, "A Guide to Fiber Bragg Grating Sensors," in Current Trends in Short- and Long-period Fiber Gratings, InTech, 2013.
[9] Hongli Li, Gang Xu, Xin Gui, Lei Liang, and Zhengying Li, "An FBG Displacement Sensor in Deformation Monitoring of Subway Floating Slab," IEEE Sens. J., 2021.
[10] Belema P. Alalibo, Wenping Cao, Adenowo Gbadebo, Lassi Aarniovuori, and Kewei Cai, "Investigation of the effect of bonding points on metal surface-mounted FBG sensors for electric machines," Prog. Electromagn. Res. C, 2019.

Effect of surface treatment materials on frost resistance of damaged concrete

Tianlun Li *, **Yonggen Wu**

Aviation Engineering School, Air Force Engineering University, Xi'an, Shaanxi, China

*Corresponding author's e-mail: litianlun2020@163.com

Abstract. In order to study the effect of silicate infiltration agent, tetraethyl orthosilicate and nano-SiO_2 surface treatment materials on the freezing resistance of damaged concrete, the peeling condition of damaged concrete after soaking in salt water was tested by single-side freeze-thaw method. The results showed that tetraethyl orthosilicate and silicon infiltration agent can effectively improve the damaged concrete frost resistance and can reduce the 7 d water quantity of concrete, the most significant is the effect of tetraethyl orthosilicate , 7 d water quantity reduced rate of 65%, after 28 times freeze-thaw cycle exfoliation volume is 679 g/m^2. Silicon infiltration agent surface treatment also has a good effect, can increase the number of freeze-thaw cycles 8 times. Nano-SiO_2 has the worst treatment effect and can not improve the number of freeze-thaw cycles of damaged concrete.

1.Introduction

In the north of China, pavement concrete is easily affected by freezing. Concrete pavement will suffer from cracks, surface shedding and other diseases under the action of freezing, resulting in pavement surface damage and reduced bearing capacity [1]. At present, in order to reduce the influence of icing on pavement surface, airport usually sprinkles deicing salt on concrete surface. Although deicing salt can eliminate snow and icing, it also accelerates surface spalling, posing new challenges to the frost resistance of pavement concrete.

At present, the main measures to improve the freezing resistance of concrete include adding different admixtures and admixtures into concrete[2]. However, for the existing concrete structure, strengthening the surface is a simple, economical and reliable method. Surface strengthening treatment can form a protective layer on the surface or block the pores of concrete, improve the compactness, prevent the invasion of water and ions in the external environment, prevent the freeze-thaw damage in the concrete, and also play a role in the resistance to sulfate, chloride ions and other erosion[3]. Therefore, it is of great significance to select suitable surface treatment materials for improving the frost resistance of airport pavement concrete.

Zhang He et al.[4] explored the influence of the composite solution of lithium silicate solution and sodium silicate on the surface performance of concrete, and the results showed that the composite sol filled the micro-pores on the surface of mortar, formed a dense microstructure, improved the surface rebound value and reduced the surface permeability. Moon et al..[5] confirmed that inorganic treatment materials with calcium silicate as the basic component can improve concrete's carbonization resistance, chloride penetration resistance and freezing damage resistance, and the effect is more obvious when multi-layer treatment is adopted.

Hou Tao[6] systematically discussed the effect of tetraethyl orthosilicate as the surface treatment material of cement concrete, and the results showed that when the volume ratio of ethyl silicate to ethanol was 1:1, the treatment effect was the best, and tetraethyl orthosilicate treatment had obvious long-term effect, which could improve the frost resistance of concrete. Barberena et al. [7] analyzed the reaction of concrete cured by tetraethyl orthosilicate using FTIR, and found that tetraethyl orthosilicate could react with hydroxytraverite to generate C-S-H gel, and could also react with the existing C-S-H gel in hydration products to generate a longer C-S-H chain than the original material chain. The surface curing treatment of tetraethyl orthosilicate reduces the porosity of cement mortar, especially the pore size larger than 1μm.

Barberena et al. [8]also added nano-SiO_2 into TEOS to treat the surface of cement mortar. Compared with the non-doped nano-SiO_2, the incorporation of nano-SiO_2 can reduce the porosity and water absorption at the same time. Scarfato et al.[9]mixed nanoparticles into epoxy resin for surface treatment of concrete. Nanofillers can improve the water permeability resistance of cured concrete by blocking the pores of concrete and reducing the diffusion performance of polymer matrix.Hou et al.[10]systematically studied the properties of cement mortar cured on the surface of nano-SiO_2 and TEOS, and the results showed that both TEOS and nano-SiO_2 can effectively reduce the water absorption, water vapor transfer coefficient and porosity of cement mortar. Comparatively, TEOS had better treatment effect.

The above studies are based on undamaged concrete, but in actual engineering, due to construction, natural erosion and other reasons will inevitably lead to different degrees of concrete damage. In this paper, silica infiltration agent, tetraethyl orthosilicate and nano-SiO_2 were selected as surface treatment materials, and the peeling condition of damaged concrete after surface treatment under salt solution immersion was tested by single-side freeze-thaw method, and the influence of three surface treatment materials on the frost resistance of damaged concrete was discussed.

2. Methods and materials

2.1 materials

Table 1 shows concrete mix ratio and basic performance. 42.5 ordinary Portland cement is used for cement, limestone is used for coarse aggregate, Ba river sand is used for fine aggregate, and ordinary tap water is used for water.

Table1 Mix proportion and properties of concrete

Cement /(kg/m^3)	Water /(kg/m^3)	Fine aggregate (kg/m^3)	Coarse aggregate /(kg/m^3)	Vebe consistometer /s	Flexural strength at 7d /MPa	Flexural strength at 28d /MPa
330	135.3	630	1328	22	5.2	5.6

Silicon infiltration agent is the product of Hunan Fenghang New Material Company,it is a mixture of lithium silicate as the main component; The tetraethyl orthosilicate adopts the products of Shandong Fengpan New Materials Company, and the nano-SiO_2 adopts the nano silicon curing agent of Shanghai Zhichuang Fine Chemical Company, as shown in Table 2-4.In this paper, B, S, N and T were used to represent concrete specimens treated with untreated, silicate infiltration agent, tetraethyl orthosilicate and nano-SiO_2, respectively.

Table 2 Performance parameters of silicate infiltration agent

Density/(g/cm^3)	pH	Solid content /%	Viscosity /Pa·s	Surface tension/(mN/m)
1.18±0.03	11.0±1.0	22±2.2	11.0±1.0	≤30.0

Table 3 Performance parameters of tetraethyl orthosilicate

Silicon content/%	Acidity/ppm	Viscosity /(CPs/25°C)	Density /(g/ml)	VOC/%
40.61	40	5.04	1.058	$<$3.0

Table 4 Performance parameters of nano-SiO_2

Particle size/nm	Solid content/%	Density /(g/cm^3)	pH	Appearance
10±1	12.5±1	1.125±0.007	11±0.5	Semitransparent

2.2 Test methods

The single-side freezing-thawing method was tested according to the "Standard for Test Methods of Long-term Performance and Durability of Ordinary Concrete (GB/T 50082-2009)". Cubic specimens of 150mm×150mm× 150mm were used for 7 days of pre-water absorption and 6 freezing-thawing cycles after curing to 28 days. At this time, the cumulative amount of spalling is about 250 g/m^2, and the concrete surface in this state is in the pockmarked stage, which is relatively common in practical engineering. The concrete specimens were taken out and dried for 1 month in the indoor environment, then the surface was cleaned with a brush and then the surface was brushed. After curing for 7 days in the indoor environment, the concrete specimens were pre-absorbent and single-side freezing-thawing cycle, with 5 specimens in each group.Concrete quality, exfoliation quality and ultrasonic propagation time were tested every 4 cycles.

The test can be stopped when one of the following conditions is reached :(1) the number of cycles reaches 28 times; (2) The total mass of exfoliation reaches 1500 g/m^2; (3) The relative elastic modulus of ultrasonic wave is reduced to 80%.

3.Results & Discussion

Figure 1 is the amount of 7d pre-water absorption of concrete specimen, and Figure 2 is the mass of exfoliation of concrete specimen after freezing-thawing cycle. During the test, it was found that the ultrasonic relative dynamic modulus of concrete after freezing-thawing cycle was far from reaching the test termination condition, so the total mass of exudates per unit surface area of the specimen was taken as the test termination index.

It can be seen from Figure 1 that the three surface treatment materials can reduce the 7d pre-water absorption of damaged concrete. Compared with B, the 7d water absorption of S, N and T specimens decreases by 31%, 9% and 65%, respectively. The treatment effect of tetraethyl orthosilicate is the best, followed by silicon infiltration agent, and the treatment effect of nano-SiO_2 is not obvious. This is because the three kinds of surface treatment materials will product and concrete hydration reaction to generate gel, gel blocking pore and micro cracks, so as to reduce concrete water imbibition, silicon infiltration agent in concrete to form a layer of transparent film on the surface, nano-SiO_2 although there will be a lot of powder material cover surface layer, but lower strength, is the penetration of ethyl silicate strongest, And the tetraethyl orthosilicate treatment can increase the contact Angle of the concrete surface, forming a hydrophobic protective layer on the surface of the concrete and the internal aperture, so the water absorption of the tetraethyl orthosilicate surface treatment specimen is the lowest. At the same time, it can be seen that after 1 day of pre-water absorption, except for the specimens treated with ethyl silicate, the water absorption of the first day can reach more than 64% of the water absorption of the seventh day. The surface curing treatment mainly improves the water absorption of the concrete specimens on the first day after soaking, and there is a small difference in the water absorption of the specimens in the following time.

It can be seen from Figure 2 that the number of freeze-thaw cycles of damaged concrete increased from 20 times to 28 times with the surface treatment of nano-SiO_2, and the total amount of exfoliation per unit area of specimens cured by ethyl silicate was only 679 g/m^2 after 28 times of cycles, far from reaching the termination conditions of the test. This is because the reduction of water absorption will reduce the depth and amount of ice in concrete, while the reduction of brine infiltration will weaken the influence of accelerating deterioration, thus enhancing the frost resistance. As the powder substance formed on the concrete surface of nano-SiO_2 falls off in the freezing-thawing cycle, the treatment effect is greatly reduced.

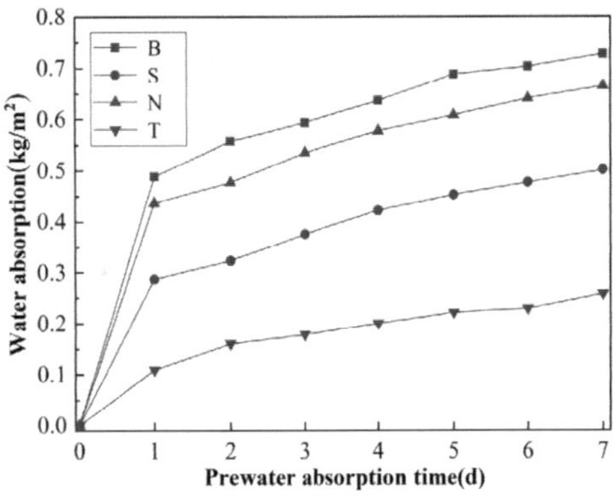

Fig.1 Prewater absorption growth curve

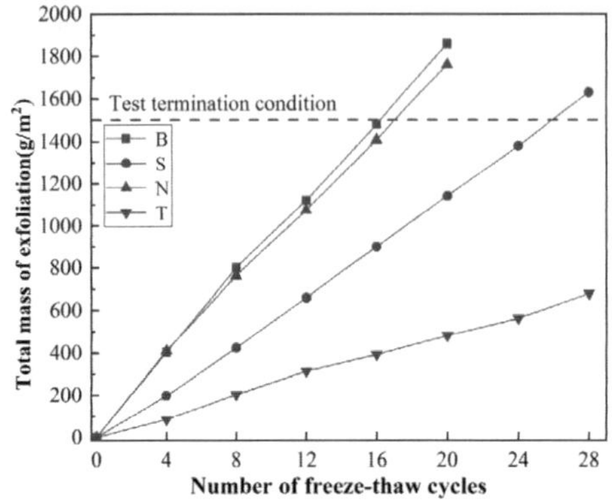

Fig.2 Quality of exfoliation after freeze-thaw cycle

4.Conclusions

According to the test, the following three conclusions can be drawn:

(1) The three surface treatment materials can reduce the 7d pre-water absorption of damaged concrete, among which, tetraethyl orthosilicate has the most obvious effect, and the water absorption reduction rate can reach 64%; The water absorption of concrete treated with nano-SiO_2 can only be reduced by 9%.

(2) The three surface treatment materials mainly improved the water absorption of concrete specimens on the first day after soaking, and there was little difference in the water absorption of specimens in the following time.

(3) Tetraethyl orthosilicate has the best improvement effect on freezing resistance of damaged concrete. After 28 freezing-thawing cycles, the total amount of exfoliation per unit area is only 679 g/m^2.Silicon infiltration agent can increase the number of freezing-thawing cycles of damaged concrete from 20 times to 28 times, while nano-SiO_2 can not improve the number of freezing-thawing cycles of concrete.

References

[1] Su,L.H,Lv,Y,Ji,X.G,et al. (2019)Research on frost resistance of airport pavement concrete enhanced by surface material. Concrete, 353:35-39+44.

[2] Zhang,A.Q,Jia,J,Liu,Z.M,et al.(2019)Influence of Epoxy Resin on Salt Scaling Resistance of Concrete. Bulletin of the Chinese Ceramic Society,38: 1278-1283.

[3] Yang,S.P. (2020)Study on the effect of different protective coatings on frost resistance of hydraulic concrete. Technical Supervision in Water Resources,4:4.

[4] Zhang,H,Zhan,S.L,Wu,H.T,et al.(2016)Study on the influence of lithium silicate composite sol on the surface properties of concrete. Rare Metal Materials and Engineering, S1:4.

[5] Moon, H.Y, Shin,D.G, Choi,D.S. (2007)Evaluation of the durability of mortar and concrete applied with inorganic coating material and surface treatment system. Construction and Building Materials, 21:362-369.

[6] Hou,T,Qi,Y,Shi,Y.G,et al. (2016)Effect of capillary crystalline reaction on water absorption and microstructure of the morta. Concrete, 03:126-131.

[7] Barberena-Fernandez, A. M , Blanco-Varela,M.T, et al. (2015)Interaction of TEOS with cementitious materials: Chemical and physical effects[J]. Cement & Concrete Composites, 55:145-152.

[8] Barberena-Fernández,A.M, Blanco-Varela,M.T, Carmona-Quiroga,P.M. (2019)Use of nanosilica- or nanolime-additioned TEOS to consolidate cementitious materials in heritage structures: Physical and mechanical properties of mortars. Cement and Concrete Composites, 95:271-276.

[9] Scarfato,P, Di Maio,L, Fariello,M.L, et al. (2012)Preparation and evaluation of polymer/clay

nanocomposite surface treatments for concrete durability enhancement. Cement and Concrete Composites, 34:297-305.
[10] Hou,P, Cheng,X, Qian, J, et al. (2015)Characteristics of surface-treatment of nano-SiO2 on the transport properties of hardened cement pastes with different water-to-cement ratios. Cement and Concrete Composites, 55:26-33.

A fault-tolerance control method of load frequency based on reverse behavior reconstruction

Fei Xue[1]*, Hanhua Zhang[2], Shuang Zhang[3], Lei Zhou[4], Ting Yang[5], Hao Li[6]

[1]Electric Power Research Institute of State Grid Ningxia Electric Power Co., LTD Yinchuan, China

[2]Electric Power Research Institute of State Grid Ningxia Electric Power Co., LTD Yinchuan, China

[3]Electric Power Research Institute of State Grid Ningxia Electric Power Co., LTD Yinchuan, China

[4]Electric Power Research Institute of State Grid Ningxia Electric Power Co., LTD Yinchuan, China

[5] Tianjin University, The School of Electrical Engineering and Automation, Tianjin China

[6] Tianjin University, The School of Electrical Engineering and Automation, Tianjin China

[*] Corresponding author: tjuxf1010@126.com

Abstract—The access of a high proportion of renewable energy and the large-scale application of open sharing communication network are the two main characteristics of modern interconnected power grid. However, the tight coupling between the cyber and physical subsystem makes the malicious network attack and sudden physical failure more likely to cause the collapse of the interconnected power grid. Therefore, a reverse behavior reconstruction strategy is proposed, which does not need to obtain a priori fault information and only depends on the input and output measurements of the system, so as to realize fault tolerance as much as possible. The simulation experiments show that the proposed strategy can achieve the frequency fault tolerance and has better practicability and applicability under complex and diverse fault conditions.

1. INTRODUCTION

In the interconnected power grid with a large number of the distributed generation, the stability between regions strongly depends on the normal operation of the cyber and power subsystem. For the physical subsystem, various faults occur in the power equipment, such as hydraulic leakage, blade fracture and other faults, causing an off-grid failure of renewable energy. For the cyber subsystem, false data injection, data packet loss and communication network delay can cause the control center make a wrong decision or the actuator cannot receive control commands in time. When the interconnected power grid has cyber and physical faults, the performance of traditional analysis and design methods will decline sharply or even fail to work normally. Therefore, it is necessary to design a new fault-tolerance control strategy to make the system run safely and stably when the corresponding fault occurs.

The existing research of active fault-tolerance control (AFTC) generally realizes fault identification by designing observers. For example, reference [1] designed a series of redundant observers and controllers for power real-time control systems with sensor faults, and selected the most matching observer and controller by defining the observed signal index. For the load frequency control (LFC) problem of interconnected power systems considering system parameter uncertainty, a distributed filtering strategy has been proposed in [2] to effectively estimate the system state. Reference [3] studies the design of multiarea LFC strategy considering cyber delay and aperiodic DoS attack.

However, the traditional observer based AFTC strategy relies too much on a priori fault information and accurate system parameters in the process of selecting the best matching controller. To solve the above problems, a reverse behavior reconstruction strategy that does not depend on a priori system parameters is proposed to realize rapid frequency fault-tolerance for cyber and physical faults. The remainder of this article is organized as follows. Section II introduces the fault library and preset controller library. Section III presents a reverse behavior reconfiguration method, which is verified by case studies in Section IV. Finally, Section V concludes the paper.

2.FAULT LIBRARY AND PRESET CONTROLLER LIBRARY

2.1.Control Model

The controlled object model under normal conditions is given as:

$$\begin{cases} \frac{d}{dt}x(t)=Ax(t)+Bu(t)+H\omega(t) \\ y(t)=Cx(t)+Du(t) \end{cases} \tag{1}$$

Where x, u, ω and y indicate the state variables, input variables, external disturbance variables and output variables, respectively. Matrix A, B, H, C, D are proper dimensional matrix.

It is noteworthy that: the model is not limited to a certain scene or system object, but has better universal representation. Although the interconnected system is nonlinear, the linearization method at the equilibrium point can be used to linearize the system. The model error caused by this linearization process can be considered in the controller design process.

2.2.Fault library

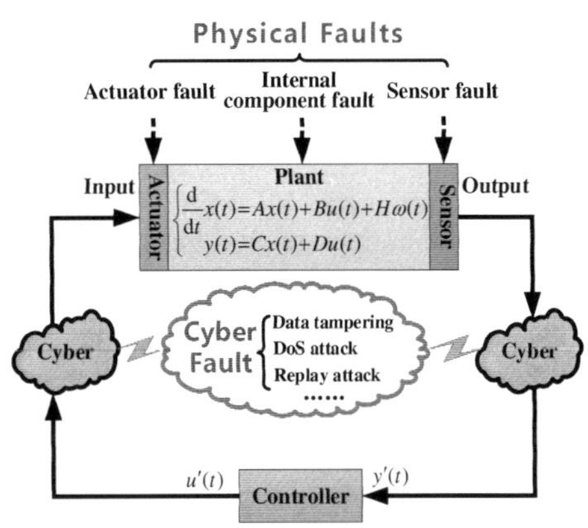

Figure 1. Closed-loop control diagram of interconnected power grid

As can be seen from Figure 1, due to the introduction of cyber subsystem, all kinds of uncertain factors (such as delay, packet loss and cyber attack) in the process of data transmission bring new security risks to real-time control systems in the interconnected power system. Although there are many fault types, they can be divided into three categories according to the impact of faults:

1) Data fault: This type of fault is mostly caused by the aging of sensors and actuators or malicious tampering of data. The measured electrical parameters seriously deviate from the real value. Under the data fault, the control parameters and measurement parameters meet the following relationship:

$$\begin{cases} u_f(t)=G^P G^C u(t) \\ y_f(t)=Y^C Y^P y(t) \end{cases} \tag{2}$$

Where subscript f indicates the fault state, Γ^P=diag(γP 1, γP 2,..., γP p) is the physical fault matrix of actuator. Γ^C=diag(γC 1, γC 2,..., γC p) refers to the cyber fault matrix between controller and actuator. Ψ^C=diag(ψC 1, ψC 2,..., ψC q) represent cyber fault matrix between sensor and controller. Ψ^P=diag(ψP 1, ψP 2,..., ψP q) indicates physical fault matrix of sensor. Take γP i and γC i for example, γP i=0 means actuator i fails completely, γp i=1 indicates actuator i works normally, γp $i\in(0, 1)$ represents actuator i has partial failure. γC i=1 indicates that the cyber transmission is normal, γC $i\in(\gamma_{\min}, \gamma_{\max})$, γC $i\neq1$ indicates data tampering attack in cyber subsystem, $\gamma_{\min}$, $\gamma_{\max}\in\mathbf{R}$ represents the ratio of the tampered value to the original real value.

2) Delay fault: Due to the transmission channel congestion or malicious cyber attack, the transmission delay of measurement data and control commands increases and then the data become uncertain or meaningless owing to the loss of its timeliness. Under the delay fault, there is a non negligible delay in the control command or measurement data from the source node to the root node. Assume that τ_{sc} and τ_{ca} respectively indicate the data delay from the sensor to the control center and from the control center to the actuator. Under normal state and delay fault state, the control commands output by the controller are $u'(t)$ and $u(t)$ respectively, the system state variables received by the control center are $y'(t)$ and $y(t)$ respectively. The following relationship can be satisfied:

$$\begin{cases} u(t)=u'(t+\tau_{ca}) \\ y'(t)=y(t+\tau_{sc}) \end{cases} \tag{3}$$

Further, Pade approximation technology is used to approximate the above pure time-delay links. For details, please refer to [4].

3) Component fault: The original structure and operation mode will be fundamentally changed by the internal component fault. The abnormality of state space description, that is:

$$\begin{cases} \dfrac{\mathrm{d}}{\mathrm{d}t}x(t)=A'x(t)+B'u(t)+H'\omega(t) \\ y(t)=C'x(t)+D'u(t) \end{cases} \tag{4}$$

Where A'、B'、H'、C'、D' represent the controlled system matrix after the component fault occurs.

2.3.Hybrid fault system modeling

Based on the mathematical modeling of three types of faults, assuming the augmented vector $X=[x, x_{ca}, x_{sc}]^T$, $x_{ca}(t)$ and $x_{sc}(t)$ are the intermediate variables introduced by Pade approximation. The state equation of real-time control system of interconnected power grid considering cyber and physical hybrid faults is established as:

$$\begin{cases} \dfrac{\mathrm{d}}{\mathrm{d}t}\begin{bmatrix} x(t) \\ x_{ca}(t) \\ x_{sc}(t) \end{bmatrix}=\begin{bmatrix} A_f & B_f C_{ca} & 0 \\ 0 & A_{ca} & 0 \\ B_{sc}C_f & B_{sc}D_f C_{ca} & A_{sc} \end{bmatrix}\begin{bmatrix} x(t) \\ x_{ca}(t) \\ x_{sc}(t) \end{bmatrix}+\begin{bmatrix} B_f D_{ca} \\ B_{ca} \\ B_{sc}D_f D_{ca} \end{bmatrix}u'(t)+\begin{bmatrix} H_f \\ 0 \\ 0 \end{bmatrix}\omega(t) \\ y(t)=\begin{bmatrix} D_{sc}C_f & D_{sc}D_f C_{ca} & C_{sc} \end{bmatrix}\begin{bmatrix} x(t) \\ x_{ca}(t) \\ x_{sc}(t) \end{bmatrix}+D_{sc}D_f D_{ca}u'(t) \end{cases} \tag{5}$$

Where $A_f=A'$, $B_f=B'\Gamma^P\Gamma^C$, $H_f=H'$, $C_f=\Psi^C\Psi^P C'$, $D_f=\Psi^C\Psi^P D'\Gamma^P\Gamma^C$.

2.4.Preset controller Library

Obviously, the real-time control system described in (5) has model uncertainty: 1) the linearization modeling of controlled system will produce errors. 2) the actual operating parameters of the controlled system deviate from the nominal value. 3) Pade approximation of network delay will also bring errors. Therefore, the controller design must be robust to model uncertainty. Based on the hybrid H_2/H_∞

method proposed in [5], this paper designs the basic controller to realize the robust control. Due to the space limitation of the article, the design process of H_2/H_∞ method is not described in detail here.

Considering that the controlled system has M normal working scenarios and N typical fault scenarios, the number of controllers in the basic library is $M+N$, assuming that $\Sigma_c=\{\Sigma_{c1}, \Sigma_{c2}, \ldots, \Sigma_{c(M+N)}\}$ represents the set of controllers designed for specific operation scenarios, which has the following general form:

$$\Sigma_{cj}: \begin{cases} \dfrac{\mathrm{d}}{\mathrm{d}t} x_{uj}(t)=A_{uj}x_{uj}(t)+B_{uj}(r(t)-y'(t)) \\ u'(t)=C_{uj}x_{uj}(t)+D_{uj}(r(t)-y'(t)) \end{cases} \tag{6}$$

Where j=1, 2, …, $M+N$, $x_{uj}(t)$ indicates the intermediate variable introduced into the controller, A_{uj}, B_{uj}, C_{uj}, D_{uj} are the controller system matrices.

2.5.Dynamic behavior modeling

The controlled system (5) and controller (6) are essentially linear differential equations with constant coefficients. The intersection of their solution sets restricts the value space of the dynamic behavior of the closed-loop system. Therefore, the signal vector is defined as $s(t)=[\omega(t), r(t), y'(t), u'(t)]^{\mathrm{T}}$, the dynamic behavior of the controlled system and the controller can be described as:

Controlled system:

$$\boldsymbol{B}_{\mathrm{P}}=\{s=(r,y',u')^{\mathrm{T}} | [-N_{\mathrm{p2}}(\xi)\ 0\ D_{\mathrm{p}}(\xi)\ -N_{\mathrm{p1}}(\xi)]s(t)=0\} \tag{7}$$

Controller:

$$\boldsymbol{B}_{\mathrm{c}j}=\{s=(r,y',u')^{\mathrm{T}} | [0\ N_{\mathrm{c}j}(\xi)\ -N_{\mathrm{c}j}(\xi)\ -D_{\mathrm{c}j}(\xi)]s(t)=0\} \tag{8}$$

Where ξ=d/dt, $N_{\mathrm{p1}}(\xi)$、$N_{\mathrm{p2}}(\xi)$、$D_{\mathrm{p}}(\xi)$、$N_{\mathrm{c}j}(\xi)$、$D_{\mathrm{c}j}(\xi)$ are proper dimensional polynomial matrix and can be strictly proved in [6]:

1) $\{N_{\mathrm{p1}}(\xi), N_{\mathrm{p2}}(\xi), D_{\mathrm{p}}(\xi)\}$, $\{N_{\mathrm{c}j}(\xi), D_{\mathrm{c}j}(\xi)\}$ are mutually prime and all revertible polynomial matrices;

2) D-1 p(ξ)$N_{\mathrm{p1}}(\xi)$, D-1 p(ξ)$N_{\mathrm{p2}}(\xi)$, and D-1 cj(ξ)$N_{\mathrm{c}j}(\xi)$ are true rational polynomial matrices.

According to (7) and (8), the controlled system and the controller are interconnected through common variables $[r(t), y'(t), u'(t)]^{\mathrm{T}}$. In addition, the preset controllers meet the H_2/H_∞ performance, that is, through the design of the controller Σ_c, the external disturbance $\omega(t)$ of the controlled system can always attenuate within the design requirements.

3.REVERSE BEHAVIOR RECONFIGURATION

3.1.Reverse reconstruction module

The core of the AFTC method is to put the best-matched controller into the controller loop. However, the traditional fault identification method requires accurate prior system parameters and a large number of state parameters, and there is a risk of mismatch between the controller and the controlled system, resulting in fault-tolerance failure.

Therefore, a AFTC strategy that does not rely on fault identification is proposed in the paper. As shown in Figure 2, the core is to use only the output $y'(t)$ of the controlled system, the control command $u'(t)$ and the existing controller library Σ_c to reversely reconstruct the dynamic behavior of the closed-loop system in a parallel mode, which is recorded as $\{\hat{B}_1, \hat{B}_2, ..., \hat{B}_{M+N}\}$. By calculating the dynamic performance index between $\{\hat{B}_j\}$ (j=1, 2, …, $M+N$), the controller with the best index result is selected to put into the control loop, so as to realize accurate switching. Therefore, the proposed AFTC strategy consists of the following three parts: 1) dynamic behavior reverse reconstruction module; 2) performance index calculation module; 3) controller switching module.

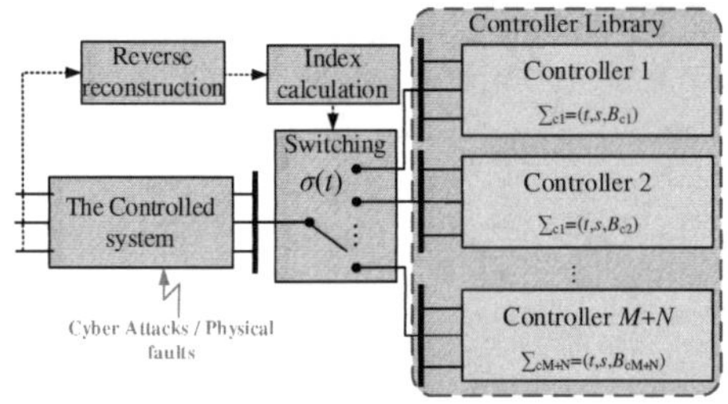

Figure 2. The proposed AFTC framework

Based on the analysis in Section II, the dynamic behavior of the controller Σ_{cj} can be completely reconstructed only through $y'(t)$ and $u'(t)$, without accurate priori system operating parameters and fault information. The dynamic behavior of the reconstructed controller is satisfied:

$$\hat{r}_j(t)=N_{cj}^{-1}(\xi)D_{cj}(\xi)u'(t)+y(t),\ t\in[(m-1)T_s,\ mT_s) \tag{9}$$

Where $\hat{r}_j(t)$ is the reconstruction reference input of Σ_{cj}.

3.2. *Performance index calculation module*

The function of this module is to calculate the performance index of the dynamic behavior of the parallel reconstructed controller in any detection cycle $[(m\text{-}1)T_s, mT_s)$. The goal of control system is that the output $y'(t)$ can track the reference input $r(t)$ in real time, so this paper considers the *MSE* (Mean Square Error) index of the error between the reconstructed reference input $\hat{r}_j(t)$ and the output $y'(t)$ as a performance index, that is:

$$MSE=\frac{1}{T_s}\int_{mT_s}^{(m+1)T_s}(\hat{r}_j(t)-y'(t))^2\,\mathrm{d}t \tag{10}$$

This index can accurately represent the control error of different controllers in the case of information uncertainty.

3.3. *Controller switching module*

The function of the controller selection module is to select the most matching controller in the next time interval according to the calculated performance index MSE, and the number value of the selected controller can be expressed as:

$$\sigma(t)=\arg\min\{MSE_i\},t\in[mT_s,(m+1)T_s] \tag{11}$$

4. SIMULATION ANALYSIS

This paper takes the three-area interconnected power system as the research object for simulation experiments. Table 1 shows the seven possible cases of the three-area interconnected system. This paper uses comparison method to verify: the passive fault-tolerance control (PFTC) strategy considering the progressive stability of all fault cases proposed in [7] and the active fault-tolerant control strategy based on fault identification (fault identification AFTC) proposed in [3].

TABLE 1. THE SCENARIOS OF OPERATING

Case	Instruction
1	Area 1, Area 2, Area 3 operate normally
2	Area 1 has data faults, Area 2, Area 3 operate normally
3	Area 2 has delay faults, Area 1, Area 3 operate normally
4	Area 3 has component faults, Area 1, 2 operate normally
5	Area 1, 2 has data and delay faults, Area 3 operate normally
6	Area 1, 3 has data and component faults, Area 2 operate normally
7	Area 2, 3 has data and component faults, Area 1 operate normally
8	Area 1, 2 and 3 has data, delay and component faults

According to Figure 3, when the PFTC strategy is adopted, the mean square error (MSE) of Δf_1 and ΔP_{tie1} is generally greater than that of AFTC, and only slightly better than the AFTC strategy in cases 2 and 6 (the MSE of Δf_1 and ΔP_{tie1} in other cases is at least high 18.40% and 220%). The performance index of the proposed AFTC strategy is roughly similar to that of the fault identification-AFTC strategy in terms of control performance, but because the proposed scheme only relies on the input and output parameters of the controlled system when selecting the best-matching controller. It does not need to rely on accurate prior knowledge of system parameters, and can achieve similar fault tolerance effects, so it is more practical.

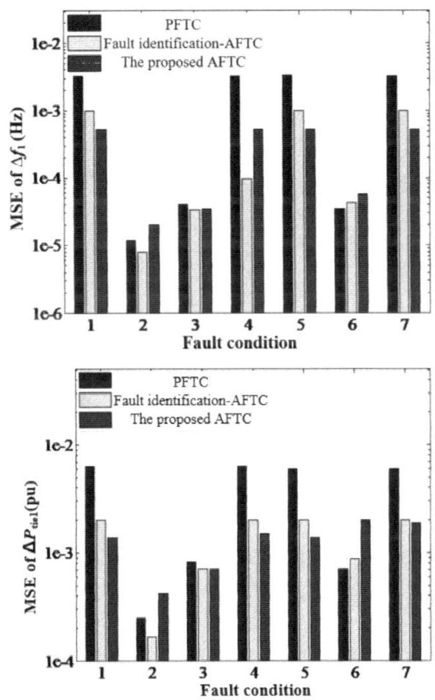

Figure 3. MSE of Δf_1 and ΔP_{tie1} in each fault case

5.Conclusion

In view of the risk that the interconnected power system is vulnerable to malicious cyber attacks and sudden physical failures, this paper proposes a reverse behavior reconstruction method. It only depends on the input and output measurements of the system, which effectively overcomes the limitation that the traditional AFTC method based on parameter identification must rely on priori system parameters. Future research will focus on further improving the controller expansion mechanism to achieve complete fault tolerance to unknown faults.

Acknowledgment

Financial support should be acknowledged here. This work was supported by the Key Research and Development Program of Ningxia Hui Autonomous Region (2021BDE931017) and the Science and technology project of State Grid Ningxia Electric Power Co.,Ltd (5229DK20005F).

References

[1] Khosravani Saeid, Moghaddam Iman Naziri, Afshar Ahmad, Karrari Mahdi, "Wide-area measurement-based fault tolerant control of power system during sensor failure," in Electrical Power Systems Research, vol. 137, pp. 66-75, 2016.

[2] Hu Zhijian, Liu Shichao, Yang Liu, Wu Ligang. "Distributed fuzzy filtering for load frequency control of non-linear interconnected power systems under cyber-physical attacks," in IET Control Theory and Applications, vol. 14, pp. 527-538, 2020.
[3] Cheng Zihao, Yue Dong, Hu Ssonglin, Xie Xiangpeng, Huang Chongxin, "Detection-based weighted H_∞ LFC for multi-area power systems under DoS attacks," in IET Control Theory and Applications, vol. 13, pp. 1909-1919, 2019.
[4] Zheng Chengde, Zhang Huaguang, Tian Hong. "Generalized homogeneous multivariate matrix Pade-type approximants and Pade approximants," in IEEE Transactions on Automatic Control, vol. 52, pp. 2160-2165, 2007.
[5] Khosrowjerdi MJ, Nikoukhah R, Sarari-Shad N. "A mixed *H-2/H*-infinity approach to simultaneous fault detection and control," in Automatica, vol. 40, pp. 261-267, 2004.
[6] T Jain, J Yame, D Sauter. "Active fault-tolerant control systems: a behavioral system theoretic perspective," in Berlin: Springer Press, pp. 15-18, 2018.
[7] Sevillal F, Jaimoukha I, Chaudhuri B, et al. "Fault-tolerant control design to enhance damping of inter-area oscillations in power grids," in International Journal of Robust and Nonlinear Control, vol. 24, pp. 1304-1316, 2013.

Development of SF_6 Gas Pressure Prediction Device in Substation

Xicheng Lin[1,a], Hua Wang[2,b]*, Zheng Qin[3,c], Wendong Zhao[4,d], Xiangchong Li[5,e], Xiaojie Yang[6,f]

[1]Substation Maintaining Centre Taizhou Power Supply Company of State Grid Taizhou, China

[2]Substation Maintaining Centre Taizhou Power Supply Company of State Grid Taizhou, China

[3]Substation Maintaining Centre Taizhou Power Supply Company of State Grid Taizhou, China

[4]Substation Maintaining Centre Taizhou Power Supply Company of State Grid Taizhou, China

[5]Substation Maintaining Centre Taizhou Power Supply Company of State Grid Taizhou, China

[6]Substation Maintaining Centre Taizhou Power Supply Company of State Grid Taizhou, China

[a]741983881@qq.com, [b]wang_hua_tz@zj.sgcc.com.cn, [c]qin_zheng_tz@zj.sgcc.com.cn, [d]zhao_wendong_tz@zj.sgcc.com.cn, [e]li_xiangchong_tz@zj.sgcc.com.cn, [f]yang_xiaojie_tz@zj.sgcc.com.cn

Abstract—SF_6 equipment has leakage due to various reasons. The longer the equipment operates, the more serious the leakage is. When the equipment gas leaks to the set air replenishment value, the air replenishment alarm signal will be sent. It can be seen from the SF_6 gas temperature pressure relationship table that the temperature change has a great impact on the pressure value. Although the equipment popularizes the SF_6 density meter or SF_6 density relay with temperature compensation, the compensation system needs to achieve temperature compensation through a long time, resulting in the distortion of the pressure value counted by the operation and maintenance personnel. Therefore, the temperature difference between winter and summer is large, and more alarm signals are obtained. At this time, there are great potential safety hazards in air replenishment. This paper introduces in detail a SF_6 gas leakage prediction device in substation, which can effectively solve the above problems in substation.

1. INTRODUCTION

The electrical equipment of GIS Substation is filled with SF_6 gas. SF_6 is an ideal insulating gas with excellent insulation performance and arc extinguishing performance. However, due to quality differences in manufacturing, installation and material aging, SF_6 gas leakage of high-voltage switchgear is a common phenomenon. Once SF_6 gas leaks from SF_6 indoor substation, it will not only cause safety problems of power grid, but also cause ground air pollution and endanger personal safety. Therefore, it is urgent to design the development of SF_6 gas leakage prediction device in substation, as in [1].

2.STATE PARAMETER EQUATION AND CURVE OF SF_6 GAS

2.1.Beattie-Bridgman Equation

Within the scope of engineering application, air or general gas can be called ideal gas when the pressure is not too large compared with atmospheric pressure and the temperature is not too low compared with room temperature. Microscopic characteristics of gas model is expressed as [2]:

- There is no interaction between molecules.
- The size of molecules, like geometric points, does not occupy volume itself.

When the gas density is less than the critical density of 0.8, Beattie-Bridgman equation is more accurate. Beattie-Bridgman equation is given by (1) to (4),

$$P = \frac{R_g T\left(1-\varepsilon_{(v,T)}\right)}{v^2}\left(v + B_{(v)}\right) - \frac{A_{(v)}}{v^2} \quad (1)$$

$$A_{(v)} = A_0\left(1-\frac{a}{v}\right) \quad (2)$$

$$B_{(v)} = B_0\left(1-\frac{b}{v}\right) \quad (3)$$

$$\varepsilon_{(v,T)} = B_0\left(1-\frac{c}{vT^2}\right) \quad (4)$$

Where A_0, B_0, a, b and c are 5 constants, whose values depend on the type of gas and can be fitted by the experimental values.

2.2.State Parameter Equation Of SF_6 Gas

SF_6 gas has large molecular weight and mutual attraction between molecules. When the gas pressure exceeds 0.3 MPa, especially 0.5 MPa, with the increase of density, the intermolecular distance is compressed, which significantly increases the attraction between molecules and weakens the force of molecules on the vessel wall, The actual gas state deviates from the ideal gas state equation: the increase of actual pressure is lower than the ideal. The higher the gas pressure, the more obvious the difference, as in [3]. The rated pressure of SF_6 gas used in high voltage apparatus is generally not less than 0.5 MPa, so it does not have the conditions of ideal gas and cannot be calculated according to Charlie's law.

The most commonly used empirical formula for calculating the state parameters of SF_6 gas is Beattie-Bridgman equation, as in (5) to (7).

$$P = 56.2\times10^{-6}\rho T(1+B) - \rho^2 A \quad (5)$$

$$A = 7.49\times10^{-5}(1-0.727\times10^{-3}\rho) \quad (6)$$

$$B = 2.51\times10^{-3}\rho(1-0.846\times10^{-3}\rho) \quad (7)$$

Where P is the pressure of SF_6 gas, ρ is the density of the gas, T is the temperature of the gas.

2.3.Curve of SF_6 Gas

The state parameter curve of SF_6 is a family of curves, in which each curve corresponds to a density value. When the gas temperature changes, the gas pressure changes along the density curve, and the intersection of the density curve and the saturated steam pressure curve is the liquefaction point, as shown in Fig. 1.

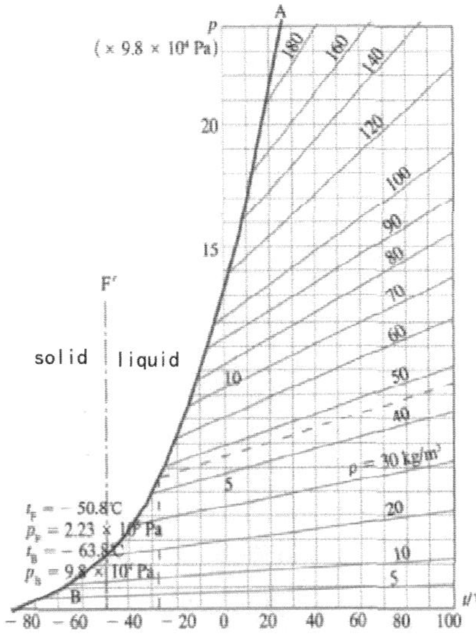

Figure 1. State parameter curve of SF_6 gas.

3.COMMON SF_6 GAS MONITORING METHODS

3.1.Traditional Measurement Method

The traditional monitoring method used by the monitoring personnel is to measure SF_6 gas with a barometer and complete the monitoring work together with the density relay measurement method. The working principle of this method is to measure the pressure generated by the gas through the barometer, and then obtain the content of SF_6 gas in the air. However, this method has many limitations and can only be used in the environment where there is no significant change in temperature and the leakage phenomenon is very obvious, as in [4]. In addition, if the substation uses this method, it also needs to be equipped with a monitoring personnel to observe the barometer, which cannot be applied to unattended substations. In addition, this method also has great disadvantages. The density relay belongs to mechanical equipment, with low accuracy and good seismic capacity. Unless the gas leakage is still insufficient to the safety line, the change of barometer is not obvious, and it is difficult for the monitoring personnel to monitor the leakage in time.

3.2.Ultrasonic Velocimetry

The monitoring personnel can use the ultrasonic velocimetry to measure the leakage of SF_6 gas. The principle is as follows: there is a difference between the molar mass of gas, and the speed of ultrasonic in it will be affected to a certain extent. Sound wave is a longitudinal wave, which propagates in an elastic medium. Ultrasonic velocity measurement method has many advantages, such as: the monitoring personnel can control the length of ultrasonic, directionally emit ultrasonic, etc. not only that, the characteristics and state of the medium are related to the propagation speed of ultrasonic. The monitoring personnel can analyze the characteristics of the medium through the measurement of ultrasonic speed. The monitoring personnel can use the difference method to eliminate the influence of temperature on the results, so the acoustic velocity is only related to the molar mass of gas. For SF_6 gas, its molar mass is about 5 times higher than that of air. Therefore, if SF_6 gas leakage occurs in the substation, the molar mass of gas in the substation is different from the previous molar mass of air, and the propagation speed of sound wave will naturally be affected. The measuring instrument can calculate the content of SF_6 gas in the air according to the change of sound wave propagation speed, so as to determine whether the leakage of SF_6 gas in the air has exceeded the standard. Compared with the traditional test method, this method is more stable, has longer service life and accurate measurement results. However, this method also has disadvantages, that is, it is expensive and is often used by most foreign companies, and it is only applied to areas with high concentration.

4. DEVELOPMENT OF SF_6 GAS PRESSURE PREDICTION DEVICE IN SUBSTATION

The SF_6 gas leakage prediction device in substation shall have the following functions:

- Calculate the gas chamber pressure at 20°C.
- Calculate the annual leakage rate of SF_6 in the gas chamber.
- Predict the remaining days from gas chamber leakage to pressure alarm value.
- Display various information of the air chamber and modify the setting parameters.
- Reliable operation of the device.

4.1. Scheme Implementation

In order to reduce the number of SF_6 low pressure alarm emergency defects caused by special weather in 220 kV substation, it needs to be realized from the following steps. Firstly, it is necessary to calculate the air chamber pressure at 20 °C. It is required to automatically collect the equipment ambient temperature and air chamber pressure at 5 a.m. every day and convert the air chamber pressure at 20 °C according to the SF_6 pressure ambient temperature curve. Secondly, calculate the annual leakage rate of SF_6 in the gas chamber. It is required to calculate the residual SF_6 gas in the gas chamber according to the pressure value of the gas chamber converted every day and calculate the annual leakage rate of SF_6 every day according to the formula. Then, according to the daily annual leakage rate of SF_6 and the residual SF_6 gas in the gas chamber, the remaining days from gas leakage in the outlet chamber to the pressure alarm value are predicted. Then, the information of the air chamber is displayed, and the information is displayed intuitively through the display. Finally, power shall be supplied to the device to meet the normal operation of the device. Based on the above analysis, the team proposed a specific module decomposition scheme. The tester of DC insulation detection device is divided into five modules: data transmission, data storage, logic control, power supply and display, as shown in Fig. 2.

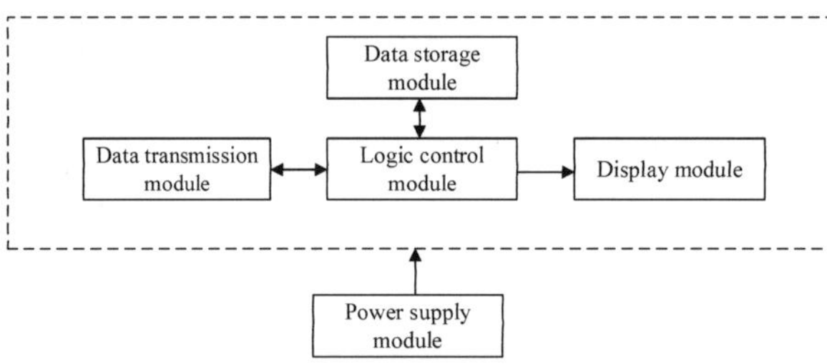

Figure 2. Device module design.

4.2. Effect Verification

The above devices are installed in 3 chambers with SF_6 low pressure alarm emergency defects in 2020. The statistics of SF_6 low pressure alarm emergency defects are shown in Table 1.

TABLE 1. SF_6 LOW PRESSURE ALARM DEFECTS IN 3 GAS CHAMBERS AFTER INSTALLATION OF DEVICE.

Number	Air chamber name	SF_6 low pressure alarm occurred
1	1878 line voltage transformer chamber	No
2	220kV switch phase a gas chamber	No
3	Auxiliary bus voltage transformer chamber	No

5.CONLUTION

The device in the substation can accurately calculate the remaining days from the leakage of the gas chamber to the low-pressure alarm of SF_6. The maintenance personnel can turn passivity into initiative, make decisions in advance and reasonably arrange the gas replenishment time, so as to avoid emergency gas replenishment in the environment with high air humidity such as rainstorm and late night. The operational stability of the equipment is improved to a certain extent. The prediction device also has the function of calculating the annual leakage rate. According to the standard of annual leakage rate of SF_6, it can judge whether there is air leakage in the gas chamber in advance. Therefore, the device can find the deterioration trend of equipment quality in advance, which greatly improve the equipment quality control ability.

ACKNOWLEDGMENT

Thanks to the Taizhou power supply company of state grid for its strong support for this project. At present, the system has been applied in the debugging and work of substation in Taizhou, which has obtained a good evaluation. It can be seen that the put into use of the system greatly simplifies the on-site work process and enables on-site operation and maintenance personnel to complete on-site work with quality and quantity while improving work efficiency.

REFERENCES

[1] Y. Hao, Z. Lu and K. Wenke, “Research on Early Warning and Recovery Technology of Gas Leakage,” New Communications. China, vol. 12, pp. 38-45, April 2019.

[2] Z. Dani, H. Weidong. “State Parameter Equation of SF_6 Gas and Its Engineering Application,” Journal of Nanjing University of Posts and Telecommunications (social science edition). China, vol. 14, pp.56-60, May 2012.

[3] P. Qingliang, Y. Shijun, L. Liumeng, et al. “Study on SF_6 Gas State Collection and Multi-parameter Evaluation Technology,”. Chinese Command and Control Institute. Proceedings of the 4th China Command and Control Congress. China, 2016, pp. 63-67.

[4] Z. Yanpei. Research on On-Line Monitoring Technology of SF_6 Leakage in Substation, Beijing: Beijing University of Posts and Telecommunications, 2018.

Parallel Simulation of The Electromechanical Transient Phenomena of Power System via Modern Software Development Technique

Yu XIE[1], Zhigang WU [1*], Erzhuan ZHOU [2]

([1.]South China University of Technology, Guangzhou 510650, China;

[2.] China Electric Power Research Institute, Beijing 100192, China

*Coresponding author: epzgwu@scut.edu.cn)

Abstract: The simulation of power system electromechanical transient phenomena is highly dependent on the knowledge and model of special domain. For this reason, traditional parallelization of the simulation must be coupled with concrete algorithm and code. In this paper, a parallelized electromechanical transient simulation program is proposed based on the open-source simulation engine called InterPSS and general open object-oriented modelling framework. The correctness and effectiveness of this framework have been verified by a large-scale simulation case.

1. Introduction

To construct a new power system with high penetrance of renewable energy, numerical simulation techniques of the system with sufficient accuracy and efficiency are required urgently, and hence, appropriate parallel computing should be implemented via modern high-performance computing techniques. The parallel computing of electromechanical transient (EMT) simulation is one of the important types of simulation tasks, which always requires specific domain knowledge.

There are three types of parallel algorithm used in EMT simulation: spatial parallel algorithms, temporal parallel algorithms, and spatial-temporal parallel algorithms. The main idea of spatial parallel algorithms is to divide the network into different partitions and define sub-tasks for each partition to simulate the corresponding part of EMT model decoupling within a single time step, finally coordinate the sub-results to get the whole result of this time step [1-3].

Temporal parallel algorithm, on the contrary, simulate the entire electrical power network without any partitioning, but try to calculate simultaneously for different time steps to speed up the calculation. Most of the proposed algorithms try to finish the job with the convergence property similar with Newton method, and sometime special task-scheduling methods such as pipeline processing technique is utilized to solve the variable coupling relationship between adjacent integration time steps [4-9].

Both spatial parallel algorithm and time parallel algorithm have limitations. If the spatial parallel algorithm uses too many subnets, the communication burden between subnets increases, which reduces the computational efficiency. The performance of the time parallel algorithm highly depends on the serial convergence of the algorithm, and its speedup tends to be saturated with the increase of the parallel scale. To further improve the technical efficiency of simulation, scholars propose spatial-temporal parallel algorithms aiming to combine the advantages of both types of algorithms. It often solves the dynamic process of power system in parallel at multiple time points according to the differential nonlinear algebraic equations. The key point is to effectively solve the contradiction between the degree of parallelism and the convergence of parallel algorithms [10]. A space-time parallel relaxed Newton method for transient stability calculation is proposed in [11], which can achieve a high degree of parallelism and maintain the convergence of Newton algorithms.

The above algorithms and parallelization techniques are based on the specific characteristics of power system simulation problems. It should be point out that such algorithm improvement based on specific domain knowledge is an essential improvement of the algorithm studied. It is an improved way to get the best performance-to-price ratio. However, this will inevitably encounter a new problem. At present, the mainstream EMT simulation parallel programs are highly dependent on professional knowledge and models in this field, and even rely on special hardware resources. It is not enough in the application of more general-purpose high-performance computing and advanced software development. To a certain extent, it restricts the expansion and upgrade of the simulation program.

In this paper, InterPSS (Internet based Power System Simulator), a power system simulation engine based on modern software engineering technology, is used to propose a general method of parallelizing EMT simulation. Firstly, with the help of InterPSS's highly open and extensible framework, the author can integrate the MATE (Multi-Area Thévenin Equivalents) algorithm that have been contributed by other key developers in the InterPSS development community. Secondly, a label propagation algorithm is proposed to achieve more effective partitioning for specific computing resources. Thirdly, the parallelization of decoupling calculation of each partition is realized by multithreading in JAVA environment, and it is coordinated effectively without a lot of deployment tasks on very general hardware platform. Finally, a large-scale EMT simulation case is used to verify the correctness and effectiveness of the proposed EMT simulation parallel framework.

2. EMT simulation engine of InterPSS

2.1. *Framework of InterPSS*

InterPSS is the most mature and open simulation engine in the power industry [12]. Its most outstanding feature is the utilization of advanced software engineering technology, distributed development mode and modular architecture. It is one of the main research directions of IEEE open-source power system simulation software development team. The system architecture of InterPSS is shown in Fig. 1.

As can be seen from Fig. 1, the core of the whole structure of InterPSS is an object-oriented power system simulation model. It is a software system that is flexible, extensible, and easy to maintain. Its most important feature is that the source code is open, and the composition is loose. According to their own requirement, one can add additional function or replace exist function in the form of plug-in. This creates favorable conditions for this paper to expand the existing functions of InterPSS and develop parallel programs for EMT simulation.

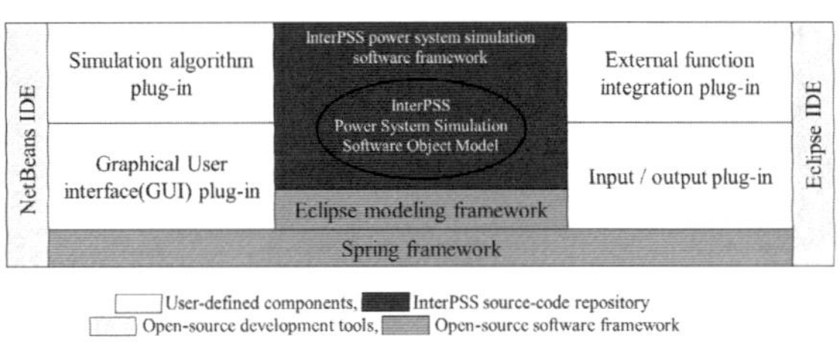

Fig. 1 Framework of InterPSS

2.2. *Object-oriented Model of EMT Simulation in InterPSS*

As all other EMT simulation programs, such simulation function provided by InterPSS also deals with the power system model formed by connecting the power network model to the generator and other dynamic models. However, unlike most programs, InterPSS follows the standard specification of object-oriented modeling and development, which is the fundamental guarantee of its openness and high scalability.

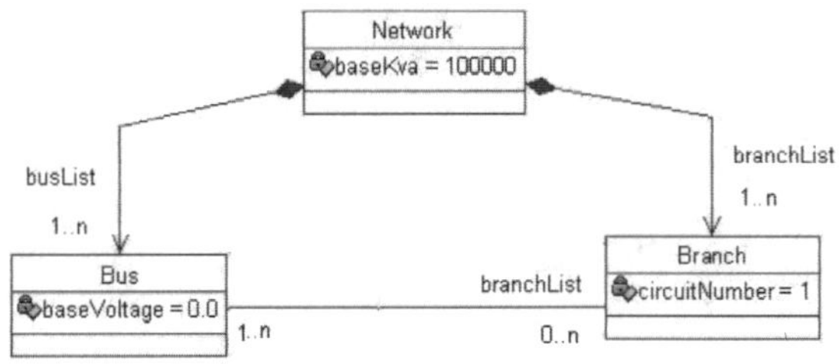

Fig. 2 Basic network model in InterPSS

The basic network model and its extension are related to EMT simulation concerned in this paper. The basic network model of InterPSS is shown in Fig. 2, in which one can find that the model includes node sets and branch sets that must be included in all power system computing tasks. The branch sets also undertakes the task of defining topological connections. In addition, the per unit value is often used in power system analysis and calculation, and it is usually necessary to specify a unified base capacity of the whole network. This is defined in InterPSS by the properties of the network.

There are three major calculations in power system: power flow calculation, short circuit calculation and stability calculation [13]. Logically, all these calculations need to establish the corresponding node admittance matrix when dealing with the network model, which is the task of the BaseAclfNetwork object in Fig. 3. Based on BaseAclfNetwork, the conventional power flow calculation network model (AclfNetwork) and the basic short circuit calculation network model (BaseAcscNetwork) can be established respectively. BaseAcscNetwork provides methods for generating sequence networks and can also be used to establish conventional short circuit calculation network model (AcscNetwork) and conventional transient simulation network model (DStabilityNetwork). In this paper, the parallelization of EMT simulation is essentially based on DStabilityNetwork.

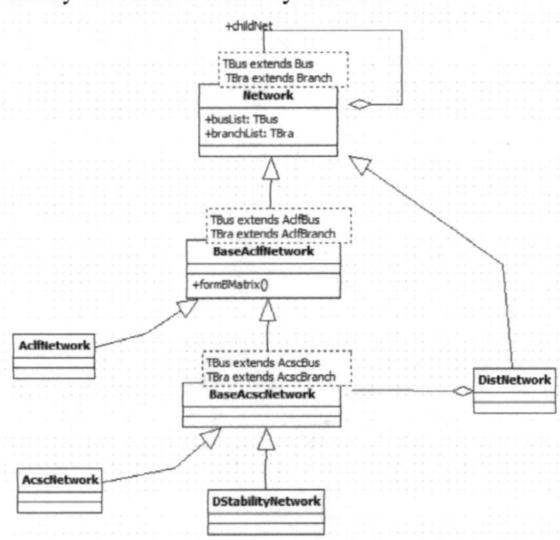

Fig. 3 Relationship between different types of networks

Finally, in InterPSS, everything is object-oriented, including all kinds of simulation algorithms. Taking the EMT simulation concerned in this paper as an example, the algorithm itself is also an object. Disturbance events, numerical integration algorithms and simulation parameters are all specific and replaceable attributes or functions of EMT simulation algorithm objects. Only in this way can we parallelize the necessary functions without changing the overall architecture of the program as much as possible, to achieve the high degree of versatility and expansibility mentioned above.

3. Realization of parallelization of EMT simulation in InterPSS

The general computing framework of EMT simulation is shown in Fig. 4. As can be seen from the figure, all dynamic models are coupled and influenced each other through the network model. To realize the parallelization of computing tasks, this paper considers that the complete computing tasks will be partitioned by an effective networking algorithm. In each integration step, the computing task is published to different computing resources for parallel calculation, and then the results of each sub-network are reasonably coordinated to obtain the result of this integration step.

To achieve the above purpose, two key problems need to be solved: (1) decoupling of the calculation and coordination of different partitions; (2) flexible and efficient partitioning algorithm. In this section, we will make an in-depth analysis to solve the two problems.

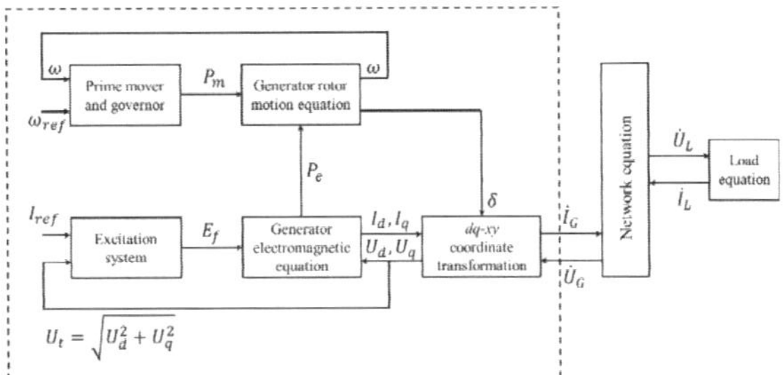

Fig. 4 General Framework for EMT Simulation

3.1 Decoupling and coordination of EMT simulation based on MATE

A method of regional decoupling calculation of large-scale power grid by MATE algorithm is introduced in [14]. Firstly, the original power grid is divided into several areas. The node set is divided into the union of several non-intersection subsets. The branches are divided into two types, one is the branch whose two ends are in the same node subset, and the other is the branch whose two ends belong to two different node subsets. The specific method of network partitioning is not discussed in the original literature, and reasonable results can be read by default. Secondly, if the current through the tie line is not considered for each subnet, the subnet can be solved independently. Thirdly, the corresponding impedance matrix, branch current vector and branch voltage drop vector are defined for the branches connecting different subnets. These quantities are used to coordinate the results of different subnets in an appropriate way to get the whole calculation results of the whole network.

Based on the MATE algorithm, a transient simulation method including simulation objects of both transmission network and distribution network is proposed in [15]. Transmission network and distribution network interact with each other via MATE. The main ideas of the partition decoupling calculation and the coordination of partition results in the parallelization of EMT simulation realized in this paper come directly from [15]. The improvement of this paper is mainly reflected in the ability to achieve a more flexible and efficient network according to the actual situation, and then truly realize the parallelization of simulation calculation, which is not realized yet in [15].

3.2 An improved label propagation algorithm appropriate for flexible partition requirements

The influence of partition results on decoupling calculation and parallel calculation is not discussed in reference [14] [15]. In fact, as a traditional problem of complex network theory, a very important starting point of Network Partitioning problem is that it can flexibly and efficiently obtain the partitioning scheme of large-scale computing task network.

The basic indicators used to evaluate the effect of parallelization of computing tasks are as follows: (1) the size of the subnet should be proportional to the computing power of a single computing resource. The purpose is to let the computing tasks allocated to all computing resources be completed as synchronously as possible, so as to avoid the mutual waiting time when the calculation results are summarized and coordinated. (2) the interaction between subnets should be as little as possible. This kind of interaction means information exchange between different computing resources in the process of parallel computing, the time required for such interaction should be reduced as much as possible. There are several mature algorithms to achieve this task, including Kernighan-Lin algorithm [16], spectral partitioning algorithm [17] and heuristic search algorithm [18].

The main goal of this paper is to provide a "universal" parallelization program framework on the premise of satisfying the best possible parallelization effect. Therefore, the networking method is also required to be universal and can adapt to a variety of hardware computing resources. For this reason, the label propagation algorithm [19] is improved in this paper according to the actual needs for "universal" purpose. The general process of the classic label propagation algorithm is as follows:

1. Each node gets a label after initialization, which can be the number or name of the node. Or only some of the known feature nodes are labeled, and the rest of the nodes are empty labels.
2. Set the propagation probability for the label. If no specific propagation probability defined, the propagation probability can be set to 1 by default.
3. Update the labels for the nodes one by one. The label with the highest propagation probability in the label of the neighbor node will be selected as the new label for that node. When the label propagation probability is set to 1, it is equivalent to selecting the labels that appear most in the neighbor nodes.
4. After updating all node labels, compare them with the previous round of labels. If the labels are different, iteration should be continued until the convergence condition is reached. If the labels are the same, the iteration is completed, and the nodes with the same labels are the same community.

However, the above classical label propagation algorithm has a disadvantage. It cannot specify the number of partitions, which greatly restricts its applicability. In addition, to achieve the goal of "general " parallelization program framework, it is also necessary to be able to set the partition proportion manually and have sufficient partition speed. Therefore, we have improved the classical label propagation algorithm as the follow way.

1. The number of nodes labeled at initialization is equal to the number of partitions specified, and the remaining nodes have no labels. Labels are propagated to all nodes through the label propagation mechanism. So that the goal of specifying the number of partitions can be achieved.
2. Nodes are weighted according to the model complexity of the electrical components corresponding to them. The partition size is defined by the sum of the weights of the nodes in the partition instead of the number of nodes, which is closer to the goal of completing the calculation of different computing resources at the same time. Multiplication times are commonly used in program development to approximate model complexity.
3. Edges are weighted according to the amount of transmission information they bear. When updating a node, the edge weights of neighboring nodes that belong to the same label are summed up. The largest label of the sum of edge weights is the new label of the node. This makes it easier for neighbor nodes with high edge weights to pass their own labels to target nodes. So that the edge weights between partitions can be minimized and the information exchange between partitions can be reduced when partitioning the network.
4. Based on the idea of node label propagation capacity in [20], when the total computing task quantity and the scale ratio of each partition are determined, the expected computing task quantity of each partition can be determined. Label propagation capacity in this paper is defined as an increasing function of the difference between the actual scale of the node and the expected computing task. As the actual size of a partition approaches the expected size, the possibility of nodes in that partition being moved to other partitions decreases. Finally, the purpose of controlling the proportion of partition scale is achieved.

It should be noted that the partition problem is in fact discrete, so it is usually unrealistic that the partition size is exactly equal to the expected size. From the point of view of engineering application, the improved label propagation algorithm introduced here can meet the requirements of the parallelization of EMT simulation.

Take the power network model with 10000 nodes and 12217 edges [21] in matPower software package as an example. We tested a variety of partition requirements and got the following results.

Test 1: Assume that there are five partitions in which the nodes and edges are not weighted, and the expected partition size ratio is 1:1:1:1:5. The partition results are shown in Table 1.

Tab.1 Results for Test 1

Expected partition ratio (%)	11.11:11.11:11.11:11.11:55.56
Actual partition ratio (%)	10.87:10.84:10.65:11.63:56.01
Maximum partition error	1.94%
Interaction edge weight ratio	13.88%
Time (s)	9.44021

Test 2: Test the situation where the computing power of computing resources varies greatly. Assume that there are 18 partitions in which the nodes and edges are not weighted, and the expected partition size ratio is 1:1:1:....:1:5:10:20. The partition results are shown in Table 2.

Tab.2 Results for Test 2

Expected partition ratio (%)	2:2:2:2:2:2:2:2:2:2:2:2:2:2:2:10:20:40
Actual partition ratio (%)	1.91:1.89:1.98:2:2.23:2.08:2.11:1.94:1.99:1.99 :2.14:2.02:2.12:1.95:1.85:10.42:19.75:39.63
Maximum partition error	2.24%
Interaction edge weight ratio	17.39%
Time (s)	35.46993

From these results, even for the ten thousand scale power grid model, the proposed network partitioning method can get acceptable results in a reasonable time. Furthermore, it has a higher adaptability to the requirements of arbitrary weight distribution and arbitrary sub-network scale. Since this kind of network partitioning method only needs to be carried out once before parallel computing, so it does not affect the effect of parallel computing procedure itself. However, the label propagation algorithm itself can also been parallized, while is suitable for the calculation tasks with the distribution scheme needs to be adjusted in real time, which is not detailed here.

3.3 Realization of the Parallelization of EMT simulation in InterPSS

Since the MATE algorithm and the label propagation algorithm are both universal, most parallel computing techniques are feasible to realize the parallelization of EMT simulation in InterPSS. In this paper, it is realized based on the multithreading technology under JAVA utilizing identical processors in traditional PCs.

Ideally, if the computing tasks are intensive, for hardware platforms with N identical processors, setting the size of the thread pool to N+1 can achieve optimal utilization efficiency. In this paper, the transient simulation program based on InterPSS needs to be parallelized for each integration step, which is a computing-intensive task. Therefore, before starting parallel computing, the original power network model should be divided into N+1 subnets.

For the parallel calculation of each time step, firstly, each subnet is solved independently without considering the injection current of the interacting tie line. The calculation of each subnet is submitted to the thread pool as an independent task, which is calculated independently in different threads.

After the tasks of each thread are completed, the results of each sub-network are coordinated and modified based on the method introduced in [14]. Because this task requires multiple subnets to be operated at the same time, it cannot be performed in a separate thread, hence it can't be parallelized further. However, the sub-network method in this paper has reduced the scale of sub-network contact as much as possible. These non-parallelizable operations account for a small proportion of the whole computing task and have a limited impact on the overall computing performance. When all the calculation tasks of this step are completed, the simulation is completed. We use the common methods in JAVA to achieve parallelization, and there are not many special operations. The parallelization method proposed in this paper is universal and does not depend on specific numerical algorithms and hardware resources. Therefore, it can be easily extended to other high-performance computing environments.

4. Case study

4.1 Generating numerical examples

For parallel computing tasks, the larger the scale of the example, the better the parallel effect. At present, the scale of power system examples that can directly carry out EMT simulation under InterPSS are limited. Therefore, similar to the method introduced in [22], the authors can generate suitable example systems with different scales based on the conventional relatively small-scale IEEE standard examples.

In this paper, the famous IEEE 3-machine and 9-bus case is used to generate large cases. The InterPSS transient simulation network object of the 3-machine 9-bus case is generated first. Then we can clone this object to get another network object, and through the object-oriented way to splice two smaller objects into larger objects. On the premise of ensuring the normal simulation of the transient simulation network, some of the nodes and edges will be deleted according to a specific mechanism to avoid the resulting network being too "uniform".

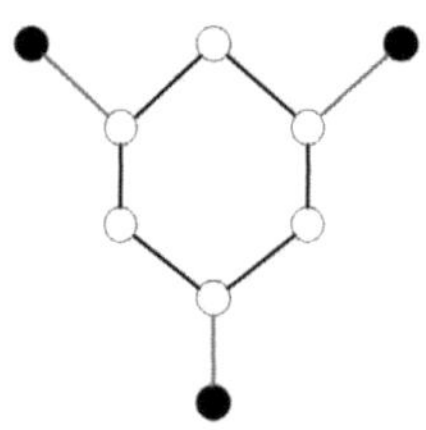

(a) 3-machine and 9-node topology

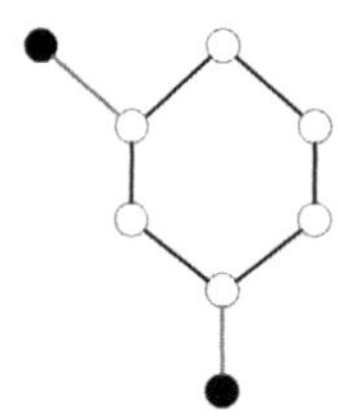

(b) Post-disturbance topology

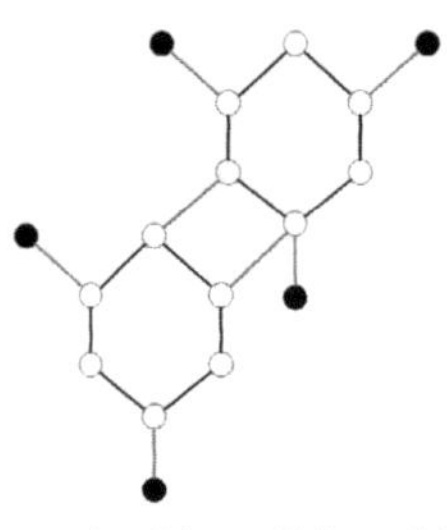

(c) The result of the splicing of (a) and (b)

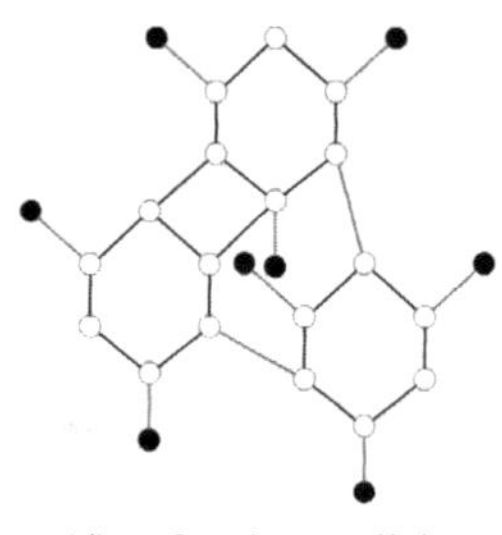

(d) Continue splicing

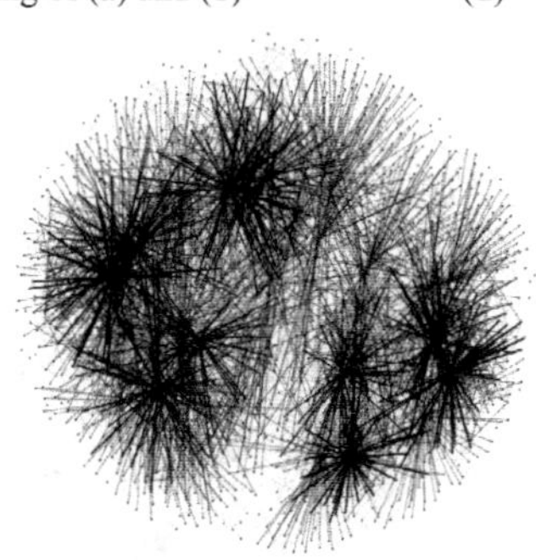

(e) Large-scale power grid topology (an 1800-node network whose layout is generated by the Frutchterman Reingold algorithm of the software Pajek)

Fig. 5 Splicing procedure for bulk power system cases

Since then, all the splicing results will be used as "material" for further expansion of network objects. For example, if the 3-machine 9-node network object is regarded as an "atomic" network, another network can be obtained after cloning and simple disturbance, and a larger network can be obtained by splicing the two together. If further expansion required, one can splice any combination of the three existing networks. Finally, more and more network objects of different sizes with differences can be obtained. The specific operation process is illustrated in Fig. 5.

As a comparison, the reference [21] is based on the standard case of 10 machines and 39 nodes to generate power network models of different sizes, and the network topology is highly regular. However, the topology of the case used in this paper is relatively irregular. Obviously, the effect of grid separation and parallel calculation may not as good as that from [21], but it is not due to the methods, but due to different test cases, while the case in this paper should be more in line with the actual power grid situation.

4.2 Design of the simulation experiments to show parallelization effect

Suitable experiments are defined to verify the accuracy of the results obtained by using this method. Using the example with the scale of 1800 nodes mentioned above, the numerical integration step is set to 0.005 seconds, and the total simulation time is 20 seconds. The specific sequence of events is as follows. At 0.1 second, the three-phase short-circuit disturbance is set for node 300, and the duration is 0.25 seconds. At 0.2 seconds, the three-phase short-circuit disturbance is set for node 820, and the duration is 0.35 seconds. At 0.25 seconds, the three-phase short-circuit disturbance is set for node 1620, and the duration is 0.55 seconds.

The curve of generator power angle at node 299 before and after parallelization is shown in Fig. 6.

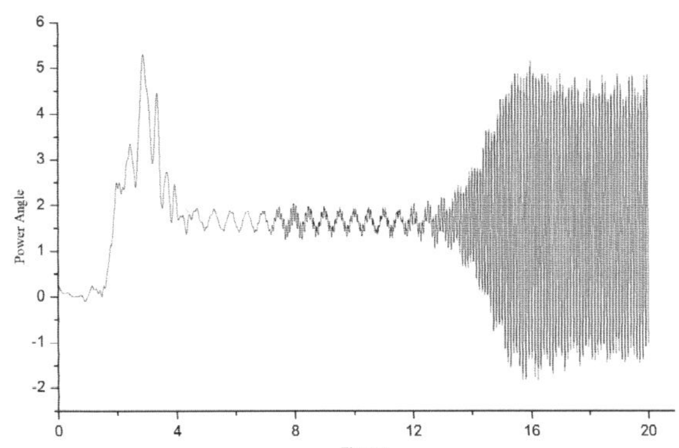

Fig. 6 Comparison of output results before and after parallelization

Fig. 6 depicts two curves corresponding to the simulation results before and after parallelization. Because the simulation results are completely coincident because the linear characteristic of MATE, even when the system has been unstable in the later stage of the simulation, only one curve can be seen in the figure.

Table 3 shows the comparison of the parallelization performance of different simulation networks with the same disturbance. The simulation was performed in a Windows Server 2012 R2 64-bit virtual machine with 16.0GB memory and up to 16 CPU.

Tab.3 Comparison of The Results of Parallelization

System scale (number of nodes)	Non-parallel computing time/ms	Number of threads	Parallel computing time/ms	Acceleration ratio
1800	88687	4	66563	1.332
4500	148662	8	74361	1.999
6200	235580	12	98234	2.398
7200	350819	16	117901	2.976

4.3 Result analysis

Firstly, the two curves in Fig. 6 are completely coincident because the decoupling calculation and the coordination of the results are linear operations. The simulation algorithm itself does not change and will not generate new errors from parallelization.

Secondly, it can be seen from Table 4 that the larger the scale of the simulated power grid, the greater the speedup obtained by parallelization. For further analysis, define

$$Normalized\_Acceleration\_Ratio = Acceleration\_ratio / Thread\_number / Network\_Scale \quad (1)$$

to characterize the effect of acceleration more accurately. Fig. 7 shows the curve of relationship between the normalized acceleration ratio and the scale of power network in Table 4. As can be seen from the figure, the effect of unit speedup decreases with the increase of the size of the network. With the increase of the size of the network, the amount of information that different networks interact with each other is also increasing, which will weaken the effect of parallelization. The existing reports on parallel computing have similar conclusions. However, as can be seen from the figure, with the increase of the scale of the power grid, the downward trend is slowing down. Limited by the available hardware resources, the conclusion has not been directly verified in larger computing resources. However, it can be expected that with the increase of the size of the power grid and the number of available threads, the parallelization effect of this algorithm will be maintained to a considerable extent.

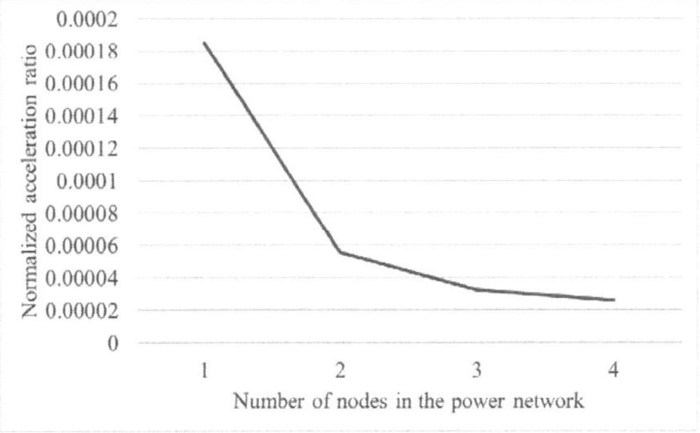

Fig. 7 Relationship between the normalized accelerative ratio and the scale of power network

5. Conclusion

In this paper, a general parallel computing framework based on the EMT simulation function of InterPSS is proposed, which can achieve flexible and efficient network partition and decoupling calculation of different subnets based on the results. It is parallelized in JAVA multithreading environment as a test to show the possibility of "general framework", and a good effect is achieved. The large-scale simulation case shows that the parallelization effect of this method is less affected by the increase of simulation scale.

The method in this paper is not limited by specific high-performance computing resources. In particular, the label propagation algorithm proposed in this paper can in fact partition the highly heterogeneous high-performance computing platform composed of hardware with great differences in computing power. Moreover, the decoupling calculation method provided has nothing to do with the specific partition scheme, which makes this method have a strong generalization ability. The methods will be more valuable for the computing tasks that need to be carried out through cloud-edge collaborative computing and communication systems in the future.

Limited by the available hardware resources, some potential values of this paper have not been fully verified, such as the slow decline of unit speedup, parallel computing under highly heterogeneous high-performance computing platforms and so on. All these can be implemented in near future.

Acknowledgement
This work is supported by Science and Technology Project of State Grid Corporation of China (Research on the parallization of electromagnetic transient and hybrid simulation for power system with HVDC and high-proportioned new energy based on heterogeneous hardware platform, No.SGSHDK00DWJS1900400).

Reference:

[1] XUE Wei, SHU Jiwu, YAN Jianfeng,*et al.* Cluster-based parallel simulation for power system transient stability analysis[J]. Proceedings of The Chinese Society for Electrical Engineering, 2003, 23(8): 38-43.
[2] Chai J S , Bose A . Bottlenecks in parallel algorithms for power system stability analysis[J]. IEEE Transactions on Power Systems, 1993, 8(1):9-15.
[3] Decker I. C., Falcao D. M., Kaszkurewic Z. E., Parllel implementation of a power system dynamic simulation methodology using the conjugate gradient method[J]. IEEE Transactions on Power System, 1992, 7(1):458-466.
[4] Alvarado F L. Parallel solution of transient problems by trapezoidal integration[J]. IEEE Traps on PowerApparatus and Systems, 1979, 98(3).
[5] Scala M L , Brucoli M . A Gauss-Jacobi-Block-Newton method for parallel transient stability analysis[J]. IEEE Transactions on Power Systems, 1990, 5(4):1168-1177.
[6] Scala M L , Sbrizzai R , Torelli F . A pipefined-in-time parallel algorithm for transient stability analysis[J]. IEEE Transactions on Power Systems, 1991, 6(2):715-722.
[7] Granelli G P , Montagna M . Relaxation-Newton methods for transient stability analysis on a vector/parallel computer[J]. IEEE Transactions on Power Systems, 1994, 9(2):637-643.
[8] Scala M L , Bose A . Relaxation/Newton methods for concurrent time step solution of differential-algebraic equations in power system dynamic simulations[J]. Circuits & Systems I Fundamental Theory & Applications IEEE Transactions on, 1993, 40(5):317-330.
[9] HONG Chao. A parallel-in-time algorithm for analysis of power system transient stability[J]. Power System Technology, 2003, 27(4): 31-35.
[10] M ahseredjian J, Dennetiere S, Dude L, *et al.* On a new approach for the simulation of transients in power systems[J]. Electric Power System Research, 2007, 77(11): 1514-1520.
[11] MU Qing, LI Yalou, ZHOU Xiaoxin,*et al.* A parallel multi-rate electromagnetic transient simulation algorithm based on network division through transmission line[J]. Automation of Electric Power Systems, 2014, 38(7): 47-52.
[12] Mike Zhou. https://github.com/interpss[OL]. 2021.
[13] He Yangzan, Wen Zengyin, Power System Analysis[M], Wuhan: Huazhong University of Science and Technology Press, 2016.
[14] Tomim M A , Marti J R , Wang L . Parallel solution of large power system networks using the Multi-Area Thevenin Equivalents (MATE) algorithm[J]. International Journal of Electrical Power & Energy Systems, 2009, 31(9):497-503.
[15] Huang Q , Vittal V . Integrated transmission and distribution system power flow and dynamic simulation using mixed three-sequence/three-phase modeling[J]. IEEE Transactions on Power Systems, 2017, 32(5):3704-3714.
[16] Kernighan B W , Lin S . An efficient heuristic procedure for partitioning graphs[J]. Bell System Technical Journal, 2014, 49(2):291-307.
[17] Fiedler M . A property of eigenvectors of nonnegative symmetric matrices and its application to graph theory[J]. Czechoslovak Mathematical Journal, 1975, 25(4):619-633.
[18] Irving M R , Sterling M . Optimal network tearing using simulated annealing[J]. Generation Transmission & Distribution Iee Proceedings C, 1990, 137(1):69-72.
[19] Zhu X , Ghahramani Z . Learning from labels and unlabeled data with label propagation[J]. Tech Report, 2002, 3175(2004):237-244.
[20] Zhao Siqi, Research and improvement of community discovery algorithm based on Spark for

large scale complicated networks[D]. Changchun: Jilin University, 2018.
[21] Birchfield A B , Xu T , Gegner K M , et al. Grid structural characteristics as validation criteria for synthetic networks[J]. IEEE Transactions on Power Systems, 2017:1-1.
[22] HUANG Yuanyang. A parallel algorithm for electromechanical transient simulation by recursive block matrix inversion[J]. Power System Protection and Control, 2019,47(24):8-14.

Research on Vibration of Mechanical System Based on MATLAB

1st Changying Feng[1,a], Chen Pang[2,b]

[1]Wuhan University of Technology, Wuhan China

[2]Wuhan University of Technology, Wuhan China

[a]987934159@qq.com

[b]2381848721@qq.com

Abstract—Mechanical vibration is the main factor affecting mechanical performance. This vibration will seriously affect the smoothness of the mechanical system and reduce the service life of mechanical system components. In addition, severe mechanical vibration may also affect the stability of the mechanical system and produce noise. Therefore, studying mechanical vibration and proposing control methods will be a significant task. This paper is mainly based on MATLAB to study the vibration of a typical mechanical system, namely the body and wheel dual-mass system, and obtain its amplitude-frequency characteristics, and focus on the influence of system parameters on the root mean square value of vibration. This research can provide a certain useful reference for reducing vehicle vibration and improving vehicle driving comfort.

1. INTRODUCTION

Mechanical vibration is the back and forth movement of an object near a certain position. This kind of vibration phenomenon is common in nature, engineering technology, and daily life. For example, the screaming of summer cicadas, the tiny vibrations of various parts when the machine is turned on, the swing of the pendulum, the beating of the heart, and other phenomena have obvious vibration characteristics. Although the vibration phenomena in different fields have their own characteristics, they have common objective laws. In order to explore the mechanism of various vibration phenomena and clarify the basic laws of vibration, it is necessary to apply mechanics knowledge to establish a mathematical model, usually a differential equation (set), and then analyze and solve the built model. This process involves tedious mathematical derivation and Solving complex differential equations is tedious, time-consuming, and error-prone to solve manually. If you use computer software to solve it, you need a certain amount of computer programming knowledge. In terms of vehicle vibration research, Hu Qiguo [1] analyzed the time-frequency characteristics of the front and rear wheel excitations of vehicles on non-stable roads. Based on the second type of Lagrangian method, the vibration dynamics equation was established and established with the help of MATLAB/Simulink. The vibration simulation model shows that under the excitation of non-stable roads, appropriately reducing the stiffness of the car seat suspension and rear suspension and increasing the damping of the car seat suspension and rear suspension can improve the ride comfort of the driver.Zhang Jianli [2] In order to study the vehicle vibration caused by the road surface unevenness, the vehicle vibration system was taken as the research object, and the numerical simulation study of the vehicle vibration under the influence of the road surface unevenness was carried out. The generalized virtual work principle is used to establish a four-degree-of-freedom vehicle vibration differential equation system to achieve related vibration research.

Jin Qijun [3] conducted vertical vibration tests under different excitation amplitudes and measured the vibration transmission from the sea floor to the contact surface with the human body to characterize the dynamic performance of the human-chair coupling system. Then, the relative amplitude estimation method is designed for subjective evaluation, and the subjective and objective correlation analysis is carried out. Finally, a parameter that can characterize the vibration isolation performance of the seat is proposed. Zhang Chunhong [4] elaborated on the phenomenon of lateral vibration of the vehicle caused by the driveshaft, explained the evaluation method of lateral vibration from both subjective and objective perspectives, and carried out an in-depth analysis of the test data to determine the mechanism of lateral vibration. Based on the universal joint structure, the force analysis is carried out to identify the causes of the third-order axial circulation force, and the influence factors of the axial circulation force are explored from the four aspects of the arrangement angle, the joint structure, the fitting clearance, and the lubricating grease. Considering the vehicle positioning and control costs, finally determine the fit clearance and the optimization of lubricating grease, and realize the control of the vehicle's lateral vibration problem. Wang Zhijun [5] established a four-degree-of-freedom vehicle vibration model in order to evaluate the impact of vertical vibration and pitch vibration on car ride comfort. Derive the mathematical model through the Laplace transform. Based on MATLAB software, the deduced mathematical model is programmed, and the influence of vehicle vertical vibration and pitching vibration on the excitation of the front and rear wheels is analyzed, and the frequency response of the front and rear axle vibration to the excitation of the front and rear wheels are analyzed. Huang Yiyang [6] used Matlab/Simulink to complete the seven-degree-of-freedom dynamics model of automobile suspension and analyzed its ride comfort with B-level road white noise as an excitation, which provided important theoretical support for subsequent ride comfort research. Tang Tiande [7] established a four-degree-of-freedom automobile two-axle suspension dynamic model and its Simulink simulation model to study the vertical vibration and longitudinal angular vibration of the vehicle body and proposed a method to solve the front and rear suspension amplitude-frequency characteristics by analyzing Simulink time-domain simulation data. A method to convert the amplitude-frequency characteristics of a general dual-axle suspension model with different suspension mass distribution coefficients not equal to 1 under dual input.

Cui Yiwen [8] simplified the structure of a light-duty vehicle into a two-degree-of-freedom vibration model and took this model as an example. In the case of using simple harmonic excitation to simulate the road surface excitation, the dynamic equation was established by the generalized Lagrangian equation, and based on Simulink Process and simulated the dynamic equations of light-duty vehicles. By setting different parameters, the simulation results of each displacement under different conditions are obtained. After analysis and calculation, the simulation results of the model are obtained. There is a certain reference for the optimization design of vehicle dynamic parameters. value. Wang Dengqiang [9] simplified the vehicle vibration analysis model by analyzing the vehicle vibration form. On this basis, Matlab/Simulink simulation software is used to establish a two-degree-of-freedom vibration simulation model of the vehicle that can be solved, and then the random B-level road spectrum model is used as the excitation source, and the acceleration response of the vehicle body, the vertical direction of the suspension, and the vertical direction of the tire are analyzed. Vehicle comfort indicators such as dynamic load. Li Jie [10] established a 1/2 vehicle 5 degrees of freedom military vehicle vibration analysis model for the random vibration analysis of military vehicles, derived the formula for vehicle random vibration analysis based on the virtual excitation method, and gave the international unit system of absorbed power Indicates that the main indicators and auxiliary indicators of military vehicle random vibration analysis have been determined, and the military vehicle random vibration analysis software has been developed using MATLAB, and the random vibration of a military vehicle has been studied. The results show that the random vibration analysis of military vehicles based on the virtual excitation method is basically consistent with the traditional method, but the method is simpler and more versatile. These studies have carried out a good analysis of the vehicle vibration. Under the MATLAB simulation, the problem has been studied thoroughly, but it is slightly complicated. This article mainly uses a simplified model to analyze the vibration of the dual-mass system of the body and the wheel for analysis. The impact of relevant parameters on vehicle vibration. Jiang Xin [11] simplified the car model into a four-degree-of-freedom dynamic model of an independent suspension vehicle. Newton's second law is used to establish the differential equation of the car in the driving process. On this basis, a vehicle suspension model based on Simulink software is built and simulated in the time domain and frequency domain, and the white noise function is used as the vibration input of the road surface. At the same time, it analyzes the amplitude-frequency characteristics and proposes a scheme to improve the suspension. At the same time, based on the time domain, the sensitivity analysis of the suspension and tire parameters is done. Hong Dekai [12] studied the vibration characteristics of light trucks based on finite element complex modal analysis and harmonic response analysis.

On the basis of summarizing and analyzing the abnormal vibration manifestations and characteristics of light trucks, according to the various parameters of the truck provided by the company, the finite element model of the truck is established, and the constraint complex modal analysis and harmonic response analysis of the entire truck model are carried out. The low-order vertical natural vibration characteristic parameters of the truck are obtained, and the vertical vibration response of the truck cab position under the action of the wheel's own radial runout excitation is explored. The above analysis results basically determine the cause of the abnormal vibration of the light truck. Based on this, the influence of suspension stiffness and damping parameters on the vibration characteristics of freight cars is further explored.

2. Modeling

This paper establishes a four-degree-of-freedom mathematical model of the automobile through the research of automobile dynamics, and transforms the vibration of the automobile into the process of establishing the model. It has a greater impact on the ride comfort and the grounding of the wheel, and is closer to the car suspension. The actual power of the system.

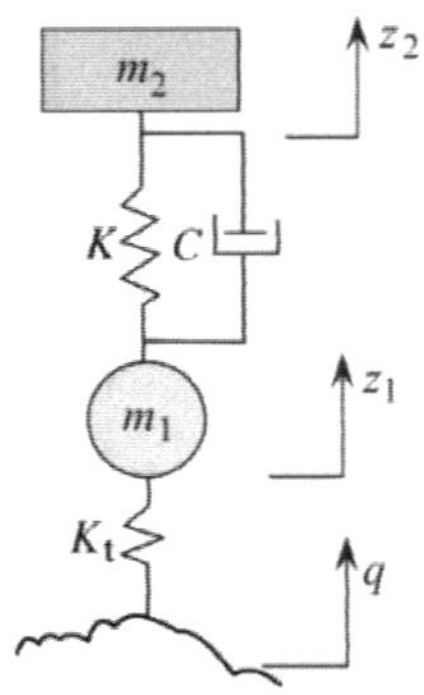

Figure 1. Two-degree-of-freedom vibration system for body and wheels

The vertical displacement coordinates of the wheel and the car body are z_1,z_2, and the coordinate origin is selected at their respective equilibrium positions. The motion equation of the car body and the car in Figure 1 is:

$$\begin{cases} m_2\ddot{z}_2 + C(\dot{z}_2 - \dot{z}_1) + K(z_2 - z_1) = 0 \\ m_1\ddot{z}_1 + C(\dot{z}_2 - \dot{z}_1) + K(z_1 - z_2) + K_t(z_1 - q) = 0 \end{cases} \tag{1}$$

In the case of undamped free vibration, the equation of motion becomes:

$$\begin{cases} m_2\ddot{z}_2 + K(z_2 - z_1) = 0 \\ m_1\ddot{z}_1 + +K(z_1 - z_2) + K_t z_1 = 0 \end{cases} \tag{2}$$

It can be seen from the equation of motion that the vibrations of m_2 and m_1 are coupled with each other. If m_1 does not move($z_1 = 0$), then

$$m_1\ddot{z}_1 + Kz_2 = 0 \tag{3}$$

This is equivalent to a single degree of freedom undamped free vibration with only the body mass m_2, and its natural circular frequency $\omega_0 = \sqrt{K/m_2}$. Similarly, if m_2 does not move ($z_2 = 0$), it is equivalent to the undamped vibration of the wheel mass m_1 with a single degree of freedom, then:

$$m_1\ddot{z_1} + (K + K_t)z_1 = 0 \tag{4}$$

The natural circular frequency of the wheel part is

$$\omega_t = \sqrt{(K + K_t)/m_1} \tag{5}$$

At this time, since the body is basically not moving, the two-degree-of-freedom system can be simplified as a single-mass system of the wheel part shown in Figure 2 to analyze the vibration of the wheel part in the high-frequency resonance region.

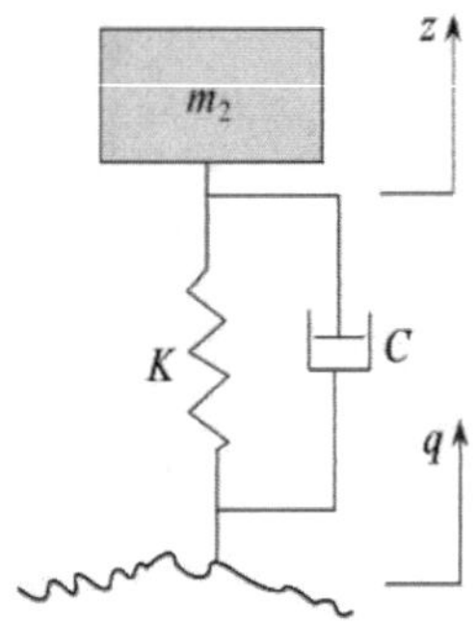

Figure 2. Single-mass system for wheel part

After simplification again, the motion equation of mass m_1 at this time is:

$$m_1\ddot{z_1} + C\dot{z_1} + (K + K_t)z_1 = K_t q \tag{6}$$

Substituting each complex amplitude into the above formula:

$$-\omega^2 m_1 z_1 + j\omega C z_1 + (K + K_t)z_1 = K_t q \tag{7}$$

The frequency response function of wheel displacement z_1 to q is:

$$\frac{z_1}{q} = \frac{K_t}{-\omega^2 m_1 + K + K_t + j\omega C} \tag{8}$$

Divide the numerator and denominator of formula (8) by $K + K_t$, and substitute the natural circular frequency ω_t of the wheel part and the damping ratio ζ_t of the wheel part into this formula, then:

$$\frac{z_1}{q} = \frac{K_t/(K + K_t)}{1 - \left(\frac{\omega}{\omega_t}\right)^2 + j2\zeta_t\omega/\omega_t} \tag{9}$$

Its amplitude-frequency characteristics are:

$$\left|\frac{z_1}{q}\right| = \frac{K_t/(K + K_t)}{\sqrt{\left[1 - \left(\frac{\omega}{\omega_t}\right)^2\right]^2 + (j2\zeta_t\omega/\omega_t)^2}} \tag{10}$$

At high frequency resonance $\omega = \omega_t$, the rms spectrum of wheel acceleration $\sqrt{G_{\ddot{z}_1}(\omega_t)}$ is proportional to the amplitude-frequency characteristics, that is,

$$\left|\frac{\ddot{z}_1}{\dot{q}}\right| = \frac{\omega_t K_t/(K + K_t)}{2\zeta_t}$$

(11)

It can be seen that reducing the tire stiffness K_t can reduce ω_t and increase ζ_t, which is an effective way to reduce the acceleration of the wheel part at high frequency resonance; reducing the non-suspension mass m_1 makes both ω_t and ζ_t increase, and when the wheel part resonates at high frequency The acceleration of is basically unchanged, but the dynamic load $m_1\ddot{z}_1$of the wheel part decreases, which is beneficial to reducing the relative dynamic load. For the dual-mass vibration system, we also need to calculate and analyze the body acceleration $\ddot{z}_2$, the suspension spring dynamic deflection and the amplitude-frequency characteristics of the wheel relative to the dynamic load to study the relative influence of the parameters of the body and the wheel dual-mass system. Body acceleration $\ddot{z}_2$The amplitude-frequency characteristic of q is:

$$\left|\frac{\ddot{z}_2}{\dot{q}}\right| = \omega\gamma\left(\frac{1 + 4\zeta^2\lambda^2}{\Delta}\right)$$

(12)

3. SIMULATION

After the research of related models, two simulation studies were mainly carried out in matlab. One is to build a related model of the dual-mass vibration system in matlab to study the root mean square value spectrum $\sqrt{G_{\ddot{z}_1}(f)}$, $\sqrt{G_{\ddot{z}_2}(f)}$ and $\sqrt{G_a(f)}$ and the amplitude-frequency characteristics of the corresponding body and wheel dual-mass vibration system $\left|\frac{z_1}{q}\right|$, $\left|\frac{z_2}{z_1}\right|$, $\left|\frac{p}{z_2}\right|$; Choose the body and wheel dual mass system parameters $f_0 = 1.5Hz$, $\zeta = 0.25$, $\gamma = 9$, $\mu = 10$, the vehicle speed is 20m/s, in matlab Code writing and simulation calculations are carried out in the process to obtain Figure 3 to Figure 4:

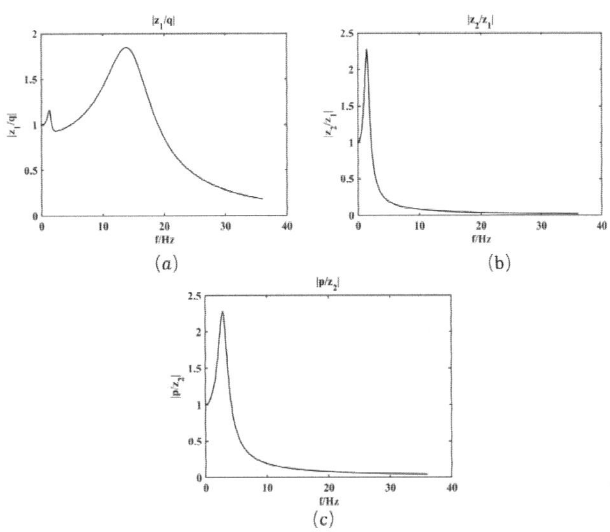

Figure 3. Amplitude-frequency characteristic diagram(a), Amplitude-frequency characteristic diagram(b), Amplitude-frequency characteristic diagram(c)

Finally, complete content and organizational editing before formatting. Please take note of the following items when proofreading spelling and grammar:

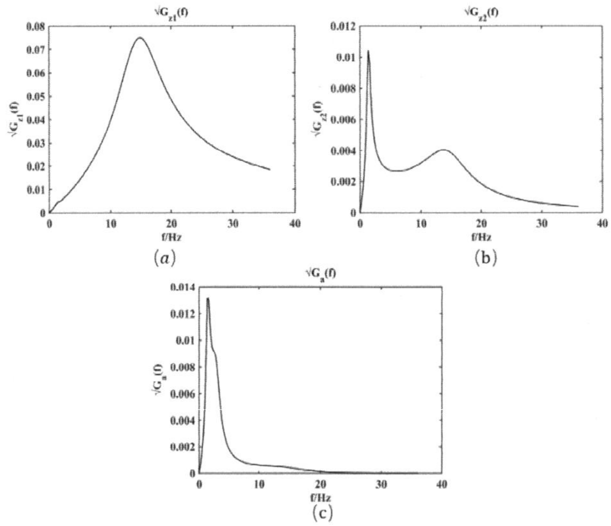

Figure 4. Root mean square spectrum spectrum of wheel acceleration(a),Spectrogram of RMS value of vehicle body acceleration(b),Human body acceleration root mean square value spectrum spectrum(c)

From the simulation results, it can be seen that the root mean square value spectrum is proportional to the corresponding amplitude-frequency characteristics, and the curve shape is highly similar, which can explain the accuracy and correctness of the related model. It can also explain that for the dual-mass system vibration system, as long as the acceleration root mean square value spectrum is studied, the corresponding amplitude-frequency characteristic curve can be obtained, and the vibration response of the vehicle body and wheels can be judged.

4. Conclusion

This paper mainly studies mechanical vibration. Through continuous simplification, a four-degree-of-freedom model is transformed into a single-degree-of-freedom model. The root mean square value spectrum $\sqrt{G_{\ddot{z}_1}(f)}$, $\sqrt{G_{\ddot{z}_2}(f)}$ and $\sqrt{G_a(f)}$ and the amplitude-frequency characteristics of the corresponding vibration system $\left|\frac{z_1}{q}\right|$, $\left|\frac{z_2}{z_1}\right|$, $\left|\frac{p}{z_2}\right|$. This method avoids the lengthy and tedious process of mathematical derivation and differential equation solving, and can easily debug model parameters, and intuitively observe the impact of related parameter changes on the vibration system. It has the characteristics of simple principles, intuitive phenomena, and strong interaction. It has certain application value in subject scientific research and engineering practice.

References

[1] Hu Qiguo, Qian Kai, Li Leike, Yue Guangjie, Si Hongjian. Modeling and simulation of vehicle vibration dynamics under non-stationary road surface excitation[J]. Journal of Zhengzhou University (Engineering Edition), 2012, 33(06): 49 -53.
[2] Zhang Jianli, Chen Rongfeng, Yao Ke. Analysis of the impact of road surface roughness on vehicle vibration[J]. Highway and Transportation Science and Technology, 2019, 36(11): 129-133.
[3] Jin Qijun, Luo Qian. Analysis of two-degree-of-freedom and four-degree-of-freedom automobile vibration model based on MATLAB[J]. Science and Technology and Innovation, 2020(17): 67-69.
[4] Zhang Chunhong. Simulation of automobile vibration system based on MATLAB[J]. Mechanical Engineering and Automation, 2008(02): 83-84.
[5] Wang Zhijun, Jing Ke, Chen Jingxin. Vehicle vibration response analysis based on MATLAB[J]. Agricultural Equipment Technology, 2016, 42(04): 54-57.

[6] Huang Yiyang, Hu Xu. Research on car ride comfort based on Simulink[J]. Equipment Manufacturing Technology, 2020(02): 112-114+127.
[7] Tang Tiande. Dynamic simulation analysis of automobile two-axle suspension based on Simulink[J]. Agricultural Equipment and Vehicle Engineering, 2010(04):24-27+30.
[8] Cui Yiwen, Liu Jun, Miao Tongchen. Light-duty vehicle vibration simulation based on Simulink[J]. Science and Technology Innovation and Application, 2015(26): 48-49.
[9] Wang Dengqiang, Dong Zhengxin, Zhang Kun. Analysis of vehicle vibration speed response based on Simulink[J]. Hubei Agricultural Mechanization, 2019(04): 62.
[10] Li Jie, Wang Wenzhu, Zhao Qi, Zhang Chuxu. Random vibration analysis of military vehicles based on virtual excitation method[J]. Automotive Engineering, 2016, 38(03): 368-372.
[11] Jiang Xin. Vehicle vibration model analysis under random road surface excitation[J]. Science Technology and Innovation, 2021(17): 7-9.
[12] Hong Dekai. Vibration analysis and damping design of light trucks [D]. Shandong University of Technology, 2020.

Journal of Physics: Conference Series **2195** (2022) 012051 doi:10.1088/1742-6596/2195/1/012051

Theory and application of gas drilling technology

Xiangbing Pei*, Jianqiu Feng, Ke Jiang, Tan Zhang, Xiaonan Wang

PetroChina Coalbed Methane Company Limited, Lvliang Shanxi 033200, China

* Corresponding author: fjq2009@petrochina.com.cn

Abstract-Gas drilling began in the 1950s. The drilling process is implemented by injecting a mixture of air and water into the well with an air compressor. Field experiments have gradually proved that air drilling in hard rock can improve the ROP by 2 ~ 3 times, especially in reservoir protection. In this paper, the gas drilling technology is analyzed and discussed. The test results of gas drilling technology show that the drilling time is reduced and the mechanical ROP is greatly improved. Compared with mud drilling, the mechanical ROP of gas drilling in the same structure and well section is generally increased by 2 ~ 10 times, the production of oil and gas wells is relatively high and the economic benefit is remarkable.

1.Introduction

Gas drilling technology is a new drilling technology with gas as circulating medium. Compared with conventional drilling, the advantages of gas drilling are mainly to discover and protect oil and gas reservoirs, reduce downhole complexity such as lost circulation, improve single well mechanical ROP and single well recovery. In recent years, gas drilling has developed rapidly in improving the ROP and protecting the reservoir. Compared with conventional drilling fluid drilling, the ROP is increased by 5 ~ 15 times and the drilling cycle is shortened. Practice shows that the influence of gas itself on borehole stability is not very serious [1]. For example, when gas drilling is used in loose sandstone formation, annulus plugging is rarely seen no matter how high the ROP is, because small-grained sandstone can smoothly pass through the key point on the upper part of the drill collar. Gas drilling is gradually becoming an important technical means to greatly improve the ROP, discover and protect oil and gas reservoirs, control serious lost circulation and reduce drilling costs. In the United States and Canada, underbalanced drilling accounts for about 40% of the total drilling workload, and the number of wells implementing gas drilling accounts for about 30% of the total underbalanced drilling workload [2].

Gas drilling has experienced two development stages in the United States. In 1953, eipaso natural gas company of the United States successfully drilled the first oil and gas well in Utah with air as drilling fluid for the first time. Until 1990, air drilling was widely used in engineering applications such as drilling speed increase, leakage prevention and control in non oil and gas bearing reservoirs. Since 1990, due to the needs of developing a large number of low-pressure and low-permeability oil and gas resources, coalbed methane, tight gas and other difficult to use oil and gas resources in the Midwest of the United States and Canada, inert gas drilling technology has been developed, especially the gas drilling technology combined with special orbital wells. China's gas drilling technology started almost at the same time as foreign countries, but its development is slow. It has experienced the early exploration stage (1950s ~ 1970s), the initial stage (late 1980s ~ 2004) and the development stage (2005 ~ now). With the rapid development of gas drilling in Sichuan basin after 2005, gas drilling shows the characteristics of revolutionary technological progress. Through the research in recent years, China's gas drilling technology and equipment have reduced the gap with foreign countries, reached the more advanced level in the world, and realized large-scale application in Sichuan, Chongqing, Xinjiang, Jilin and other regions [3].

At present, foreign countries have mature downhole information transmission technologies such as wireline MWD, downhole storage measurement and electromagnetic MWD, as well as gas screw drilling tools and pneumatic hammers for horizontal wells, but China is still in the research stage in these aspects. Many domestic institutions have carried out EM-MWD research. The current MWD technology of Chuanqing drilling and production Institute has been tested to a well depth of 4300m, and mature products will be available soon. Many domestic institutions have gas screw drilling tools, but there are problems such as high speed, low torque, small load stall and short service life. The technology is imperfect and has not been popularized and applied in a large area. Chuanqing company has developed the guided air hammer, which is currently in the test stage [4]. The successful test of this product will promote the gas drilling horizontal well technology to a new level. Domestic Chuanqing gas drilling company has successfully carried out gas drilling tests of two wells in Guang'an area, Sichuan, with the longest horizontal section drilled to 537.9m, creating a domestic record of gas drilling horizontal wells.

According to the application at home and abroad, combined with the characteristics of gas drilling technology, the advantages are mainly manifested in discovering and protecting the reservoir, improving the ROP and reducing lost circulation. The main characteristics and mechanisms are as follows: (1) protecting and discovering the reservoir. Reservoir damage is irreversible [5]. As long as damage occurs around the wellbore, it will be difficult to eliminate these damage. In the process of gas drilling, the gas phase (mixed gas phase) is used as the circulating medium, and the bottom of the well is in an unbalanced state to prevent drilling fluid from entering the formation. Therefore, gas drilling can play an important role in discovering and protecting oil and gas reservoirs; (2) improve the mechanical ROP. The substantial reduction of fluid column pressure in gas drilling wellbore not only improves the rock breaking efficiency at the bottom of the well, but also helps to reduce the "holding effect", so that the bit can continue to cut new rock instead of rolling broken rock cuttings, so as to improve the mechanical ROP; (3) Reduce or avoid lost circulation. The problem of lost circulation greatly increases the cost of drilling engineering, and will cause serious losses caused by formation damage. Gas drilling can effectively eliminate the problem of lost circulation. The bottom hole equivalent density of gas drilling is 0.001 ~ 0.90g/cm3, the window range is large, and underbalanced drilling can be realized in various pressure formations, so gas drilling can eliminate lost circulation; (4) Extend bit life. During gas drilling, the bottom hole is under negative pressure, which is beneficial to blow the broken rock away from the bottom of the well, effectively prevent the repeated breaking of the rock, and prolong the service time of the bit; (5) Reduce complex situations such as sticking. Gas drilling machinery has high ROP, shortens drilling time and reduces the wear of drilling tools and centralizers; Reduce the drilling tool load and prolong its service life; At the same time, the vertical pressure of gas drilling is low, reducing the probability of penetration and leakage. Therefore, the occurrence of complex underground accidents is reduced; (6) Reduce well completion and stimulation measures. Due to the underbalanced drilling in the process of gas drilling, the solid particles and liquid phase can not enter the formation, so that the reservoir can be protected. Therefore, no further stimulation measures are required and the relevant costs are reduced.

2.FLOW PATTERN THEORY OF GAS DRILLING

The theory and technology of gas drilling technology are mainly based on the study of different fluid forms. Through experimental study, the flow pattern of gas-liquid two-phase flow in annulus tube is divided into five types by Caetano: bubble flow, slug flow, dispersed bubbly flow, annular flow and bubbling flow. Under normal temperature and normal pressure, the flow patterns of gas-liquid two-phase flow in annulus can be divided into four types: bubbly flow, slug flow, stirring flow and annular fog flow. The specific description is as follows: bubbly flow. The gas phase is distributed in the form of smaller discrete bubbles in the continuous liquid phase of the annular space, and the smaller spherical bubbles move upward in an approximate linear trajectory. In a fully eccentric annular tube, small bubbles tend to move to a larger annular space in the region of small annular gap; slug flow. The gas phase mainly exists in the liquid phase in the form of large bubbles, namely Taylor bubbles. Bubbles are separated from each other by a liquid column containing small bubbles. The shape of Taylor bubble in annular tube is different from that in circular tube. It is wound on the outer wall of inner tube and occupies almost the whole section of annular tube; stirring flow. It is exhausted with slug flow, but it is very chaotic and unstable. The bubbles become very narrow and distorted into bubble flow; annular fog flow. There is a liquid film between the outer wall of the inner tube and the inner wall of the outer tube, and the annular space between the two films is occupied by the gas core containing small droplets. The four fluid forms are shown in Figure 1.

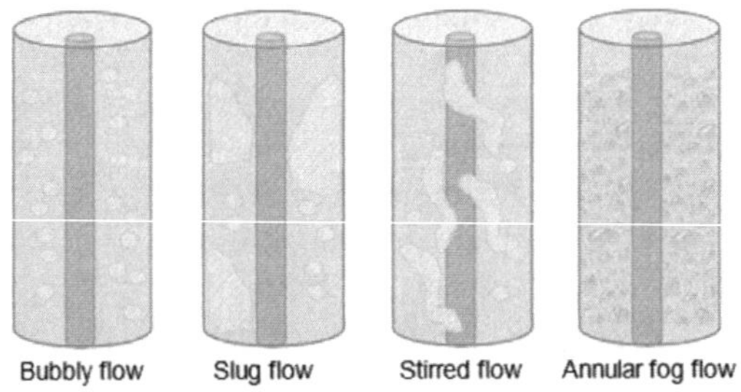

Fig. 1 flow pattern of gas-liquid two-phase flow in vertical annulus

Gas (including air or inert) drilling fluid is injected into the well at the same time to reduce the pressure of annulus fluid column. Inflatable drilling technology is a drilling circulating fluid with liquid phase as continuous phase and gas phase as dispersed phase. The bottom hole equivalent density can be adjusted by inflation and liquid injection. According to the different injection methods, it can be divided into riser injection, parasitic injection and concentric injection. The advantages of riser injection method are simple, do not affect the wellbore structure design, and can reduce the density of the whole circulation system. The disadvantage is that MWD and LWD measurement while drilling systems cannot be used, and the optimization of downhole motor parameters is limited due to the limitation of displacement. Parasitic pipe injection method and concentric string injection method are less used due to the disadvantages of high cost, complex process (special wellhead device is required), reduced wellbore (concentric string method), etc. According to the actual situation of Liaohe Oilfield, the riser (drill pipe) inflation injection method is selected as the nitrogen filling method. Therefore, the research and field test of aerated drilling are mainly aimed at this way. The process flow of aerated drilling is shown in Figure 2.

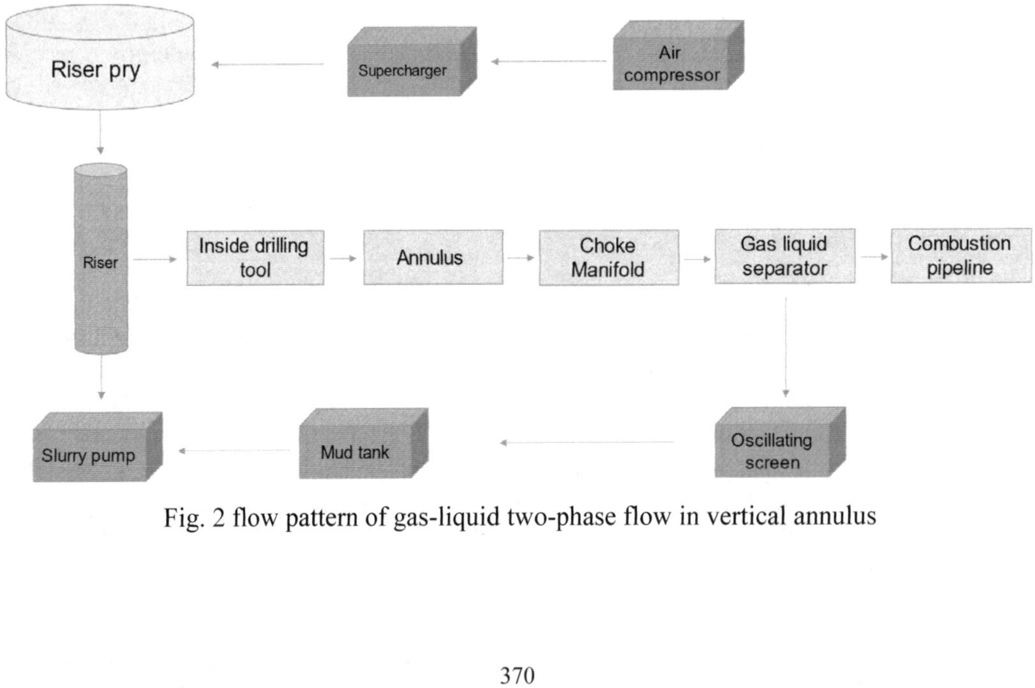

Fig. 2 flow pattern of gas-liquid two-phase flow in vertical annulus

3. PRINCIPLES OF PARAMETER DESIGN FOR AERATED DRILLING

The negative pressure value design of nitrogen filled drilling parameters must pay attention to the following problems:

(1) The separation capacity of the surface separation equipment is greater than the fluid flow of the formation invading the wellbore. The greater the negative pressure difference during nitrogen filled drilling, the greater the oil and gas output. Therefore, the designed negative pressure difference can not make the oil and gas output exceed the separation capacity of surface equipment, especially in high-yield reservoirs. The output of oil and gas can be estimated according to the production pressure difference and maximum initial production of adjacent wells.

(2) The rated working pressure of wellhead equipment meets the requirements of safe drilling. The more oil and gas produced during nitrogen filled drilling, the higher the casing pressure controlled by the wellhead. In principle, the maximum allowable casing pressure shall not exceed the minimum of the rated working pressure of the wellhead device, the internal pressure resistance of the casing (calculated as 80%) and the formation fracture pressure. The influence of oil and gas production on casing pressure is very complex, which needs to be calculated by gas-liquid two-phase flow numerical simulation software.

(3) Maintain wellbore stability. Sidewall stability is a major factor affecting the use of nitrogen filled drilling. Sidewall instability is mainly in mud shale formation, fracture zone and piedmont structural zone. Nitrogen filled drilling in Liaohe Oilfield is mainly Archean buried hill formation. The contact surface of this formation is unconformity ancient contact surface, mainly ancient buried hill weathering crust buried hill. There is fragmentation zone on the contact surface, so formation stability analysis should be carried out. Therefore, for underbalanced drilling in multiple layers, it is necessary to analyze the wellbore stability and determine the formation collapse pressure to determine the allowable maximum negative pressure difference.

(4) Instability of gas-liquid two-phase flow. The instability of gas-liquid two-phase flow leads to periodic instantaneous positive pressure difference and production formation damage, which is an important problem in gas filled drilling through drill string. Due to the instability of gas-liquid two-phase flow, the negative pressure difference shall not be too small to avoid periodic instantaneous positive pressure difference caused by pressure fluctuation.

(5) Optimization of gas injection rate. The relationship between bottom hole pressure and gas flow under a certain liquid flow rate. It can be divided into two areas: static pressure control area and friction control area. In the static pressure control area, hydrostatic pressure is dominant and the influence of gas flow is very sensitive; in the friction control area, the friction pressure drop is dominant, and the increase of gas flow has no obvious impact on the bottom hole pressure, which is relatively stable. In hydraulic design, the impact of formation gas should be fully considered to make the inflation volume the best.

4. EQUIVALENT DENSITY CONTROL METHOD FOR AERATED DRILLING

For conventional drilling, aerated drilling can adjust the bottom hole equivalent density by adjusting the gas-liquid ratio (liquid displacement, gas displacement) or casing pressure without changing the density of drilling fluid base fluid. In this way, the density of drilling fluid base fluid needs to be increased only when the bottom hole pressure is greater than the density of drilling fluid base fluid. It can be seen from Figure 3 that the density is affected by the displacement of drilling fluid. In the same case, changing the fluid injection volume can significantly affect the bottom hole equivalent density. The bottom hole equivalent density can be increased by increasing the drilling fluid flow. On the contrary, the bottom hole equivalent density can be reduced by reducing the drilling fluid flow. However, it should be noted that excessive displacement of drilling fluid will lead to excessive vertical pressure, and too small displacement of drilling fluid will affect the return of rock cuttings.

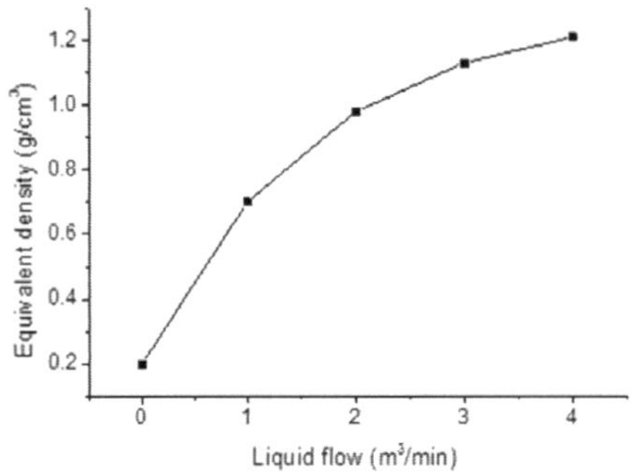

Fig. 3 Effect of adjusting liquid flow on bottom hole equivalent density

For the convenience of analysis and comparison, the data of pressure and well depth in the annulus and drill string in the wellbore are drawn in one diagram, as shown in Figure 4. The well depth at the turning point of the two curves is roughly the same, which is exactly the starting point of the drill collar tappet, that is, the position at the connection between the drill string and the drill collar. And the two curves are discontinuous, which is mainly due to the pressure drop caused by the air flow passing through the air hole of the bit.

Generally speaking, the annulus pressure increases gradually from the wellhead to the bottom of the well, and the curve slope decreases at the connection between the drill string and the drill collar, indicating that the pressure gradient increases rapidly, which shows that the pressure value increases rapidly. This is because the annular cross-sectional area at the junction of drill pipe and drill collar suddenly decreases, resulting in the rapid increase of flow rate and the increase of pressure loss, resulting in the sudden change of annular pressure in drill collar section and the sharp increase of pressure. The pressure in the drill string also increases gradually from the bottom of the well to the wellhead, which is similar to the distribution law of annular pressure. At the connection between the drill pipe and the drill collar, the inner diameter of the drill collar becomes smaller, which reduces the overflow cross-sectional area, and the pressure gradient increases rapidly, resulting in the rapid increase of pressure. Because the gas flow in the drill string is single-phase flow, and the fluid density is less than the density of the mixed fluid in the annulus, its pressure gradient is less than the pressure gradient in the annulus, making the pressure value change less. It can be seen from the above analysis that the sudden change of pressure gradient in the wellbore in gas drilling is at the intersection of drill pipe and drill collar, and the value of pressure gradient is large in the drill collar section, so the loss mainly occurs in the annulus drill collar section. Because the pressure gradient is closely related to the flow cross-sectional area, and the pressure gradient increases significantly with the decrease of the flow cross-sectional area. Therefore, where the cross-sectional area changes in gas drilling is the dangerous section where the pressure changes suddenly, and the pressure discontinuity in annulus and drill string is due to the pressure drop of bit.

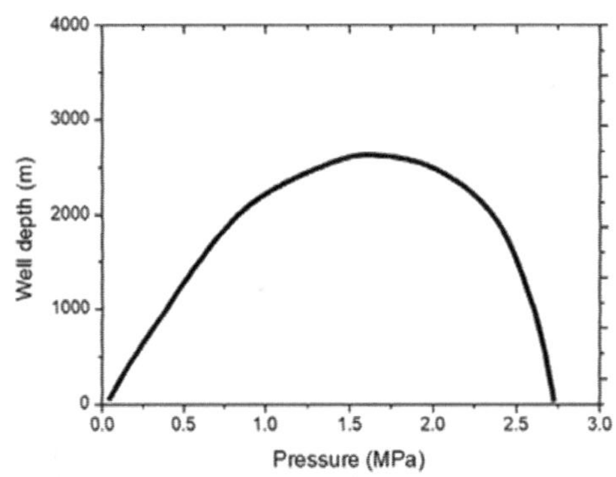

Fig. 4 relation curve between wellbore pressure and well depth in gas drilling

5.CONCLUSION

The field test data show that the drilling rate of the gas injection designed according to the minimum kinetic energy method or the minimum rock cuttings carrying method gradually increases with the increase of the gas injection. The analysis shows that it is mainly due to the increase of the gas annulus return velocity, the decrease of the rock cuttings settlement speed, the cleaning better of the rock cuttings at the bottom of the well, the avoidance of repeated cutting of the rock cuttings, the improvement of the rock breaking efficiency of the bit, and the acceleration of the drilling speed,However, too much gas injection will greatly erode the wellbore and drilling tools. Therefore, appropriately increasing the injection gas is conducive to accelerating the drilling speed.

ACKNOWLEDGMENTS

This work was not supported by any funds. The authors would like to show sincere thanks to those techniques who have contributed to this research.

REFERENCES

[1] Guo, X.L., Wang, Z.M., Long, Z.H. (2010) Study on three-layer unsteady model of cuttings transport for extended-reach well. Journal of Petroleum Science & Engineering. 73: 171-180.
[2] Jiang, G., Li, L., Xu, W., Li, Y. (2013) Research and evaluation on bubble sweep used for extended reach well. Drilling Fluid and Completion Fluid. 30: 5-8.
[3] Schneider, M., Berthe, L., Fabbro, R., Muller, M., Nivard, M. (2013) Gas investigation for laser drilling. Journal of Laser Applications. 19: 165-169.
[4] Zhu, H., Zhou, F., Wang, X., Fan, X., Tang, H., Lian, Z. (2012) Design and application of anti-fatigue sub for gas drilling. Natural Gas Industry. 32: 43-45.
[5] Zhen, N., Xia, B., Zhou, L., Deng, J. (2011) Modeling of wellbore stability for gas drilling. Natural Gas Industry. 31: 71-76.

Power output control method of small power photovoltaic grid connected power generation under complex illumination

Yun Luo[1]*, WenLing Lu[2]

[1,2]School of Intelligent Manufacturing, NanNing College of Vocational Technology, Nanning, China

* Corresponding author: 81737208@gmail.com

Abstract—Due to the irregular change of lighting conditions, there is a large error in the control of photovoltaic grid connected power output. Therefore, this paper proposes proposes a power output control method for low-power photovoltaic grid-connected power generation considering complex illumination. Based on the analysis of the basic characteristics of photovoltaic grid connected power generation, the output characteristics of photovoltaic array under complex lighting conditions are studied. Combined with its variation characteristics, based on fuzzy control strategy, the output of photovoltaic grid connected energy storage system is adjusted by filter control with variable time constant, and economic operation and stable operation are taken as constraints. The test results show that the design method has a high degree of fit between the control of photovoltaic grid connected power output and the actual results, and the error is within a reasonable range, which is obviously due to the two groups of comparison methods.

1. INTRODUCTION

The traditional primary energy such as coal, oil and natural gas has been used by human beings for a long time, but the energy on our earth is actually very limited and can not be consumed by human beings without limit, and will be exhausted one day [1-2]. There is only one feasible way: vigorously develop new energy. Since the end of last century, countries all over the world have paid more and more attention to the development and utilization of new energy [3-5]. Solar energy is a green and clean alternative energy. However, today's solar photovoltaic power generation still has some defects, such as low efficiency, high power generation cost and high on grid electricity price. These defects lead to the inability of large-scale production of solar photovoltaic power generation and limit its large-scale promotion and utilization, which is greatly constrained in economic benefits[6]. Therefore, taking measures to greatly improve the utilization rate of light energy and reduce the power generation cost is the top priority of photovoltaic power generation research. It is assumed that when the light intensity of the sunlight perpendicular to the surface of the silicon wafer and the temperature of the silicon wafer are constant, the output volt ampere characteristics of the solar cell are theoretically determined for the same photovoltaic module, and there is a maximum power value when the photovoltaic panel is not aged and covered. Therefore, in order to maximize the use of solar energy and reduce the power generation cost, it is necessary to make the photovoltaic cell run near the maximum power output node as far as possible, for example, by adjusting the equivalent impedance of the load. At the same time, it will also affect the stable output of power, and even make the power curve fluctuate seriously. If no measures are taken, the photovoltaic module will enter an unstable working state. Photovoltaic power generation is a new power generation mode that directly converts solar radiation energy into electric energy by using the photovoltaic effect of solar cell semiconductor materials. However, photovoltaic power generation is random and intermittent, which will affect the quality of power and aggravate the fluctuation of power grid. Equipped with certain energy storage in the light energy system, it can effectively suppress power fluctuation and improve the safety and stability of power grid operation. Therefore, the smooth control of the output power of the optical energy motor unit has become the focus of research.

2. ANALYSIS OF BASIC CHARACTERISTICS OF PHOTOVOLTAIC GRID CONNECTED POWER GENERATION

2.1. *Analysis of basic properties of photovoltaic cells*

The basic model of photovoltaic cell based on light energy is shown as Figure 1.

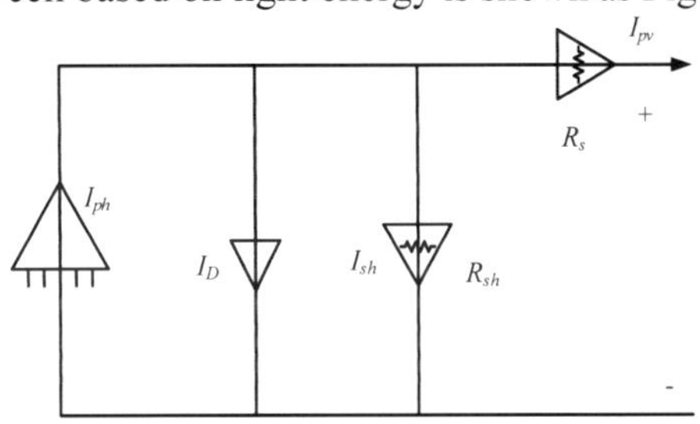

Figure 1. Equivalent circuit diagram of photovoltaic cell

According to the equivalent circuit diagram, the KCl current equation can be obtained

$$I_{pv} = I_{ph} - I_D - I_{sh} \quad (1)$$

among,

$$I_{ph}=[I_{sc} + K(T - T_{ref})]\frac{S}{S_{ref}} \quad (2)$$

$$I_D=I_{rs}[\exp\frac{q}{AkT}(U + i_{pv}R_s) - 1] \quad (3)$$

Where I_{pv} represents the output current of the photovoltaic panel, I_{ph} represents the photogenerated current, I_D represents the short-circuit current of the photovoltaic cell, U represents the external voltage of the photovoltaic cell, R_s represents the series resistance, R_{sh} represents the parallel resistance, I_{sh} represents the forward current of the PN junction, I_{sc} represents the leakage current of the PN junction, i_{pv} represents the reverse saturation current of the diode, K represents the temperature coefficient, k is the Boltzmann constant (1.38 × 1033J/K), q represents the charge of electrons (1.6x101 ℃), A represents the diode characteristic factor, T_{ref} represents the standard cell temperature, T represents the photovoltaic cell temperature, S_{ref} represents the standard light intensity, S represents the light intensity.

In practical application, it is found that the series resistance R_s is very small, while the parallel resistance R_{sh} is very large. Therefore, we can further simplify the analysis of the above formula, so we can obtain the characteristic equation representing the ideal solar cell:

$$I_{pv} = I_{ph} - I_{rs}[\exp\frac{qU}{AkT} - 1] \quad (4)$$

On this basis, the photovoltaic cell model based on MATLAB / Simulink is constructed.

3. Study on Output Characteristics of Photovoltaic Array under Complex Illumination

When the illumination is uneven, the illumination intensity of each battery panel is different, and the photovoltaic voltage and current generated are different. The diodes connected in parallel to the battery panel may form a positive voltage and be in the on state, which changes the $I-V$ characteristics and $P-V$ characteristics of the series array.

(1) When the output current of the series array is within $[0, I_{sc}]$, the bypass diode is in the reverse bias state without conduction. The photovoltaic panels $S1$ and $S2$ flow the same current and can output power externally. The voltage of the series array is equal to the sum of the output voltages of $S1$ and $S2$, i.e

$$V = V_{S1} + V_{S2} = \frac{nKT}{q}[(\ln\frac{I_{sc}-I}{I_0}+1)] - 2IR_S \quad (5)$$

The current $\max I_{S2} \in (0, I_{sc})$ at the maximum power point of photovoltaic panel $S2$ can be obtained from the output characteristics of photovoltaic cells. When the output current of the series array is within $[0, \max I_{S2}]$, with the increase of the current, the voltage of the photovoltaic panel $S2$ decreases within the interval $[\max V_{S2}, V_{oc}]$, and the output power of the corresponding photovoltaic panel $S1$ and $S2$ increases, resulting in the increase of the output power of the series array, in which the output power of $S2$ increases from 0 to the maximum power $\max P_{S2}$. When the output current of the series array is in $[\max I_{S2}, I_{oc}]$, with the increase of the current, the voltage of the photovoltaic panel $S2$ decreases in the interval $[0, \max V_{S2})$, and the output power of the photovoltaic panel $S2$ decreases. Therefore, when the output current of the series array is within $[0, I_{S2}]$, the output power of the array is the sum of the output power of photovoltaic panels $S1$ and $S2$. With the increase of current, it first increases and then decreases, reaching a local peak near $\max I_{S2}$.

(2) When the output current I of the series array is within $[I_{S1}, I_{S2}]$, the current generated by the photovoltaic panel $S1$ is greater than the current generated by $S2$, the bypass diode of $S2$ is in the forward bias state and turned on, and the current flowing through is $[I - I_{S2}]$. In this case, the photovoltaic panel $S1$ outputs power to the outside. The voltage of the series array is equal to the sum of $S1$ and the bypass diode, i.e

$$V = V_{S1} + V_{S2} = \frac{nKT}{q}[(\ln\frac{I_{PH}-I}{I_0}+1)] - IR_S - V_D \quad (6)$$

Thus, the series array in $[I_{S1}, I_{S2}]$, series array output voltage and photovoltaic cell $S1$ voltage difference diode on-voltage drop, their output characteristics are basically the same. The maximum power point of photovoltaic cell $S1$ is within this range, so the series array has a peak power point in this range.

4. Research on Power Output Control of Photovoltaic Grid-Connected Generation

4.1. Smooth output control of energy storage system

The power fluctuation of smooth renewable energy is mainly divided into short time power fluctuation and long time power fluctuation, and the time scale of short time fluctuation is generally S-class. Long-term power fluctuation time scale for the general min level, generally applied to renewable energy and load short-term power mismatch between the following analysis of photovoltaic grid-connected smooth output control.

4.1.1 Filtering control of variable time constant
Because of the dynamic change of illumination condition, the fixed filter can not meet the output requirement. Therefore, the variable time constant is added in this paper, and the filter time constant is adjusted at the same time to keep the SOC of the energy storage system in a certain range, so as to achieve the goal of smoothing the output and taking into account the life of the storage battery. The control structure is shown in Figure 2.

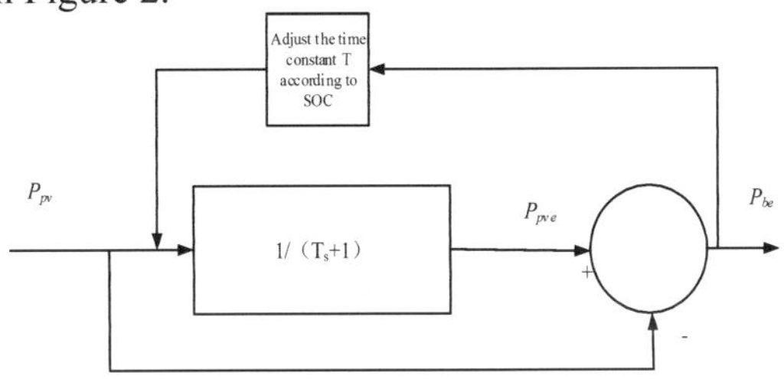

Figure 2. Filter control structure of variable time constant

Among them, T_s is the filtering time constant, according to the principle of system filtering function, the output power of energy storage system is proportional to the time constant T_s. Therefore, when the SOC of energy storage battery is high, the charging power of energy storage system can be reduced by decreasing the time constant T_s, which can effectively prevent the SOC from exceeding the upper limit. Similarly, when the energy storage battery is in the low position and the battery is in the state of discharge compensation, reducing the time constant T_s can effectively prevent the SOC from exceeding the lower limit. Similarly, when the energy storage battery is in the low position and the battery is in the state of discharge compensation, reducing the time constant AA can effectively prevent the SOC from exceeding the lower limit.

4.2.Control constraints
On the basis of the above, but also on the photovoltaic grid-connected power output control constraints, this paper mainly from the following two aspects.

4.2.1 Photovoltaic grid-connected economic operation
In the process of battery operation, large charging and discharging current and over charging and over discharging will affect the battery life, so it is necessary to make the battery work in a certain bound range. To this end, we must first control the current and voltage.

(1) Current constraint conditions, when the charging and discharging current of the battery exceeds the limit, the service life of the battery will be affected.

$$I_{sc} < \max I_{sc} \qquad (7)$$

(2) Voltage constraints, when the battery voltage is lower or higher than a certain value will affect the battery life.

$$V_{\min} < V < V_{\max} \qquad (8)$$

Among them, $V_{\min}$ and $V_{\max}$ represent the minimum and maximum of the charging and discharging voltage of the battery respectively.

(3) SOC constraints for storage batteries: the storage batteries must be operated between the minimum and maximum SOC values, and the service life of the storage batteries will be affected if they exceed the limits.

$$SOC_{\min} < SOC < SOC_{\max} \quad (9)$$

Under the condition of economic operation, the control strategy of energy storage system searches for fuzzy rules according to the SOC status of storage battery, PV power forecast value, load forecast value and P_l exchange status of current PV power interconnection line, and achieves the control of battery charging and discharging to achieve the goal of economic operation. The fuzzy rules table is shown in Table 1.

TABLE 1. FUZZY CONTROL TABLE FOR GRID CONNECTED ECONOMIC OPERATION OF PHOTOVOLTAIC BATTERY

P_l	SOC	P_{pv}			
+	$[0, SOC_{\min}]$	[0,0.25]	[0.25,0.50]	[0.50,0.75]	[0.75,1.00]
	$[SOC_{\min}, SOC_{\max}]$	[0,0.25]	[0.25,0.50]	[0.50,0.75]	[0.75,1.00]
-	$[0, SOC_{\min}]$	[0,0.25]	[0.25,0.50]	[0.50,0.75]	[0.75,1.00]
	$[SOC_{\min}, SOC_{\max}]$	[0,0.25]	[0.25,0.50]	[0.50,0.75]	[0.75,1.00]

In the table, P_l for "+" means positive power status: photovoltaic grid-connected to the distribution network to sell electricity; For "-" means negative power status: means photovoltaic grid-connected distribution network to buy electricity.

4.2.2 Photovoltaic grid-connected stable operation

On the premise of guaranteeing the service life of the battery, the fuzzy control of the storage battery is designed according to the power fluctuation and the continuous power supply capacity of the system. Based on the current SOC status of the battery, the predicted photovoltaic power and the predicted load value, the system can run stably by querying the fuzzy rule table. The fuzzy rules table is shown in Table 2.

TABLE 2. FUZZY CONTROL TABLE FOR GRID CONNECTED STABLE OPERATION OF PHOTOVOLTAIC BATTERY

SOC	P_{pv}
$[0, SOC_{\min}]$	[0,0.25]
$[SOC_{\min}, SOC_{\max}]$	[0.25,0.50]
$[SOC_{low}, SOC_{high}]$	[0.50,0.75]
$[SOC_{high}, SOC_{\max}]$	[0.75,1.00]

On this basis, to ensure the reliability of the control.

4.3.Fuzzy control output

On the basis of the above, the output of defuzzy control is the process of converting the fuzzy quantity of fuzzy inference and decision output into the definite control quantity, which corresponds to the fuzzy quantity of the system. According to the symmetry of the output fuzzy set in this paper, the output defuzzy quantity is obtained by weighted average method. The calculation formula is shown in formula (10).

$$P^* = \frac{\sum_{i=1}^{n} \mu(P_i) P_i}{\sum_{i=1}^{n} \mu(P_i)} \quad (10)$$

In the equation, P_i and $\mu(P_i)$ represent the center of mass and the value of membership function of symmetric membership function respectively. The actual control quantity can be obtained by solving P^* according to the above formula and multiplying by the corresponding scale factor

5. Simulation of Power Control for Photovoltaic Power Station

5.1.Experimental environment design

Using Matlab/Simulink to build 20 photovoltaic power plant models and active power control models with the rated power of 1MW, and through the simulation of the photovoltaic power plant output power control to verify the distribution strategy proposed in this paper, the control model is shown in Figure 3, in simulation, it is assumed that each photovoltaic inverter is in normal state.

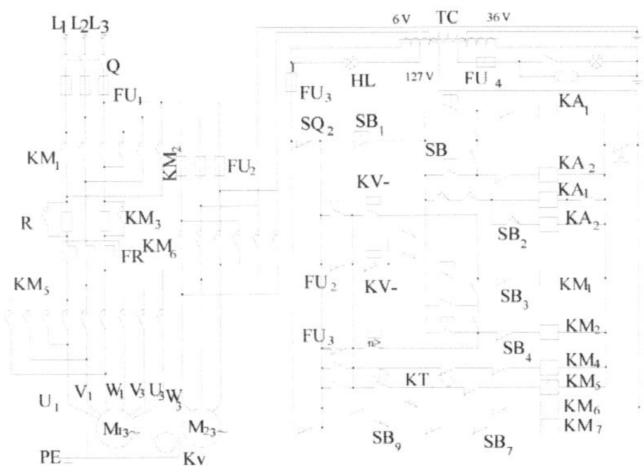

Figure 3. Photovoltaic array model

5.2.Test results

The output power of photovoltaic power station is set to 12 MW, 11 MW, 10 MW, 9 MW, 8 MW, 9 MW, 10 MW and 11 MW respectively at different time periods. The output power curve of photovoltaic power station is shown in Figure 5 by using the method proposed in literature [8] and literature [9] as a comparison.

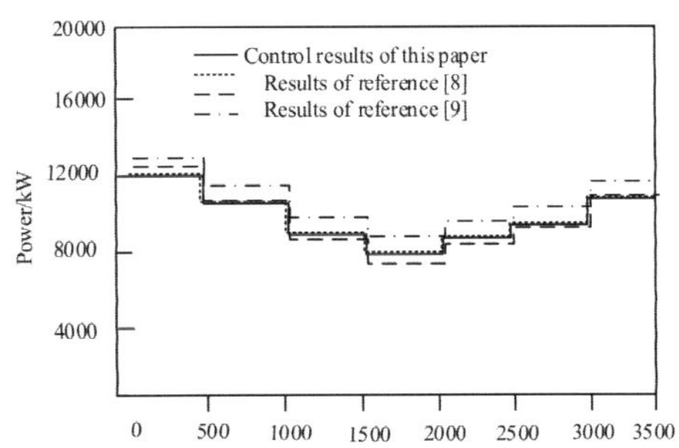

Figure 4. Output power control curve of photovoltaic power station

As can be seen from the figure 4, the set power of the photovoltaic power plant is lower than the predicted power of the photovoltaic power plant. When the set value is constant, the actual output power of the PV plant can ensure the stable output of the set power. When the set power is changed, the PV plant can complete the tracking of the new set power in a short time.

6. Conclusion

With the rapid development of solar photovoltaic power generation technology in the world, more and more large scale photovoltaic power plants begin to be connected to the power grid, and the stability of the power grid caused by this problem is becoming more and more serious. In order to improve the stability and dispatchability of output power of photovoltaic power station and promote the acceptance of large scale photovoltaic power station, this paper studies the control strategy. The simulation results show that the active power control method studied in this paper can meet the requirements of active power control of photovoltaic power plants.

ACKNOWLEDGMENT

The study was supported by “Young and middle-aged project of Guangxi Department of Education, China (Grant No. 2019KY1234)”.

REFERENCES

[1] Yadav S , Kumar V , Dhok S B , et al. (2019). Energy-Efficient Design of MI Communication-Based 3-D Non-Conventional WSNs. IEEE Systems Journal, PP(99):1-4.
[2] Hu B , Sun Y , Shao C , et al. (2021). A Decentralized Market Framework for Procurement of Operating Reserves From District Energy Systems. IEEE Transactions on Sustainable Energy, PP(99):1-1.
[3] Boschi F, Basso PR, Corridori I et al (2019) Weak biophoton emission after laser surgery application in soft tissues: analysis of the optical features. J Biophoton 12(9):1–9.
[4] Altowilib A , Alsaihati A , Alhamood H , et al. (2020). Reserves Estimation for Coalbed Methane Reservoirs: A Review. Sustainability, 12(24):1-26.
[5] Crow D , Balcombe P , Brandon N , et al. (2019). Assessing the impact of future greenhouse gas emissions from natural gas production. The Science of the Total Environment, 668(JUN.10):1242-1258.
[6] Vilanova A , Kim B Y , Chang K K , et al. (2020). Linear-Gompertz Model-Based Regression of Photovoltaic Power Generation by Satellite Imagery-Based Solar Irradiance. Energies, 13(4):1-12.

Characterization and Modeling of SiC MOSFET Modules for High-voltage Applications

Yi Hao[1,a], QingPing Li[2]*, YunFei Xu[3], Lü Cui[4], WeiGuo Li[5], JuanJuan Lu[6]

[1,2,3,5,6]Power Electronics Department, Global Energy Interconnection Research Institute Co., Ltd. Beijing, China

[4]Science and Technology Department, State Grid Shanghai Municipal Electric Power Company. Shanghai, China

[a]389233314@qq.com

[*]liqingping2017@163.com

Abstract—This paper presents a detailed characterization and modeling of 6.5kv SiC MOSFET half-bridge module. Based on the equivalent circuit topology, the model parameters are extracted from the data manual and test data, and the behavior model is constructed to describe the static and dynamic characteristics of the device. This paper used behavioral model functions, and carries out parameter extraction and model fitting for new devices. The model is verified by double pulse test under inductive load. The verification results show that the modeling work is correct and effective, it can predict the switching performance of devices in high voltage applications.

1. INTRODUCTION

As a wide band gap material, Silicon Carbide (SiC), has the advantages of high breakdown field strength, high saturation electron drift rate, high thermal conductivity, high melting point and low on-state impedance. It can be used to realize new power semiconductor devices for high voltage, high power, high frequency and high temperature applications. Compared with traditional Silicon (Si) materials, SiC materials have attracted extensive attention in the application fields of high-power switching circuits and power systems [1].

Although some manufacturers provide simulation models, most of them do not consider the temperature characteristics, and the model has convergence problems. Therefore, it is easy not to converge in practical simulation applications, and can't accurately predict the circuit characteristics. In recent years, SiC MOSFET modeling has received more and more attention. Some models are constructed based on the Si MOSFET model. However, due to the difference in physical structure, the static and dynamic characteristics of SiC devices can't be better adapted [2]. Some models focus on the internal physical structure, the model accuracy is high, but because the physical model parameters are more and have a certain physical meaning, parameter extraction is more difficult, the simulation model is more complex, which limits its application in circuit simulation [3]. Some models are fitted by building mathematical models by introducing voltage controlled current source, which is convenient to extract model parameters. In the voltage controlled current source model, the cutoff, linear and saturation regions of static characteristics are described by segmented equations [4]. Because the discontinuity of the function easily leads to the non-convergence of the model, based on this, some models describe and fit the device through non-segmented functions, which improves the simulation efficiency and convergence of the model, but the model is mainly used for medium voltage device [5].

In this paper, mathematical formula modeling methods are used to construct behavior models, using device characteristic curves to describe static characteristics of the high-voltage half-bridge module. At the same time, a non-segmented, continuously and differentiable equation is used to model the junction capacitance. And through the 6.5kV SiC MOSFET device and its double pulse circuit for simulation verification.

2. SiC MOSFET MODULES CHARACTERIZATION

The static characteristics of SiC MOSFET are mainly the static characteristics of MOS core. The MOS core is modeled by voltage controlled current source. The dynamic characteristics of SiC MOSFET mainly refer to its switching characteristics. The essence of the turn-on and turn-off process of SiC MOSFET is the charge and discharge process of junction capacitance. The body-diode affects the loss of devices in different circuits and the realization of soft switching. Therefore, the static characteristic modeling of SiC MOSFET is mainly the voltage controlled current source modeling of MOS core, and the dynamic characteristic modeling of SiC MOSFET is mainly the modeling of junction capacitance and body-diode, as shown in Fig. 1.

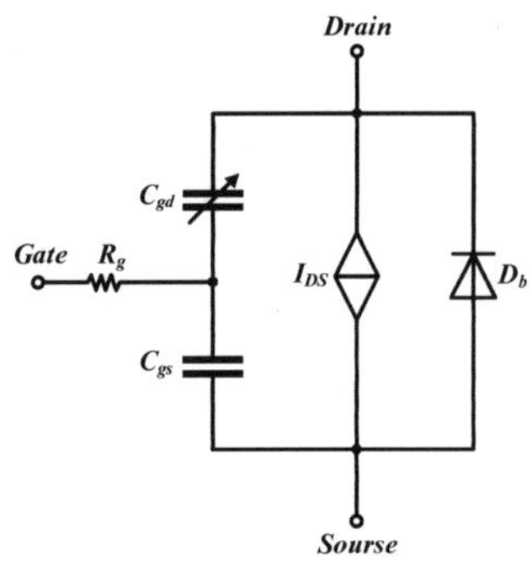

Figure 1. The equivalent circuit diagram of SiC MOSFET.

3. BEHAVIORAL SPICE MODEL FOR 6.5KV/400A SiC MOSFET

Since the hyperbolic tangent function y=tanh(x) is close to the trend of the transfer characteristic curve, this paper combines the hyperbolic tangent function to model the output characteristics of the device by mathematical formulas. At the same time, the hyperbolic tangent function is continuous and differentiable. Conducive to the convergence of the model.

The established transfer characteristic model is shown in formula (1):

$$I_{transfer} = k \cdot \left[1 + \tanh(a \cdot V_{GS} + m)\right]^n \tag{1}$$

k, a, m and n —transfer characteristic parameters. The established output characteristic model is shown in formula (2):

$$I_{output} = V_{DS} \cdot \left[1 + (e \cdot V_{GS} + r)^t \cdot V_{DS}\right]^b \tag{2}$$

e, r, b and t —output characteristic parameters. In order to fit the transfer characteristics and output characteristics at the same time, formula (1)-(2) are multiplied to obtain a voltage-controlled current source I_{DS} model, as shown in formula (3):

$$I_{DS} = k \cdot \left[1 + \tanh(a \cdot V_{GS} + m)\right]^n \cdot V_{DS} \cdot \left[1 + (e \cdot V_{GS} + r)^t \cdot V_{DS}\right]^b \quad (3)$$

The general global optimization algorithm is used to fit the output characteristics and transfer characteristics data at 25℃, 100℃ and 150℃ respectively through 1stOpt fitting software. If each fitted parameter changes with temperature, the model will be too complicated, and setting too many constants will greatly reduce the accuracy of the model. Therefore, after multiple iterations of fitting, considering the accuracy and simulation convergence, some parameters *a*, *b*, *e*, and *t* are set as constant values. The specific values are shown in Table 1. The parameters *r*, *m*, *n*, and *k* are the temperature sensitivity parameters, specific parameter values that vary with temperature are shown in Table 2. At the same time, consider the continuity under multiple temperatures, set the parameter range, and ensure the monotonicity of parameter changes, so that the trend of output characteristics and transfer characteristics with temperature changes is monotonous.

TABLE 1. TEMPERATURE INDEPENDENT MATHEMATICAL MODEL PARAMETERS

Parameter	Value
a	0.45914
b	-1.93542
c	0.330955
d	-2.60371

TABLE 2. MATHEMATICAL MODEL PARAMETERS THAT VARY WITH TEMPERATURE

Temperature	Parameter			
	r	*m*	*n*	*k*
25℃	0.100000306	-5.75388158	0.173127602	120.5953208
75℃	0.749047901	-4.54493487	0.228032175	85.56266646
150℃	3.161626804	-3.20124584	0.961353693	31.10238451

According to the model formula of equation (3) and the model parameters in Table 1 and table 2, the model of SiC MOSFET is constructed. The static characteristic curves at multiple temperatures are compared with the experimental static characteristic data. The results are shown in Fig. 2-7.

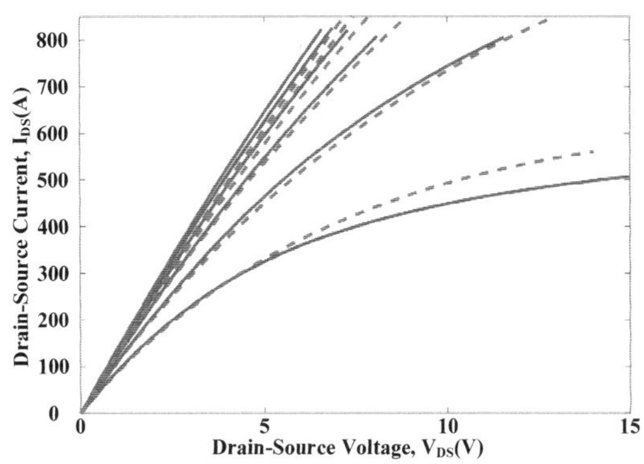

Figure 2. Output characteristic curve of mathematic model (dotted) and experiment (solid) at Tj=25℃.

It can be seen from Fig.2-7 that the fitting degree of output characteristic curve is good at 25 ℃, 100 ℃ and 150 ℃; At 25 ℃ and 150 ℃, the fitting effect of transfer characteristic curve is good, and the fitting effect of transfer characteristic curve at 100 ℃ is general, but in general, the fitting results of transfer characteristic and output characteristic at each temperature are good.

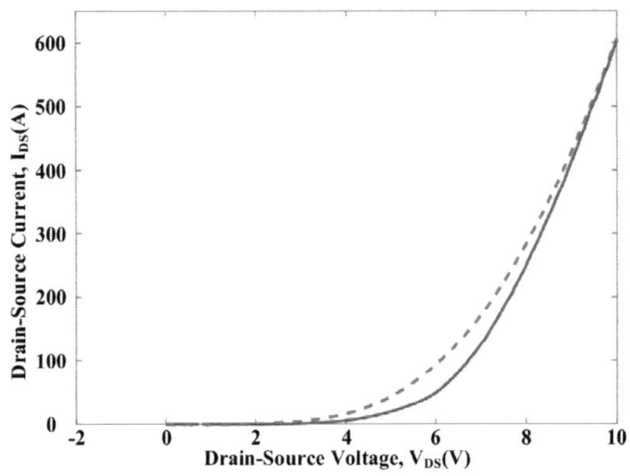

Figure 3. Transfer characteristic curve of mathematic model (dotted) and experiment (solid) at Tj=25℃.

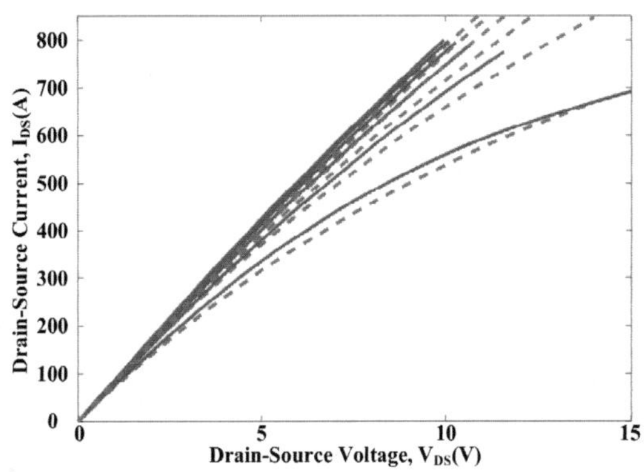

Figure 4. Output characteristic curve of mathematic model (dotted) and experiment (solid) at Tj=100℃.

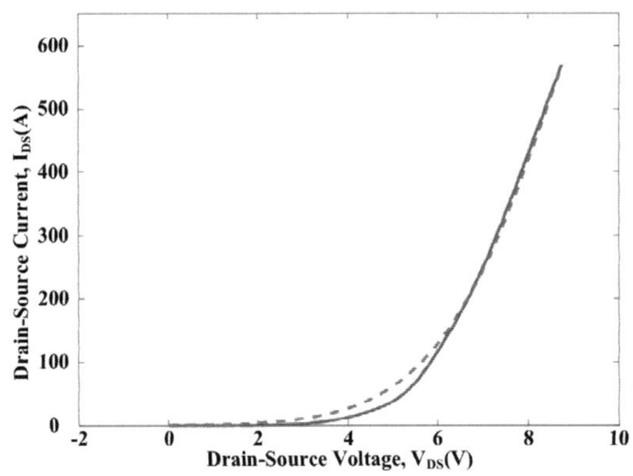

Figure 5. Transfer characteristic curve of mathematic model (dotted) and experiment (solid) at Tj=100℃.

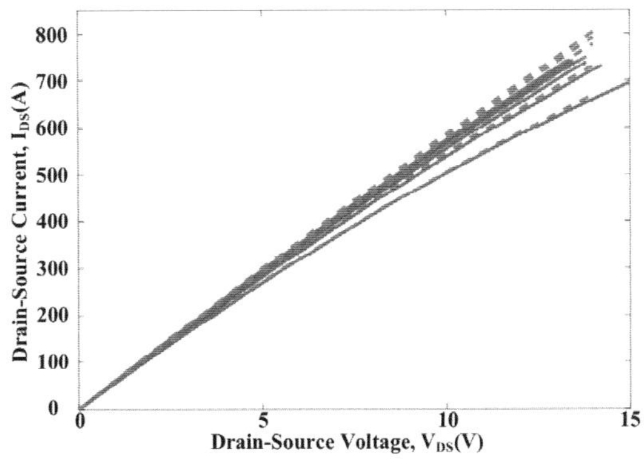

Figure 6. Output characteristic curve of mathematic model (dotted) and experiment (solid) at Tj=150℃.

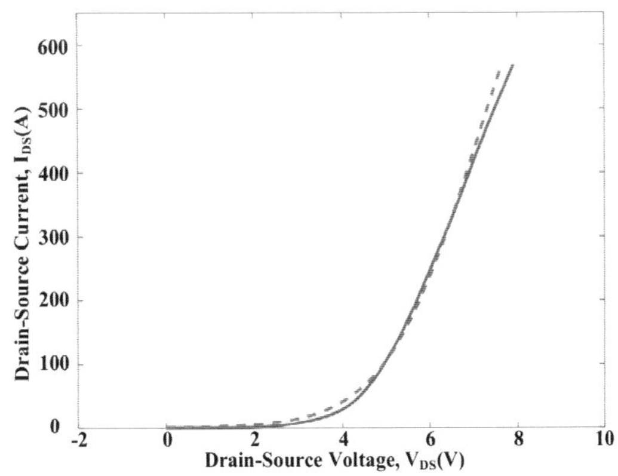

Figure 7. Transfer characteristic curve of mathematic model (dotted) and experiment (solid) at Tj=150℃.

4. DEVICE APPLICATION AND DISCUSSION

In order to verify whether the dynamic characteristics of the high-voltage SiC MOSFET model at the switching moment are correct, the simulation result waveform of the model is compared with the actual test waveform of the module under two temperature conditions of 25°C and 150°C. The circuit is shown in Fig. 8, where V_{DC}=3600V is the DC bus voltage, L_{load}=350uH is the load inductance, D is the continuous current diode (body diode), R_g=10Ω is the gate drive resistance, and the drive signal is a double-pulse drive signal.

Through the double-pulse test comparison, the timing characteristics of the circuit gate drive, drain-source current, and drain-source voltage can be analyzed. The gate-source voltage and drain-source current at 25°C are shown in Fig. 9-10, and the gate-source voltage and drain-source current at 150°C are shown in Fig.11-12.

It can be seen from the results of waveform comparison in Fig. 9-10 that the double-pulse simulation results of SiC MOSFET model at 25℃ are basically consistent with the experimental results, but there are still some deviations due to the comprehensive consideration of parasitic parameters in the circuit.

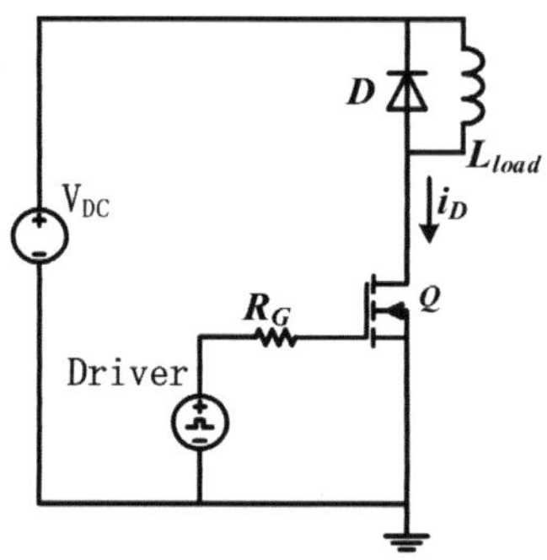

Figure 8. SiC MOSFET double-pulse test circuit.

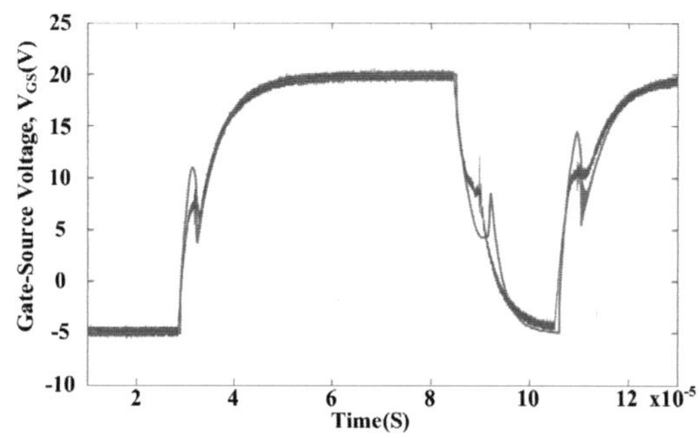

Figure 9. Simulation waveform (red) and experiment waveform (blue) of Gate voltage at 25℃ & 100A.

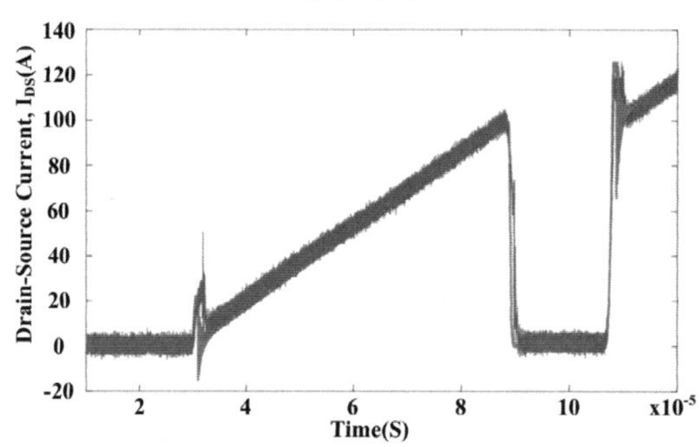

Figure 10. Simulation waveform (red) and experiment waveform (blue) of Current at 25℃ & 100A.

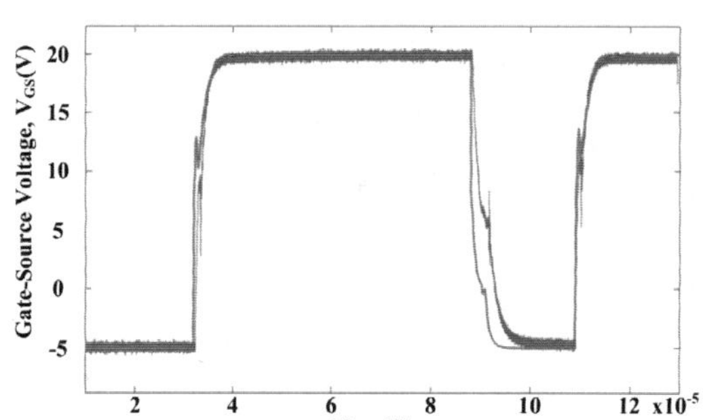

Figure 11. Simlulation waveform (red) and experiment waveform (blue) of Gate voltage at 150℃ & 60A.

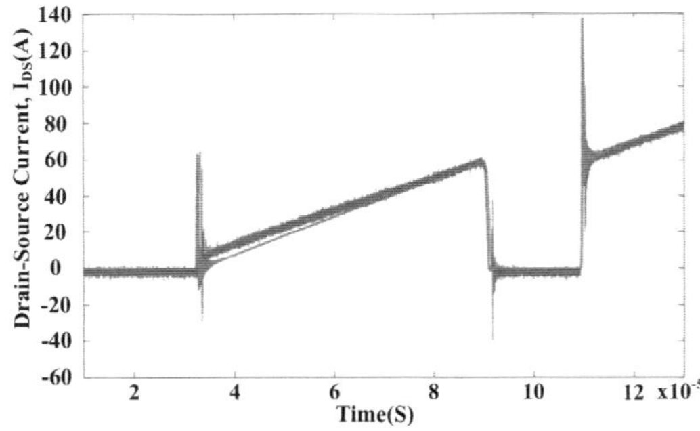

Figure 12. Simlulation waveform (red) and experiment waveform (blue) of Current at 150℃ & 60A.

It can be seen from the comparison waveform results in Figure 11-12 that the double-pulse simulation results of the SiC MOSFET model at 150°C are basically consistent with the experimental results.

By comparing the simulation and experiment waveform, it can be seen that the established device simulation model can accurately reflect the circuit characteristics of the device, predict the dynamic behavior of the circuit, provide simulation analysis for the practical application of subsequent SiC MOSFET devices, and guide the engineering application of optimizing and improving the circuit.

5. Conclusion

In this paper, the SiC MOSFET model based on mathematical formula method is proposed, including static core model and junction capacitance model. The Spice simulation model is constructed by fitting the output and transfer characteristic data of the device at 25℃, 100℃ and 150℃. The comparison of test and simulation results at different temperatures shows that the simulation model can better reflect the device characteristics. Finally, a double pulse simulation circuit is built. The results show that the simulation waveform is basically consistent with the test waveform and has high accuracy.

Acknowledgment

State Grid Corporation Headquarters Science and Technology Project Subsidy, "Research on Modeling Technology and Applicability of High Voltage SiC MOSFET Module (SGGR0000DLJS1900546)".

References

[1] J. Biela, M. Schweizer, S. Waffler and J. W. Kolar, "SiC versus Si-evaluation of potentials for performance improvement of inverter and DC–DC converter systems by SiC power semiconductors," IEEE Trans. Ind. Electron., vol. 58, no. 7, pp. 2872-2882, July 2011.

[2] J. Wang et al., "Characterization, Modeling, and Application of 10-kV SiC MOSFET," IEEE Transactions on Electron Devices, vol. 55, no. 8, pp. 1798-1806, 2008.

[3] L. Di Benedetto, G. D. Licciardo, T. Erlbacher, A. J. Bauer, R. Liguori and A. Rubino, "A model of electric field distribution in gate oxide and JFET-region of 4H-SiC DMOSFETs," IEEE Trans. Electron Devices, vol. 63, no. 9, pp. 3795-3799, Sept. 2016.

[4] Y. Cui, M. Chinthavali, and L. M. Tolbert, "Temperature dependent Pspice model of silicon carbide power MOSFET," in Proc. 27th Annu. IEEE Appl. Power Electron. Conf. Expo. (APEC), 2012, pp. 1698–1704.

[5] R. Pratap, R. K. Singh, and V. Agarwal, "SPICE model development for SiC power MOSFET," in IEEE International Conference on Power Electronics, Drives and Energy Systems, 2013, pp. 1-5.

A Flexible Control Strategy of Air-Conditioning Cluster Participating in Primary Frequency Modulation

Lei Li[1], Chaoliang Wang[2*], Songsong Chen[3] and Lingyu Chen[3]

[1]State Grid Zhejiang Electric Power Co. ,Ltd. , Hangzhou, Zhejiang, 310000, China

[2]State Grid Zhejiang Marketing Service Centre, Hangzhou, Zhejiang, 311100, China

[3]Southeast University, Nanjing, Jiangsu, 210096, China

*Corresponding author： email: wang_chaoliang@zj.sgcc.com.cn

Abstract. As renewable energy resources are gradually connected to the grid, demand response gradually develops and expands. As a response resource with a huge proportion of the load, the role of air-conditioning load is gradually being valued. The paper first describes the equivalent thermal parameter model of the air-conditioning load. Then, an air-conditioning cluster aggregation model is established based on Monte Carlo with uncertainties considered. Finally, the paper proposes a flexible trigger frequency setting method to provide a relatively consistent amount of adjustment in different situations. And the simulation proves the effectiveness of the method.

1. Indroduction

With the increasing proportion of renewable energy resources in the power grid, the frequency regulation and power balance of the power grid are facing new challenges. It is difficult to deal with it only by the generation side, and the cost is also high. Therefore, the use of demand-side resources is a new way out[1]. The air-conditioning load accounts for a huge proportion of the load in the demand side. It is reported that the air-conditioning load in summer can account for 49.2% in Beijing. In addition, since the air-conditioning is a thermal energy storage load, a short-term shutdown will not have a big impact on user's comfort[2]. Therefore, it is necessary and urgent to study how air-conditioning loads participate in grid interaction.

The first step is to establish air-conditioning models to measure the total amount of air-conditioning that can be adjusted. There have been some researches on the single model and aggregation model. Based on data-driven, linear regression is used to learn the data relationship for prediction in [3]. Through this method, the more accurate models are not needed to perform optimization. The SOC model of battery is used to simulate the air-conditioning cluster in [4], with the power limits and energy capacity expressed by SOC parameters and random exogenous. Reference [5] uses the state queueing (SQ) model to describe the aggregated thermostatically controlled loads considering the uncertainties. Based on the existed work, a new regulation capacity evaluation method is proposed in [6] with the air-conditioning switch cycle constraints considered.

Droop control is widely adopted for the air-conditionings to participate in the primary frequency modulation. Whether to act is determined by comparing the trigger frequency and monitoring the system frequency[7][8]. However, the traditional trigger frequency setting method is to set the trigger frequency once and keep it fixed. When the total power of the air-conditioning cluster fluctuates, the response amount provided is also different, which brings more uncertainties factors to grid dispatch.

The contribution of the paper is to evaluate the adjustable capacity of air-conditioning clusters through establishing an air-conditioning cluster aggregation model based on Monte Carlo and equivalent thermal parameter(ETP) model. And a flexible trigger frequency setting method is designed to provide a relatively consistent amount of adjustment in different situations.

The remainder of the paper is organized as follows. Chapter 2 describes the ETP model of a single air conditioner. Chapter 3 establishes the aggregation model based on Monte Carlo. Chapter 4 describes how to set the trigger frequency of the air-conditioning cluster. And Chapter 5 is the case study.

2. The equivalent thermal parameters (ETP) model

As shown in figure 1(a), the indoor temperature T_{in} of the space where the air conditioner is located can be equivalently expressed by the thermal resistance R, the heat capacity C, and the outdoor temperature T_{out} and the cooling power P. The change of T_{in} can be expressed by (1)-(3), where Δt is the interval of time t and time $t+1$, s indicates the ON/OFF status of the air conditioner[7].

$$T_{in}^{\ t+1} = T_{out}^{\ t+1} - \eta PR - (T_{out}^{\ t+1} - \eta PR - T_{in}^{\ t})\varepsilon, s = 1 \tag{1}$$

$$T_{in}^{\ t+1} = T_{out}^{\ t+1} - (T_{out}^{\ t+1} - T_{in}^{\ t})\varepsilon, s = 0 \tag{2}$$

$$\varepsilon = e^{-\frac{\Delta t}{RC}} \tag{3}$$

According to the working principle of the air conditioner, it can be known that its ON/OFF status is constantly changing as shown in figure 1(b). When the air conditioner is ON, the indoor temperature drops and the power is equal to the rated power. When the air conditioner is OFF, the indoor temperature rises, and the power is equal to 0.

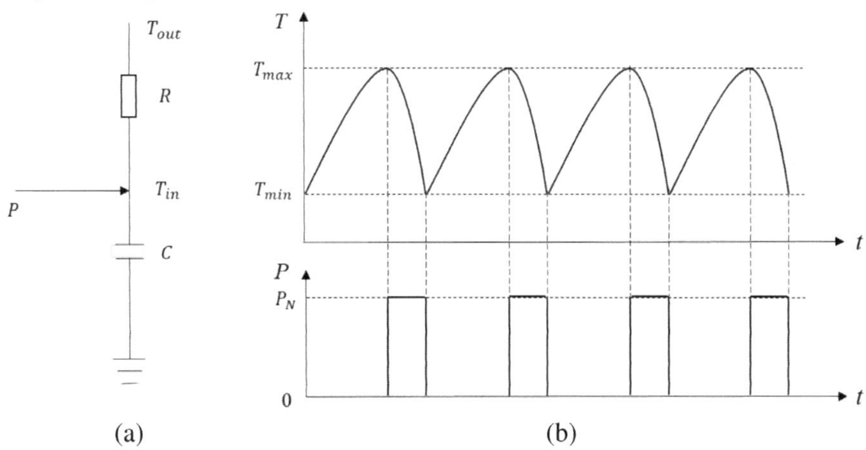

Figure 1. Air-conditioning equivalent model

3. Air-conditioning cluster aggregation model based on Monte Carlo

3.1. Monte Carlo method

The operation of each air conditioner is an independent process. The parameters and initial state of each air conditioner are also independent and random, but there is a probability distribution about them. Based on this, this paper uses the Monte Carlo method to study the air-conditioning aggregation model. Monte Carlo method, also known as random sampling method, is a calculation method based on probability and statistics theory. The basic solution steps are as follows:

- Construct a parameter probability distribution model. If the problem studied is random, it only needs to accurately describe the random process; if the problem studied is deterministic, it needs to artificially construct a random process and probability distribution model.

- Sample based on parameter distribution. After constructing the parameter probability model, random variables are extracted from the known probability distribution, which is the basic method to realize the Monte Carlo method.
- Establish estimates. After constructing a probability model and sampling, a random variable must be determined as the solution to the problem studied.

3.2. The aggregation model of the air-conditioning cluster

Firstly, according to the ETP model of a single air conditioner described in Chapter 2, the parameters needed to establish the aggregation model clearly defined. Secondly, the probability distribution of the above parameters is determined, and sampling is carried out according to the probability distribution. Each sampling result can act as the parameters of a single air conditioner. Finally, the operating conditions of each air conditioner are added together to obtain the aggregation characteristics of the air-conditioning cluster. For the air-conditioning cluster, its aggregate power and working status can be described by (4) and (5).

$$s_c(t) = \frac{1}{M}\sum_{i=1}^{M} s_i(t) \tag{4}$$

$$P_{con\_all}(t) = \sum_{i=1}^{M} P_i(t)\cdot s_i(t) \tag{5}$$

where M is the number of air conditioners in the cluster, $P_i(t)$ and $s_i(t)$ are the rated power and ON/OFF status of air conditioner *i* respectively.

4. Flexible trigger frequency setting method

Suppose that the trigger frequency of the air conditioner is set to fc. That is, when the system frequency is less than trigger frequency, the air conditioner is turned off. The trigger frequency is evenly distributed in the range [fcmin, fcmax]. The upper limit is usually fixed according to frequency modulation rule. When the system frequency drops to fsystem, the air-conditioning load response is:

$$P_{response} = P_{con\_all} \times \left(\frac{f_{c\max} - f_{system}}{f_{c\max} - f_{c\min}}\right) \tag{6}$$

As shown in the figure 2, under different temperature and climatic conditions, the total amount of air-conditionings is different. If the response frequency is set between the same frequency response range, maybe it will lead to over-regulation or under-regulation. Therefore, the frequency response range must be set according to the total amount of air-conditioning to provide a relatively consistent response of the air-conditioning.

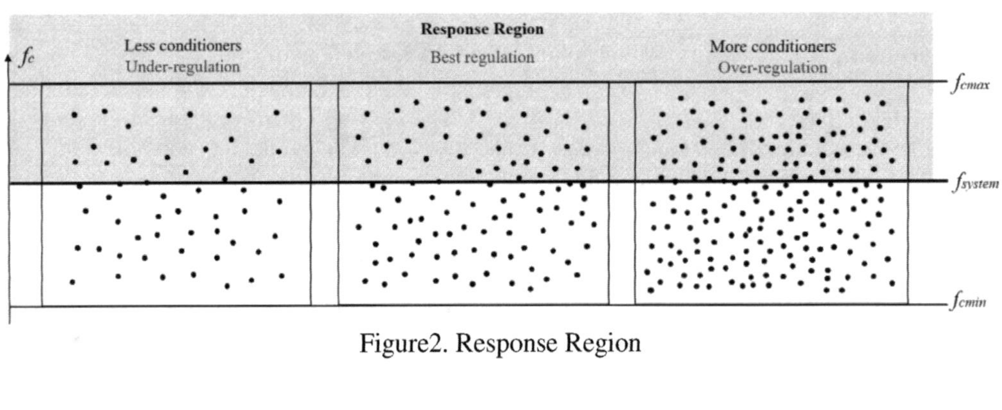

Figure2. Response Region

In order to keep the air-conditioning load response consistent under various conditions, it can be obtained from (6):

$$(f_{c\max} - f_{c\min}) \propto P_{con\_all} \tag{7}$$

5. Case Study

5.1. Parameter setting

The parameter settings of the air-conditioning cluster are shown in table 1. In the simulation examples in this chapter, the parameters that have not been re-declared are set according to table 1.

Table 1. Parameter settings

Parameter	Meaning	*Description/Value*
M	Number of air conditioners	10000
$[T_{\min}, T_{\max}]$	The upper and lower limits of indoor temperature set by the user	[24℃, 26℃]
T_{out}	Outdoor temperature	36℃
R	Equivalent thermal resistance	N(3,0.2^2) ℃/kW
C	Equivalent heat capacity	N(1.5,0.2^2) kWh/℃
P_N	Rated power	N(3.5,0.2^2) kW

5.2. System frequency under different frequency response ranges

In this part, different air-conditioning load total power (p.u.) is used to characterize the different number of air-conditioning loads in working condition. The frequency response interval is set to 0.2Hz, 0.4Hz, 0.6Hz in the three scenes of air-conditioning load total power (p.u.) respectively 0.1p.u., 0.2p.u., and 0.3p.u. Simulations are performed in the three scenes respectively to demonstrate that different frequency response ranges should be set at different outdoor temperatures to obtain better frequency regulation effect. The simulation results are shown in Figure 3. Observation shows that:

- In the scene where the total power of the air-conditioning cluster is 0.1p.u., when the frequency response interval is set to 0.2Hz and 0.4Hz, the number of air-conditioning loads participating in the demand response is small, the system frequency drops significantly. When the frequency response interval is 0.6Hz, the air-conditioning loads participating in the response are the most, and the frequency modulation effect is the best.

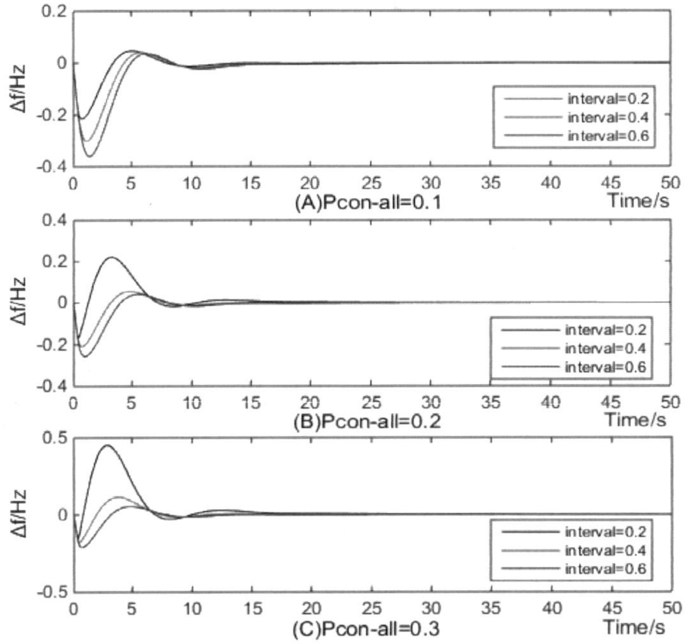

Figure3. System frequency fluctuation

- In the scene where the total power of the air-conditioning cluster is 0.2p.u., the air-conditioning load is over-adjusted when the frequency response interval is set to 0.2Hz, and the system frequency overshoot is as high as 0.2Hz, which endangers the stability of the system. Compared with the frequency response interval of 0.4Hz, when it is 0.6Hz the overshoot is basically the same, the minimum system frequency is lower, and the air-condition load is under-adjusted. So the frequency response interval should be set to 0.4Hz.
- In the scene where the total power of the air-conditioning cluster is 0.3p.u., there are too many air-conditioning loads in the open state at this time. As a result, the air-conditioning load is over-adjusted when the frequency response interval is set to 0.2Hz and 0.4Hz, and the overshoot of system frequency is large. So it should be set to 0.6 Hz.

6. Conclusion

The paper establishes a Monte Carlo-based aggregation model and proposes a trigger frequency setting method. Under different conditions of adjustable capacity, setting the trigger frequency according to the method proposed in the paper can provide a relatively consistent amount of adjustment. The simulation verifies that the method is more conducive to system frequency stability when air-conditioning cluster participates in primary frequency modulation. The paper conducts research on the cluster level. On the basis of the paper, how to set the trigger frequency of each air-conditioner in consideration of various factors will be the focus of the next stage.

Acknowledgments

This work was supported by the "Research and Demonstration Application of Key Technologies for Resource Assessment, Planning Operation, and Optimization Adjustment of Large-scale Adjustable Loads." The program was funded by State Grid Zhejiang Electric Power Co., Ltd. under Grant 2021ZK52.

References

[1] Callaway D S. Tapping the energy storage potential in electric loads to deliver load following and regulation, with application to wind energy[J]. *Energy Conversion and Management*, 2009, 50(5): 1389-1400.

[2] Kamgarpour M, Ellen C, Soudjani S E Z, et al. Modeling options for demand side participation of thermostatically controlled loads[C]. *Bulk Power System Dynamics and Control-IX Optimization, Security and Control of the Emerging Power Grid (IREP),* 2013 IREP Symposium. IEEE, 2013: 1-15.

[3] X. Qin, S. Lysecky and J. Sprinkle. A Data-Driven Linear Approximation of HVAC Utilization for Predictive Control and Optimization[J]. in *IEEE Transactions on Control Systems Technology*, vol. 23, no. 2, pp. 778-786, March 2015.

[4] H. Hao, B. M. Sanandaji, K. Poolla, and T. L.Vincent. Aggregate flexibility of thermostatically controlled loads[J]. in *IEEE Transactions on Power System*, vol. 30, no. 1, pp. 189–198, Jan. 2015.

[5] Ning Lu, D. P. Chassin and S. E. Widergen, "Modeling uncertainties in aggregated thermostatically controlled loads using a state queueing model," *IEEE Power Engineering Society General Meeting, 2005*, 2005, pp. 1336 Vol. 2-.

[6] X. Hu and J. Nutaro, "A Priority-Based Control Strategy and Performance Bound for Aggregated HVAC-Based Load Shaping," in *IEEE Transactions on Smart Grid*, vol. 11, no. 5, pp. 4133-4143, Sept. 2020.

[7] Z. Xu, J. Ostergaard and M. Togeby, "Demand as Frequency Controlled Reserve," IEEE Trans. Power Syst., vol. 26, no. 3, pp. 1062-1071, Aug. 2011.

[8] Molina-García, A., Bouffard, F., Kirschen, D.S.: 'Decentralized demand-side contribution to primary frequency control', IEEE Trans. Power Syst., 2011, 26, (1), pp. 411–419.

Evaluation Method of 500kV Transformer Bushing Insulation State Based on Active Power Loss

Zhaoliang Gu[1, 2*], Ran Xu[1, 2], Zhen Li[1, 2], Xiaolei Li[3], Yaguan Wang[4], Wei Xu[1, 2], Feng Yi[1, 2] and Zhixin Gao[1, 2]

[1]Shandong Electric Power Research Institute, Jinan 250003, China

[2]State Grid Shandong Electric Power Research Institute, Jinan 250003, China

[3]State Grid Shandong Jinan Power Supply Company, Jinan 250000, China

[4]State Grid Shandong Maintenance Company, Jinan 250000, China

[*]Corresponding author's e-mail: 792372434@qq.com

Abstract. The bushing is one of the most important components of power transformer, and it is very important to find the internal defects of bushings as soon as possible, effectively and accurately diagnose its insulation state information. In this paper, the active power loss of 500kV transformer bushings have been analyzed, and the insulation state evaluation method based on active power loss is proposed. To verify the feasibility of evaluation method, one abnormal 500kV transformer bushing was selected to analyze its insulation state. The result shows that this method can be used as a powerful complement to tanδ.

1. Introduction

The bushing is one of the most important components of power transformer. With the continuous expansion of power grid scale and the continuous improvement of power transformer voltage level, higher requirements are put forward for the insulation performance of important components, such as bushings. During the long-term operation of bushings in complex environment, due to the combined action of mechanical pressure, chemical corrosion, surface pollution, over-voltage, humidity and other factors for a long time, the insulation of equipment will continue to decline, which may eventually lead to insulation breakdown and transformer fault [1-3]. Therefore, it is helpful to reduce the probability of equipment accident and ensure the safe operation of equipment to find the internal defects of bushings as soon as possible, effectively and accurately diagnose its insulation state information, and formulate a scientific operation and maintenance plan for bushings [4-5].

At present, the dielectric loss tangent, tanδ, is the most commonly used characteristic parameter to characterize the overall insulation condition of bushings. But this single evaluation method still can not comprehensively and effectively evaluate the insulation condition of Bushings. This is mainly because , tanδ is the intermediate parameter of insulation dielectric loss. It can not truly reflect the insulation aging and power loss of the medium without effectively unifying the factors, such as insulation size and capacitance [6-7]. In addition, Most of tanδ are tested under power frequency voltage, which only reflects the inherent insulation characteristics of insulating medium to power frequency voltage. In some cases, bushings will bear the effect of complex voltage waveform, the insulation characteristics are different from those at power frequency, and the applicability of power frequency detection parameters is greatly limited [8].

Therefore, by analyzing the essential causes of bushings insulation aging, it can be determined that the insulation aging process itself is gradually developed with the long-term active power loss process of insulating medium in this paper. The electrical and thermal properties of dielectric are closely related, and the thermal state of bushings will directly affect its electrical properties. Therefore, active power loss is an important characteristic parameter to characterize the insulation condition of bushings, which can well reflect the insulation condition of bushings. In addition, in some cases, bushings will bear composite voltage, including power frequency, DC, pulse and other voltage components, which aggravates the dielectric electro-thermal effect inside the insulation. Therefore, starting from the essential physical process that the active power loss reflects the insulation condition, the active power loss of bushings can be analyzed respectively, which can lay a foundation for the effective evaluation of the insulation condition of bushings.

2. Analysis of Active Power Loss of Bushings

Dielectric will produce energy loss under the action of voltage. These losses can be divided into two types, one is the loss caused by conductivity, the other one is the loss caused by polarization. The energy consumed by a dielectric per unit time, that is, the energy generated by dielectric heating, is called dielectric loss. Under a constant electric field, there is no periodic polarization process in the dielectric, so the dielectric loss is only caused by conductivity [9-10]. Under AC voltage, in addition to the loss caused by conductivity, there is periodic relaxation polarization related to thermal motion, resulting in energy loss [11].

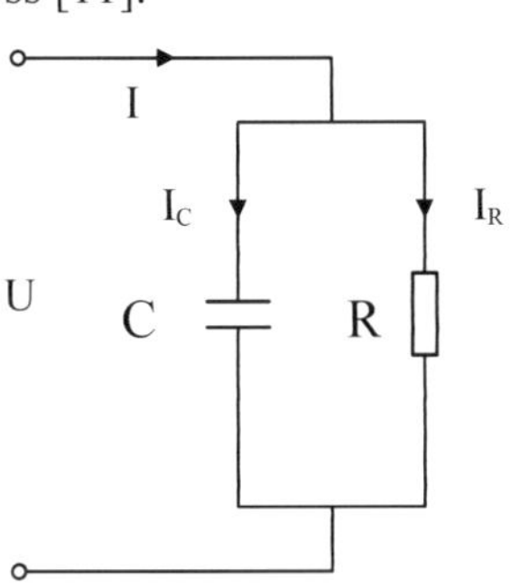

Figure 1. Equivalent circuit of dielectric

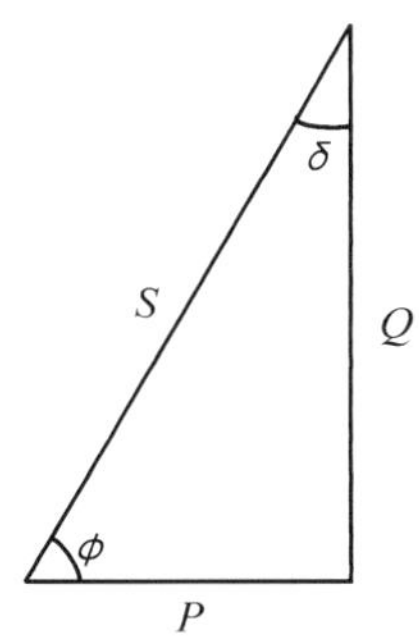

Figure 2. Power loss triangle

Voltage U is applied at both ends of the dielectric in the bushings. Due to the loss of the dielectric, the current is not a pure capacitive current, but includes active and reactive components I_R and I_C. Therefore, the apparent power supplied by the power supply is

$$S = P + jQ = UI_R + jUI_C. \tag{1}$$

Its dielectric loss is

$$P = Q\tan\delta = \omega CU^2 \tan\delta \tag{2}$$

Although the active power loss of bushings is related to capacitance, voltage and frequency, it is mainly referred to the value of tanδ. This leads to the neglect of the main factor of capacitance in the evaluation of bushings insulation state. Even if the capacitance factor is considered, it is not related to the specific capacitance. Under different voltage levels, even if the active loss of bushings is the same, the difference of bushings insulation volume will lead to different active loss per unit insulation volume. In other words, the power loss density of bushings insulation medium is also different, and the insulation condition of bushings is also different. Therefore, the insulation volume of bushings is also an important factor affecting its insulation condition.

For 500kV transformer bushings, their capacitance is often greater than those of 110kV of 220kV transformer bushings. So, even their tanδ are similar, it also shows that the active power loss of 500kV transformer bushing is much greater than that of 110kV or 220kV transformer bushing. However, in this case, the insulation state of 500kV bushings can not be simply said to be poor, because their volume are also larger than 110kV or 220kV transformer bushings. At this time, the influence of bushings insulation volume should be fully considered. Active power loss per unit volume is a more direct, effective and essential diagnostic parameter for insulation evaluation of bushings, which is directly related to the insulation state of bushings. The active loss per unit volume can be defined as p=P/V. The active loss density p of bushings is small, indicating that the bushing insulation is in good condition. When the p value is larger, it indicates that the active power loss per unit volume of the bushing increases, the heating is more serious, and there may be insulation defects.

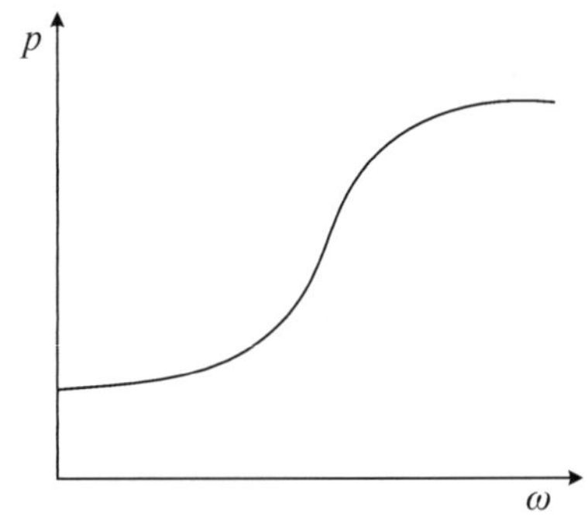

Figure 3. The relationship between power loss and frequency

When the frequency is low, the polarization of dielectric can be established in time. Meanwhile, the dielectric constant ε_r is also similar to the static dielectric constant ε_s. Therefore, p can be calculated by

$$\begin{aligned} p &= (\gamma + g)E^2 = (\gamma + \frac{\varepsilon_0(\varepsilon_s - \varepsilon_\infty)}{1+(\omega\tau)^2}\omega^2\tau)E^2 \\ &= (\gamma + \frac{\varepsilon_0(\varepsilon_s - \varepsilon_\infty)}{1+(\omega\tau)^2}(\omega\tau)\omega)E^2 \\ &\approx \gamma E^2 \end{aligned} \quad . \tag{3}$$

When the frequency is high, the relaxation polarization can not be established in time. As a result, the polarization mainly consist of displacement polarization. Besides, the dielectric constant ε_r is more similar to the optical frequency dielectric constant ε_∞. meanwhile, no relaxation loss can be produced and loss caused in each week decreases. p can be described by

$$\begin{aligned} p &= (\gamma + g)E^2 = (\gamma + \frac{\varepsilon_0(\varepsilon_s - \varepsilon_\infty)}{1+(\omega\tau)^2}\omega^2\tau)E^2 \\ &= (\gamma + \frac{\varepsilon_0(\varepsilon_s - \varepsilon_\infty)}{\tau})E^2 \\ &\approx (\gamma + g_i)E^2 \end{aligned} \quad . \tag{4}$$

It can be seen from the above formula that the bushings active power loss will surge at high frequency, thus accelerating the insulation deterioration.

3. Test Data and Discussions

The active power loss of 500kV bushings are analyzed in this paper. Firstly, the tanδ and capacitance C are obtained through manual preventive test or on-line monitoring data. Then, the active power loss of the bushing, and then converted to the power loss corresponding to the unit volume. It is assumed that the insulation volume of bushings at the same voltage level is approximately equal. According to the specific insulation structure of bushings, one 500kV capacitive bushings (phase A) data are selected for analysis in Shandong power grid, and the insulation volume is about $0.279m^3$. As shown in Table 1, the test data of these two groups of bushings from 2006 to 2021 are selected for analysis, and their active loss and active loss density are calculated.

Table 1. Calculation and analysis of 500kV transformer bushing (phase A).

	Rated voltage (kV)	Tempe-rature (°C)	Rated capacitance (pF)	Measured capacitance(pF)	Deviation (%)	tanδ (%)	Active power loss (W)	Active power loss density (W/ m^3)
2006	500	23	679	675	-0.59%	0.21	56.3	201.8
2009	500	21	679	681	0.29%	0.23	59.4	212.9
2012	500	19	679	673	-0.88%	0.22	55.6	199.3
2015	500	21	679	673	-0.88%	0.22	55.6	199.3
2018	500	18	679	675	-0.59%	0.24	60.1	215.4
2021	500	17	679	652	-3.98%	0.87	121.3	434.8

At present, tanδ is generally recognized as an important characteristic parameter to characterize the overall insulation condition of bushings, and is used to judge the insulation of equipment. By measuring the tanδ of bushings to roughly judge whether the main insulation of the bushing is good and how the insulation strength of the bushing is. We can see that the capacitance deviation of the bushing data in 2021 was -3.98%, which is much higher than before. Besides, its tanδ is also more than standard value. And its active power loss and active power loss density have also increased significantly compared with previous years. In order to further verify that there are problems in the insulation of the transformer bushing (phase A), the data of phase B transformer bushing data are calculated and analyzed, which is the same batch and model to phase A transformer bushing, as shown in table 2. We can see that there is no obvious fluctuations in the data of phase B transformer bushing in 2021.

Table 2. Calculation and analysis of 500kV transformer bushing (phase A).

	Rated voltage (kV)	Tempe-rature (°C)	Rated capacitance (pF)	Measured capacitance(pF)	Deviation (%)	tanδ (%)	Active power loss (W)	Active power loss density (W/ m^3)
2006	500	23	679	673	-0.88%	0.22	58.4	209.3
2009	500	21	679	680	0.15%	0.21	57.5	206.1
2012	500	19	679	674	-0.74%	0.22	59.6	213.6
2015	500	21	679	675	-0.59%	0.23	62.8	225.1
2018	500	18	679	673	-0.88%	0.22	58.5	209.7
2021	500	17	679	674	-0.74%	0.23	60.9	218.3

The main cause of abnormal data is analyzed, and it is considered that the insulation of bushing is damp and aging. The poor quality of the sealing ring of the bushing leads to the lax sealing of the bushing, resulting in the gradual infiltration of water into the capacitor core through the outer insulation. After the paper fiber in the insulation absorbs water, the conductivity of the fiber increases and the mechanical properties become worse, which reduces the main insulation and outermost insulation resistance of the bushing, increases the conductivity of the insulation system and increases the conductivity loss. At the same time, water will participate in the chemical degradation of polymer materials such as oil, paper cellulose and so on. This increases the number of particles participating in polarization and polarization loss per unit volume of insulating medium. The dielectric loss of the insulation system consists of conductivity loss and polarization loss. The increase of the total active power loss will cause the dielectric to heat and further accelerate the deterioration of the insulation. The degradation of insulating paper cellulose and insulating oil will generate water, which further participates in the chemical degradation reaction of polymer materials such as oil and paper cellulose.

Thus, playing a 'positive feedback' role in insulation aging. Therefore, the active power loss analysis can effectively find the aging and moisture of bushings insulation. The greater the active power loss value, the worse the insulation medium state.

In addition, the active power loss fully considers the factors of bushing capacitance and insulation volume, which makes up for the conventional parameter tanδ in the effectiveness of bushing insulation state detection. Based on the analysis of abnormal bushings data of Shandong power grid, it is preliminarily considered that when the active power loss density of capacitive bushings exceeds 200 W/m^3, the insulation may be aged and damp.

4. Conclusions

In this paper, the method based on active loss has been proposed and used for evaluating the insulation state of transformer bushings. One abnormal 500kV transformer bushing test data is used for the analysis of bushing insulation evaluation in Shandong power grid. And the feasibility of the evaluation method based on active loss has been verified. Based on the analysis of a large number of data, if the active loss density of bushings is more than 400 W/m3, there may be some problems in the bushings insulation. Using the method of active power loss evaluation, transformer bushings can be evaluated uniformly, and can be used as a powerful tool rather than tanδ.

Acknowledgments

This work was supported by the Technical Innovation Project of Shandong Electric Power Research Institute (ZY-2020-02).

References

[1] Zhou, Q. , Zhi, X. U. , Liao, R. , Zhang, Y. , & Zheng, B. (2013). Insulation condition assessment of power transformer bushing based on cloud model and kernel vector space model. Gaodianya Jishu/High Voltage Engineering, 39(5), 1101-1106.

[2] Suwanasri, T. , & Suwanasri, C. (2015). Development of on-line monitoring for power transformer bushing. Applied Mechanics and Materials, 781, 384-387.

[3] Wang, Z. (2013). New-style dielectric loss of transformer bushing on-line monitoring apparatus. Guangdong Electric Power, 4(23), 215-223.

[4] Gebelein, J. P. , & Elzi, J. A. (2013). Analysis of bushing-current-transformer performance. Electrical Engineering, 59(9), 541-546.

[5] Hashemnia, N. , Abu-Siada, A. , & Islam, S. . (2016). Detection of power transformer bushing faults and oil degradation using frequency response analysis. IEEE Transactions on Dielectrics and Electrical Insulation, 23(1), 222-229.

[6] Ye, T. , Bo, Q. , Ran, Z. , Fu, M. , & Li, C. (2016). Locating partial discharge source occurring on transformer bushing by using the improved TDOA method. 2016 International Conference on Condition Monitoring and Diagnosis (CMD). IEEE.

[7] Jiang, J. , Chen, J. , Li, J. , Yang, X. , & Schwarz, H. (2021). Partial discharge detection and diagnosis of transformer bushing based on uhf method. IEEE Sensors Journal, PP(99), 1-1.

[8] Kim, S. H. , Jung, K. J. , Kim, Y. J. , Shin, B. I. , & Lee, H. B. (2020). Correction to: the reduction method of a stray load loss of distribution transformer on the bushing flange plate with sts. Journal of Electrical Engineering & Technology, 15(1), 515-515.

[9] Zhang, L. , Sun, L. , Bin, F. U. , Li, M. A. , Han, Y. H. , & Shen, W. (2019). On-line monitoring technology of transformer bushing insulation based on dielectric response characteristics. Transformer.

[10] Yang, Z. , Ruan, J. , Huang, D. , Du, Z. , & Zhou, T. (2019). Calculation of hot spot temperature of transformer bushing considering current fluctuation. IEEE Access, PP(99), 1-1.

[11] He, C. , Xie, Q. , Yang, Z. , & Xue, S. (2019). Seismic evaluation and analysis of 1100-kvuhv porcelain transformer bushings. Soil Dynamics and Earthquake Engineering, 123(AUG.), 498-512.

Design and Realization of Intelligent Aero-engine DDPG Controller

Rong-rong Qian[1], Yimin Feng [2], Mengman Jiang [2], Lijun Liu [2*]

[1]AECC Commercial Aircraft Engine Co., LTD;

[2]School of Aerospace Engineering, Xiamen University, Xiamen 361000, China

* School of Aerospace Engineering, Xiamen University, liulijun@xmu.edu.cn

Abstract: As the artificial intelligence technology advances, intellectualization has become an important development trend for the future aero-engine industry. The intelligent aero-engine will bring a new round of technology revolution in aviation industry. Since an aero-engine is a complex system with strong nonlinearity and coupling, it is difficult to achieve the optimal control performance with traditional PI control. Therefore, the DDPG (Deep Deterministic Policy Gradient, DDPG) algorithm belonging to the family of Deep Reinforcement Learning is hereby proposed for designing aero-engine controller. Based on the non-linear polynomial state-space mathematical model of JT9D turbofan engine, the intelligent DDPG controller is designed and then compared with the performance of PI controller. The proposed controller can achieve the optimal control effect for aero-engine through the training the neural network. The simulation results demonstrates that compared with PI control, the control method proposed in this paper achieves superior control performance in response speed, overshoot and anti-interference ability.

1. Introduction

An aero-engine is a complex system with strong nonlinearity, and its performance directly affects the safety, reliability and economy of the aircraft. For the aero-engine development, it is necessary to design an effective control system in order to ensure that aero-engine can work within stability margins with desired performance. At present, the traditional aero-engine control method is generally based on proportional integral(PI) controller, which has a simple structure, strong robustness, and easy adjustment[5]. However, the turbofan engine is a complicated plant with wide operating range and long parameter-varying features, traditional PI controller can only realize the desired control performance on the single point or within a small operating range[6]. Therefore, the control structure of traditional PI controllers with limit protections, showing in Fig.1, is the most popular control method for the aero engine.

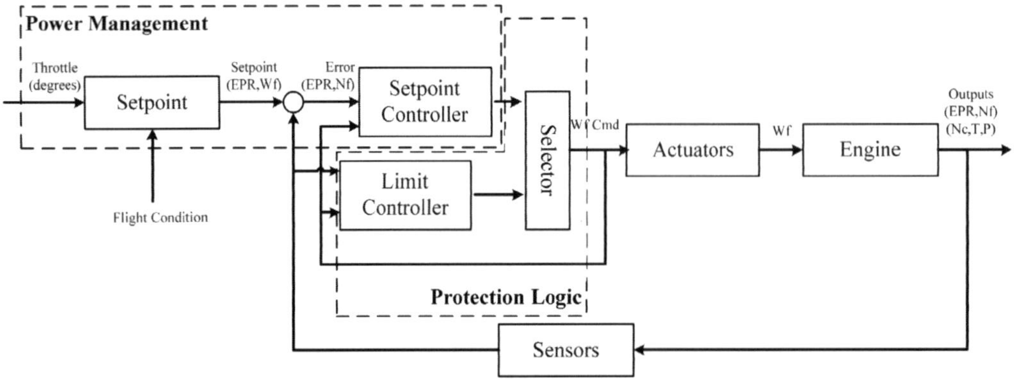

Fig. 1 Diagram of an aircraft engine control system

Although the above structure has been widely applied in the aero-engine control, it can not satisfy the increasing demand of aero engine control performance such as high efficiency, low fuel consumption with high level safety. On account to these disadvantages of PI controller, many improved control methods for aero engine are developed. Recent examples of such efforts include applying dispatch control[1], linear quadratic Gaussian/ loop transfer recovery (LQG/LTR)Control[3], adaptive control[4], sliding mode controllers to improve the dynamic engine response. In order to maximize the potential of aero-engines, more advanced control strategies must be taken into the aero-engine control. With the development of artificial intelligence technology and the evolution of computing processing power, the research and development of intelligent aero engine controller can be available for the requirements of high performance, high reliability and low cost, etc., for future aero engine. With respect to the artificial intelligence controllers, literature[8] proposes a new reactive control method through the combining reinforcement learning with fuzzy logic, which realizes different levels of robots' control in an unknown environment. In reference to the complexity and variability of the environment in automatic driving control, literature[9] applies the online learning of reinforcement learning to train neural network using the interactive learning methods. At the same time, it improves the generalization ability of the model by clustering and sampling the state-space. However, the DRL (Deep Reinforcement Learning, DRL) algorithm is rarely used in the domain of aero-engine control.

Based on mentioned above, a kind of reinforcement learning, the deep deterministic policy gradient (DDPG) algorithm as a steady-state aero-engine controller, is designed and realized in this paper by training the nonlinear polynomial state space(NLPSS) model of JT9D aero-engine. In order to accelerate the rate of convergence and avoid the divergence and over-fitting of training process for DDPG controller based on NLPSS model, the DDPG controller based on linear state-space model(LSS) of JT9D has been designed as the pre-training and the acquired controller parameters are used as the initial parameters for DDPG controller based on NLPSS model. The pre-training provides the better model initialization and improves the generalization capability of network. The simulation results show that the proposed method has better control effect on the big working range of JT9D aero-engine in the aspect of response speed, overshoot and anti-interference ability.

2. Establishment of aero-engine state-space model

In this paper, the engine model of JT9D from T-MATS is used to design the intelligent controller. The T-MATS module is a thermodynamic system simulation library based on Matlab/Simulink platform developed by NASA Glenn Research Center. The aero-engine model structure of JT9D is analyzed in Matlab. Taking the cruise phase as an example, the aero-engine operating environment is as follows:

W = 661.26 pps, h_t = 130 BUT/(lbm*R), T_t = 448.46 degR, P_t = 5.528 psia, Pamb = 3.626 psia.

Where W is the gas path flow of fan inlet, h_t and T_t are the total enthalpy and total temperature of fan inlet air flow, Pamb is the ambient pressure.

Supposed that the aero-engine works at the certain steady point $X_0=[n_L\ n_H]^T$, where n_L is the low pressure turbine speed, n_H is the high pressure turbine speed. Δn_L and Δn_H, the variation of n_L and n_H, are considered as the state variables of aero-engine state-space model. ΔFAR, the variation of fuel-air ratio *FAR* of aero engine, is the control variable. In this paper, the steady speeds of n_L and n_H are specified as 3350.9 rpm (round per minute) and 7189.7 rpm and the corresponding fuel-air ratio is 0.0165.

The proposed NLPSS model of engine is designed as follows.

$$\begin{cases} \Delta \dot{n}_L = \sum_{n=1}^{N} a_{1(2n-1)} \Delta n_L^N + \sum_{n=1}^{N} a_{1(2n)} \Delta n_H^N + b_1 \Delta FAR \\ \Delta \dot{n}_H = \sum_{n=1}^{N} a_{2(2n-1)} \Delta n_L^N + \sum_{n=1}^{N} a_{2(2n)} \Delta n_H^N + b_2 \Delta FAR \end{cases} \tag{1}$$

Where, N is represented as the model order. When N=1, equation (1) is the LSS of JT9D aero-engine. $a_{1(2n-1)}, a_{1(2n)}, a_{2(2n-1)}, a_{2(2n)}$ are the coefficients of state variables, b_1, b_2 are the coefficients of input variable.

It has been proved in the previous research[10] that the NLPSS model of aero-engine could be used in much broader operating range around steady point than the linear model. In order to reduce the switching times of controllers, the NLPSS model is used for the DDPG controller design. However, as the model order is getting bigger, the convergence of DDPG controller based on NLPSS model is hard. The initial values of controller parameters have enormous impact on the convergence of training process. Due to the simple structure and small working range of linear model, the parameters training for DDPG controller based on linear model has good astringency with quick convergence and stability. Therefore, in order to reduce the training time and avoid trapping in the local optimum, the DDPG controller design based on linear model are used as the pre-training for the final controller and the acquired controller parameters are used as the initial values for DDPG controller based on NLPSS model.

The final LPSS and NLPSS model of JT9D aero-engine after normalization at a certain steady-state operating point are given in Equation (2) and (3) . The outputs of fitting results are shown in Fig.2.

$$\begin{cases} \begin{bmatrix} \Delta \dot{n}_L \\ \Delta \dot{n}_H \end{bmatrix} = \begin{bmatrix} -2.5172 & 1.9325 \\ 2.0162 & -3.6242 \end{bmatrix} \begin{bmatrix} \Delta n_L \\ \Delta n_H \end{bmatrix} + \begin{bmatrix} 0.6337 \\ 1.3420 \end{bmatrix} \Delta FAR \\ \begin{bmatrix} \Delta n_L \\ \Delta n_H \end{bmatrix} = \begin{bmatrix} 0 & 1 \\ 1 & 0 \end{bmatrix} \begin{bmatrix} \Delta n_L \\ \Delta n_H \end{bmatrix} \end{cases} \tag{2}$$

$$\begin{cases} \begin{bmatrix} \Delta \dot{n}_L \\ \Delta \dot{n}_H \end{bmatrix} = \begin{bmatrix} -4.181 & 2.714 \\ -0.872 & -0.213 \end{bmatrix} \begin{bmatrix} \Delta n_L \\ \Delta n_H \end{bmatrix} + \begin{bmatrix} 5.410 & -2.825 \\ 2.062 & -1.180 \end{bmatrix} \begin{bmatrix} \Delta n^2_L \\ \Delta n^2_H \end{bmatrix} \\ \qquad + \begin{bmatrix} -2.804 & 1.234 \\ -1.214 & 0.537 \end{bmatrix} \begin{bmatrix} \Delta n^3_L \\ \Delta n^3_H \end{bmatrix} + \begin{bmatrix} 0.01073 \\ 0.01610 \end{bmatrix} \Delta FAR \\ \begin{bmatrix} \Delta n_L \\ \Delta n_H \end{bmatrix} = \begin{bmatrix} 0 & 1 \\ 1 & 0 \end{bmatrix} \begin{bmatrix} \Delta n_L \\ \Delta n_H \end{bmatrix} \end{cases} \tag{3}$$

The final single-input linear and nonlinear state-space models of the JT9D aero-engine after normalization at a certain steady-state operating point are given in Equation (2) and (3) .

3. Design of DDPG controller

3.1 Design scheme of aero-engine DDPG controller

The design scheme of the DDPG controller proposed in this paper is shown in Fig.2. The main design process is as follows.

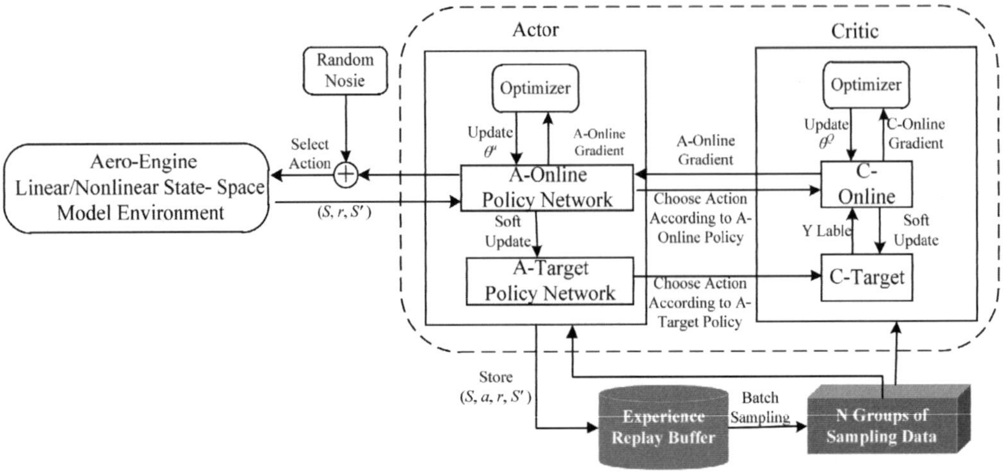

Fig.2 Design Scheme of Aero-engine DDPG Controller

1) To build the training environment for the controller training based on the previously established JT9D engine linear/non-linear model;
2) To design the intelligent DDPG controller based on the linear model, including the construction of network structure, reward function design, experience replay buffer design, etc.
3) Based on the obtained linear controller parameters, the nonlinear model environment is optimized and trained.
4) Finally, the nonlinear single-input intelligent controller of the JT9D aero-engine is constructed.

It can be seen from equation (2) and (3) that the control input of engine is the change of fuel-air ratio ΔFAR, which is used as the action *a* given by the agent in the reinforcement learning. The output of the engine model is Δn_L and Δn_H, which is used as the state *S* in the reinforcement learning. In the control process of aero-engine, the agent is desired to acquire the ability to control the engine through self-learning. When the expected value of low pressure turbine speed of aero-engine changes to a certain value, in order to achieve the best control effect, the agent is supposed to accurately give the corresponding action value, which is the control variable ΔFAR. The training process of the engine controller is shown in Fig.3.

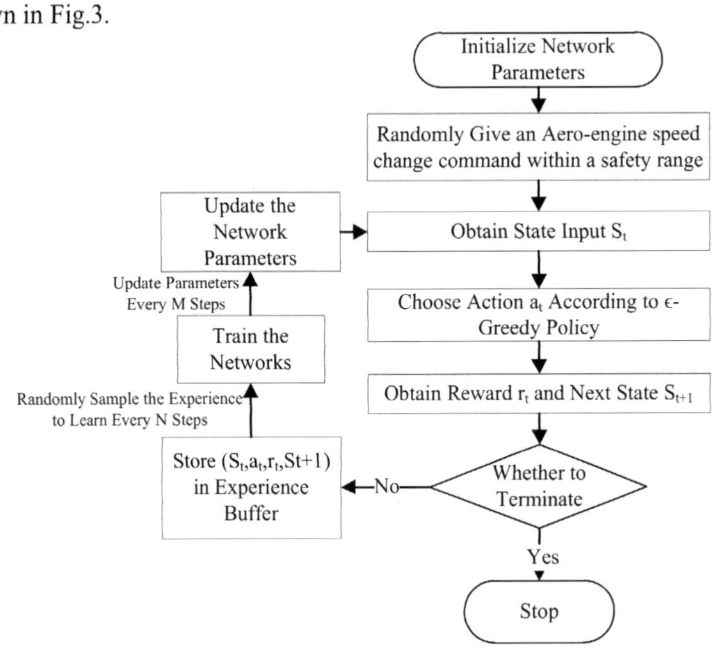

Fig.3 DDPG Controller Training Flowchart

3.2 Network framework design

The aero-engine intelligent controller is mainly composed of the modules of aero-engine mathematical model environment, policy networks, value networks, experience buffer, as shown in Fig.2.

A-Online is an online policy network whose function is to update the parameters of the policy network. The input of the network is the current state of the engine *S*, that is, the current low-pressure turbine speed change value Δn_L of the engine. The output of the network is action *a*, that is ΔFAR. Through the interaction with the environment, the next state S_{t+1} and the reward *r* can be obtained. The loss function can be expressed as equation (4):

$$L_a = J(\theta) = -\frac{1}{m}\sum_{i=1}^{m} Q_{\text{targ}}(s_i, a_i) \qquad (4)$$

A-Target is a target policy network. Its function is to sample the actions through the experience buffer as the input to C-Target network and improve the convergence speed of the network. The input of A-Target is the next state of the state *S* from the online actor network. Then the action of the A-online network is transferred to the aero-engine simulation environment and executed to obtain the next step state S_{t+1} of the engine. The output of the A-target network is the action a_{t+1}. The update mode of network parameter is to copy regularly from the A-Online network, that is so-called the soft-update. The update equation is shown in equation (5):

$$\theta_{\text{targ}} \leftarrow \rho\theta_{\text{targ}} + (1-\rho)\theta \qquad (5)$$

C-Online is an online critic network. Its function is to update the parameters of the A-online policy network and optimize the update direction for the actor network. The inputs of the C-online network are the current state of the engine *S*, the actual execution action *a*. The output of the network is the Q value of the current state *S*. The loss function can be expressed in equation (6):

$$L_c = J(\phi) = \frac{1}{m}\sum_{j=1}^{m} (y_i - Q_{\phi_{t\arg}}(s_i, a_i))^2 \qquad (6)$$

C-Target is the target critic network. Its function is to calculate the target Q value, to optimize the iterative direction for A-online policy network, and to improve the convergence speed of the network. The input of the C-Target network is the next state of the engine S_{t+1} and the next state a_{t+1}. The update mode of C-Target network is similar to that of A-Target network. It also uses a soft-update method to periodically obtain the parameters from the C-Online network. The update equation is as follows in equation (7):

$$\phi_{t\arg} \leftarrow \rho\phi_{t\arg} + (1-\rho)\phi \qquad (7)$$

3.3 Reward function design

In the reinforcement learning algorithm, the design of the reward function is significant which directly influences the control effect and the astringency. The agent in the reinforcement learning algorithm must be told which decision is preferred and which decision is not encouraged by the reward function. Therefore, in the aero-engine reinforcement learning control system, the difference between the current engine speed change value and the given command value is considered as the most important indicator. However, in the agent decision process, the reward becomes greater when the engine speed is getting closer to the command value and it would increase the amplitude of the action. This would result in the big overshoot of control effect. In order to avoid the overshoot, $\Delta\dot{n}_L$ is introduced in the reward function. $\Delta\dot{n}_L$ represents the acceleration of the low pressure turbine speed. The larger the value of $\Delta\dot{n}_L$ grows, the greater the stride range of the agent in the previous step to the target value becomes. The introduction of $\Delta\dot{n}_L$ can regulate the amplitude of the agent's decision to a limited extent.

In this paper, we design two reward functions respectively for DDPG controller based on LPSS and NLPSS model.

3.3.1 Reward function design for DDPG controller based on LPSS

In this paper, the single-input linear reward function of the aero-engine is designed as shown in equation (8):

$$r = |a(\Delta n_L^* - \Delta n_L)| + |b(\Delta n_H^* - \Delta n_H)| + |c\Delta \dot{n}_L| \tag{8}$$

Where, *a*, *b* and *c* are the arbitrary constants. Δn_L^* and Δn_H^* are the command change values of low and high pressure turbine speeds. Δn_L and Δn_H are the current change values of low and high pressure turbine speeds. $\Delta \dot{n}_L$ represents the low pressure turbine speed acceleration of the previous moment.

3.3.2 Reward function design for DDPG controller based on NLPSS

For the NLPSS model, the nonlinear state order is up to 3 according to equation (2). The increase of model order would result to the explosion of reward function during the training process of the network. Therefore, this paper limits the maximum value of the reward function. When the reward value is larger than or equal to a certain set value, the reward value will be forcibly assigned to the set value to improve the convergence speed of the algorithm. A nonlinear reward function is designed for DDPG based on NLPSS model is shown in equation (8):

$$r = \begin{cases} |a(\Delta n_L^* - \Delta n_L)| + |b(\Delta n_H^* - \Delta n_H)| + |c\Delta \dot{n}_L| + |d\Delta \dot{n}_H| & r < X \\ k & r > X \end{cases} \tag{9}$$

Where, *a*, *b*, *c* and *d* are the arbitrary constants. Δn_L^* and Δn_H^* are the command change values of low and high pressure turbine speeds. Δn_L and Δn_H are the current change values of low and high pressure turbine speeds. $\Delta \dot{n}_L$ and $\Delta \dot{n}_H$ respectively represent the low pressure and high pressure turbine speed accelerations of the previous moment. *k* is the preset value and *X* is the switching threshold of the reward value.

4. Simulation and analysis

4.1. Control effect of DDPG controller based on LPSS

The step demand signal of Δn_L is used to test the control effort of the proposed DDPG controller based on LPSS model. Fig.4 (a) and (b) are respectively the low-pressure and high-pressure turbine speed response curves of the engine DDPG controller based on LPSS model in the normal operation (without disturbance) on the JT9D simulation model. The result shows that the proposed controller can acquire the fast response of engine low-pressure turbine speed control without overshoot.

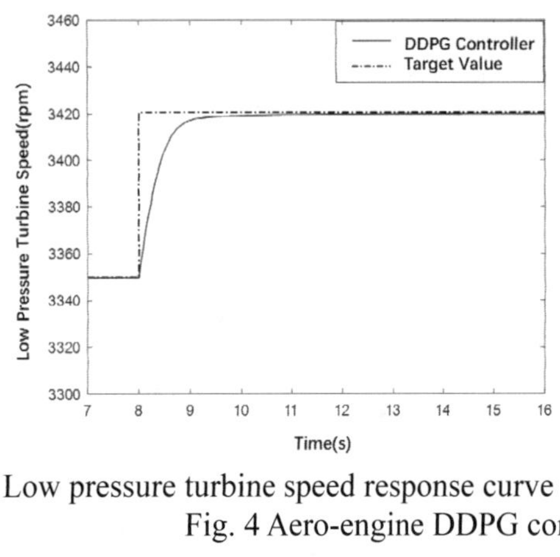

(a) Low pressure turbine speed response curve

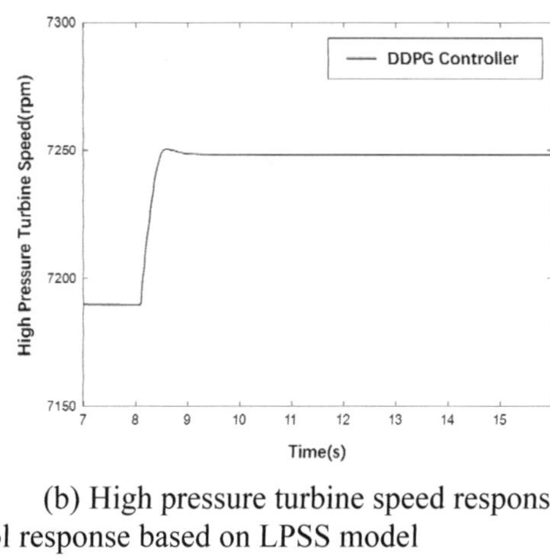

(b) High pressure turbine speed response curve

Fig. 4 Aero-engine DDPG control response based on LPSS model

4.2. Control effect diagram of DDPG controller based on nonlinear model

Fig.5 (a) and (b) are the response curves of the engine DDPG controller based on NLPSS model in normal operation on the JT9D simulation model. Compared to the DDPG controller based on LPSS model, the DDPG controller based on NLPSS can acquire better dynamic response of high pressure turbine speed which has no overshoot. Fig.5 also shows that the DDPG controller based on nonlinear model can be used to control the larger change value of low pressure turbine speed which means less switch in the full envelop control of aero-engine.

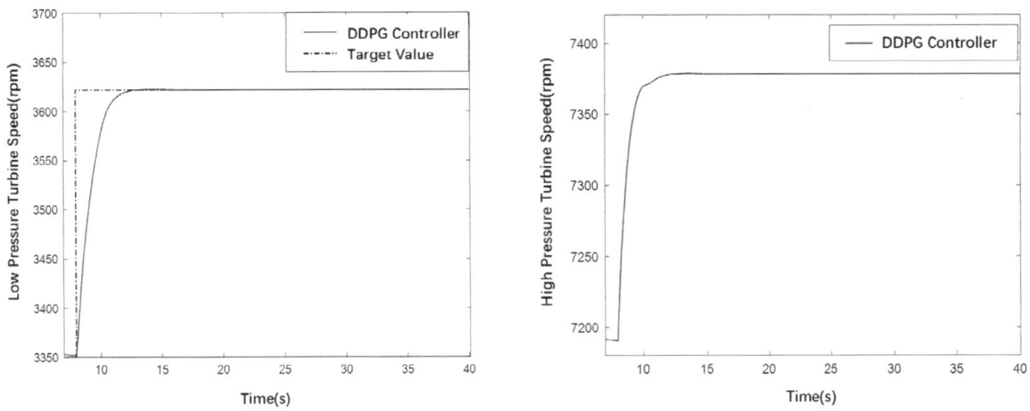

(a) Low pressure turbine speed response curve (b) High pressure turbine speed response curve
Fig.5 Aero-engine DDPG control response based on NLPSS model

4.3. Comparison between DDPG and PI controller

Aiming at the speed control in the cruise phase of aero-engine, Fig.6(a) and (b) show the simulation comparison of DDPG controller based on NLPSS model and PI controller without interference. It can be clearly seen from Fig.6(a) that given the same control command, the stabilization time of DDPG controller is smaller than PI controller in low pressure turbine speed response.

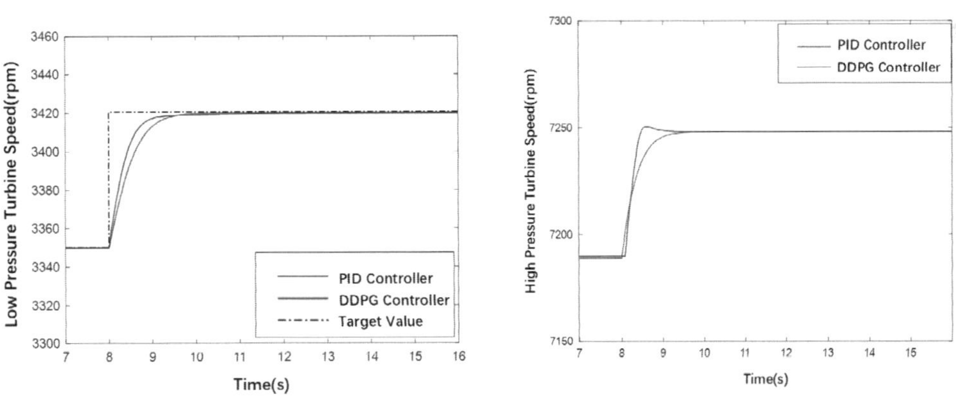

(a)low pressure turbine speed response curve (b) high pressure turbine speed response curve
Fig.6 Comparison of control response between PI controller and single-input DDPG controller

Fig.7(a) and (b) show the simulation comparison between the intelligent DDPG controller based on NLPSS model and PI controller when the total pressure P_t of the aero-engine is added with a strong disturbance of 10%. It can be seen from Fig.7 that the overshoot of the intelligent DDPG controller is significantly smaller than the PI controller whose response curve jitters more violently. It means that the DDPG controller in aero-engine control has a better anti-disturbance ability.

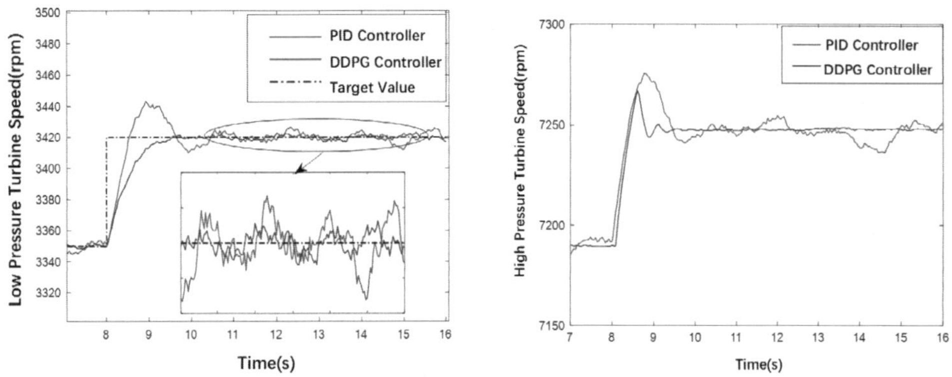

(a)low pressure turbine speed response curve (b) high pressure turbine speed response curve
Fig.7 Comparison of control response between PI controller and DDPG controller after adding disturbances of d=±10%

4.4. Simulation summary and analysis

It can be seen from Figure 6-7 that no matter whether the controlled variable is $\triangle n_L$ or $\triangle n_H$, the DDPG controller performs with a faster response speed than the PI controller. At the same time, the response curve of the proposed controller is smoother, without overshoot, and has a good control performance. In addition, when ±10% strong interference is added to the total pressure of aero-engine P_t, as shown in Fig.7, the intelligent DDPG controller has better steady state and dynamic response and anti-disturbance ability than PI controller. Therefore, the intelligent DDPG controller has better control effect than the PI controller in terms of the speed control problem during the cruise phase of the aero-engine.

5. Conclusion and Future Work

In this paper, we proposed a new control method, DDPG, for the aero-engine design. Developing the aviation intelligent engine using industrial big data, the work presented in this paper is the first application of intelligent tools in the aero-engine control area. Aiming at optimizing the aero-engine rotational speed control, the JT9D engine in the high-precision aero-engine simulation platform T-MATS is chosen as an example. The linear and nonlinear state-space models, normalized at a certain steady-state operating point, are established in this paper. Then, the DDPG algorithm is applied to design the intelligent controllers. After training the neural networks in DDPG algorithm, two model-based intelligent controllers are successfully obtained. The advantages and feasibility of the method proposed are demonstrated by the comparison with PI controller. The simulation results have shown that, compared with PI control, the proposed DDPG controller in this paper has a better control effect in response speed, overshoot and anti-interference ability.

Although the DDPG controller designed in this paper performed well in the simple simulation environment with the single work condition, further research on the modeling and construction for the full envelop operation and multivariable control with more uncertainties should be conducted in future. Furthermore, we must develop a new simulation and experiment test theory and system to verify the improvement of the aero-engine performance by new control strategy. It would be a new technology trip for us and it is really worthy of our more in-depth study.

Funding Statement

Sponsored by National Science and Technology Major Project (2017-I-0001-0001) and Shanghai Sailing Program (20YF1454300).

References

[1] R. Zhang, G. A. Alleyne, and D. E. Carter, "Generalized multivariable gain scheduling with

robust stability analysis," ASME J. Dyn. Syst., Meas.,Control, vol. 127, no. 4, pp. 668–687, Jan. 2005.

[2] R. R. Sumar, R. A. R. Coelho, and L. dos Santos Coelho, "Computational intelligence approach to PID controller design using the universal model,"Inf. Sci., vol. 180, no. 10, pp. 3980–3991, Mar. 2010.

[3] Y.-T. Wang, Y.-Q. Guo, and H.-Q. Wang, "Design of non-fully recover LQG/LTR control system of aeroengine," J. Aerosp. Power, vol. 22, no. 3, pp. 485–489, Mar. 2007.

[4] I. D. Landau and B. Courtiol, "Design of multivariable adaptive model following control systems," Automatica, vol. 10, no. 4, pp. 483–494,Sep. 1974.

[5] H. Richter, Advanced Control of Turbofan Engines. New York, NY, USA: Springer Science & Business Media, 2011.

[6] L. C. Jaw and J. D. Mattingly, Aircraft Engine Controls: Design, System Analysis, and Health Monitoring. Reston, VA, USA: American Institute of Aeronautics and Astronautics, 2009.

[7] A. Kwiatkowski, H. Werner, J. P. Blath, A. Ali, and M. Schultalbers, "Linear parameter varying PID controller design for charge control of a spark-ignited engine," Control Eng. Pract., vol. 17, no. 11, pp. 1307–1317, 2009

[8] C. C. Chen. Autonomous learning and navigation control of mobile robots based on reinforcement learning[D]. University of Science and Technology of China, 2006.

[9] W. Xia, H. Y. Li. Training Method of Automatic Driving Strategy Based on Deep Reinforcement Learning [J]. Journal of Integration Technology, 2017(5): 29-40.

[10] Cheng D, Liu L, Yu Z. A novel multivariable nonlinear robust control design for turbofan engines. Transactions of the Institute of Measurement and Control. October 2021.

Cooperative Control Mechanism of Key Objects and Key Nodes in Container Terminals

Hankun Shi[1], Xuelin Wang[2*], Hongmao Zhou[3]

[1] China Waterborne Transport Research Institute, Beijing, China

[2] China Waterborne Transport Research Institute, Beijing, China

[3] China Waterborne Transport Research Institute, Beijing, China

[*]Corresponding author's e-mail: wangxuelin@wti.ac.cn

Abstract. This research starts from the operation process of the container terminal, constructs a technical route for the analysis of key objects and key nodes, summarizes the specific distribution of key objects and key nodes of the container terminal, and analyzes the process mechanism and interactive information of the container terminal. The cooperative control mechanism of key objects and key nodes creates a certain theoretical framework for the automation and intelligent design of container terminals.

1. Introduction

There are more than 30 automated container terminals built around the world. The ECT terminal in the Port of Rotterdam, the Netherlands was completed in 1993, marking the entry of the container terminal into the era of automation[1]. Subsequently, the Port of London in the United Kingdom, the Port of Kawasaki in Japan, the Port of Singapore, and the Port of Hamburg in Germany have successively built automated container terminals[2]. Since the Xiamen Yuanhai Automated Terminal was completed in March 2016[3], several newly built or renovated automated container terminals such as Shanghai Port, Qingdao Port and Tianjin Port have been built and put into operation one after another. Tianjin Port integrates the application of unmanned automated rail bridges, unmanned electric trucks, and remotely controlled unmanned and autonomous quay cranes[4-5]. The self-developed "intelligent horizontal transportation system" achieves the best overall efficiency. The intelligent production and real-time equipment scheduling system of Shanghai Port Yangshan Phase IV Automated Container Terminal has realized unmanned supervision of equipment operation execution, and the intelligent test of dual-cantilever ARMG container area has realized automatic container sorting, container turning and pre-turning Work, manual work intensity is greatly reduced[6-7]. The Qingdao Port Automated Container Terminal is highly integrated with new technologies such as the Industrial Internet, Internet of Things, and 5G[8-9]. It is the world's leading fully automated container terminal and the world's first "hydrogen +5G" smart green terminal[10], with a maximum operating efficiency of 52.1 natural containers/hour , Compared with the traditional terminal, the operating efficiency is increased by 30%, and the labor cost is reduced by 70%.

In short, the current problems in the direction of intelligent and automation of container terminals mainly include the following two aspects: the first problem is the uneven development of intelligent and automation technology of container terminals. It is mainly reflected in the unbalanced development of key technologies. For example, the upgrading of control technology can not catch up with the development of information technology, and the coordinated development between single machine equipment automation and system informatization is insufficient. The second problem is the lack of deep integration of emerging technologies. The integration of emerging technologies such as artificial intelligence, big data, blockchain, information physics system and automatic driving with port business scenarios and key equipment systems is far from enough to support the future development needs of container terminal intelligence and automation. Therefore, starting from the basic theory and overall architecture, it is particularly necessary to find the key objects and key nodes of the container terminal, and deeply analyze the cooperative control mechanism between them.

2. The technical route of analyzing key objects and key nodes of the container terminal

The characteristics of the operation process of the container terminal include: the loading and unloading operations in the container terminal are completely reversible; the operation machinery has cross-regional operations in the operation process; the operation machinery has obvious cooperation and correspondence in the operation process. Based on a comprehensive survey of major domestic automated container terminals such as Qingdao Port, Tianjin Port, Shanghai Port, etc., it analyzes the existing container moving operation process, sorts out the main ship loading and unloading processes, and studies the matching transportation methods. Finally, the key nodes and key objects in the container operation process are clarified. The key object and key node analysis technology roadmap is shown in Figure 1.

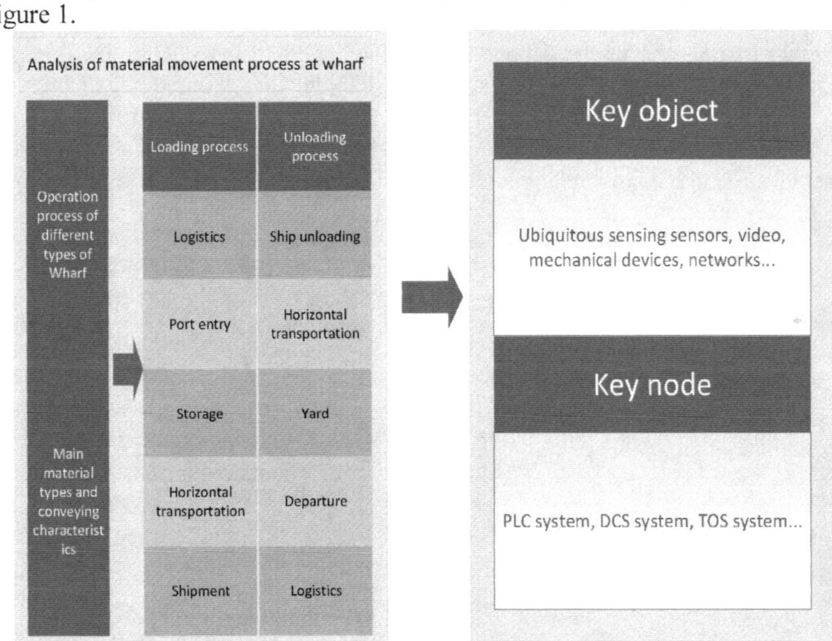

Figure 1. The key object and key node analysis technology roadmap

3. Distribution of key objects and key nodes in container terminals

The key objects of the automated container terminal are divided into four categories in terms of equipment types, namely terminal frontier loading and unloading equipment, terminal horizontal transportation equipment, terminal yard equipment, and in and out terminal loading and unloading equipment. The equipment under these four types of equipment together constitute the container terminal. The key objects are shown in Table 1. The organic cooperation between them determines the normal and orderly operation of container terminal production operations.

Table 1. Key Objects of Automated Container Terminal

Wharf type	Equipment type	Equipment
Container	1.Wharf apron handling equipment	Single trolley double 20ft box shore bridge
		Single trolley double 40ft box shore bridge
		Double trolley double 20ft box shore bridge
		Double trolley 40ft box shore bridge
	2.Wharf horizontal transportation equipment	Horizontal transportation truck
		AGV/IGV
		Straddle carrier
		Rail shuttle car
	3.Wharf yard equipment	Rail mounted gantry crane
		Rubber tyred gantry crane
		Straddle carrier
	4.Loading and unloading equipment in and out of the wharf	Container truck
		Driverless container truck

The key nodes of the container terminal are divided into four categories from the core system to which they belong, namely the terminal operating system, the support system, the equipment control system, and the remote control system. The subsystems under these four types of systems constitute the key nodes of the container terminal. As shown in Table 2. The coordination and interconnection between them assisted the normal operation of the key equipment of the container terminal, realized the business operation and the exchange and sharing of information and data, and improved the work efficiency and intelligence of the terminal to a certain extent.

Table 2. Key node of container terminal

Wharf type	Core system	Subsystem
Container	1.Terminal Operation System	System management
		Configuration management
		Basic information management
		Appointment acceptance
		Documentation and electronic data processing
		Operation plan
		Dispatching and monitoring
		Gate control
		Container information management
		Billing and settlement
		Statistical analysis
		Work terminal system
		Dangerous goods management system
		CFS management
	2.Support system	Online business hall
		Video monitoring system
		Electronic chart system
		Tally system
		Ship safety information retrieval
		Meteorological information platform
		Remote monitoring system for

Wharf type	Core system	Subsystem
		refrigerated containers in storage yard
		Electronic map of port area
		External truck navigation system
		Data center
		Large screen system
		Access control system
	3.Equipment control system	Quayside container crane management system
		Container gantry crane management system
		Horizontal transportation equipment management system
	4.Remote control system	Remote control system of quayside container crane
		Remote control system of container gantry crane

4. Cooperative Control Mechanism of Key Objects and Key Nodes in Container Terminals

Based on the above research content, combined with the container terminal material movement process mechanism and the container terminal material movement interaction information, from the frontier loading and unloading equipment of the terminal, the horizontal transportation equipment of the terminal, the terminal yard equipment, the loading and unloading equipment in and out of the terminal, and the level of key objects and the terminal operating system, Starting from the level of key nodes such as the equipment control system, remote control system, and support guarantee system, we sort out the control logic structure required to achieve the remote real-time online material movement control function on the basis of ensuring the safe and stable operation of the terminal, as shown in Figure 2.

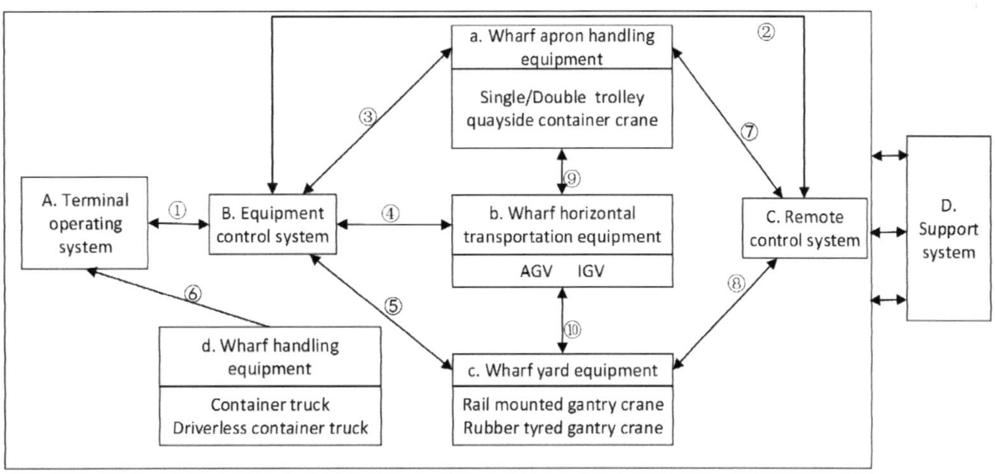

Figure 2. Logical structure of remote real-time material movement control

Combining the above control logic, the control mechanism of the coordination and cooperation between the key nodes and key objects of the container terminal is shown in Table 3. The interaction mechanism between the systems can open up the circulation path of decision-making information between the systems during the terminal operation. The interaction mechanism between devices can integrate the internal relationship between information flow and physical flow, and the interaction mechanism between devices can determine the optimal structure of the device in the geographical space layout.

Table 3. Cooperative control mechanism between key objects and key nodes of the container terminal

Interaction category	**information flow**		**Interactive information**
Interaction between systems	1	A-B	Work order
		B-A	Work order execution status
	2	B-C	Work order
			Work order execution status
		C-B	Work status
			Work intervention request
Interaction between system and equipment	3	B-a	Detailed work instructions
		a-B	Fault information
			Work status
			Work order execution status
			Abnormal operation instruction execution
	4	B-b	Detailed work instructions
		b-B	Fault information
			Work status
			Work order execution status
			Abnormal operation instruction execution
	5	B-c	Detailed work instructions
		c-B	Fault information
			Work order execution status
			Work status
			Abnormal operation instruction execution
	6	d-A	Driver identification information / vehicle number
	7	C-a	Remote operation instruction
		a-C	Work status
	8	C-c	Remote operation instruction
		c-C	Work status
Interaction between devices	9	a-b	3D scanning signal of wharf horizontal transportation equipment
		b-a	Positioning signal of wharf horizontal transportation equipment
	10	c-b	3D scanning signal of wharf horizontal transportation equipment
		b-c	Positioning signal of wharf horizontal transportation equipment

5. Conclusions

The analysis of the key objects and key nodes of the container terminal and the study of the control mechanism between them can lay a theoretical foundation for the construction of the container terminal control model and digital twin. On the basis of focusing on the material movement process mechanism and comprehensively grasping the material movement interactive information, combined with the actual operating conditions of each container terminal, in the future, we can create a system design plan and engineering application guidance specifically for automated container terminals.

This paper belongs to the research results of the national key research and development project "Theories and methods of super-large comprehensive port operation prediction based on cyber-physical integration" (2020YFB1710801)

References

[1] Liu Zhaoqing. Strengthening the implementation of ISPs measures at the ECT terminal of Rotterdam port [J]. World shipping, 2004, 27 (2): 1
[2] Qiu Huiqing, Lu Kailiang. Review of international automated container terminal technology development [C] / / independent innovation to achieve sustainable and scientific development of logistics engineering -- Proceedings of the 8th Annual Academic Conference on Logistics Engineering. 2008
[3] Zhang Yiying, Wu Qing. Analysis of the impact of automated terminal construction on China's ports -- Taking Xiamen offshore automated terminal as an example [J]. Navigation, 2016 (2): 4
[4] Sun Fuqi. Tianjin port, building a world-class smart port [J]. China storage and transportation, 2020 (8): 2
[5] Xu caifei. Research on integrated management of Tianjin Port Container Logistics Information System [D]. Tianjin University of technology, 2008
[6] Zhang Chuanlong, Feng Dinggen. Application of automation technology in traditional container terminals [J]. Containerization, 2019, 30 (7): 2
[7] Luo Xunjie. Key equipment technology and development of fully automated container terminal [J]. Port handling, 2019 (1): 5
[8] Yuan Zheng, Yu Qingshuang. Overview of key technologies of container automation terminal in Qingdao port [J]. Port engineering technology, 2020, 57 (5): 4
[9] Zhang Yulong. Business process design of automated container terminal [J]. Water transportation engineering, 2019 (10): 6
[10] Gong Congcong, Han Lei, Li Qiang. Shandong Port Group innovates Chinese port "black technology" [J]. Shandong State owned assets, 2019 (12): 2

AUTHOR INDEX

Institute of Physics
Dirac House, Temple Back
Bristol BS1 6BE UK

ISSN: 1742-6588
ISBN 978-1-7138-5162-2

CURRAN ASSOCIATES INC.
proceedings.com